R. Wetherbee, R. A. Andersen,
and J. D. Pickett-Heaps (eds.)

The Protistan Cell Surface

Springer-Verlag Wien New York

Dr. Richard Wetherbee
Dr. Jeremy D. Pickett-Heaps
School of Botany
University of Melbourne
Parkville, Victoria, Australia

Dr. Robert A. Andersen
Provasoli-Guillard Center for Culture of Marine Phytoplankton
Bigelow Laboratory for Ocean Sciences
West Boothbay Harbor, Maine, U.S.A.

Softcover reprint of the hardcover 1st edition 1994

Printed on acid-free and chlorine-free bleached paper

With 566 Figures

Library of Congress Cataloging-in-Publication Data

The protistan cell surface / R. Wetherbee, R. A. Andersen, and J. D. Pickett-Heaps, eds.
p. cm.
Includes bibliographical references.
ISBN-13: 978-3-7091-9380-8
1. Protista. 2. Cell membrances. 3. Fungal cell walls.
I. Wetherbee, R. (Richard). II. Andersen, Robert A. (Robert Arthur). III. Pickett-Heaps, Jeremy D.
QR74.5.P76 1994 94-33622
576'.14–dc 20

Special Edition of "Protoplasma", Vol. 181, 1994

ISBN-13: 978-3-7091-9380-8 e-ISBN-13: 978-3-7091-9378-5
DOI: 10.1007/978-3-7091-9378-5

Preface

The protists represent an immensely diverse group of organisms comprised of algae, fungi and protozoa. Historically studied in isolation within the specialized disciplines of phycology, mycology and protozoology, respectively, recent phylogenetic analyses by a variety of techniques, most notably ultrastructural and gene sequence comparisons, have shown these classical separations to be artificial. In the search for true phylogenetic classifications, the need for a common vocabulary that reflects the relationship between protists, rather than the field of study that produced the terms, is imperative. This communal vocabulary will allow modern protistologists to identify and compare homologous structures and to break down the historical barriers between disciplines. To help address this problem, a number of Task Forces were formed in 1987 to help create a common vocabulary for the Protists as a whole. The history of the origin and function of Task Forces is discussed in the Preface to the report of the Task Force on the Cytoskeleton, which was published both as an issue of the international journal "Protoplasma" (Vol. 164) and subsequently as a book by Springer-Verlag: Melkonian, M., Andersen, R. A., Schnepf, E. (eds.), The Cytoskeleton of Flagellate and Ciliate Protists, Springer-Verlag, Wien New York (1991).

The nature of the protistan cell surface is as diverse as the terminology that has evolved to describe the various surface components. This terminology is defined and discussed in the report of the Task Force on the Protistan Cell Surface chaired by Dr. Hans Preisig. The Task Force members were representative of mycology (Dr. M. J. Powell and Dr. R. W. Roberson), phycology (Dr. O. Moestrup and Dr. R. Wetherbee), and protozoology (Dr. O. R. Anderson and Dr. J. O. Corliss). Although the Task Force made no attempt to dictate a common vocabulary, several obvious discrepancies in terminology are cited and recommended for change (e.g., the common and inconsistent use of the terms "periplast" and "theca"). The remaining contributions to this book represent the proceedings of two symposia, entitled "The Cell Surface in Protistan Taxonomy and Systematics" and "The Structure, Development and Function of Protistan Cell Surfaces", or were solicited by the editors to fill obvious gaps in the symposia program. Both the Task Force and the Symposia met at the Phycological Society of America meetings held in conjunction with the AIBS at Iowa State University in August of 1993. Despite the efforts of the editors, not all of the contributors to the symposia were able to produce their required written papers. These proceedings therefore have some obvious gaps, most notably in the coverage of protistan cell walls (fungal and algal) and the organic coverings of many protistan groups (e.g., the periplast, pellicle, and theca).

We are grateful to Dr. James Rodman and the Systematic Biology Program of the National Science Foundation of the U.S.A. for funding both the Task Force and the Symposia that lead to this book (NSF grant DEB-9304086). We are also grateful to Dr. Randall Alberte and the Office for Naval Research (U.S.A.) for providing additional funds in support of the Editors and the Symposia. This is the second Task Force to report, and several others (e.g., on mitosis/cytokinesis: paleobiology) are scheduled to meet and report in the near future. The common goal of each Task Force is the same, to promote and enhance the study of protistan biology through a common terminology and a better understanding of the interwoven evolutionary relationships of algae, fungi, and protozoa.

R. Wetherbee, R. A. Andersen, and
J. D. Pickett-Heaps

Contents

Protoplasma (1994) 181: 1–28

PROTOPLASMA

Terminology and nomenclature of protist cell surface structures

H. R. Preisig[1,] *, **O. R. Anderson**[2], **J. O. Corliss**[3], **Ø. Moestrup**[4], **Martha J. Powell**[5], **R. W. Roberson**[6], and **R. Wetherbee**[7]

[1] Institute of Systematic Botany, University of Zurich, Zurich, [2] Department of Biological Oceanography, Lamont-Doherty Earth Observatory, Columbia University, Palisades, New York, [3] Albuquerque, New Mexico, [4] Botanical Institute, Department of Mycology and Phycology, University of Copenhagen, Copenhagen, [5] Department of Botany, Miami University, Oxford, Ohio, [6] Department of Botany, Arizona State University, Tempe, Arizona, and [7] School of Botany, University of Melbourne, Parkville, Victoria

Received September 28, 1993
Accepted January 2, 1994

Summary. The use of a precise terminology is important to the unambiguous exchange of information in the multidisciplinary area of protistology. In this paper we attempt to establish clear definitions, give illustrations, and comment on the different terms used for cell surface structures of protists and related organisms.

Keywords: Protists; Algae; Fungi; Protozoa; Cell surface structures; Terminology; Nomenclature.

Introduction

Cell surface structures are scientifically important for a number of reasons including species identification, cell structural support and protection, evidence of cellular biosynthetic activity and stages of development, as well as cell-cell interactions and antigenic properties of the cell.

The terminology used to describe cell surface structures of protists is sometimes confusing. This is mainly because protists have traditionally been studied by workers from different disciplines of biology (mycology, phycology, protozoology), in which special vocabularies have been developed and used over scores of years. In order to assist the flow of information amongst the different scientific fields, to address the terminological problems and to provide definitions of terms used in protistology, a "Task Force" has been established with expertise in all three protistological disciplines (for background and rationale of the Task Force concept, see a previous account on "Terminology and Nomenclature of the Cytoskeletal Elements in Protists", Andersen et al. 1991).

The task of defining cell surface features for protists is complex for several reasons:

(1) The limits of this kingdom (if it *is* indeed a single assemblage; see Corliss 1993) are not clear, and the "Kingdom Protista" most likely encompasses a paraphyletic assemblage of eukaryotic organisms with relatively simple body structure (i.e., no highly differentiated tissues, multicellular organs, or complex embryological development). We have been inclusive in the organisms considered and have incorporated terms which apply to all taxa of algae, protozoa and fungi. The reason for this strategy is that "lower" organisms in the Kingdoms Plantae, Animalia, and Fungi intergrade into Protista. For example recent molecular sequence studies (Bowman et al. 1992, Wainright et al. 1993) support the view that chytridiomycetes, with progenitor and close phylogenetic affinities with higher fungi, are also phylogenetically related to protozoa such as choanoflagellates. The term fungal-protists refers to the "lower" plasmodial and zoosporic "fungi", that are organisms resembling fungi but which are phylogenetically diverse, including those with affinities to Chromista (Cavalier-Smith 1986; labyrinthulids, thraustochytrids, plasmodiophoromycetes, hyphochytriomycetes, oomycetes) and to fungi (chytridiomycetes). In preparing this glossary, we have unanimously adopted a "protist perspective" in order to demonstrate the need today to bridge the lamentable gap between

* Correspondence and reprints: Institut für Systematische Botanik, Universität Zürich, Zollikerstrasse 107, CH-8008 Zürich, Switzerland.

the conventionally separate fields of mycology, phycology and protozoology.

(2) Widely used terms may not always apply to homologous structures. Common structures, such as cell walls and stalks, probably arose and evolved independently many times among organisms, and a single term applies to analogous structures in different protists, just as wings do in animals. Cell surface features are highly adaptive, and chances of similarity of form due to similar selective pressures are great. It is beyond the scope of this effort to suppose or propose homologies, but hopefully the terms listed will suggest structures present in diverse groups of protists which can be explored for potential homologies and as indicators of phylogenetic relationships and organellar evolution.

(3) As it is difficult to resolve homologous versus analogous functions for structures defined with the same term among protists, the situation is even more perplexing when terms also apply to structures produced by other eukaryotes, as well as prokaryotes. We have tried to identify these possible confusions, and again emphasize that the same term may in practice be used for analogous and unrelated structures.

(4) Deciding the limits of extracellular material becomes difficult when demarcations are not certain and when material may have the superficial appearance of extracellular material. In the strictest sense some terms included in our list are not actually non-living components of the cell surface, but because they superficially appear to be and because they may contain non-living material, we have included them (e.g., alveoli, amphiesma). Plasma membrane, cortex, epiplasm and pellicle are main constituents of the outer layers of the living cytoplasm and have also been included, but otherwise living cell-surface structures and appendages have been omitted from direct consideration, even though they do protrude or extend from the surface of the protist in some cases (e.g., cilia, flagella, haptonemata, pseudopodia, tentacles, undulating membranes, etc.). Extrusive organelles have also been excluded here (see Kugrens et al. 1994).

(5) There is a multiplicity of terms used in protistological literature with reference to components of cell surface structures, but in this account we limited ourselves to major components and not all their "parts". For example, we dealt with "frustule" of diatoms, but not with subparts of this structure such as mantle, girdle bands, labiate processes, etc.

Literature citations have not been included in the definitions because so many are possible. A great number of pertinent citations can be found in accompanying papers in this volume and in the following references to major sources of protistological glossaries embracing appropriate mycological, phycological, and protozoological terms (overviews that naturally go beyond our own coverage restricted to cell surface structures): Alexopoulos and Mims (1979), Anderson (1988), Berner

---

Abbreviations used in all legends: BF (bright field microscopy), DIC (Nomarski differential interference contrast microscopy), FM (fluorescence microscopy), PC (phase contrast microscopy), SEM (scanning electron microscopy), TEM (transmission electron microscopy)

Fig. 1. Adhesions. Extracellular adhesions (*A*) released from the germ tube (*GT*) of a germling of *Entophlyctis variabilis* (Chytridiomycetes). High pressure cryofixed and freeze substituted material, TEM, × 25,000

Figs. 2 and **3.** Adhesion pad

Fig. 2. Oomycete cystospore (*S*) attached to the surface of a glass coverslip at the site indicated with the arrow and germinating with a germ tube (*GT*). The arrow denotes the location of the adhesion pad, which is not visible with DIC but is detected with staining, as seen in Fig. 3. DIC, × 1,250

Fig. 3. Binding pattern of fluorescein isothiocyanate labelled-wheat germ agglutinin lectin to N-acetylglucosamine sugar residues within extracellular material on oomycete cystospore. The arrow points out the adhesion pad which adheres the cystospore to the glass cover slip. Although cystospore walls do not bind the lectin, sheath material coating germ tube surfaces does. FM, × 1,250

Fig. 4. Amphiesma. Transverse section through the cell surface of the dinoflagellate *Katodinium rotundatum* showing the amphiesma consisting of the plasma membrane (*PM*) and underlying flattened amphiesmal vesicles (= alveolar sacs) (*AS*). A thin plate (*P*) is visible within the vesicle. Groups of microtubules (*Mt*) are situated beneath the vesicles; *Sc* body scales. TEM, × 45,000

Fig. 5. Appressorium ring. Basidiospore (*Bs*) of *Gymnosporangium juniper-virginianae* (Basidiomycetes). Note the appressorium (*A*) and the appressorium ring (*AR*). DIC, × 1,250

Figs. 6 and **7.** Auxospore wall

Fig. 6. Siliceous scale from the auxospore wall of the diatom *Melosira nummuloides*. TEM, × 10,000 (courtesy of R. M. Crawford)

Fig. 7. Auxospore pair of the diatom *Rhoicosphenia curvata* showing the numerous siliceous bands (perizonium) of the secondary auxospore wall. The arrow points out the initial epivalve developing within the perizonium. SEM, × 10,000 (courtesy of D. G. Mann)

(1993), Bold and Wynne (1985), Grell (1973), Harrison and Corliss (1991), Hawksworth et al. (1983), Lee et al. (1985), Margulis et al. (1990, 1993), Moore-Landecker (1990), Olive (1975), Ross et al. (1979), Sleigh (1989), van den Hoek et al. (1994).

For the taxa of higher rank we have generally used vernacular names. In most cases the taxonomic composition of the group will be clear to the reader; there is debate concerning the most correct scientific name for some taxa and the hierarchical level of classification (hence the suffix of the name) for some taxa (see Corliss 1993).

The terms are arranged in alphabetical order and are not organized around a theme (such as function or spatial relationships), as one might expect. Thus, we assume cross-references will be easier to find in the text. Furthermore, functional or other groupings of terms could be out-of-date at any point in the near future if additional functions or relationships were discovered necessitating reassignment. For cross-references the ab-

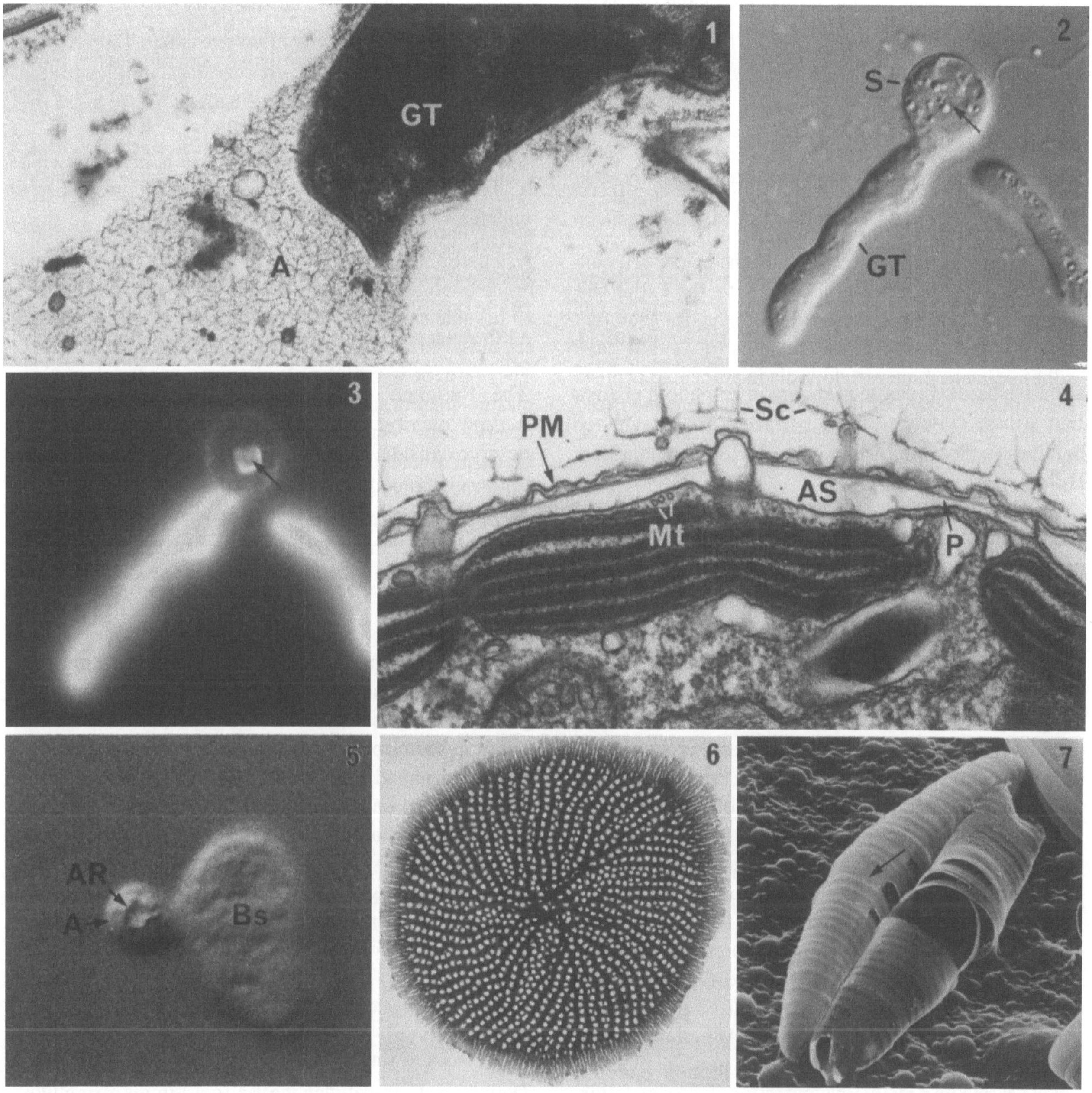

breviation q.v. ("quo vide") has been used. Also note that some "lesser" synonyms have not been defined independently but entered with a notation to see the chosen "major" synonym (e.g., for "acellular stalk" see "stalk").
We hope that this account serves to identify specific terminological problems and to suggest specific solutions that will be acceptable to the scientific community.

Results

One hundred terms (including 34 synonyms) are dealt with below.

Acellular stalk, see Stalk.

Adhesions

Definition: Morphologically and chemically diverse classes of compounds which function in attachment of cells to substrates or hosts (Fig. 1).
Comments: Distributions of adhesions at the cell surface range from adhesive junctions to the entire cell surface. Symphoriontic ciliated protozoa, for example, attach to host carapaces with stalks (q.v.) that have adhesive bases. Fine pseudopodia in amoeboid protists that adhere closely to the substratum are categorized as adhesive. See also, Adhesion pad, Holdfast adhesion, Holdfast organelle.

Adhesion pad

Definition: A type of extracellular matrix (q.v.) which spreads at the cell or spore substrate interface, binding the two surfaces together (Figs. 2 and 3).
Comments: "Adhesion pad", "adhesion plaque", "mucous pad" and "spore tip mucilage" are used as synonyms. The polarized distribution of this material distinguishes it from an extracellular matrix or sheath (q.v.). Among oomycetes it attaches encysting zoospores to a host or substrate; among pennate diatoms it denotes distinct, localized regions of polysaccharide adhering cells together into colonies; among pedinellid flagellates (such as *Apedinella*) it is a surface associated structure that spine-scales attach to. Adhesion pads may have diverse origins including: (1) discharge from pre-existing organelles such as K-bodies as oomycete zoospores are induced to encyst and (2) selective secretion of polysaccharide from localized pores on the frustule (q.v.) surface of diatoms. Adhesion pads of diatoms are a specialized part of the non-silicified, organic casing and maintain connections between daughter cells following division. The exact position of these pads determines the overall morphology of the colony. An adhesion disc made of denticles (ring) immediately beneath the pellicle (q.v.) occurs in the ciliate *Trichodina*. See also, Adhesions, Holdfast adhesion, Plaque, Stalk.

Adhesion plaque, see Adhesion pad.

Alveoli

Definition: Membrane-bounded flattened vesicles or sacs underlying the plasma membrane (q.v.) in certain protists, most notably, the (majority of the) ciliates, the dinoflagellates [where they are called "amphiesmal vesicles" (q.v.)], some sporozoan (apicomplexan) parasitic protists, and possibly the glaucophyte algal protists (Figs. 4 and 17).
Comments: A single alveolar vesicle is also known as a "lacuna", and the entire system in ciliates has been called a "lacunar system" in older literature. In more recent usage, it is called pellicular alveoli. A similar system found in the Sporozoa is known as "inner membrane complex". In some actinopods (e.g., radiolaria) an organic capsular wall (q.v.) surrounding the central cell mass is deposited within lacunae resembling alveoli. The flattened peripheral cisternae underlying the plasma membrane of oomycete zoospores may well deserve consideration as a kind of alveoli. In diatoms, the term alveoli is used for elongated chambers of the valve (q.v.) opening to the inside of the cell by a large opening and with a perforate outer layer. See Conclusions and recommendations; see also, Cortex, Epiplasm, Pellicle.

Amphiesma

Definition: Term describing the plasma membrane (q.v.) and the underlying flattened vesicles of dinoflagellates (amphiesmal vesicles = "thecal vesicles") which in some species contain plates (= thecal plates). In certain naked dinoflagellates the amphiesmal vesicles appear to be without plate-like structures (Fig. 4).
Comments: Apparently the same system of vesicles underneath the plasma membrane occurs in ciliates, some apicomplexans, glaucophytes, and perhaps other protists (see Alveoli), but the term amphiesma is presently restricted to dinoflagellates.

Amphiesmal vesicles, see Amphiesma.

Appressorium ring

Definition: Extracellular matrix (q.v.) in an o-ring like arrangement through which the penetration peg from

the appressorium of some pathogenic fungi, such as rust fungi, passes just prior to entering host cells (Fig. 5).
Comments: Believed to assist in attachment of appressorium to host cell and preventing dislodgement as host colonization occurs.

Auxospore wall

Definition: The cell wall (q.v.) of a diatom auxospore (a cell resulting from sexual fusion or autogamy, or one formed purely by vegetative processes that swells to maximum size) (Figs. 6 and 7).
Comments: The auxospore wall differs from that of the normal vegetative cell. Its primary wall consists of polysaccharide material, which in most centric diatoms contains overlapping siliceous scales (q.v.) (Fig. 6). Auxospores may also develop a secondary wall system consisting of siliceous hoops or bands, collectively termed the "perizonium" (in pennate diatoms) and "properizonium" (in centric diatoms), respectively. Unlike the perizonium, the properizonium is not separated spatially from the primary auxospore wall, but is physically and developmentally continuous with it. The individual siliceous components of the perizonium (properizonium) resemble modified girdle bands, and are secreted sequentially to create the secondary auxospore wall. The new frustule (q.v.) of the enlarged initial cell is deposited within the perizonium (properizonium), which then ruptures to release the cell (Fig. 7).

Basal disc

Definition: A general term describing the attachment of the stalk (q.v.) (or occasionally the entire aboral surface of the body) to a substrate in a wide range of protists.
Comments: See also, Hypothallus.

Basket

Definition: (1) An enclosing structure composed of loosely arranged components [spicules, costae (q.v.), etc.] surrounding a cell as in some choanoflagellates and silicoflagellates (Fig. 8); (2) a type of complex scale (q.v.) on the surface of some flagellates; (3) the conical array of microtubules forming a feeding apparatus in nassulid-type ciliates.
Comments: The feeding apparatus in nassulid-type ciliates is also known as a "feeding basket", "cytopharyngeal basket", or "cyrtos" and is an intracellular structure with the narrow end of the conical array of microtubules pointed toward the interior of the cell. It is a dynamic structure for grasping and ingesting food particles, including filamentous algae in some species. See also, Lorica.

Bristle

Definition: Siliceous elongated structure that is tucked under, and cemented to, the distal portion of a scale in the synurophyte genus *Mallomonas* (Fig. 9). The term "bristle" is also widely used to describe diverse stiff hair-like appendages on cell walls and other surfaces (e.g., in some chlorococcalean algae). In ciliates it is a common name for stiff cilia of several kinds and functions.
Comments: The terms "bristle" and "spine" (q.v.) are sometimes used as synonyms (e.g., in ciliates). Bristles of *Mallomonas* radiate outwards from the cell and may change their position relative to the longitudinal axis of the cell. They are streamlined when the cell is actively swimming and positioned closer to 90° at rest. Movable bristle-like components, possibly homologous to bristles of *Mallomonas*, also occur on the valve (q.v.) of the diatom *Corethron*. Euplotid ciliates have modified cilia known as bristles on the dorsal surface; they may be sensory.

Capillitium

Definition: Sterile threads, which are often coiled and bear ornamentation, intermixed with spores of fruiting bodies of some Mycetozoa. Differentiated prior to complete spore cleavage, these threads aid in spore dispersal (Fig. 10).
Comments: Among higher fungi the term is not restricted to extracellular material, but can describe specialized hyphae in frutifications such as those in puffballs. The term "elaters" is sometimes used for capillitial strands that are short and unbranched.

Capsular wall

Definition: An organic wall in actinopods, especially radiolaria, composed of closely spaced plates deposited within cytoplasmic lacunae or alveoli (q.v.) at the periphery of the central, more dense cytoplasm (Figs. 11 and 12). Cytoplasmic projections protrude through pores in the wall. In myxozoan spores, the polar capsular wall consists of two layers with varying sensitivities to chemical digestion which can be observed with transmission electron microscopy; the inner electron-lucent layer resists alkaline hydrolysis and the outer electron-dense layer is digested with proteases.
Comments: The thickness and organization of the wall

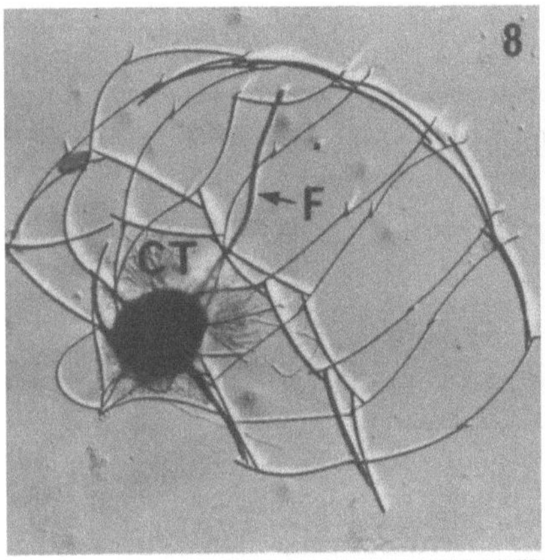

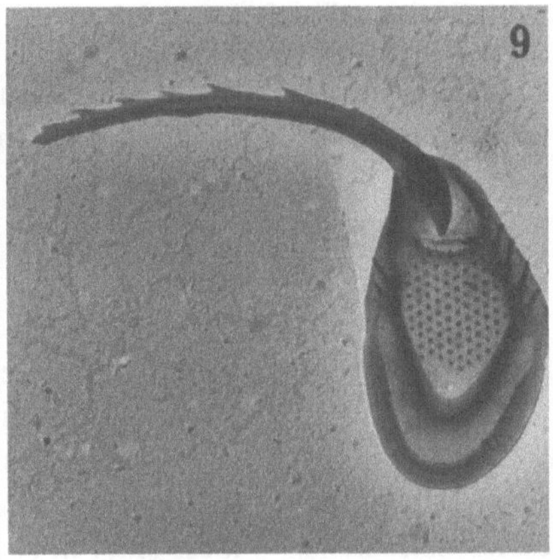

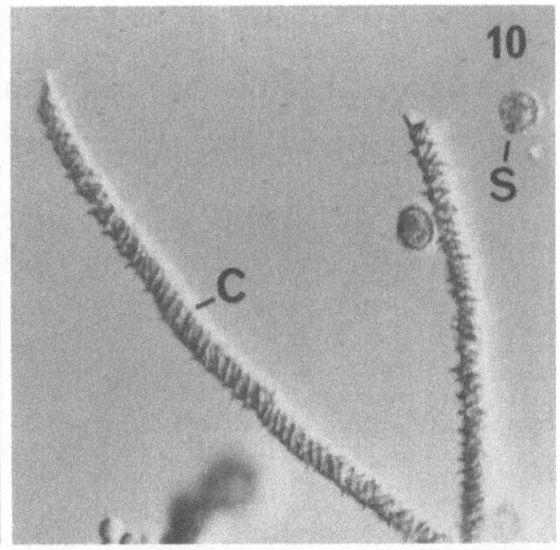

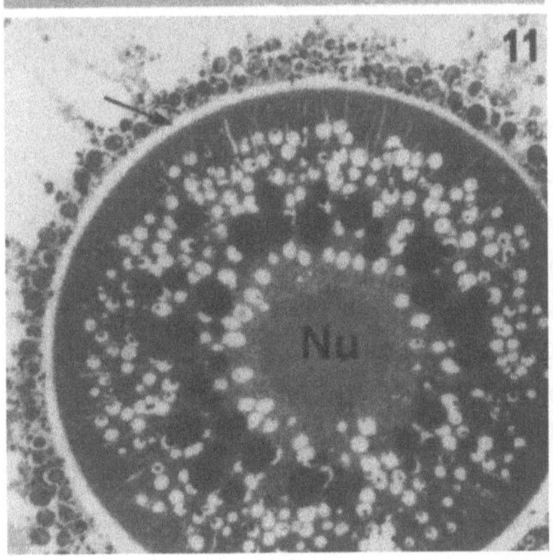

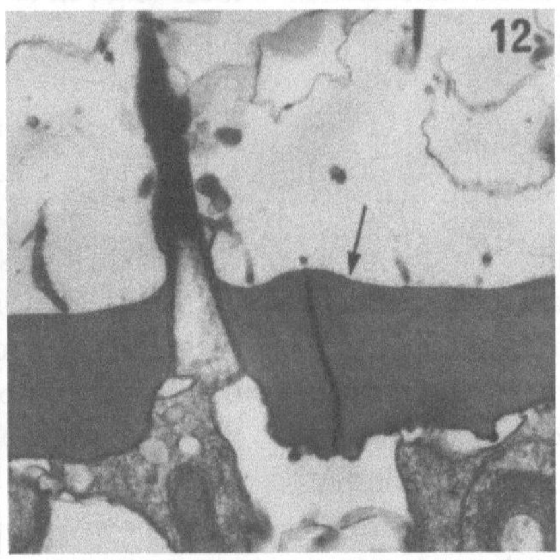

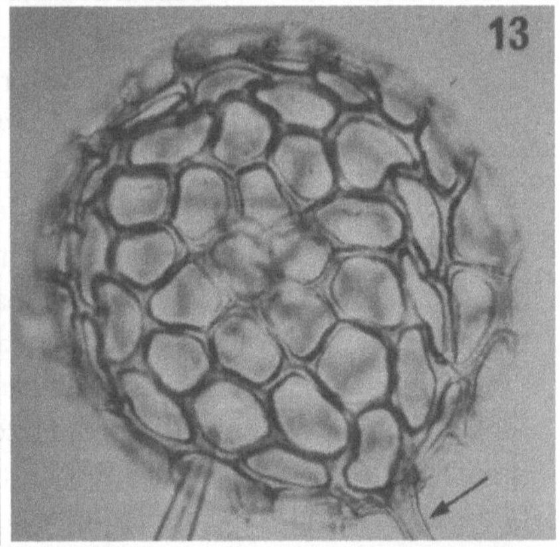

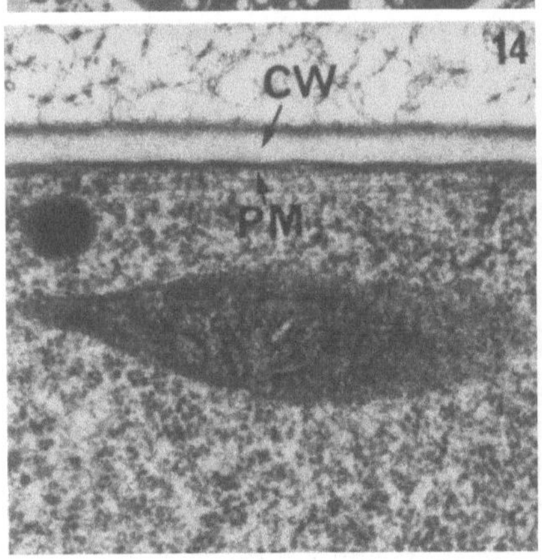

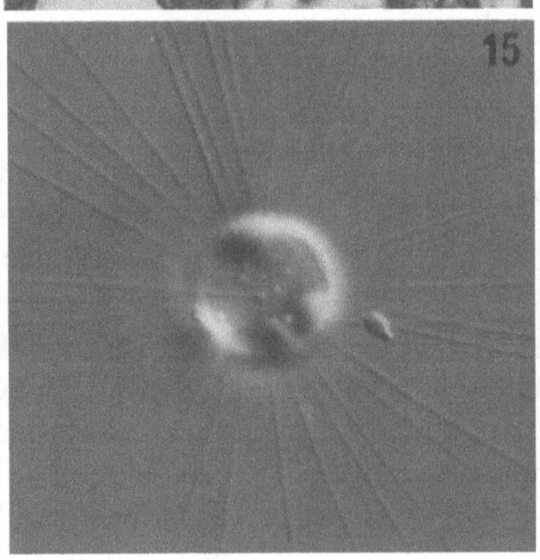

Fig. 8. Basket (lorica) of the choanoflagellate *Cosmoeca* aff. *ventricosa* composed of costae (costal strips); *F* flagellum, *CT* collar tentacles; see also Fig. 18. TEM, × 3,000

Fig. 9. Bristle. Serrate bristle articulating at the anterior end of a detached scale of *Mallomonas papillosa* (Synurophyceae). TEM, × 10,000

Fig. 10. Capillitium. A few spores (*S*) remain among the thread-like and sculptured strands of capillitium (*C*) dissected from the sporangium of a slime mold. DIC, × 400

Fig. 11. Capsular wall. Section through the central capsule of a radiolarian showing the capsular wall (arrow) surrounding the central cytoplasmic mass and nucleus (*Nu*). BF, × 120

Fig. 12. Capsular wall (arrow) of a radiolarian. TEM, × 14,000

Fig. 13. Capsule. Stalked latticed capsule of the heliozoan *Clathrulina elegans*. The stalk (arrow) may be several times longer than the diameter of the capsule. BF, × 700

Fig. 14. Cell wall. Cross section through the cell wall (*CW*) of a fungal hypha (*Allomyces macrogynus*, Chytridiomycetes). Note plasma membrane (*PM*). TEM, × 66,000

Fig. 15. Chitinous threads emanating from pores in the silica valve of the diatom *Thalassiosira fluviatilis* (valve in face view). DIC, × 700 (courtesy of L. Fritz)

Fig. 16. Coccolith. Overlapping coccoliths in coccosphere (*Emiliania huxleyi*). SEM, × 8,700

material varies among actinopod species, and the penetrating cytoplasmic strands vary in complexity and function among major groups. See also, Capsule.

Capsule

Definition: A perforated organic layer surrounding the cell body of heliozoa, but not necessarily enclosed within a cytoplasmic envelope (Fig. 13).
Comments: Capsule and capsular wall (q.v.) are sometimes used synonymously for the organic wall surrounding radiolarian or clathruline heliozoan bodies, but it is preferable to separate them based on morphological differences between the wall of radiolaria (capsular wall) and the more loosely organized structure of heliozoa (capsule). A special type of extrusive organelle found in myxozoa is termed "polar capsule". Some bacteria and yeast are surrounded by a protective organic wall, also known as a "capsule".

Case, see Lorica.

Cell coat, see Glycocalyx.

Cell membrane, see Plasma membrane.

Cell wall

Definition: A mostly rigid, often multilayered structure consisting of discrete microfibrillar polysaccharides embedded in an amorphous matrix composed of polysaccharides, lipids, and proteins which together comprise an outermost layer of the cell proper. Architectural complexity and composition may vary among taxonomic groups, stages of cell development and environmental conditions (Fig. 14).
Comments: The cell wall confers shape to the cell and constrains the turgid protoplast. In addition, the cell wall (1) influences the cellular uptake of water and other molecules thereby regulating intracellular ionic and solute concentrations, (2) passively regulates secretions, (3) contains enzymes involved in the conversions of nutrients into metabolizable forms prior to entry into the cytoplasm, (4) serves as a store for carbon reserves, and (5) is involved in cell-cell interactions/recognitions. See also, Cyst wall, Frustule, Spore wall, Wall fimbriae.

Cellular stalk, see Stalk.

Chitinous threads

Definition: Stiff, elongate fibrils of β-chitin (sometimes called chitan) that emanate from pores in the silica valves of some centric diatoms, and may extend a considerable distance from the cell surface (Fig. 15).
Comments: Chitinous threads (chitin fibrils) in diatoms originate from specialized invaginations of the plasma membrane that appear involved in fibril formation. One or occasionally two fibrils emit from a single pore, and appear to be replaced if dislodged from the surface. In some species chitin fibrils may be involved in colony formation. Chitin is also a major component of cell walls (q.v.) in many protists (especially fungal protists) and also occurs in loricae, thecae, and cyst walls (q.v.) of some algal and protozoan protists (chrysophytes, choanoflagellates, ciliates).

Ciliary membrane, see Plasma membrane.

Coccolith

Definition: Calcareous scale (q.v.) or plate-like particle deposited on the surface membrane of some prymnesiophytes (the coccolithophorids), varying in complexity and surface decoration according to species. Coccoliths are synthesized within cytoplasmic vesicles and deposited by exocytosis on the surface of the cell in an overlapping arrangement forming a coccosphere (Fig. 16).
Comments: Between 10 and 100 coccoliths comprise the coccosphere, which may range in size from 5 to 50 µm. The function of the coccosphere is not fully understood. Under certain growth conditions coccoliths are not secreted and the cells are naked. Coccoliths with little modification of the calcite crystal form are termed holococcoliths and are apparently developed at the surface of organic scales. Heterococcoliths are composed of identifiable component elements and are produced within the cell. A special terminology has been developed for various subtypes of holococcoliths (e.g., calyptroliths, crystalloliths) and heterococcoliths (e.g., cricoliths, placoliths, etc.).

Cortex

Definition: The outer portion or layer(s) of a protist cell, including the plasma membrane (q.v.) but excluding secreted nonliving structures that may lie outside that membrane. In ciliates, the cortex exhibits considerable thickness, like the ectoplasm (q.v.) of many rhizopod amoebae, and contains the pellicular alveoli (q.v.) and the epiplasm (q.v.) as well as infraciliary organelles, notably the kinetosomes and their associated microfibrillar and microtubular structures. It also bears the cilia (Fig. 17).
Comments: In recent literature, the term "cortex" is

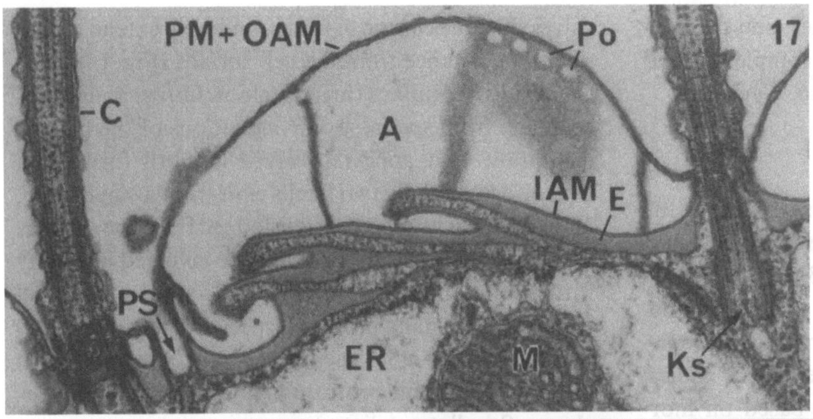

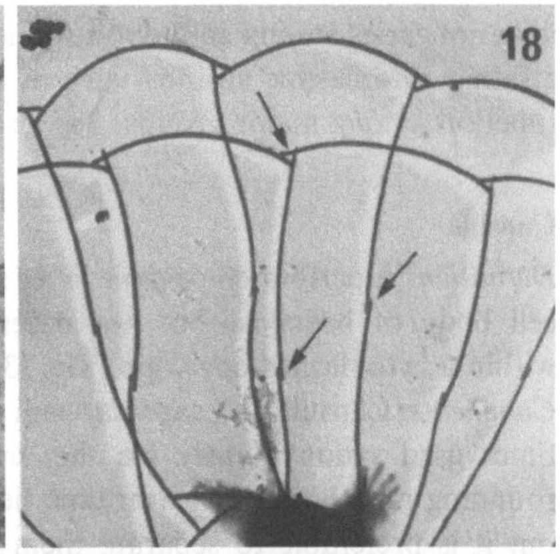

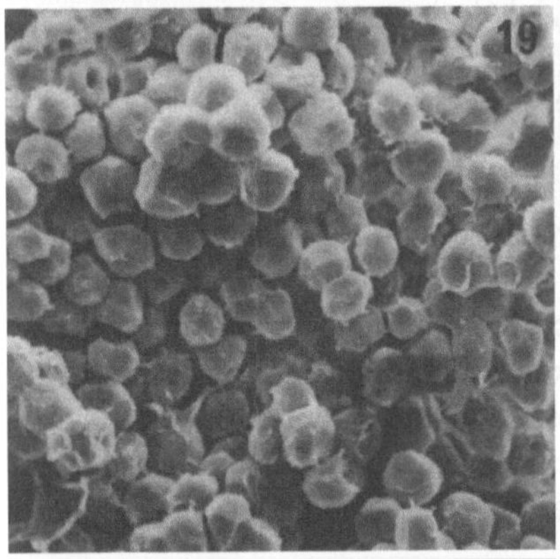

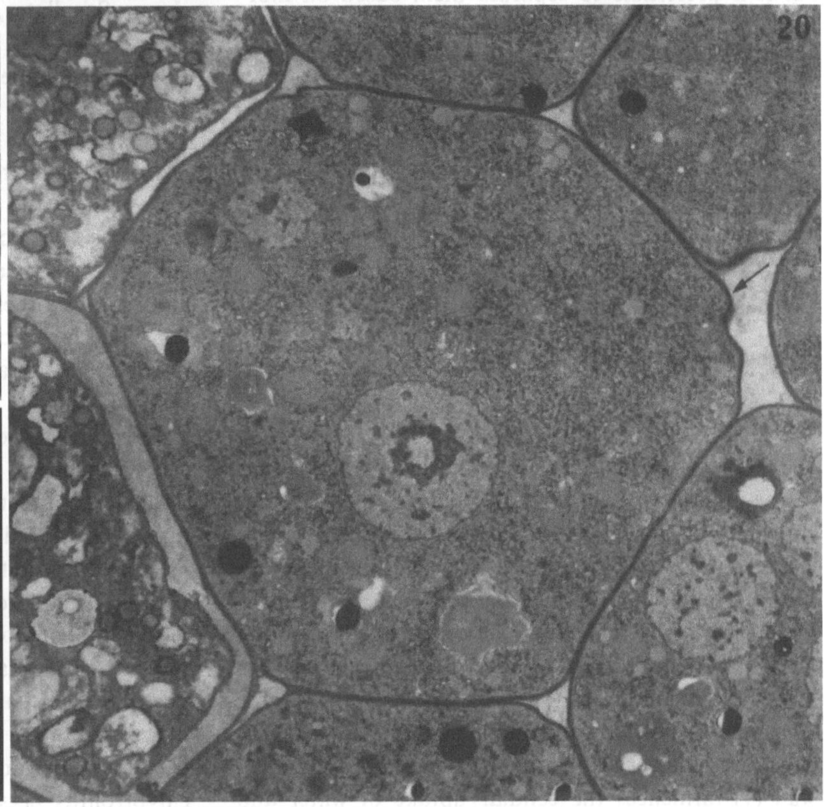

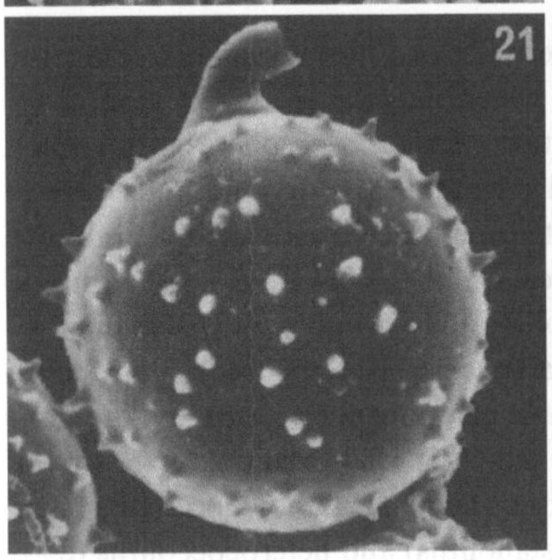

Fig. 17. Cortex. Cross-section through the cell surface (cortex) of the ciliate *Lembadion bullinum*; *A* alveolus, *C* cilium, *E* epiplasm, *ER* peripheral endoplasmic reticulum, *IAM* inner alveolar membrane, *Ks* kinetosome, *M* mitochondrion, *OAM* outer alveolar membrane, *PM* plasma membrane, *Po* pores in the lateral alveolar septum, *PS* parasomal sac. The large pores connect the alveoli with adjacent alveoli. TEM, × 37,500 (courtesy of C. F. Bardele)

Fig. 18. Costa. Longitudinal and transverse costae of a basket of the choanoflagellate *Cosmoeca phuketensis* (cf. Fig. 8). Note the end-to-end joints between component costal strips (arrows). TEM, × 8,000

Fig. 19–21. Cyst wall

Fig. 19. Closely packed cysts with organic cyst walls in a sclerotium of *Physarum polycephalum* (Myxomycetes). SEM, × 430

Fig. 20. Cross-section through a cyst of *Physarum polycephalum*. Note the organic cyst wall (arrow). TEM, × 6,460

Fig. 21. Cyst (stomatocyst) of the chrysophyte *Dinobryon cylindricum*. Note the ornamented siliceous cyst wall. SEM, × 4,000

used almost exclusively in reference to ciliates. See also, Cuticle, Pellicle.

Costa

Definition: Siliceous rib made up of costal strips which join and make up a spicular basket or basket-like lorica (q.v.) in members of the choanoflagellate family Acanthoecidae (Fig. 18). Term also used for an elongated, solid thickening of the diatom valves (q.v.), the "attachment band" connected at both ends to coiled filaments that confer elasticity to the cortex (q.v.) in acantharian actinopods, and a ridge or rib serving as an external ornament on the calcareous test (q.v.) of certain foraminifers, running parallel to the length or axis of the test.

Comments: Within a lorica of choanoflagellates, costae are usually positioned in a plane parallel or perpendicular to the long axis of the protoplast and are called longitudinal and transverse costae, respectively. Some species may have helically or more irregularly arranged costae. For many taxa the number and organization of costae are used taxonomically.

In addition to the above usage, the term "costa" is also used in certain parasitic flagellates (trichomonads) for a structure which is strictly speaking not a cell surface structure, namely a slender, rib-like structure or rod (a modified flagellar root, but not microtubular), which subtends the undulating membrane and which is often highly contractile and used in locomotion of the entire organism. The term "costa" is also widely used to describe diverse non-homologous rib-like structures in non-protist organisms, e.g., the ribs of vertebrates, the anterior vein on an insect wing, or the single midrib of a plant leaf, frond, or moss thallus.

Cuticle

Definition: In the broadest sense, a thin, cellular or non-cellular hydrophobic layer deposited on the outermost surface of the cell wall (q.v.). In euglenoid flagellates this term is sometimes used synonymously with pellicle (q.v.). In some protists also used as a synonym of cortex (q.v.).

Comments: The terms "cuticle", "cutis", and "pellis" are synonyms. The cuticle functions as an effective barrier to the movement of water, leaching of nutrients, and adds protection from injuries due to wind and physical abrasion. It may be present as either a non-cellular waxy layer on hyphae and spores or as a cellular layer of compressed hyphae on fruiting bodies. Multiple layers may be identified in the latter where the uppermost is referred to as the epicuticle and the next below as the subcuticle. Third and fourth layers, if present beneath the subcuticle, are the hypodermial and the subhypodermial layer, respectively.

Cutis, see Cuticle.

Cyst wall

Definition: The cell wall (q.v.) of a cyst (a nonmotile, often dehydrated, resistant, inactive, dormant stage in the life cycle of many protists). Cyst walls or cyst membranes, sometimes multiple in number and quite complex, may bear sculpturing or protrusions of a striking nature, and sometimes may be equipped with an emergence pore (plugged or unplugged) (Figs. 19–21).

Comments: Cysts are generally considered to serve in protection, propagation, and/or dispersal of the species. Many kinds of cysts have been described for protozoan protists alone, and very likely they are not all homologous structures. They are characteristic of numerous free-living as well as parasitic groups of protists and are useful in taxonomy. Planktonic foraminifera (e.g., *Globigerinita glutinata*) are occasionally enclosed by a thickened protective wall, surrounding the shell (q.v.), forming a "feeding cyst" containing undigested remains of prey such as diatom frustules. Some cysts contain a cyst plug, a non-living substance as a protective closure within the opening. The term cyst (cyst wall) is sometimes used synonymously with spore [spore wall (q.v.)]. See also, Pellicle.

Cytokalymma

Definition: A thin cytoplasmic sheath surrounding the siliceous endoskeleton (q.v.) of some actinopods (e.g., radiolaria) that supplies the silicate during skeletal deposition; the geometry of its internal space determines the shape of the skeleton (Fig. 22).

Comments: The cytokalymma is a dynamic structure that enlarges with the growing skeleton. By cytoplasmic streaming and changes in form, it determines the ontogenetic sequence of stages during skeletal growth.

Diatotepum

Definition: A distinct, continuous organic layer found between the siliceous frustule (q.v.) and the plasma membrane (q.v.) in diatoms (Figs. 23 and 24).

Comments: Previously referred to as the "pektin" or "diatopectin" layer, this terminology was confusing due to the possible chemical interpretations. The diatotepum stains for acidic polysaccharide and has been

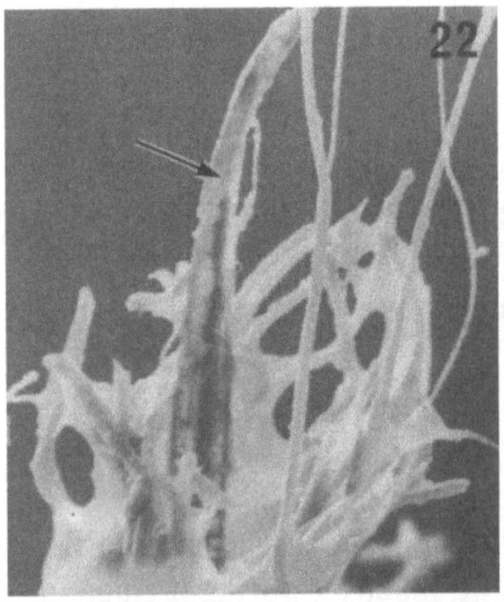

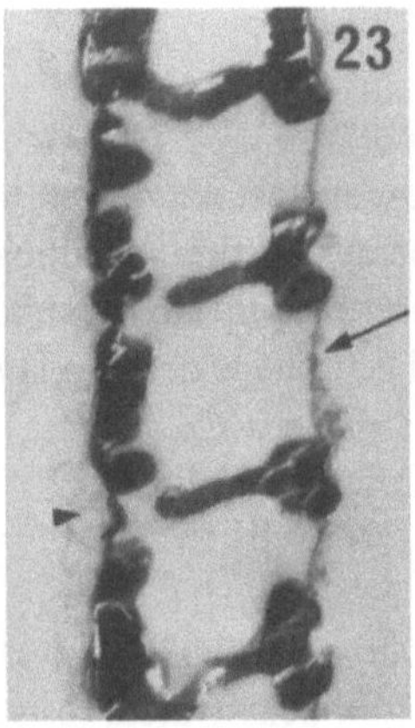

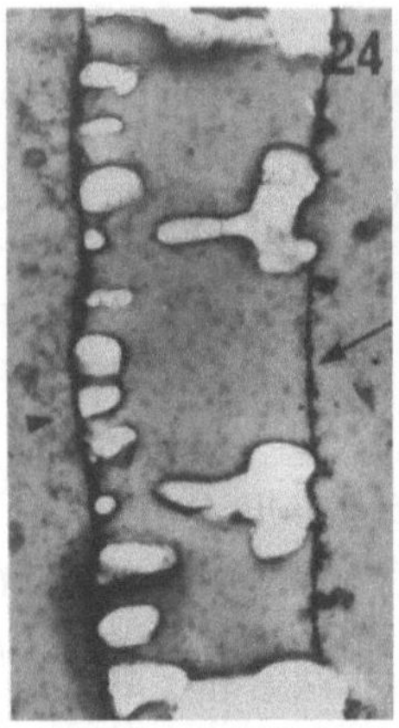

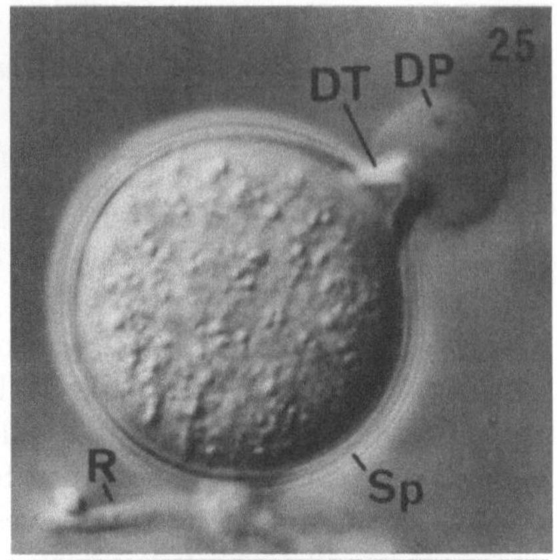

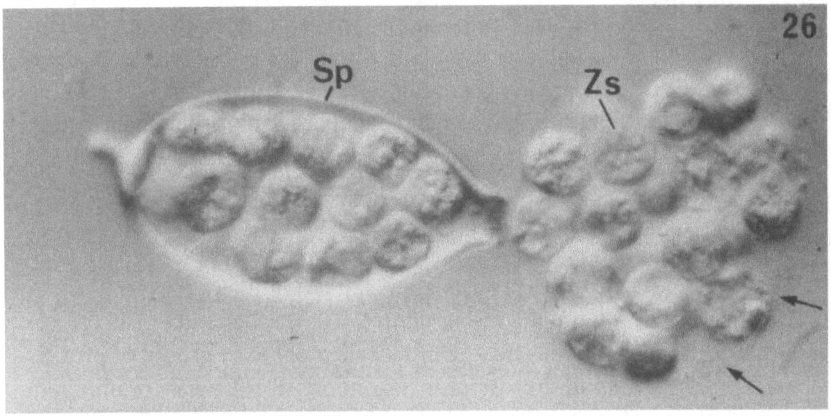

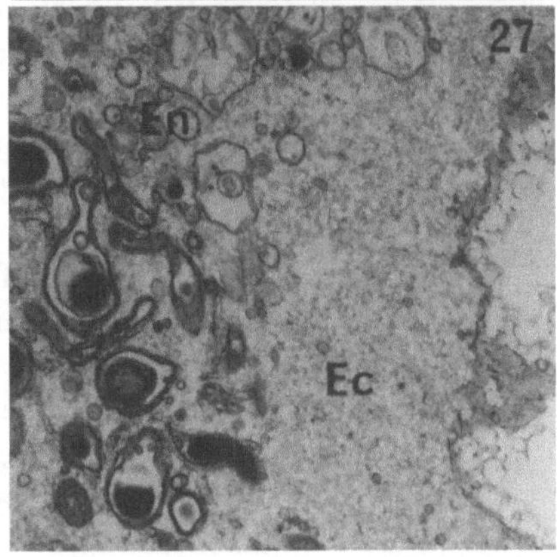

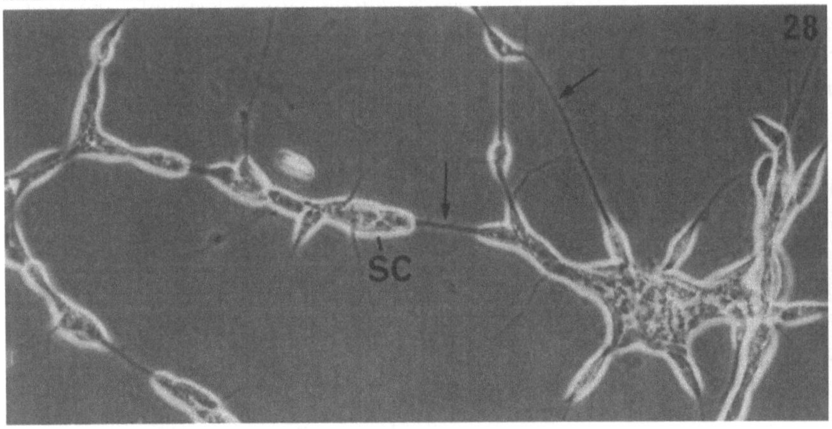

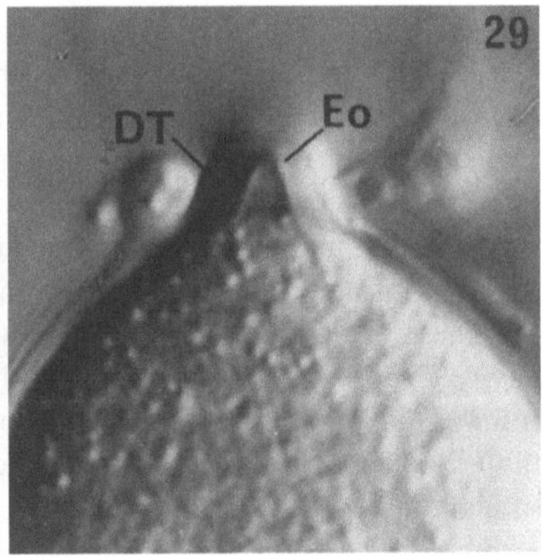

Fig. 22. Cytokalymma. View of a radiolarian cytokalymma (arrow) enclosing the growing tip of a spine, and surrounded by rhizopodial extensions of the extracapsular cytoplasm. SEM, × 1,000

Figs. 23 and **24.** Diatotepum

Fig. 23. Cross-section through the valve of the diatom *Coscinodiscus wailesii* showing chambered construction (outer layer to the left). Note organic coat (arrowhead) and inner organic layer (= diatotepum) (arrow). TEM, × 10,000 (courtesy of A.-M. M. Schmid)

Fig. 24. The same as in Fig. 23 after treatment with hydrofluoric acid; silica has dissolved, organic layers remain as in Fig. 23. TEM, × 10,000 (courtesy of A.-M. M. Schmid)

Fig. 25. Discharge plug. Prior to release of zoospores from the sporangium (*Sp*) of a chytridiomycete, a highly refractive, globose discharge plug (*DP*) emerges through a pore in an inoperculate discharge tube (*DT*). The rhizoidal system (*R*) of this simple thallus is visible. DIC, × 850

Fig. 26. Discharge vesicle. From the deciduous sporangium (*Sp*) of an oomycete a thin film, the discharge vesicle (arrows), distends and restrains the emerging zoospores (*Zs*) as a spherical mass. Eventually the discharge vesicle is broken, and zoospores disperse. DIC, × 1,000

Fig. 27. Ectoplasm. Edge of an amoeba cell showing the ectoplasm (*Ec*) and endoplasm (*En*). TEM, × 4,500

Fig. 28. Ectoplasmic net. Spindle-shaped cells (*SC*) of a labyrinthulid within interconnected strands of an ectoplasmic net (arrows). PC, × 800 (courtesy of D. Porter)

Fig. 29. Endooperculum (*Eo*) within discharge tube (*DT*) of sporangium of a chytridiomycete. DIC, × 1,300

described for a number of diatoms, mostly centric. The diatotepum is considered a distinct layer (= "diatotepic layer") in addition to the organic coat (= organic casing). The latter envelopes the surface of all siliceous components of the frustule.

Discharge plug

Definition: Gelatinous, highly refractive material secreted between the plasma membrane (q.v.) and sporangial wall at early stages of spore formation in chytrids (Fig. 25).
Comments: Found among diverse chytrids including those with operculate and inoperculate zoospore discharge.

Discharge vesicle

Definition: Extracellular material produced during zoosporogenesis in fungal and algal protists, which at spore or protoplast discharge, evaginates into an external sac, constraining spores until they rupture the vesicle (Fig. 26).
Comments: Involved in zoospore discharge of many algae and oomycetes such as *Pythium* and chytridiomycetes such as *Chytriomyces*. Also appropriately applied to the polaroplast of Microsporidia. See also, Gun cell.

Ectoplasm

Definition: Outer layer of the cytoplasm of a cell, bounded by the cell or plasma membrane (q.v.). In many rhizopod amoebae, it may be of considerable thickness, clear, non-granular, and more viscous than the underlying endoplasm (Fig. 27).
Comments: In ciliates, the special name "cortex" (q.v.) is often used for essentially this layer of protoplasm. See also, Ectoplasmic net, Pellicle.

Ectoplasmic net

Definition: Slime filaments interconnected into a fine network (Fig. 28) generated from a specialized membrane aggregate, the sagenogen. The sagenogen was first termed the bothrosome in the labyrinthulids (meaning pit body), and later the sagenogenetosome (meaning net-producing body) in the thraustochytrids. Sagenogen is a modification of the later term. In labyrinthulids, spindle-cells glide within the net; and in thraustochytrids, the net extends from the sporangium as rhizoidal-like structures. In both groups, the net is actually continuous with the plasma membrane (q.v.) which protrudes from a pore within the layer of cell-covering scales (q.v.).
Comments: At the light microscopy level, the ectoplasmic net has the appearance of extracellular material, but electron microscopy has revealed that in strict terms it is not, but rather is simply an elaboration of the plasma membrane.

Elaters, see Capillitium.

Endooperculum

Definition: Wall-like layer deposited between the plasma membrane (q.v.) and spore discharge area. In chytrids, it is typically secreted between the plasma membrane and discharge plug material just prior to zoosporogenesis (Figs. 29 and 30).
Comments: Distinguished from an exooperculum [= "operculum" (q.v.)] because it is distinct from the original sporangial wall and is an additional layer deposited late in sporangial development.

Endoskeleton, see Skeleton.

Envelope

Definition: A general term in protistology used variously for such structures as plasma membranes, cystic membranes, pellicles, coats, walls, sheaths and gelatinous coverings, and even shells or loricae.
Comments: The range of cell coverings loosely called "envelopes" precludes their consideration as homologous structures. See also, Lorica.

Epiplasm

Definition: A peripheral, cytoplasmic layer of varying thickness directly underlying the innermost cell membrane; thus, in ciliates, it is positioned just beneath the pellicular alveoli (q.v.). The epiplasm is part of the cortex (q.v.) of ciliates and in some species can be subdivided into a discontinuous set of plates (Fig. 17).

Exooperculum, see Operculum.

Exoskeleton, see Skeleton.

Extracellular matrix

Definition: Mucilaginous glycoproteins external to the plasma membrane (q.v.) or to the compact, microfibrillar cell wall (q.v.), characteristically appearing flocculent, loosely fibrillar, finely granular, or amorphous. It may totally surround a cell or be localized to specific regions at the cell surface. Certain zones may not be preserved or visualized with standard conventional fix-

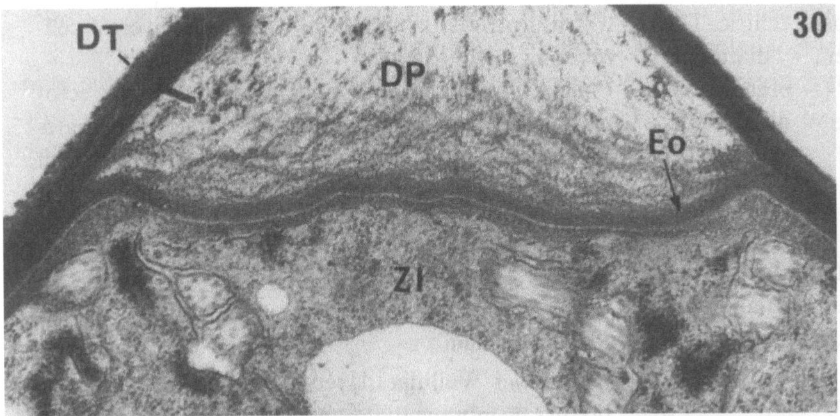

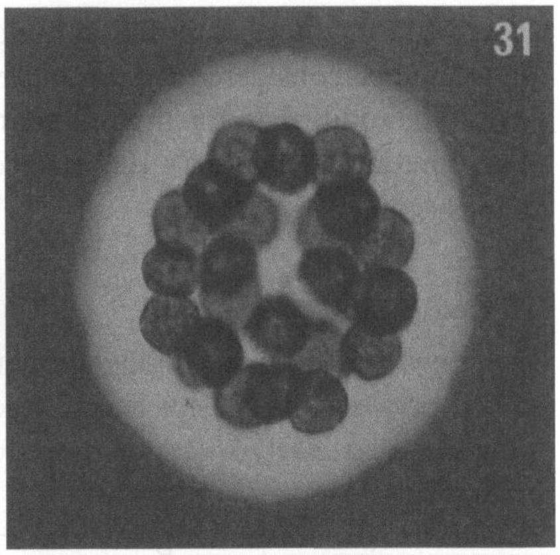

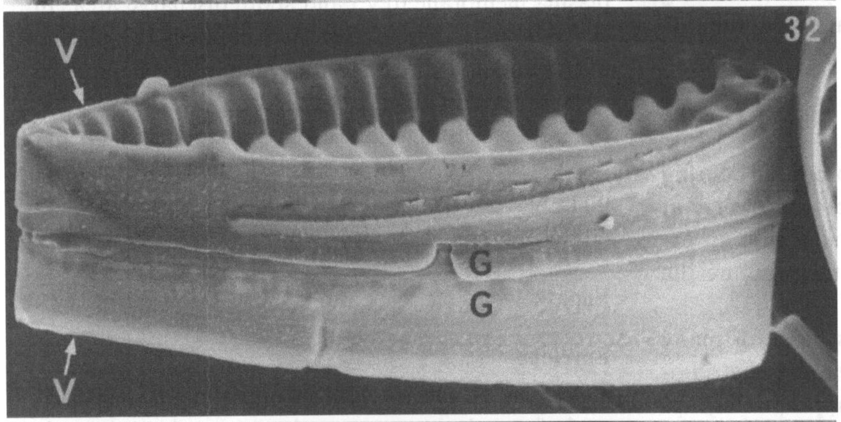

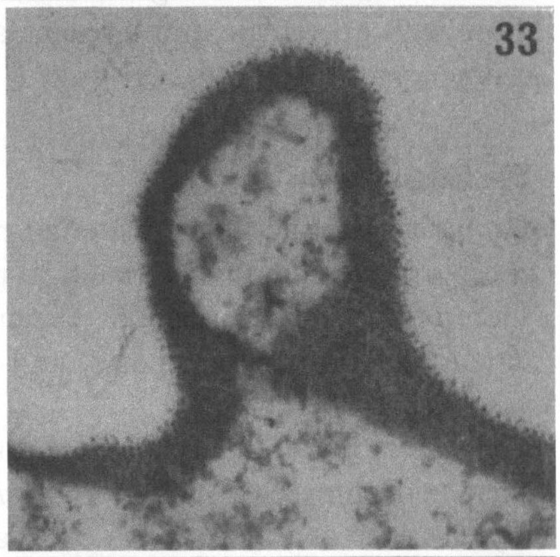

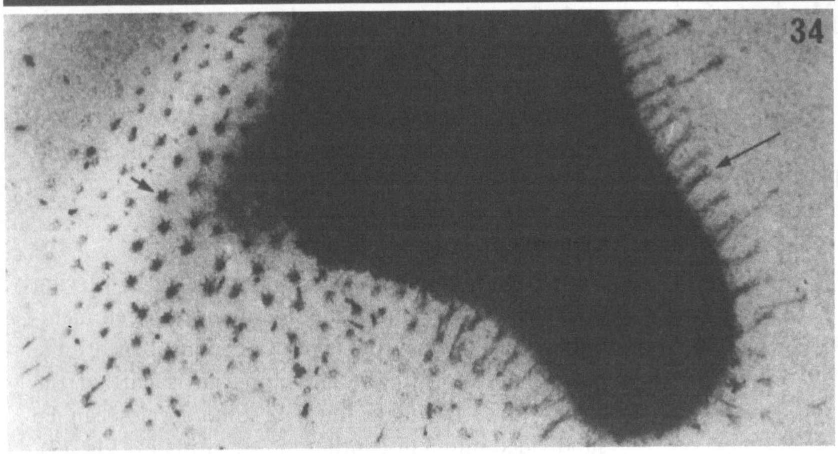

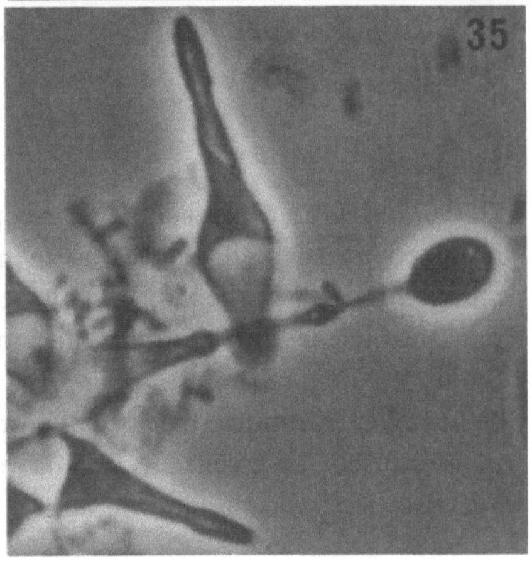

Fig. 30. Endooperculum. Section through sporangium of a chytridiomycete in which zoospore initials (*ZI*) have formed. Notice that finely granular material fills the interstitial spaces between zoospore initials and coats their plasma membrane surfaces. Within the discharge tube (*DT*), there is a discharge plug (*DP*). Located at the interface between the discharge plug and the interstitial extracellular material around zoospore initials, there is an endooperculum (*Eo*). When zoospores are released, the endooperculum is broken open. TEM, × 40,000

Fig. 31. Extracellular matrix. India ink preparation of the coenobic colonial green flagellate *Eudorina elegans*. The individual cells in the coenobium are held together by an extracellular matrix of hydroxyproline-rich glycoproteins that are similar to the cell wall of *Chlamydomonas*. BF, × 400

Fig. 32. Frustule of the diatom *Surirella biseriata* consisting of two valves (*V*) and associated girdle elements (*G*) (see also Fig. 80). SEM, × 1,200

Fig. 33. Glycocalyx. Prominent glycocalyx on the surface of an amoeba plasma membrane. TEM, × 55,100

Fig. 34. Glycostyle. Longitudinal (long arrow) and tangential (short arrow) sections of glycostyles attached to the surface membrane of the freshwater amoeba *Vannella platypodia*. TEM, × 50,000 (courtesy of F. C. Page)

Fig. 35. Gun cell of *Haptoglossa mirabilis* (Oomycetes) before and after firing. PC, × 2,500 (courtesy of G. L. Barron)

ation for electron microscopy, but may be revealed using cationic dyes such as ruthenium red or cryofixation techniques (Fig. 31).
Comments: This is a broadly descriptive term which encompasses different functional categories. "Cell coat" and "glycocalyx" (q.v.) refer to extracellular matrix adjacent to the plasma membrane. "Adhesion pad", "appressorium ring", "hyphal sheath", and "spore mucilage" (q.v.) describe extracellular matrix adjacent to the cell wall. Reported functions of this material include adhesion, collection of functional degradative enzymes and cell protection either with prevention of cell desiccation or binding of toxic compounds. See Conclusions and recommendations; see also, Holdfast adhesion, Sheath, Slime track.

Exudates, see Spore mucilage.

Frustule

Definition: All siliceous components of the vegetative cell wall (q.v.) of diatoms, comprising two valves (q.v.), the epivalve and hypovalve, and the girdle (or cingulum). The epivalve plus epicingulum comprises the epitheca, which is that component of the frustule inherited from the parent cell during division. The hypovalve plus hypocingulum constitutes the hypotheca, or that component of the frustule deposited by daughter cells following division (Fig. 32).
Comments: The frustule plus any additional organic component(s) (see Diatotepum), regardless of the composition, are together referred to as the diatom cell wall. It has been reported that silicification of the frustule may occur on an organic matrix (template?), though this has not been conclusively demonstrated. Frustule components, particularly the valves, are normally adorned with a vast range of morphological features defined by an equally vast and precise terminology. See also, Auxospore wall.

Fuzzy coat: see Glycocalyx.

Glycocalyx

Definition: An organic coat, covering, sheath (q.v.) or extracellular matrix (q.v.) extending from the outer surface of the plasma membrane (q.v.); e.g., the organic layer coating the surface membrane of many "naked" rhizopod amoebae (Fig. 33), the mucilaginous cell coat covering the plasma membrane in some "naked" algal protists (e.g., the green alga *Dunaliella*), the cell coat on zoospores of some oomycetes (e.g., *Phytophthora, Lagenidium*) and chytridiomycetes (e.g., *Polyphagus, Chytriomyces*), or the surface covering of other protists ranging from trypanosomes to ciliates (Fig. 78).
Comments: The surface coat, among other surface membrane components, confers antigenic cellular specificity. Variations among "naked" amoebae include surface coats that: (1) are amorphous and thin or thick, (2) consist of various configurations that are not resolvable into separate structures (fuzzy, with a suggestion of an organized pattern, or containing hexagonal or tubular elements), or (3) contain more discrete structures, varying in geometry, arranged in a definite pattern, and not separable intact from the surface membrane (see Glycostyle).

Glycostyle

Definition: Complex, flexible, discrete structures attached to the surface of cells (especially "naked" amoebae), each separate from its neighbors, arranged in a definite pattern, and not removable intact from the cell surface (Fig. 34).
Comments: In "naked" amoebae, the glycostyles vary among genera (e.g., pentagonally symmetrical in *Vannella*, and hexagonal in *Vexillifera*). Glycostyles are distinguished from other discrete surface structures such as scales, plates, and spicules (q.v.), by being flexible, sometimes erect, projections from the cell surface.

Gun cell

Definition: Developing from an encysting zoospore of the oomycete *Haptoglossa*, this apparatus consists of a cell with: (1) an extracellular plug and projectile located in an invaginated chamber of the cell surface and (2) an intracellular complex system of vacuoles. These components are used to forcibly discharge the missile-shaped projectile into rotifer or nematode hosts which collide with the cell, injecting infective sporidia through the host cuticle and into the host body where the fungal-protist grows into cylindrical thalli (Fig. 35).
Comments: This mechanism of host infection is similar to one found among some members of the plasmodiophoromycetes infecting plants (see Stachel), but differs in its explosive trigger and discharge of the structure used to pierce the host surface. The discharge vesicle (q.v.) of the Microsporidia can be considered a type of gun cell.

Holdfast adhesion

Definition: Extracellular matrix (q.v.), mucilaginous secretion, cementing one end of a filament, typically via a morphologically distinct holdfast cell, to a substrate.

Comments: Found with single algal filaments, particularly in shallow aquatic habitats. In actual use the term holdfast sometimes applies to a cell, to extracellular matrix, or to both a cell and secreted adhesive matrix. This term is also used among fungi such as with trichomycetes inhabiting guts of arthropods as commensals. The sessile state of the ciliate *Stentor* and many other sessile protists also produces an adhesive attachment at the posterior end. See also, Adhesions, Adhesion pad, Holdfast organelle, Stalk.

Holdfast organelle

Definition: Term broadly used in protozoology in reference to any structure(s) by which a given organism can affix or attach, temporarily or permanently, to some living or inanimate substrate. A wide range of organelles, often specialized and involving secreted or otherwise formed non-living and/or living structures, are implicated: e.g., cilia (individual or in thigmotactic fields), flagella, pseudopodia, tentacles, tails, suckers, hooks, spines, epimerites, sucking discs, threads, filaments, stalks of diverse kinds and loricae (q.v.).
Comments: As widely used by protozoologists, these diverse holdfast or adhesive organelles go beyond the rather restricted concepts of adhesions (q.v.) and holdfast adhesion (q.v.) as defined above.

House, see Lorica.

Hyphal sheath, see Sheath.

Hypothallus

Definition: A thin clear crust at the base of fruiting bodies of some Mycetozoa.
Comments: This structure is continuous with the acellular stalk (see Stalk) and appears as a disc. In protostelids this structure is also called a "basal disc" (q.v.) and in certain slime molds the hypothallus may be a continuous structure interconnecting the bases of multiple fruiting bodies. In coralline red algae the term "hypothallus" (= "hypothallium") is used for the lower part of the thallus composed of large cells.

Idiosome

Definition: An inorganic surface particle in the wall of testate amoebae or other biomineralizing protists, secreted within cytoplasmic vesicles and deposited on the surface of the newly formed cell during binary fission (Figs. 36 and 50).
Comments: Idiosomes, secreted by the cell, are distinguished from particles collected from the environment and cemented into the wall [= xenosomes (q.v.)]. The shape and chemical composition of idiosomes vary among species of testate amoebae and include pearl-

Fig. 36. Idiosome. Peg-shaped idiosome (*Id*) in test of the testate amoeba *Netzelia tuberculata*. SEM, × 8,400

Fig. 37. Keel (arrow) at the edge of a non-spinose planktonic forminiferan shell. SEM, × 150

Figs. 38 and **39.** Lorica

Fig. 38. Lorica of *Codonella cratera*, a common tintinnid (oligotrichid) ciliate in the plankton of freshwater lakes. The lorica consists of tiny sand grains and some diatom frustules (arrow). SEM, × 960 (courtesy of W. Foissner)

Fig. 39. Lorica of the colourless heterocont *Bicosoeca epiphytica*. TEM, × 1,700

Fig. 40. Operculum. A monocentric sporangium (*Sp*) of a chytridiomycete is growing on pollen (*Po*) and releasing zoospores (*Zs*) into a discharge vesicle. An operculum (*Op*) has been pushed from the discharge pore in the sporangium wall and is visible in profile as a curved hinge to the right side of the mass of zoospores. BF, × 1,000 (from slide collection of the late Dr. Charles Miller)

Fig. 41. Outer rodlet layer. Rotary-shadowed preparation of the outer rodlet layer of conidia of *Aspergillus flavus* (Ascomycetes). TEM, × 110,000 (courtesy of W. M. Hess)

Fig. 42. Pellicle. Cross-section through the cell surface of a euglenoid flagellate showing the ridge and groove construction of the pellicle. Each pellicle strip is bounded by the plasma membrane (*PM*), supported by a proteinaceous layer (*PL*) and underlain with microtubules (*Mt*). TEM, × 59,000 (courtesy of R. E. Triemer)

Figs. 43 and **44.** Peridium

Fig. 43. The peridium (*Pe*) covering the mass of spores contained in the sporangium of a zygomycete. The expanded apex of the sporangiophore (*Sp*), a special hyphal branch, forms the columella (*Co*) which protrudes into the sporangium. DIC, × 400

Fig. 44. In the sporangium of a slime mold, the peridium partially disappears leaving a network of threads connected with expanded nodes at the upper portion of the sporangium and a cup-like base at the lower portion of the sporangium. An acellular stalk (*St*) joins to the cup-like base of the peridium. DIC, × 250

shaped, globose, ovate, peg-shaped, or plate-like structures that may be either siliceous or calcareous.

Inner membrane complex, see Alveoli.

Keel

Definition: Any ridge or flange-like extracellular thickening resembling the longitudinal plate or timber (keel) on the bottom of the hull of a ship (Fig. 37).
Comments: In foraminifera, some species are reinforced by a thickened ridge along one edge of the shell; in some pennate diatoms, it occurs on the valves (q.v.); in euglenoid flagellates, the term "keel" is used to describe prominent pellicular ridges.

Lacuna, see Alveoli.

Lorica

Definition: A structure surrounding or enveloping the body of some ciliates (Fig. 38) and flagellates (Fig. 39); it fits loosely over the body proper of the organism, with a wide opening at one, occasionally both end(s),

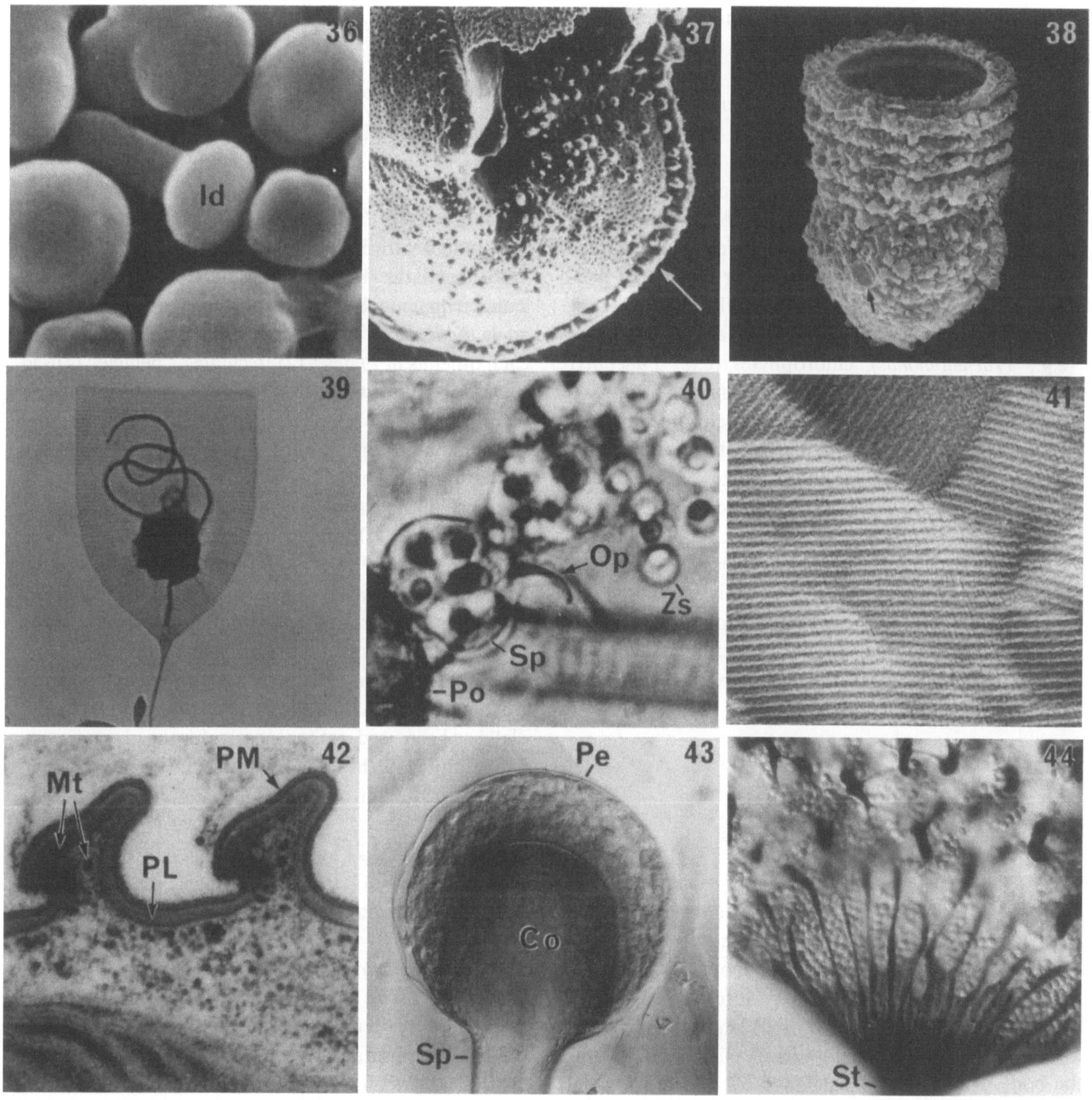

and is sometimes attached to a substrate at the other pole (see Holdfast organelle). In tintinnid oligotrichs and flagellated protists it is often carried about by the organism freely swimming in the water column. Its composition may be of a calcareous, chitinous, pseudochitinous, proteinaceous, siliceous, or tectinous nature, or it may be mineralized by iron and manganese compounds, or it may be made up of sand grains, coccoliths, diatom frustules, or simply debris held together by a secreted "glue".
Comments: Broadly speaking, the following terms are sometimes synonymized with "lorica": basket, case, envelope, house, sheath, shell, test, theca, tube, valve. Loricae are secreted or assembled by their occupants; when not solitary, they may appear in an arboroid "colony" arrangement. When an occupant has undergone cellular division during asexual reproduction in the lorica, one of the filial organisms often departs to secrete its own house elsewhere, the other remaining, often attached, in the parental abode. Peritrichs, folliculinids, and tintinnines are the major ciliate taxa in which loricae are to be found. As for flagellated protists, loricae occur in some choanoflagellates (acanthoecids), some (fossil) ebriids, euglenoid flagellates (e.g., *Trachelomonas*), chlorophytes (e.g., *Phacotus*) and heterokonts (e.g., *Bicosoeca, Dinobryon, Epipyxis*). *Epipyxis* produces a lorica composed of imbricate plates or scales (q.v.) of fibrillar construction. The loricae of acanthoecid choanoflagellates are uniquely basket-like (Fig. 8), composed of costal strips (see Costa, Skeleton). A fine membranous or fibrillar investment ("lorica membrane") lines part of these basket-like loricae and an extension of this attaches to the protoplast, holding it firmly in position within the lorica. The lorica of fossil ebriids is an inflated chamber-like structure which generally develops at the anterior end of the skeleton (q.v.).

Mucous pad, see Adhesion pad.

Nectar, see Spore mucilage.

Operculum

Definition: Literally, lid or bowl-shaped flap. Term used for diverse protist structures including (1) the "plug" of an emergence pore of a cyst or spore, (2) the whole or partial covering of certain peritrich loricae (q.v.), (3) the so-called "epistomial disc" (an anterior part of the body of certain other peritrich ciliates), or (4) the circumscissile bowl-shaped flap dissolved from a fungal sporangial wall allowing spore discharge (Fig. 40).
Comments: In fungi, "operculum" is sometimes synonymized with "exooperculum". Operculate (= exooperculate) discharge occurs on sporangia releasing zoospores among some chytrids. This term is also used among fungi describing asci-releasing ascospores.

Outer rodlet layer

Definition: Distinct zones of rods scattered in spore walls (q.v.) of ascospores and conidia; best visualized with freeze-etch techniques (Fig. 41).
Comments: In hydrophobic conidia the structure initiating interface with host insect cuticle of invertebrate fungal pathogens.

Pedicel, see Stalk.

Peduncle, see Stalk.

Pellicle

Definitions: (1) Proteinaceous plate or plates located just underneath the plasma membrane (q.v.) in euglenoid flagellates (Fig. 42). The individual plates (pellicular strips) may be capable of sliding along each other as in *Euglena*, or the plates may be firmly attached to each other as in *Phacus*. (2) In ciliates, "pellicle" is used to describe the outer part of the cell, comprising plasma membrane, alveoli, and the epiplasm (q.v.), but it is only the outer portion of the entire ciliate cortex (q.v.). (3) In dinoflagellates, it is used as a term for the "wall" of temporary cysts developing after the theca (q.v.) is shed by ecdysis. This type of pellicle comprises the fused inner amphiesmal vesicle (q.v.) membranes, an internal probably cellulosic layer and an external thin cellulosic (sometimes sporopollenin-containing) layer which originates from material present in the amphiesmal vesicles. (4) In various other protists (e.g., apicomplexans, opalinids, acantharians, scattered other taxa), "pellicle" has been used, rather loosely, for the plasma membrane plus any thickened material just below it, reminiscent of the terms "cortex" (q.v.) and "ectoplasm" (q.v.). (5) The term "pellicle" has also been used to denote the skin-like aggregations of zygotes developing after clumping of gametes of opposite mating types in chlamydomonadalean flagellates.
Comments: The pellicle of some euglenoid flagellates (e.g., *Euglena spirogyra*) is decorated by ferric hydroxide "warts". The term "pseudopellicle" has sometimes been used for euglenoid flagellates to describe the plasma membrane and the underlying proteinaceous plates. See Conclusions and recommendations; see also, Cuticle, Periplast.

Pellis, see Cuticle.

Peridium

Definition: A covering enclosing spores within fruiting bodies (Figs. 43–45).

Comments: In Mycetozoa the peridium originates from material secreted as plasmodia differentiate into fruiting bodies. In zygomycetes the peridium is the original sporangial wall enclosing spores after they are cleaved from a multinucleate sporangium and produce spore walls (q.v.). With fruitifications of higher fungi, the term is not restricted to extracellular material. Among rust fungi peridium pertains to the cup composed of hyphae surrounding aeciospores of fruiting bodies called aecia.

Perilemma, see Plasma membrane.

Periplast

Definition: In many protists, a colloquial term for a covering on the outside of the plasma membrane (q.v.), e.g., "a scaly periplast". In cryptomonad flagellates, a trilaminate structure consisting of proteinaceous inner and surface components that are closely associated with (i.e., "sandwich") the plasma membrane (Fig. 46). In euglenoid flagellates, sometimes used as a synonym of "pellicle" (q.v.).

Comments: In cryptomonad flagellates, the periplast covers the entire surface except for a subapical vestibulum from which the flagella emerge. Components are often composed of highly organized plate areas that correspond in size and position on either side of the plasma membrane. There is great variation on this simple theme; the inner component may consist of a single sheet or discrete plates that have variable relationships with the plasma membrane. The surface component, the details of which can only be viewed following freeze-fracture/etch, displays an even greater variation in structure; consisting of discrete plates or scales or fibrils or mucilage and combinations of any or all of these structures. See Conclusions and recommendations.

Perizonium, see Auxospore wall.

Plaque

Definition: (1) An organic plate-like structure cemented among the inorganic particles in the wall of testate amoebae; discoidal, with complex areolate or reticulated pore pattern in the central space (Fig. 47). (2) In ciliates, a rectangular arrangement of intramembranous ultramicroscopic particles in the ciliary membrane covering a cilium, at the proximal end of the cilium.

Comments: The plaques are secreted within Golgi-derived vesicles in the cytoplasm of testate amoebae, and are deposited by exocytosis on the surface of the newly formed cell during binary fission. Organic cement, secreted by other Golgi-derived vesicles, forms the matrix of the wall containing the plaques and test particles. Note that the term "adhesion plaque" [= "adhesion pad" (q.v.)] has a different meaning. "Plaque" is also used in bacteriology to describe a clear zone in a bacterial lawn on a culture plate. See also, Pore plate.

Plasma membrane

Definition: With rare exceptions, the outermost *living* membrane of all protistan cells; therefore, the homologue of the universal limiting unit "cell membrane" of the cytological literature, possessing the same ultrastructure and general functions (Fig. 14).

Comments: While "cell envelope", "cell membrane", "cytoplasmic membrane", and "plasmalemma" are often used as synonyms of "plasma membrane", certain other terms may relate to structures that either sometimes *include* the cell membrane, e.g., the cortex (q.v.) and, in ciliates, but not in euglenoid flagellates, the pellicle (q.v.), or are positionally in juxtaposition to it, e.g., the perilemma, an additional living membrane that lies just *out*side the plasma membrane (ciliary membrane), in some oligotrich ciliates and certain stichotrichous hypotrichs (Figs. 48 and 49), and the alveoli (q.v.), with their own unit membrane, lying just *in*side (under) it. Although we are not including motile appendages in our treatment, some remarks are made on clarifying the use of the term "membrane" and "membranelle". The plasma membrane (often called the "ciliary membrane" by ciliatologists) of the cell body also encloses, without a break, the central axoneme of the protruding flagellum or cilium; whether such structures are present individually or, as in many ciliates, in some compound form. The latter are sometimes called a membrane or membranelle (e.g., the paroral or undulating membrane and the adoral zone of membranelles or polykineties of the oral or buccal area and the cirri on the ventral somatic surface of hypotrichs). See also, Glycocalyx.

Plate

Definition: A relatively thin, flattened structure deposited on the surface of cells, usually inorganic (calcareous or siliceous) or organic as in the case of thecal plates in dinoflagellates; varying in geometry, circular, elliptical, quadrangular, or irregular in outline (Figs. 50 and 51).

Comments: Plates are secreted within cytoplasmic vesicles and deposited by exocytosis on the cell surface in some amoebae and centrohelidian heliozoa (or cysts in actinophrydian heliozoa), or enclosed in amphiesmal vesicles (q.v.) in dinoflagellates. A layer of skeletal plates is secreted within alveoli (q.v.) of the ciliate *Euplotes*. The plates may be planar to slightly concave, concavo-convex, or biconvex; perimeters variable, round to polygonal. Plates can be considered a form of scale (q.v.) or a synonym, but in general a plate is much thinner than broad, whereas a scale may be much more elaborate in three-dimensions. See also, Lorica, Periplast, Pore plate.

Pore plate

Definition: An organic or mineralized septum containing pores (Fig. 52).

Comments: In foraminifera, some pores in the shell contain septa, spanning the opening usually at a position below the rim, and penetrated by micropores

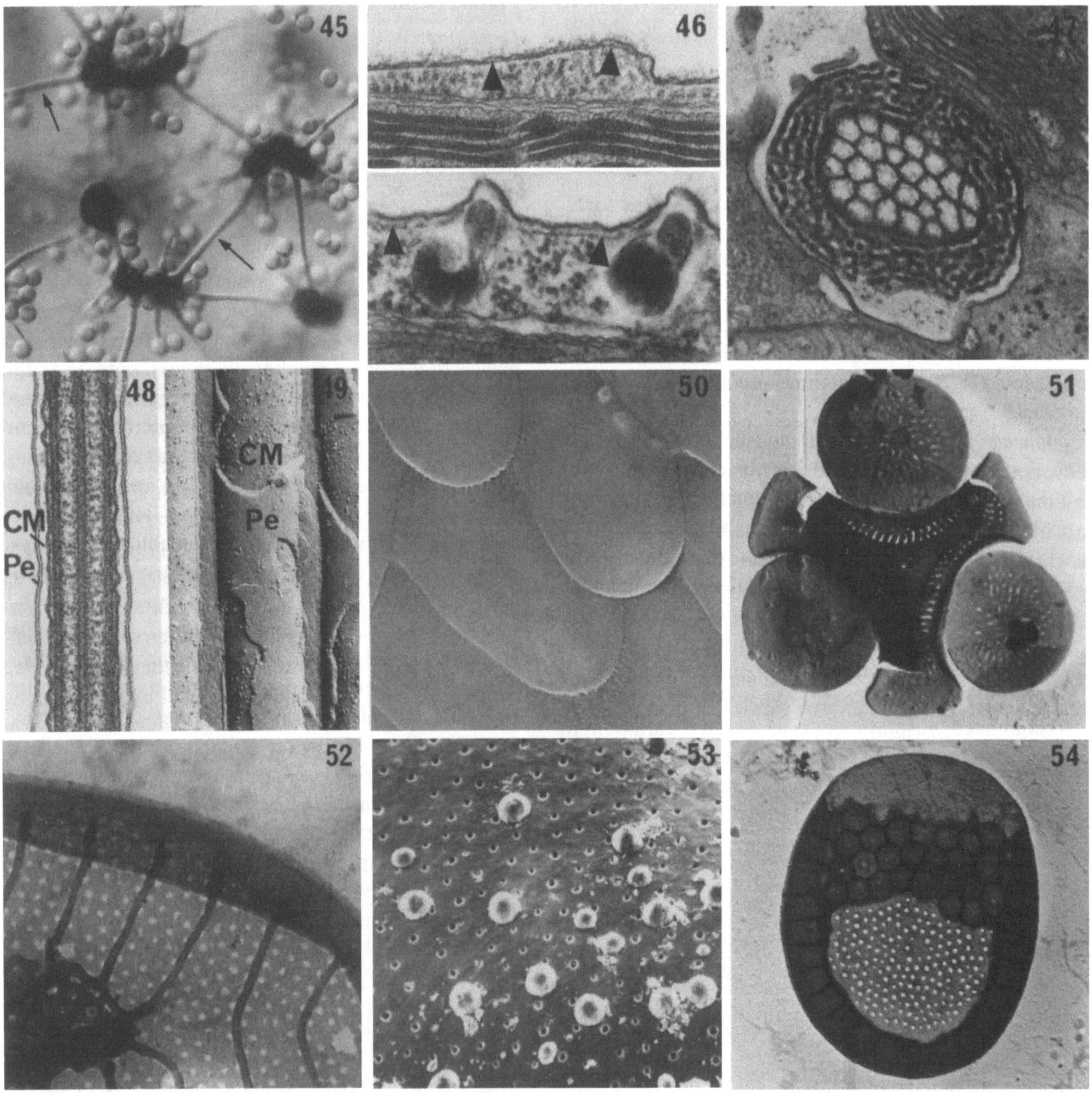

that often contain thin, cytoplasmic projections. A fine silicified plate with pores that span the areola of the frustule (q.v.) of many diatoms. An organic, sometimes partially silicified, areolate plate or plaque (q.v.) is deposited in the test (q.v.) of some testate amoebae.

Properizonium, see Auxospore wall.

Pseudopellicle, see Pellicle.

Pustule

Definition: Blister-like, papillate, or knob-like projection on the surface of a cell enclosure, or eruptive spot or spore mass in fungi (Fig. 53).
Comments: Pustules occur as calcified projections on the outer surface of foraminiferal shells, or as ornamentations on the surfaces of frustules and thecae (q.v.) of some algae.

Sagenogen, see Ectoplasmic net.

Scale

Definition: An organic or inorganic cell-surface element of variable geometry, distributed individually or arranged in a pattern sometimes forming an envelope around the cell. The scale may be thin and plate-like (solid and ornamented or perforated with varying patterns), composed of meshwork or latticed structures, or complex, three-dimensional shapes forming basket-shaped, boat-shaped, crown-shaped, tower-like, or spine-bearing structures (as in some flagellates, amoebae and heliozoa) (Figs. 6, 9, 52, 54, 55, and 68).
Comments: Surface scales are secreted by a wide range of protist taxa spanning diverse algae, flagellates, sarcodinians, ciliates, labyrinthulids, thraustochytrids and protostelids. Bristles, coccoliths, and plates (q.v.) may be considered special forms of surface scales. The fine structural features of scales, as well as the position and manner of their formation and deployment, are important taxonomic characteristics. See also, Auxospore wall, Lorica.

Sheath

Definition: A general term for some sort of outer covering or extracellular matrix (q.v.) of a protist cell or of an extension of such a cell: e.g., the gelatinous investment outside the cell wall of some coccoid and filamentous algae (Fig. 56); the non-living outer covering, or annulus, of the stalk (q.v.) of many peritrich ciliates; the material surrounding a germ tube of directly germinating oomycete cystospores (Fig. 3).
Comments: Sheaths also commonly occur in cyanobacteria. The sheath in most organisms is composed of mucopolysaccharides or glycoproteins. The degree of development of a sheath is sometimes associated with the type of the substrate the cell contacts. "Hyphal

Fig. 45. Peridium. Arrows point out details of the persistent threads of the peridium of a slime mold. The round cells are spores which remain in the sporangium. DIC, × 500

Fig. 46. Periplast. Thin sections through the periplast of the haplomorph (top) and diplomorph (bottom) of the cryptomonad *Proteomonas sulcata*. The inner periplast component (arrowheads) is located either just beneath the plasma membrane (bottom) or tightly appressed to it (top). The surface periplast component is not obvious. TEMs, × 50,000

Fig. 47. Plaque. Fenestrated organic plaque secreted in the cytoplasm and eventually deposited in the wall of a testate amoeba. TEM, × 53,900

Figs. 48 and **49.** Perilemma

Fig. 48. Thin section through a cilium of the ciliate *Urostylis grandis* showing the perilemma (*Pe*) which covers each individual cilium. The perilemma is slightly thicker than the ciliary membrane (*CM*) and clearly shows a unit membrane character as in other biological membranes. TEM, × 50,000 (courtesy of C. F. Bardele)

Fig. 49. Freeze-fracture aspect of the perilemma (*Pe*) in a cilium of the cilate *Uroleptus caudatus*; *CM* ciliary membrane. TEM, × 50,000 (courtesy of C. F. Bardele)

Figs. 50 and **51.** Plate

Fig. 50. Siliceous plates (idiosomes) in the wall of a testate amoeba. SEM, × 8,000

Fig. 51. Isolated plates from a broken plate case of *Triparma* cf. *retinervis* (Parmales, Chrysophyceae?). TEM, × 12,000 (courtesy of H. A. Thomsen)

Fig. 52. Pore plate in the scale of *Synura petersenii* (Synurophyceae). TEM, × 38,000

Fig. 53. Pustules on the surface of a non-spinose planktonic forminiferan shell. SEM, × 600

Fig. 54. Scale. Siliceous scale of *Synura uvella* (Synurophyceae). TEM, × 13,000

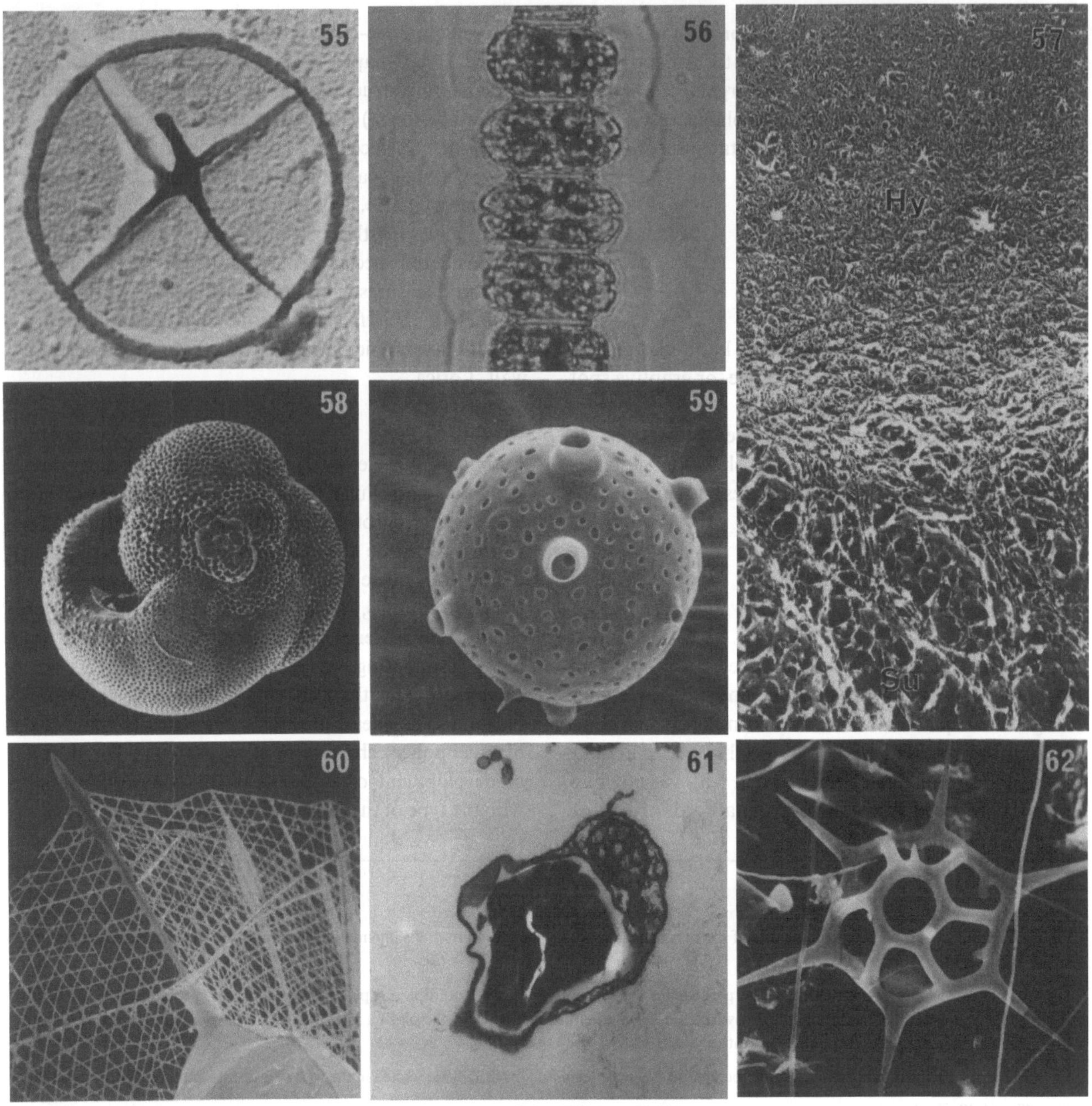

Fig. 55. Scale of *Pseudodendromonas insignis* (Pseudodendromonadida). TEM, × 50,000

Figs. 56 and **57.** Sheath

Fig. 56. Filamentous desmid (*Desmidium swartzii*) surrounded by a copious gelatinous sheath. BF, × 600

Fig. 57. Hyphal sheath of *Uromyces appendiculatus* (Basidiomycetes). Note its tight fibrillar nature over hypha (*Hy*) while over the substrate (*Su*) it is more loosely organized. Rotary-shadowed deep-etch preparation, TEM, × 28,000

Figs. 58 and **59.** Shell

Fig. 58. Calcitic shell of a planktonic foraminiferan. SEM, × 100

Fig. 59. Siliceous shell of a colonial radiolarian. SEM, × 700

Figs. 60–62. Skeleton

Fig. 60. Siliceous framework of a radiolarian skeleton. SEM, × 1,000

Fig. 61. An ultrathin section of a segment of siliceous skeleton (endoskeleton) of a radiolarian enclosed by cytoplasm. TEM, × 50,000

Fig. 62. Siliceous skeleton of the silicoflagellate *Dictyocha speculum*. SEM, × 2,000 (courtesy of H. A. Thomsen)

sheath" (Fig. 57) refers to extracellular material along hyphae of oomycetes as well as "higher" fungi. Hyphal sheaths are particularly prominent with aquatic hyphomycetes. The degree of development of a sheath is sometimes associated with the type of substrate the hypha contacts. See also, Adhesion pad, Envelope, Glycocalyx, Lorica, Slime track.

Shell

Definition: A shell is a protective inorganic or organic enclosure, deposited directly by the organism or built-up by cementing foreign particles into a wall matrix. Its wall is usually solid and smooth or variously ornamented. One or more openings (apertures, pores, etc.) provide access to the surrounding environment (Figs. 58 and 59). Also used for the cellular enclosure known as a "spore membrane" enclosing myxosporidian spores; the myxosporidian shell is composed of two to six valves (q.v.) or sections with thickened edges and often bearing sculpturing, markings, or processes on the outer surface.
Comments: The shell geometry is often used as a taxonomic indicator. "Shell" and "test" (q.v.) are sometimes used synonymously, and a clear distinction is not easily maintained; however, in general, a shell may be considered to be a rather thick, wall-like or framework enclosure with one or more major openings. See also, Envelope, Lorica.

Skeleton

Definition: A hardened non-living, protective or supportive structure, enclosed by, or attached to, cytoplasmic structures. In a strict usage, it includes only non-living frameworks and spine-bearing structures that form anchorage or support for cytoplasm. This excludes smooth-surfaced, solid-walled shells, tests, or loricate structures lacking supportive spines or open frameworks. An endoskeleton is partially or completely enclosed by cytoplasm; an exoskeleton lies totally outside the cytoplasm, though cytoplasmic structures may be attached to it (Figs. 60–62).
Comments: Few protists have a true exoskeleton, since most of the skeletal frameworks are at least partially surrounded by cytoplasm at some time (e.g., in diatoms). The elaborate siliceous skeletal frameworks of radiolaria and the spine-bearing shells of planktonic foraminifera may be considered endoskeletons since they are partially surrounded by, and form a supportive framework for, cytoplasm. External skeletons occur in silicoflagellates, but evidence indicates that they are formed internally and then extruded by the cell (detailed ultrastructural studies on the origin of the skeleton are missing). The skeletal elements of the basket-like lorica (q.v.) of choanoflagellates are also produced intracellularly and extruded to the cell surface; when a full complement has been produced, they are rapidly assembled to produce a basket (q.v.) (Figs. 8 and 18).

Slime sheath, see Slime track.

Slime track

Definition: Extracellular matrix (q.v.) sloughed off plasmodia of slime molds and most amoebae, particularly amoebae as pseudoplasmodia of cellular slime molds migrate prior to culmination and differentiation of sorocarps.
Comments: This material is most prominent in cellular slime molds that do not lay out a horizontal stalk during migration. The term "slime sheath" is most widely used to describe the continuous, extracellular material that remains around the pseudoplasmodia.

Somatonemes

Definitions: Surface, hair-like structures on the soma (body) of some flagellates such as proteromonads (Figs. 63 and 64).
Comments: The somatonemes are synthesized in the endoplasmic reticulum and migrate to the Golgi vesicles before they are discharged on the cell surface. There they appear to be linked to cortical microtubules by anchoring proteins which are part of the plasma membrane (q.v.).

Spicule

Definition: A rod-like, spindle-shaped, stellate, or variously curved and ornamented structure, usually siliceous or calcareous, with blunt or tapered tips deposited individually on the cell surface or distributed throughout the peripheral cytoplasm (Figs. 65 and 66).
Comments: The term spicule is occasionally used to mean a tapered rod-like extension attached to a skeletal framework as in radiolaria, but preferentially the term spicule is used for unattached elements. The geometry of the scattered spicules is a taxonomic characteristic of some actinopod species. In some solitary and colonial radiolaria, the spicules may be lunate, tri-radiate to hexa-radiate, s-shaped, or variously curved, needle-like elements scattered within the gelatinous coat surrounding the cell. Among other examples, curved spi-

cules occur in centrohelidian heliozoa (produced within specialized organelles called "spicule-generating organelles"), and calcareous spicules are secreted by the marine amoeboid testacean *Trichosphaerium* spp. Endoskeletal siliceous spicules occur in some dinoflagellates (e.g., in the genus *Plectodinium*). The framework of thin, siliceous rods surrounding some choanoflagellates (e.g., *Diaphanoeca* spp.) is known as a spicular basket (see Costa). A wide variety of siliceous spicules are also secreted by sponges and form supportive structures for the soft body.

Spine

Definition: A non-living (e.g., organic, calcareous or siliceous), rod-like or tapered, sometimes ornamented, elongate structure attached to a scale, shell, wall, or skeletal framework (Figs. 67 and 68).

Comments: Spines can serve as supportive structures for cytoplasm and as protection. For example, in radiolaria and planktonic foraminifera the spines are attachment sites for feeding rhizopodia and provide a mechanical advantage during capture of large prey. The cross-section may be round, elliptical, triradiate, or of

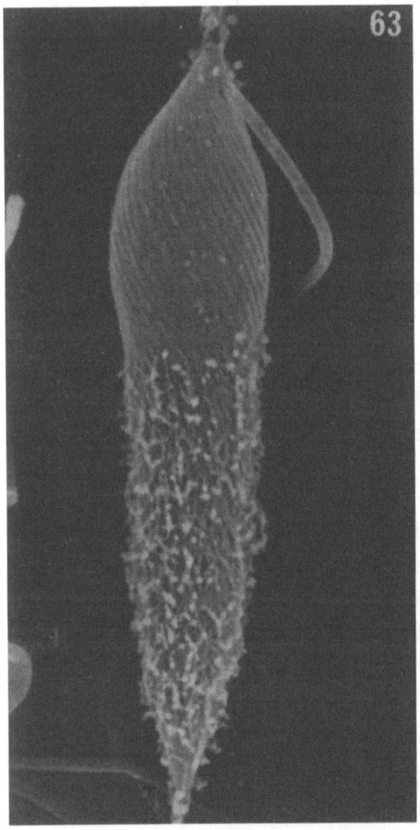

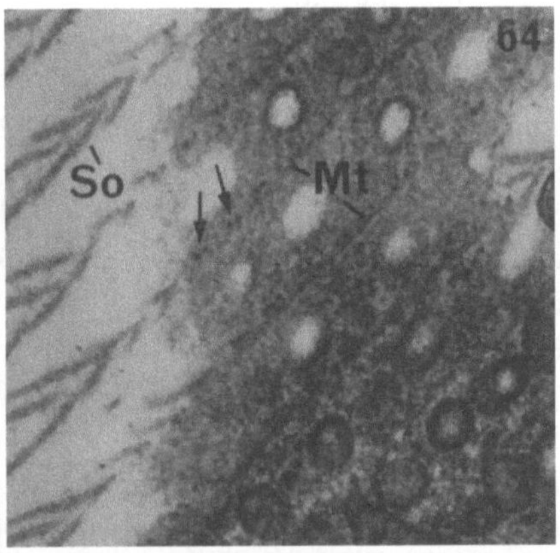

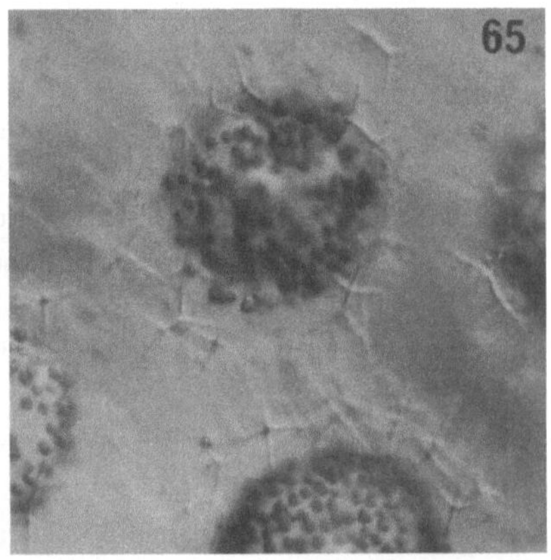

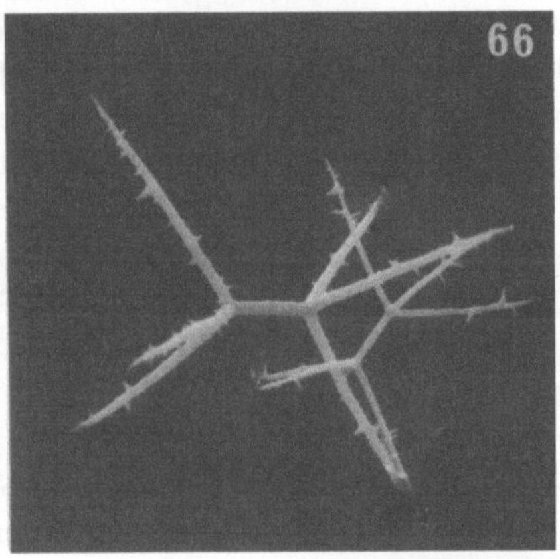

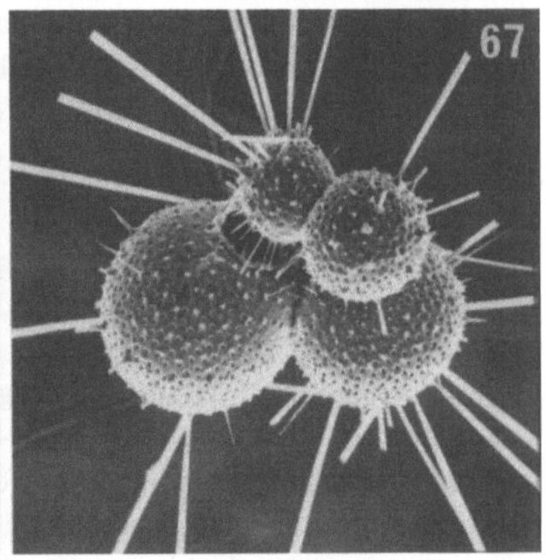

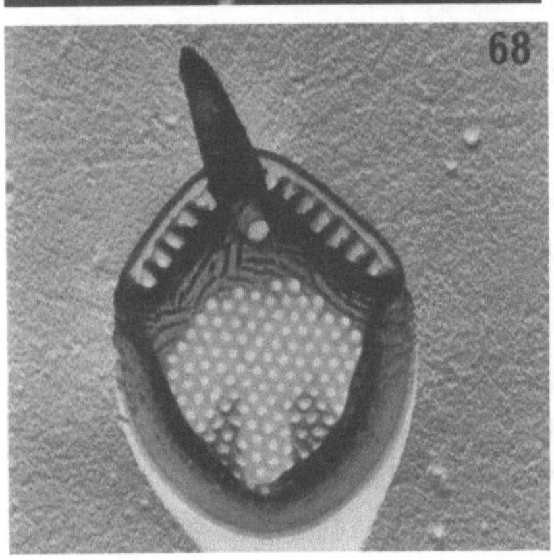

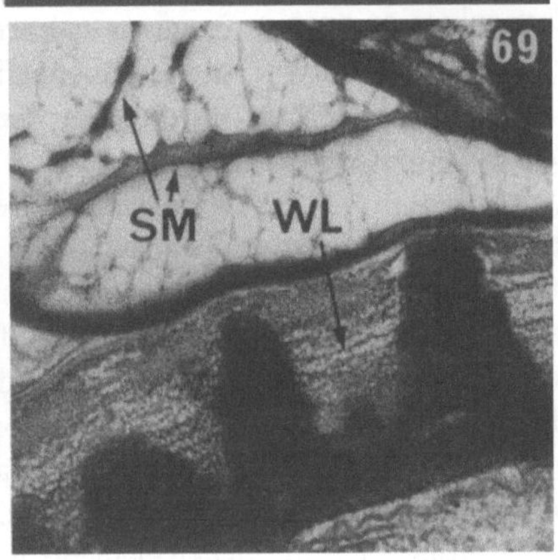

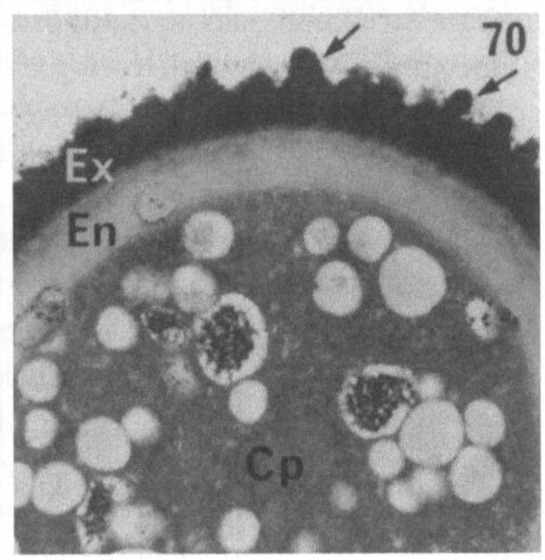

complex design and the surface may be smooth, supplied with short spines or embellished with a variety of ornamentations. Prominent spines occur on frustules (q.v.) of some diatoms, on scales (q.v.) of some chrysophytes and synurophytes, and on dinoflagellate cysts. In the synurophyte genus *Mallomonas*, stiff spines on the scales should not be confused with movable bristles (q.v.), which are elongate structures developed intracellularly separate from scales, joining together with scales only after their release on the cell surface.

Spore mucilage

Definition: Extracellular matrix (q.v.) commonly believed to be rich in carbohydrates, preventing spore desiccation or attracting and attaching to animal vectors. Spores released from generative structures typically cluster in the mucilage as a globular mass (Fig. 69).
Comments: "Exudates" and "nectar" are synonyms of "spore mucilage". Examples include spermatia of black-stem rust of wheat fungus on barberrry host, conidia of Dutch elm disease fungus, conidia of chestnut blight fungus, conidia of ergot fungus, oidia of inky cap mushroom, and ascospores of yeast *Dipodascopsis uninucleatus*.

Spore tip mucilage, see Adhesion pad.

Spore wall

Definition: The cell wall (q.v.) of a spore (a microscopic propagule capable of giving rise to a new individual(s) or transferring genetic information). Surface features, architectural complexity and chemical composition of the spore wall vary among taxonomic groups (Fig. 70).
Comments: The fungal spore wall serves as a protective barrier from environmental stresses (i.e., cold, heat, desiccation, radiation) and microorganism attack and may aid in spore dispersal. The surfaces of spores may be smooth or ornamented. Examples of surface ornamentation include appendages (conidia of *Pestalotia macrotricha*), warts (ascospores of *Genea hispidula*), spines (uredospores of *Puccinia graminis*), reticulations (ascospores of *Tuber rufum*), verrucae (aeciospores of *Cronartium quercuum*), and ridges (ascospores of *Neurospora crassa*). Spore surfaces may also have scars, pegs, tubes (all marking the points of attachment), or germ pores. Many spores have a fine ornamentation that is detected only with high resolution electron microscopy (see Outer rodlet layer). The spore wall is typically composed of more than one wall layer and may have as many as five or more distinct layers when examined with transmission electron microscopy. Layer terminology varies between authors. The most commonly used terms, starting from the inner wall layer, are: (1) endosporium (= endospore), (2) episporium (= epispore), (3) exosporium (= exospore, epitunica, trachytectum), (4) perisporium (= mucostratum, myxosporium), and (5) ectosporium. Some parasitic protozoa (e.g., apicomplexans, microsporidians, and myxosporeans) produce a small dispersal, infective stage known as a "spore". This, however, is not strictly equivalent to the environmentally resistant or quiescent dispersal stages cited for fungi and other protists. The "spores" of these parasitic protozoa are sometimes enclosed by valves (q.v.) with ridges or thickenings at the point of contact (sutures), and may have additional surface features such as a mucus coat, filamentous

---

Fig. 63. Cell surface of the proteromonad *Proteromonas lacertae*. Note the corrugated surface of the anterior half of the cell and the hairy posterior zone covered with somatonemes. SEM, × 3,400 (courtesy of G. Brugerolle)

Fig. 64. Grazing section of the posterior part of the cell of the proteromonad *Proteromonas lacertae*. Note the pairs of dense dots (arrows) along the cortical microtubules (*Mt*) to which the somatoneme pairs (*So*) are anchored. TEM, × 38,000 (courtesy of G. Brugerolle)

Figs. 65 and **66.** Spicule

Fig. 65. Clusters of siliceous spicules surrounding the central capsule of a colonial radiolarian. DIC, × 250

Fig. 66. Siliceous radiolarian spicules. SEM, × 1,000

Figs. 67 and **68.** Spine

Fig. 67. Spines on a calcitic shell of a planktonic foraminiferan. SEM, × 200

Fig. 68. Spine on a scale of *Synura echinulata*. TEM, × 13,000

Fig. 69. Spore mucilage. Outer wall layer (*WL*) of a teliospore of the smut fungus *Ustilago bullata* (Basidiomycetes) and associated spore mucilage (*SM*). TEM, × 39,000

Fig. 70. Spore wall (teliospore wall) of *Ustilago bullata* (Basidiomycetes). Note endosporium (*En*), exosporium (*Ex*), spines (arrows) and poorly preserved spore cytoplasm (*Cp*). TEM, × 13,000

fleece, elongated caudal appendage, or warts that aid in their dispersal. The quiescent stages of free-living protozoan protists are known as cysts rather than spores. See also, Auxospore wall, Cyst wall.

Stachel

Definition: An osmiophilic knife-shaped structure which develops outside the cell in a cavity, the tubular invagination of the plasma membrane (q.v.) termed "Rohr", in encysted zoospores of plasmodiophoromycetes. The Stachel is surrounded by adhesive material and after contact with a plant host, the Stachel pierces the cell wall, allowing entry of the parasite protoplast (Fig. 71).

Comments: Although this structure might be considered an example of an extrusive organelle, it is assembled external to the plasma membrane and is not an intracellular organelle.

Stalk

Definition: An elongate structure specifically formed (1) to attach an organism to a living or non-living substrate or (2) to raise spores or group of spores (e.g., sporangia) above a substrate or host, a strategy which increases spore dispersal potential by wind or animal vectors in terrestrial habitats or by turbulence in aquatic environment (Figs. 72–74).

Comments: Broadly speaking, the following terms are sometimes synonymized with "stalk": pedicel, peduncle, and stipe. These are general terms widely used for both non-living and living structures. The various kinds of non-living stalks may have different origins. The acellular stalk (Figs. 72 and 73), as found among some myxogastria, protostelids, and chytrids, arises as a mound or tube from secretion of extracellular material. In some protostelids and chytrids the acellular stalk is hollow, and in protostelids the microfibrillar stalk tube is covered with a sheath (q.v.). Algae (e.g., diatoms) often attach to substrates with polysaccharide acellular stalks. Living stalks can be merely extensions of the body in many algal and protozoan protists. Cellular stalks of dictyostelids (Fig. 74) form as cells aggregate and raise the sorocarp above the substrate surface and die at culmination of fruiting body formation, leaving their cell walls as a dictyoid, non-living stalk. The cellular stalks of dictyostelids may be homologous to those of the acrasids which are composed of living cells. Multicellular stalks (stipes) also attach some algae (e.g., benthic, macroscopic brown algae) to substrates. In ciliates (e.g., peritrichs and suctoria), stalks commonly arise by extension of (living parts of) the body of the organism or by secretion of specific non-living materials or by some combination of these means. The stalk is sometimes produced from a specialized region known

---

Fig. 71. Stachel. Primary cyst of *Plasmodiophora brassicae* (Plasmodiophoromycetes) showing the location of the Stachel (*Sta*) within the Rohr (*Ro*) and the nucleus (*Nu*). TEM, × 36,700 (courtesy of J. R. Aist)

Figs. 72–74. Stalk

Fig. 72. In the fruiting body of a slime mold, a solid finely granular acellular stalk (*St*) supports the sporangium. DIC, × 130

Fig. 73. A hollow acellular stalk (*St*) elongates from a basally located cyst (*Cy*) and supports at its tip an aerial sporangium (*Sp*) of a chytridiomycete. DIC, × 1,000

Fig. 74. Cell walls, devoid of cellular contents, remain as a scaffold, the cellular stalk (*St*) supporting a sporangium of a dicytostelid. Spores (*S*) and sheath (*Sh*) of extracellular material coating the stalk are visible. BF, × 500

Fig. 75. Test of *Netzelia tuberculata* (see also Figs. 36 and 81). SEM, × 700

Figs. 76–79. Theca

Fig. 76. Theca of the choanoflagellate *Salpingoeca gracilis*. TEM, × 2,500

Fig. 77. Theca of the dinoflagellate *Protoperidinium thorianum*. SEM, × 750 (courtesy of G. Hansen)

Fig. 78. Section showing thecae (*T*) of two daughter cells of the prasinophyte *Scherffelia dubia* at different stages of development. The theca of the left cell is at an early stage of assembly showing its composition of scales. Secondary cytoskeletal microtubules subtend the plasma membrane (arrowheads) during scale aggregation. The theca of the right cell is almost complete. TEM, × 70,000

Fig. 79. Rotary-shadowed deep-etch of theca of *Scherffelia dubia*. TEM, × 140,000

Fig. 80. Valve view of the diatom *Surirella biseriata* (cf. Fig. 32). SEM, × 1,100

Fig. 81. Xenosome (arrow), a particle collected from the environment, silica coated, and deposited in the test of *Netzelia tuberculata*. SEM, × 2,600

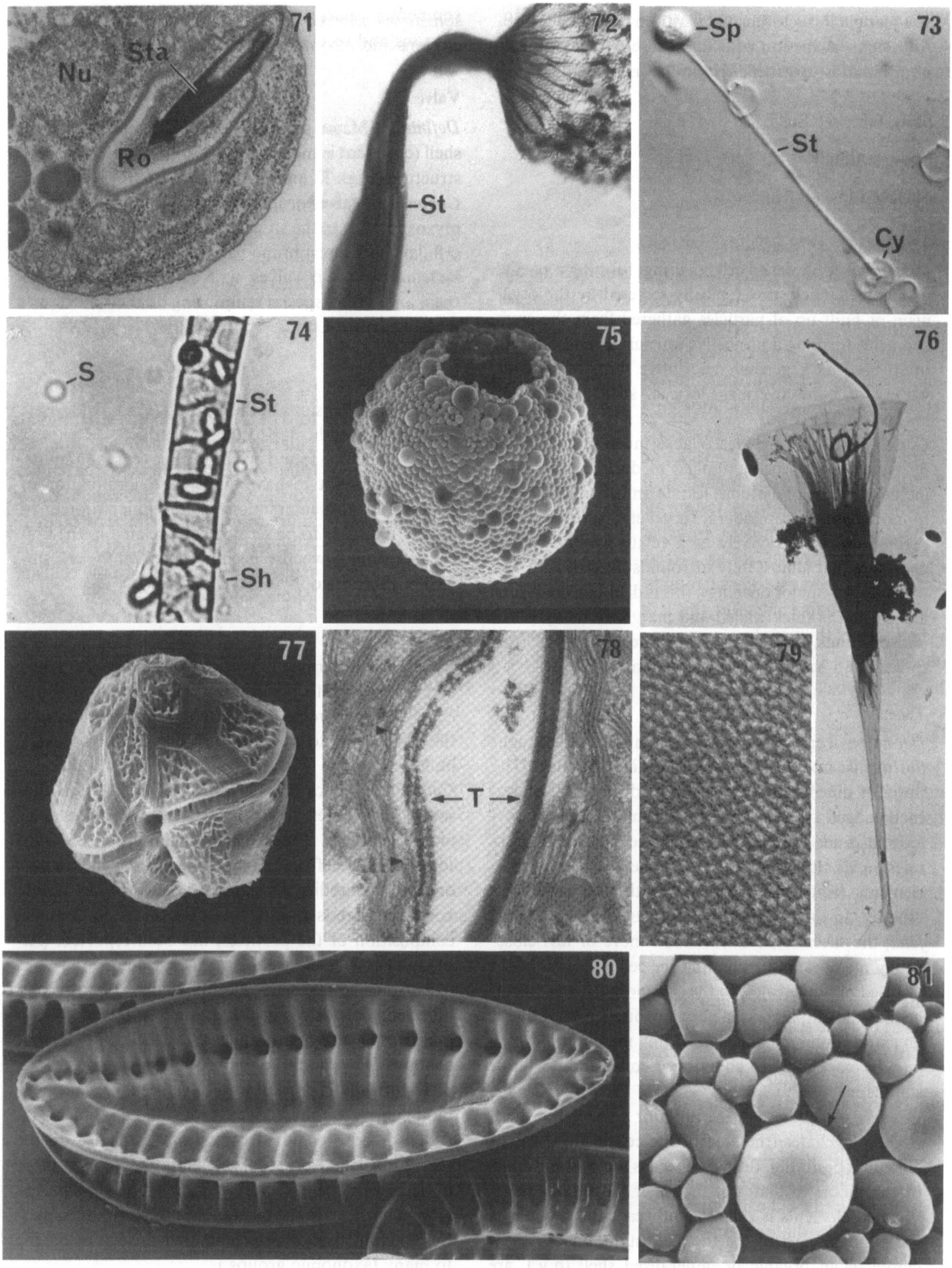
71
Nu
Sta
Ro
72
St
73
Sp
St
Cy
74
S
St
Sh
75
76
77
78
T
79
80
81

as a scapula in suctorian and peritrich ciliates. See also, Adhesions, Adhesion pad, Basal disc, Holdfast adhesion, Holdfast organelle, Hypothallus, Sheath.

Stalk tube, see Stalk.

Stipe, see Stalk.

Surface coat, see Glycocalyx.

Test

Definition: A hardened cell covering, forming a partial or complete enclosure, typically secreted by the organism or built up of particles gathered from the environment, forming a protective barrier around the cell; often used as a general descriptive term for a wide variety of hard-coverings, loricae, shells, thecae, or valves (q.v.) (Fig. 75).

Comments: The term "test" lacks much specificity in its typical usages being applied to diverse enclosures including those of diatoms, testate amoebae, and some monothalamic foraminifera. In general, a test may be defined more restrictively as a relatively thin-walled enclosure that almost fully surrounds the cell, with one or only a few major openings; this is distinguished from a shell that is thick-walled and may have many pores and openings, or a lorica that more loosely encloses the cell and has few to many openings.

Theca

Definition: Term used in various connections throughout protist taxa to describe a wall-like structure. The theca of dinoflagellates is composed of thecal plates, each lodged in a vesicle [amphiesmal vesicle (q.v.)] located inside the plasma membrane. In diatom frustules (q.v.), the valve (q.v.) together with the girdle elements (cingulum) associated with it is called the "theca". In the prasinophytes *Tetraselmis* and *Scherffelia*, the theca is an extracellular more or less close-fitting membrane surrounding the cell. It is composed of fused organic scales. In the choanoflagellates, members of the Salpingoecidae are surrounded by a theca, an extracellular usually close-fitting membrane, which may be extended into an anterior collar and posterior stalk (Figs. 76–79).

Comments: Epitheca is the anterior part of the theca in dinoflagellates (the area of the cell in front of the cingulum or girdle), the hypotheca is the posterior part of the theca. In diatoms, the older valve and its associated elements is called "epitheca", the hypotheca is the newer part of the theca. Testate amoebae, enclosed in an organic or mineralized shell (q.v.), are sometimes referred to as thecate amoebae. See Conclusions and recommendations; see also, Lorica, Test.

Valve

Definition: Major section of a frustule, theca, test, or shell (q.v.) that is more or less recognizable as separate structure (Figs. 32 and 80).

Comments: Valves occur, for example, in diatoms and myxosporidia. In the so-called shell valves of the living cellular "spore membrane" of myxosporidia, there may be from two to six valves, with suture planes between them and, often, special sculpturing on their outer surfaces. See also, Lorica, Spore wall.

Wall fimbriae

Definition: Extensions of the cell wall (q.v.) that radiate as fibers perpendicular to the cell surface.

Comments: Often found on pathogenic yeast and smuts, they are thought to have a role in determination of fungus and host interaction. Extracellular protein fibrils that appear to be serologically related to the fimbriae or fungi have also been observed on the cell surface of prymnesiophyte flagellates.

Xenosome

Definition: A particle collected from the environment and cemented into the organic matrix of a wall. Typically used for foreign mineral matter incorporated into the wall of testate amoebae or other agglutinate-wall-building protists (Fig. 81).

Comments: Xenosomes are distinguished from idiosomes. The latter are secreted de novo within cytoplasmic vesicles. The chemical composition of xenosomes, collected from the environment, depends in part on the availability of particles; however, in testate amoebae, species-specific selection, based on chemical composition of the particles, has been reported. (Although not in reference to a surface structure, the word "xenosome" has recently been introduced into the evolutionary protistological literature as a name for an intracytoplasmic or intranuclear organelle or body that contains DNA and is membrane-bounded and that is, or historically is *presumed* to have been, an endosymbiotic entity within a given protistan cell, having invaded or been engulfed by the host organism in the recent or distant past, subsequently becoming a more or less permanent inclusion in the host body, e.g., Kappa particles in the ciliate *Paramecium* and the plastids and mitochondria so common in protists belonging to many taxonomic groups.)

Conclusions and recommendations

Many terms associated with the cell surface of protists go back to the time before the advent of electron microscopy. Present knowledge, based on detailed studies on ultrastructure, has provided evidence that the terminology used is often unsatisfactory, the same term being used in the different groups of protists for different (non-homologous) structures. This may be illustrated in the thecate amoebae, in which the "theca" has been termed a test, shell, lorica, and theca and (in *Cochliopodium*) also stiffened pellicle and tectum (see Bark 1973). Electron microscopy has shown that the "theca" of the latter is a scaly periplast as found in many flagellates.

Terms used in widely different contexts which need to be reconsidered are: alveoli, extracellular matrix, pellicle, periplast and theca. These will be briefly discussed below.

Alveoli

If it could be proven that the structure in the three so-called "alveolate" groups of protists (the ciliates, dinoflagellates, and apicomplexans) is truly a homologous one, then there would be justification for such a wide usage of the term. But, at the very least, it should be retained for the ciliates where it has long been used.

Extracellular matrix

As discussed above, this term is used to identify numerous categories of cell surface structures (adhesion pad, appressorium ring, cell coat, glycocalyx, holdfast adhesion, hyphal sheath, sheath, slime track, spore mucilage) which may differ, sometimes significantly, in function, organization, and chemical composition. We recommend its usage only in the most general sense, preferring the use of descriptive/functional terms when applicable.

Pellicle

This term is also used in widely different contexts. Its use should be restricted to one only or it should be abandoned entirely. The "pellicle" in euglenoids is a proteinaceous intracellular plate system not to be confused with the "pellicle" of ciliates, which comprises both plasmalemma, alveoli and epiplasm. The "pellicle" surrounding the dinoflagellate cyst probably deserves a separate term. Considering the meaning of the word pellicle ("little skin") it is perhaps to be preferred to restrict the use of this term to the skin-like aggregation of zygotes that develop during sexual reproduction in chlamydomonad flagellates.

Periplast

We suggest restriction of this word to serve as a colloquial term for a covering on the outside of the plasmalemma (e.g., "scaly periplast"). A new term should be coined for the "periplast" of cryptomonads, a structure which comprises also cytoplasmic constituents.

Theca

This term is used in widely different contexts. Its use in prasinophytes such as *Tetraselmis* is superfluous and should be discontinued. The "theca" in *Tetraselmis* is a periplast of fused scales.

The "theca" in salpingoecid choanoflagellates is a lorica and we recommend that specialists working on choanoflagellates reconsider the terminology presently used in this group. Is the "lorica membrane" of acanthoecid choanoflagellates the homologue of the "theca" (lorica) of salpingoecid choanoflagellates? "Basket" may be an alternative for the silicified part of the lorica in acanthoecid flagellates.

The term "theca" is also used in dinoflagellates and diatoms. This use should be discouraged, a difficult recommendation considering the widespread use of "thecate" or "athecate" dinoflagellates. The term is superfluous in dinoflagellates, and "amphiesma" (containing amphiesmal plates) should take its place. "Epitheca" and "hypotheca" may be replaced with "epicone" and "hypocone". Whether special terms need to be coined to replace "thecate dinoflagellates" and "athecate dinoflagellates" is less clear, considering that separation of the dinoflagellates into these two groups is artificial.

The use of the term "thecate amoeba" needs to be carefully reconsidered, since present usages for the term "theca" do not conform to the diverse enclosures found among this group.

While we agree that the present confusion must be resolved by the establishment of new terms, we have refrained from doing so here. The rationale has been that specialists in the different groups of organisms need to be consulted and, if necessary, should be involved in establishing the new terminology. It is the intent of our effort to stimulate discussion and research on cell surface features, particularly of protistan and protistan-like organisms. Much still awaits discovery about functional roles, mechanisms of assembly, biochemical composition, and potential homologies of extracellular material. The members of this Task Force hope to convene again in the future to reconsider this terminology, revising it based on input from new results, insights, and synthesis.

Acknowledgements

We appreciate the scholarly input of many colleagues who aided our efforts in preparing this paper, and in particular we thank C. F. Bardele and M. A. Farmer for their comments on the manuscript. We also express our thanks to J. R. Aist, C. F. Bardele, G. L. Barron, D. Breger, G. Brugerolle, U. Christen, R. M. Crawford, W. Foissner, L. Fritz, G. Hansen, W. M. Hess, C. Kinkade, D. G. Mann, G. I. McFadden, F. C. Page, D. Porter, A.-M. M. Schmid, H. A. Thomsen, R. E. Triemer, and the late Charles Miller slide collection for providing some of the figures. The assistance with the photographic work by A. Zuppiger is also gratefully acknowledged.

References

Alexopoulos CJ, Mims CW (1979) Introductory mycology, 3rd edn. Wiley, New York

Andersen RA, Barr DJS, Lynn DH, Melkonian M, Moestrup Ø, Sleigh M (1991) Terminology and nomenclature of the cytoskeletal elements associated with the flagellar/ciliary apparatus of protists. Protoplasma 164: 1–8

Anderson OR (1988) Comparative protozoology: ecology, physiology, life history. Springer, Berlin Heidelberg New York Tokyo

Bark AW (1973) A study of the genus *Cochliopodium* Hertwig and Lesser 1874. Protistologica 9: 119–138

Berner T (ed) (1993) Ultrastructure of microalgae. CRC Press, Boca Raton

Bold HC, Wynne MJ (1985) Introduction to the algae, 2nd edn. Prentice-Hall, Englewood Cliffs

Bowman BH, Taylor JW, Brownlee AG, Lee J, Lu S-D, White TH (1992) Molecular evolution of the fungi: relationship of the basidiomycetes, ascomycetes, and chytridiomycetes. Mol Biol Evol 9: 285–296

Cavalier-Smith T (1986) The Kingdom Chromista: origin and systematics. Prog Phycol Res 4: 309–347

Corliss JO (1993) Should there be a separate code of nomenclature for the protists? BioSystems 28: 1–14

Grell KG (1973) Protozoology. Springer, Berlin Heidelberg New York

Harrison FW, Corliss JO (eds) (1991) Protozoa, vol 1. Wiley-Liss, New York [Harrison FW (ed) Microscopic anatomy of invertebrates]

Hawksworth DL, Sutton BC, Ainsworth GC (1983) Ainsworth and Bisby's dictionary of the fungi, 7th edn. Commonwealth Mycological Institute, Kew, Surrey

Kugrens P, Lee RE, Corliss JO (1994) Ultrastructure, biogenesis, and function of extrusive organelles in selected non-ciliate protists. Protoplasma 181: 164–190

Lee JJ, Hutner SH, Bovee EC (eds) (1985) An illustrated guide to the Protozoa. Society of Protozoologists, Lawrence, KS

Margulis L, Corliss JO, Melkonian M, Chapman DJ (eds) (1990) Handbook of Protoctista. Jones and Bartlett, Boston

– McKhann HI, Olendzenski L (eds) (1993) Illustrated glossary of Protoctista. Jones and Bartlett, Boston

Moore-Landecker E (1990) Fundamentals of the fungi, 3rd edn. Prentice-Hall, Englewood Cliffs

Olive LS (1975) The Mycetozoans. Academic Press, New York

Ross R, Cox EJ, Karayeva NI, Mann DG, Paddock TBB, Simonsen R, Sims PA (1979) An amended terminology for the siliceous components of the diatom cell. Nova Hedwigia Beih 64: 513–533

Sleigh MA (1989) Protozoa and other protists. Edward Arnold, London

van den Hoek C, Mann DG, Jahns HM (1994) Algae. Cambridge University Press, Cambridge

Wainright PO, Hinkle G, Sogin ML, Stickel SK (1993) Monophyletic origins of the Metazoa: an evolutionary link with fungi. Science 260: 340–342

Protoplasma (1994) 181: 29–42

PROTOPLASMA

Siliceous structures and silicification in flagellated protists

H. R. Preisig*

Institute of Systematic Botany, University of Zurich, Zurich

Received October 14, 1993
Accepted January 16, 1994

Summary. Flagellated protists produce a diverse range of siliceous structures, such as internal and external skeletons, scales, spines, bristles, cell walls, cyst walls, and loricae. The different groups of silica-depositing flagellates, i.e., chrysophytes/synurophytes, choanoflagellates, dinoflagellates, ebriids, silicoflagellates, thaumatomastigids, and the genus *Petasaria* are reviewed. Brief mention is also given to those algal groups in which silicification is uncommon and rare (i.e., chlorophytes, euglenophytes, haptophytes/prymnesiophytes, xanthophytes/tribophytes), but in which silicified structures nevertheless occur in few flagellate genera. Special attention is given to aspects of morphology and development of the different siliceous structures as well as on aspects of systematics and taxonomy.

Keywords: Protists; Flagellates; Algae; Siliceous structures; Morphogenesis; Taxonomy.

Introduction

The utilization of silica, one of the most abundant minerals on earth, in the formation of skeletal structures is well-known from many organisms (see, e.g., Simpson and Volcani 1981, Evered and O'Connor 1986). Several groups of protistan organisms have made especially wide use of this mineral which, in the majority of them, is deposited intracellularly as amorphous silica within membrane-bounded vesicles known as silica deposition vesicles (SDVs). Sometimes silica is also deposited extracellularly in association with cell walls or loricae. Among the protists, morphogenesis and silicification have been studied most extensively in diatoms and sarcodines (Schmid 1994, Anderson 1994). A diverse range of siliceous structures is also known from many flagellated protists. Many of these structures are characteristically patterned and ornamented and have fascinated microscopists with their intricacy and diversity, and almost all serve as the basis for the taxonomy of their groups. This paper presents a survey of these taxonomically important structures in flagellated protists and describes what is known about the mechanisms involved in the formation of these structures.

* Correspondence and reprints: Institut für Systematische Botanik, Universität Zürich, Zollikerstrasse 107, CH-8008 Zürich, Switzerland.

Chrysophytes/Synurophytes

The distinctive resting cysts (stomatocysts) and scales (including scale-like structures such as spines and bristles) are the main siliceous structures found in Chrysophyceae and Synurophyceae. These two algal classes have only recently been separated based on pigment and ultrastructure data (Andersen 1987, 1989; Moestrup 1994). Nevertheless they have many features in common and even now, as before 1987, the "synurophytes" (orders Synurales/Mallomonadales) are sometimes included in the Chrysophyceae (e.g., Kristiansen 1990). In synurophytes (i.e., the 120+ described species of *Mallomonas, Synura, Chrysodidymus, Tessellaria*) siliceous scales are always present, and stomatocysts are probably also produced by all species, though for many species these have not yet been described. Chrysophytes sensu stricto (see Preisig 1994), on the other hand, comprise about 160 genera and more than 900 species, the majority of which are capable of producing stomatocysts (Bourrelly 1963), whereas siliceous scales only exist in about 65 species (Preisig et al. 1991), i.e., members of the family Paraphysomonadaceae (genera

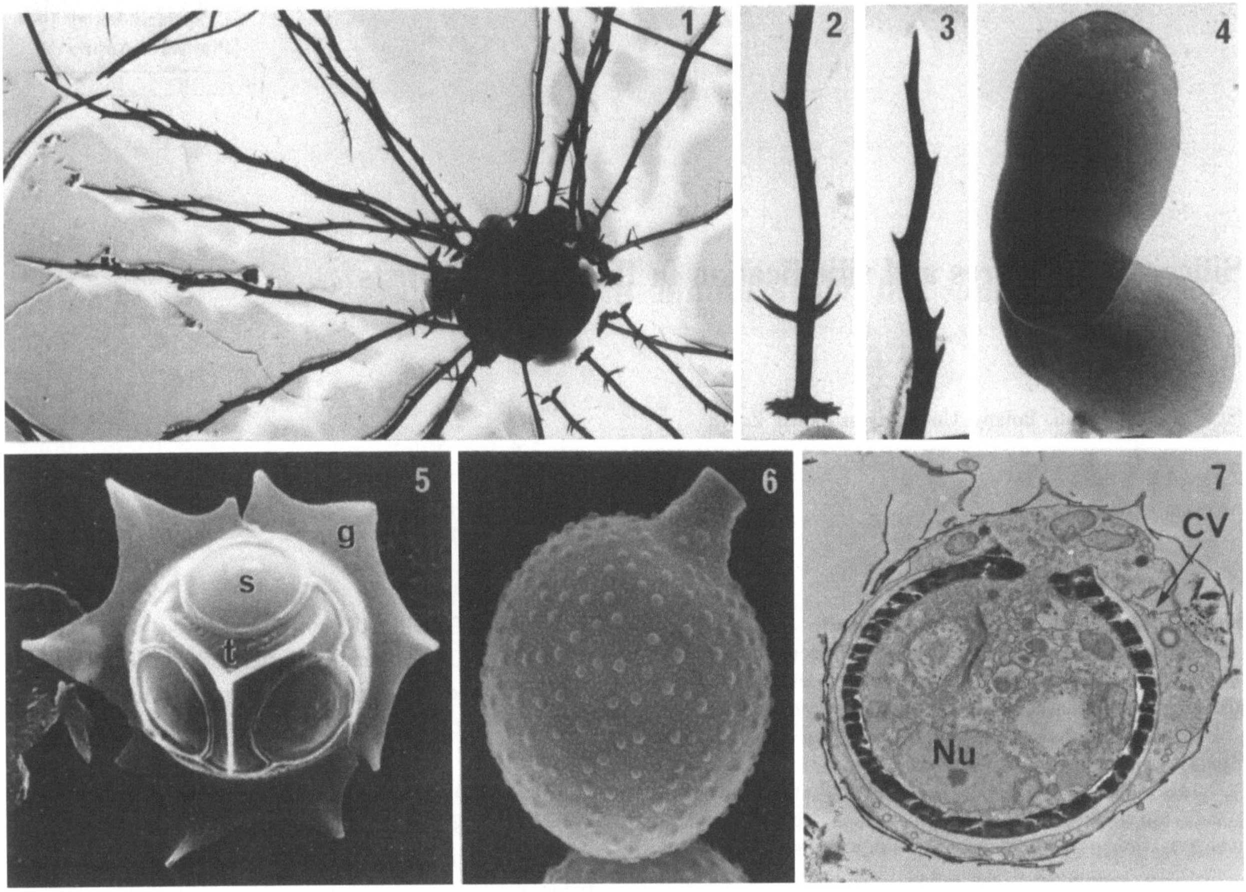

Figs. 1–4. *Meringosphaera mediterranea*. Cell covered by siliceous scales and spines (Fig. 1, × 2,650); spine base (Fig. 2, × 7,600), spine tip (Fig. 3, × 7,600) and scales (Fig. 4, × 15,000) at higher magnification. TEM (courtesy of H. A. Thomsen)

Fig. 5. *Triparma laevis* ssp. *mexicana* (Parmales, Chrysophyceae?). In this view, seven of the eight siliceous plates covering the cell are visible: one triradiate (*t*) plate, three shield (*s*) and three girdle (*g*) plates (ventral plate not visible). SEM, × 11,500 (courtesy of H. A. Thomsen)

Fig. 6. *Syncrypta pallida*. Stomatocyst. SEM, × 4,050

Fig. 7. *Paraphysomonas corynephora*. Section passing vertically through the pore region of a developing stomatocyst inside the cell; *CV* contractile vacuole; *Nu* nucleus. TEM, × 7,500

Chrysosphaerella, Paraphysomonas, Polylepidomonas, and *Spiniferomonas*).

Siliceous scales and stomatocysts are important taxonomic characters (see below), and their fossils have been used in paleoecological studies following recognition that they may be reliable environmental indicators (Smol 1988, 1994). So far, most of these studies have been restricted to Holocene sediments, but stomatocysts are known from at least the Cretaceous (Tappan 1980). Siliceous scales which appear to have chrysophycean (synurophycean) affinities have even been recorded from 800 million year old Lower Cambrian cherts (Allison 1981).

Silicification may also occur in loricae (envelopes) of some chrysophytes (e.g., species of *Chrysococcus*, see Bourrelly 1963), or extraneous particles of primarily siliceous composition can be accreted to form a cover outside the lorica as in *Bicosoeca accreta* (Hibberd 1978). These species are poorly known and will not be discussed further here. Only brief mention will be given to the siliceous scales and spines of the coccoid genera *Aurosphaera* and *Meringosphaera* (Figs. 1–4) (Leadbeater 1974) and the siliceous cell walls of Parmales (Fig. 5) which consist of a number of plates arranged in symmetrical patterns (Booth and Marchant 1987, Mann and Marchant 1989, Kosman et al. 1993). All these taxa are poorly understood and have an uncertain taxonomic position, though they have often been assigned to the Chrysophyceae. Nothing is known about the mode of formation of their siliceous structures.

Stomatocysts

Stomatocysts (also called statospores) generally range in diameter from ca. 3 µm to 35 µm. They are usually spherical or oblong in shape (Fig. 6) and have a single pore which may be surrounded by a collar. The pore is sealed by an organic plug containing little or no silica. Stomatocysts show wide variation in structure and ornamentation and may be smooth, reticulated or ornamented with markings such as spines, ridges, punctae and depressions of various design. Ultrastructure and life history strategies of stomatocysts have been comprehensively reviewed by Sandgren (1983 a, 1991).

Stomatocyst formation

For a survey on morphogenesis of stomatocysts the reader is referred to the studies of Sandgren (1983 a, 1989). So far there is no evidence for fundamental differences between synurophytes and chrysophytes sensu stricto regarding stomatocyst formation (Fig. 7). During early stages of encystment the Golgi apparatus is very active producing vesicles which appear to fuse forming a large, spherical silica deposition vesicle (SDV) analogous to the SDV of diatoms. In contrast to the diatoms, however, the SDV appears to be fully formed prior to silica deposition. The pore of the stomatocyst appears to result from a preformed aperture in the SDV or, alternatively, by a rupture in the SDV during an early stage in wall silicification. These differences concerning pore formation together with some related features of encystment may have potential phylogenetic significance within the chrysophytes/synurophytes (Sandgren 1983 a). Silicification of the stomatocyst wall is a continuous process, believed to occur in two phases. Primary silicification within the SDV proceeds rapidly from a silicification site in the cell posterior towards the anterior leading to a thin primary wall. This wall is subsequently thickened and the collar and surface ornamentation added during a slower and more controlled secondary phase of silicification. This secondary phase resembles the development of areolate valves and complex ornamentation in diatoms, where intimate contact between the SDV membrane and the growing siliceous structures is usually observed. Primary stomatocyst silicification, on the other hand, differs from diatom valve silica deposition in that it appears to proceed without recourse to templating on membrane surfaces. Further research on stomatocyst formation is certainly needed to confirm these differences.

Stomatocysts as taxonomic characteristics

Stomatocysts often have species-specific morphology. Even within the same genus different species may produce stomatocysts with unique characteristics (Cronberg 1986). On the other hand, many species produce stomatocysts that are difficult to differentiate (e.g., simple unornamented stomatocysts are formed in different chrysophyte families). Some species produce stomatocysts with a large variation in form, size and external morphology (Cronberg 1988). Immature stomatocysts that lack a full complement of ornamentation can also be found. The mature pore-collar complex is of high taxonomic value, but only when fully developed. In addition, the degree of ornamentation of the stomatocyst is known to be influenced by the physiological state of the organism during encystment (Sandgren 1983 b). Clearly, more research is needed in order to fully document the degree of variation in the structure of the stomatocyst that exists for individual taxa.
It is also important to note that thus far more than 600 stomatocyst types have been described with light or electron microscopy, but only about 45 have been related to the species that produced them. A variety of artificial classifications have attempted to provide a temporary nomenclature for stomatocysts (for a proposal for the standardization of stomatocyst nomenclature, see Cronberg and Sandgren 1986). Fossil silicified cysts thought to be of chrysophyte origin, but with no taxonomic or ecological information about the growing phase, have often been assigned to the families Chrysostomataceae (freshwater forms) and Archaeomonadaceae (marine forms) in paleontology (see Tappan 1980, Mitchell and Silver 1986, Lipps 1993).

Scales, spines, and bristles

Synurophytes

Siliceous scales occur in all synurophytes (Figs. 8–11), either encasing individual cells (e.g., *Mallomonas, Synura, Chrysodidymus*) or surrounding the entire colony (*Synura lapponica, Tessellaria volvocina*). Scales may range in length from about 1 µm (e.g., caudal scales of *Mallomonas dickii*) to over 10 µm (body scales of *M. pseudocoronata*). They are generally oval or elongate, with clear anterio-posterior differentiation of ornamentation (Asmund and Kristiansen 1986, Kristiansen 1986, Siver 1991). Most scales exhibit essentially bilateral symmetry, though they are usually slightly asymmetric due to the shape of the surface structures. Scales on the extreme ends of a cell are generally more asymmetric. Ornamentation of the scales includes pat-

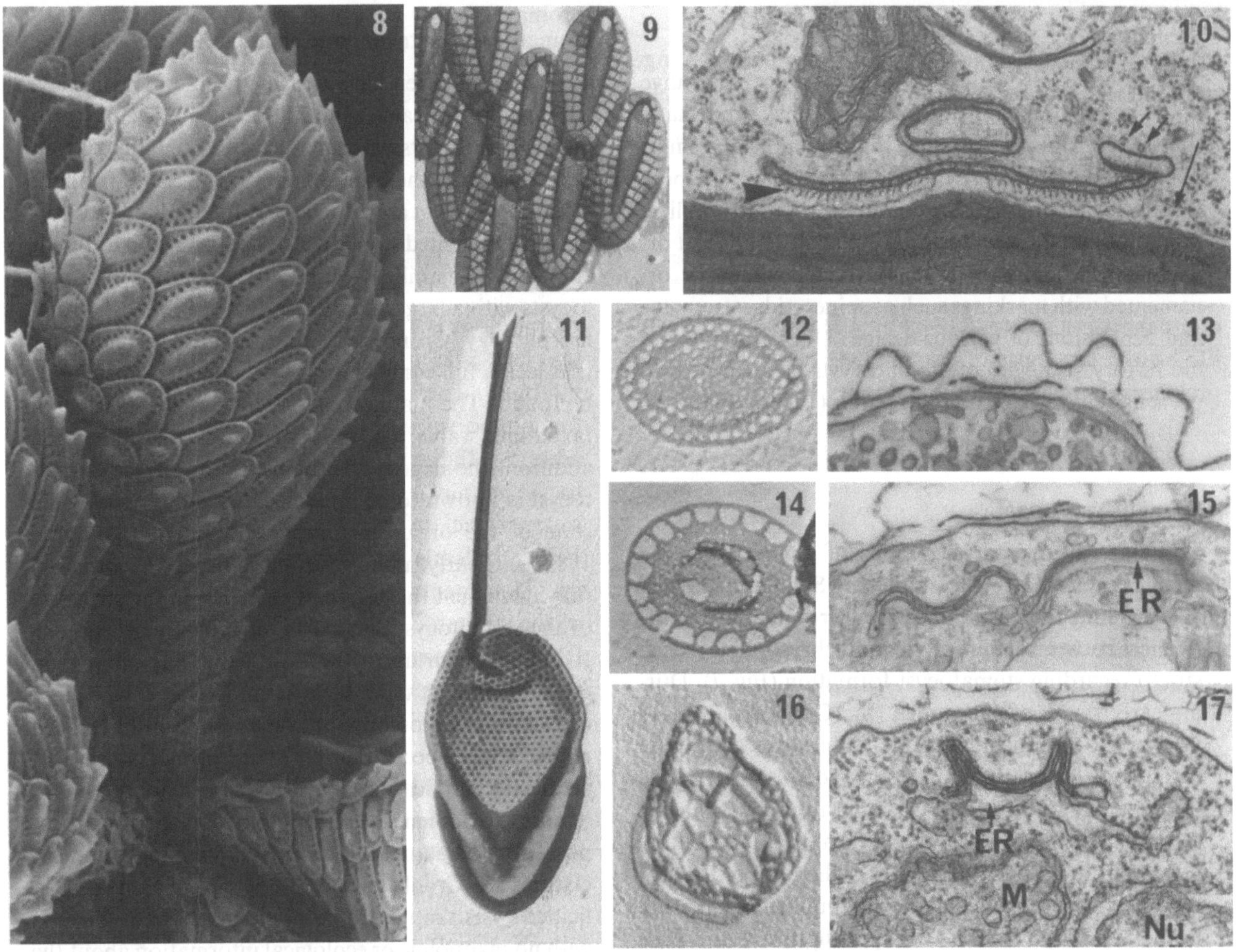

Fig. 8. *Synura petersenii.* Note overlapping scale pattern. SEM, × 4,700 (courtesy of R. Wetherbee)

Fig. 9. *Synura petersenii.* Group of detached scales. TEM, × 4,700

Fig. 10. *Synura glabra* cell showing scale vesicle being formed and guided by microtubules (short arrows) and actin-like microfilaments (arrowhead). Flagellar hairs are cross-sectioned in the chloroplast endoplasmic reticulum (long arrow). TEM, × 33,800 (courtesy of R. A. Andersen)

Fig. 11. *Mallomonas paxillata.* Scale with articulating bristle. TEM, × 7,800

Figs. 12–15. *Paraphysomonas undulata* (TEM). Whole mounts of plate-shaped scale (Fig. 12, × 28,000) and undulated scale (Fig. 14, × 20,000); both types of scales in section on cell surface (Fig. 13, × 23,000) and scales developing in scale vesicles associated with cisternae of endoplasmic reticulum (*ER*) (Fig. 15, × 28,000)

Figs. 16 and **17.** *Paraphysomonas morchella* (TEM). Whole mount of meshwork scale (Fig. 16, × 44,000) and a scale developing in a scale vesicle associated with a cisterna of endoplasmic reticulum (*ER*); *M* mitochondrion, *Nu* nucleus (Fig. 17, × 40,000)

terns of perforation, papillae, ribs, ridges, thickened marginal rims and sometimes anterior spines. Size and morphology of scales are species-specific. In some taxa (e.g., *Synura petersenii*), most scales are of similar form and patterning whereas in others (e.g., *S. spinosa*) three or four different scale types may occur on a cell. Even if the scales are all of the same basic type there may be great variation in morphology, with the anterior scales often being best developed, and the scales towards the posterior end of the cell often being gradually reduced (Kristiansen 1986).

Individual cells have their scales arranged in overlapping spiral rows forming a precise and consistent imbrication (for more information on scale cases, see Leadbeater 1990, Siver and Glew 1990, Lavau and Wetherbee 1994). The precise imbrication and arrangement of scales in a scale case together with distinctive scale morphology are features unique to the synurophytes (Andersen 1987).

In *Tessellaria* plate-shaped and spine-shaped scales form a multilayered investment on the surface of the colony (Pipes and Leedale 1992). The plate-scales are biradial in symmetry and are not imbricated but appear to be stuck together by diffuse material.

In addition to scales, most species of *Mallomonas* possess elongate siliceous bristles formed independently from the scales (see below), but becoming secondarily attached to the anterior end of some scales, often articulating there within a dome-like structure (Fig. 11). In some species bristles are confined to a region around one or both poles, while in others they may be universally distributed over the cell surface (Asmund and Kristiansen 1986, Siver 1991).

Chrysophytes (Paraphysomonadaceae)

Scales of *Paraphysomonas* (Figs. 12–17) and related genera are generally smaller and more delicate than those of synurophytes, and they are usually radial or biradial in symmetry (as opposed to the bilaterally symmetrical scales of most synurophytes). Many species have solid or perforated plate-scales; sometimes the scales are undulated in cross-section or they have elongated central processes forming spines or raised, three-dimensional mesh networks of siliceous elements (Preisig and Hibberd 1982, 1983). Up to three different types of scales may occur on the same cell; usually these types are not restricted to certain cell areas as in synurophytes, but are regularly distributed on the cell. The arrangement of the scales appears tobe loose and without any clearly defined pattern and thus differs from the precise construction of the scale cases in synurophytes.

Scale formation in synurophytes

Morphogenesis of scales and scale cases in this algal group are discussed elsewhere (Wetherbee et al. 1989, 1992, 1994; Lavau and Wetherbee 1994), but a few significant matters will be briefly mentioned here. Scales of *Mallomonas* and *Synura* are produced within a silica deposition vesicle (SDV) which forms along the distal surface of the periplastidial endoplasmic reticulum (ER) on one of the two chloroplasts (Fig. 10). The origin of the SDV is unclear, though it has been suggested to arise from the fusion of vesicles transported anteriorly from the Golgi body. The SDVs are moulded into the shape of the mature scales before silica is deposited in them. Initially, microtubules and actin microfilaments are intimately associated with the SDV and are probably involved in the shaping of the SDV, thereby controlling the morphology of the future scale. Mitochondria have also been suggested to be possibly involved in morphogenesis of SDVs (Andersen 1989). Neither the ontogeny of microanatomical ornamentation features nor the mechanism for silicon transport into the scale vesicles are known. When mature, the scales are extruded onto the surface of the plasma membrane and are incorporated into the overlying scale case (Lavau and Wetherbee 1994).

The bristles of *Mallomonas* are formed in a similar way to the scales in a SDV, but they arise separately and become attached to the scales outside the cell (Wetherbee et al. 1989, Beech et al. 1990).

Scale formation in *Tessellaria* differs from that of *Synura* and *Mallomonas* in that the chloroplast ER plays no role in scale formation. The SDVs are distributed throughout the cell and are moulded by closely appressed cisternae of cytoplasmic ER, as well as by microtubules and microfilaments (Pipes and Leedale 1992). Plate-scales and spine-scales are not produced simultaneously in one cell.

Scale formation in chrysophytes (Paraphysomonadaceae)

SDVs in Paraphysomonadaceae have never been found to be associated with a chloroplast (or leucoplast). Initially the SDVs lie close to the Golgi apparatus, but direct contact has never been observed and it is not clear whether the SDV is of Golgi origin or not.

A flattened cisterna of ER is always closely appressed to the SDV and this association also remains when the SDVs migrate away from the Golgi apparatus to the peripheral cytoplasm (Figs. 15 and 17). As silica is deposited in the preshaped SDV, the ER cisterna is reduced in size and later it disappears. It seems that the ER cisterna is involved in the shaping of the SDV, possibly serving as a template (Lee 1978, Preisig and Hibberd 1983, Preisig 1986). Ontogeny of the specific fine-scale ornamentation features such as perforations and the raised meshwork elements is unknown. In *Chrysosphaerella* mitochondrial profiles and membranous structures are also closely associated with the SDV during spine-scale formation (Andersen 1990), but microtubules and microfilaments have so far never been observed in association with developing scales in this genus and other Paraphysomonadaceae. As in the synurophytes, the SDVs ultimately fuse with the plasma

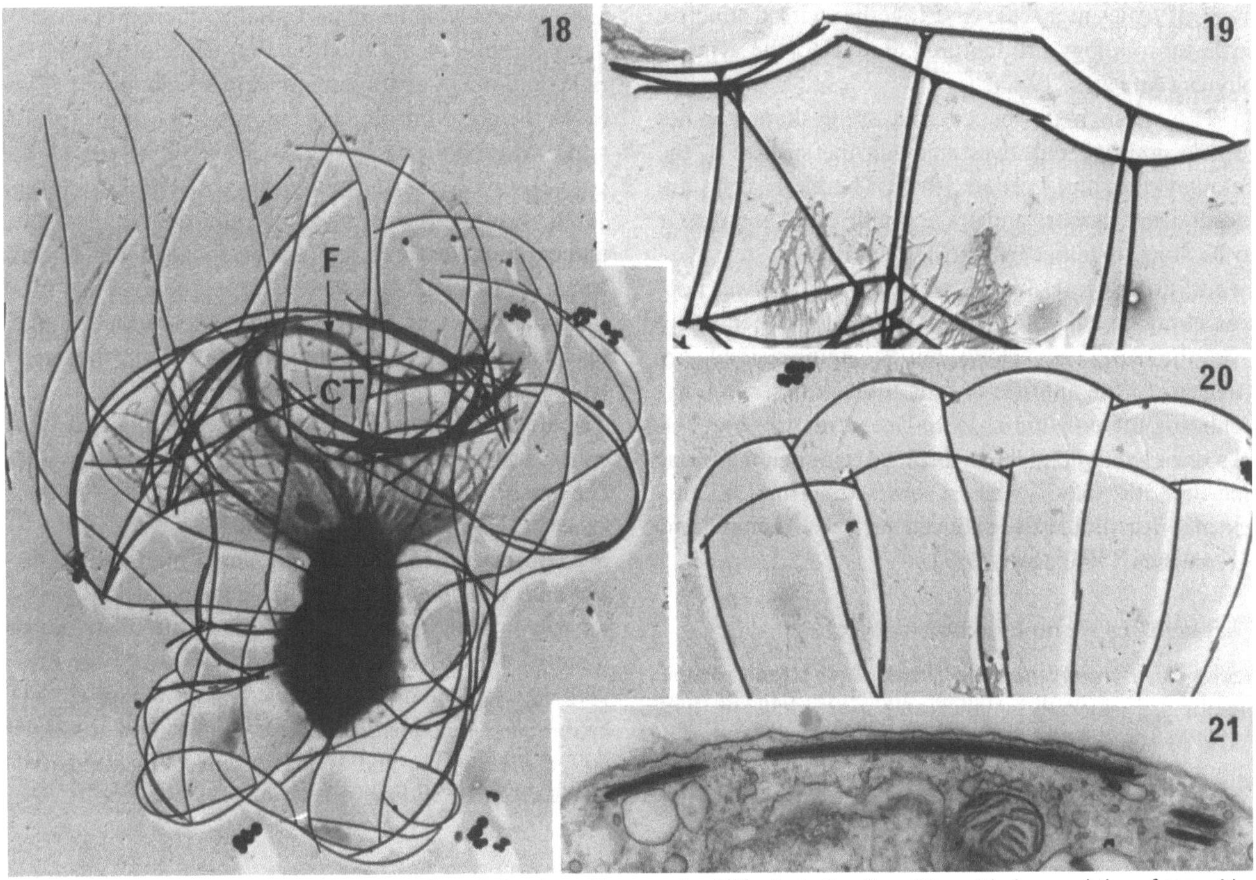

Fig. 18. *Acanthocorbis apoda.* Whole mount of a cell showing collar tentacles (*CT*), single flagellum (*F*), and lorica consisting of some 16 longitudinal and 3 transverse costae. The arrow denotes a junction of two component costal strips of a longitudinal costa. TEM, × 6,600 (courtesy of Ø. Moestrup)

Fig. 19. *Pleurasiga tricaudata.* T-junctions of longitudinal and transverse costae. TEM, × 6,000 (courtesy of Ø. Moestrup)

Fig. 20. *Cosmoeca phuketensis.* End-to-end junctions of longitudinal and transverse costae. TEM, × 8,850 (courtesy of Ø. Moestrup)

Fig. 21. Longitudinal section of a developing costal strip (from a mixed culture of *Acanthocorbis apoda* and *Diaphanoeca grandis*). TEM, × 16,500 (courtesy of Ø. Moestrup)

membrane and the mature scales are extruded onto the cell surface where they are interlocked in some unknown way.

Scales as taxonomic characteristics

The taxonomic importance of scale morphology is well established, and scales are important characters in the differentiation between the classes Chrysophyceae and Synurophyceae (see above). However, it should be noted that populations of actively growing, completely scale-free cells can exist under conditions of extreme silicon starvation (Sandgren and Barlow 1989, Leadbeater and Barker 1994). Small details in scale morphology have sometimes been used to describe new taxa, but it is generally not known to what extent scale structure can be influenced by environmental factors (for preliminary studies, see Sandgren and Barlow 1989, Gutowski 1993). There is a clear need for experimental studies for a better understanding of variability in scale morphology. Such studies together with phenetic and cladistic studies on scale morphology (Siver 1991) and scale biogenesis may have a great impact on taxonomy (Wee et al. 1992).

Choanoflagellates

Of the approximately 50 genera and 180 species of choanoflagellates known at present, some 30 genera and 90 species have been described to produce siliceous, basket-like loricae. All choanoflagellates possessing such loricae have been accommodated in a single family, Acanthoecidae. Many of these species form a prominent part of marine and brackish water nanoplankton.

However, the siliceous loricae have only low preservation potential and fossil records are not known so far. In the choanoflagellate family Salpingoecidae, the cells are encased in thecae, which may also be silicified to some extent as suggested by De Saedeleer (1929), but this needs to be re-analysed using modern methods of investigation. For general information on choanoflagellates, see Buck (1990).

Lorica structure

The loricae of acanthoecid choanoflagellates are composed of costae which, themselves, are constructed of rod-like units, the costal strips (Figs. 18–20). The total number of costal strips may be as low as 5 (*Monocosta fennica*) to 300 or more (e.g., *Stephanoeca norrisii*) and the range in overall lorica size can also be considerable (e.g., 3–4 µm in diameter in *Monocosta fennica*, more than 100 µm long in *Bicosta spinifera*; for a survey of lorica diversity, see Thomsen and Buck 1991). In most species the costal strips are of similar morphology throughout the lorica, but in a few taxa several distinctive morphological types of costal strip may occur, each located in a different part of the lorica. Within a lorica, costae are usually arranged parallel or perpendicular to the long axis of the protoplast, and they are called longitudinal and transverse costae (Figs. 18–20). Taxa with helically or more irregularly positioned costae, or with projections made up of costal strips, are also known. Junctions between one costa and another may be confined to the ends of costal strips or to midpoints along one or both costal strips involved (Figs. 19 and 20). In the majority of species, costal strips are narrow rods with an approximately circular profile in transverse section, while some species have broad flattened strips, sometimes with elaborate patterns of pores and markings. The loricae are lined partly by a fine membranous or fibrillar investment, and an extension of this attaches to the protoplast to hold it firmly in position within the lorica. In most species the protoplast is located at the base of the lorica, in others it is more central and, in species with small loricae, it can even be partly projecting beyond the anterior fringe of the lorica.

Lorica formation

Morphogenesis of costal strips and their assembly to form a lorica has been studied in considerable detail in a few species of acanthoecid choanoflagellates (mainly *Stephanoeca diplocostata* and *Diplotheca costata*; for reviews, see Leadbeater 1986, 1989, 1991). The formation of costal strips takes place individually in the peripheral cytoplasm (Fig. 21) in SDVs. The origin of the SDVs is not clear, but early stages are frequently found near the Golgi apparatus and the endoplasmic reticulum. At early stages of development, the SDVs are associated with microtubules, which may serve in the shaping of the developing costal strips. Silicification proceeds without clear evidence of an organic matrix being involved in the control of this process. When mature, costal strips are extruded sideways onto the cell surface, and are accumulated in groups that are transferred to the top of the collar. The positioning of the accumulated costal strips apparently anticipates their ultimate position in the mature lorica. Lorica assembly takes only a few minutes and involves a sliding of the accumulated costal strips over each other until they reach their final position. A group of tentacles outside the collar provides the movement necessary for sliding to occur. There is no information how costal strips stick to each other when costae are fully formed. There is evidence, however, that silicification in the costal strip is not uniform. Mann et al. (1983) believe that the outside of the strip is of quite dense silica, while the inside contains more hydrated silica which easily dissolves producing hollow tubes. Initially the join between strips may be made by organic material, but some of this silica may flow into the join and may assist in gluing the components of the skeleton together.

The initiation of lorica production varies in relation to cell division between the species. In one group of species (e.g., *Acanthoeca spectabilis*), one of the two siblings formed after cell division remains in the parent lorica, while the other is naked, starting to produce and subsequently assembling new costal strips once it has moved away from the parent lorica (nudiform replication). In a second group of species (e.g., *Stephanoeca diplocostata, Diplotheca costata*), a new set of costal strips is produced and accumulated at the top of the collar of the parent cell prior to cell division. After cytokinesis one of the two siblings keeps the parent lorica, whereas the other leaves this lorica backwards and immediately assembles a new lorica from the preformed, accumulated costal strips (tectiform replication).

Taxonomy

Protoplast morphology in acanthoecid choanoflagellates appears to be quite uniform and is not used so far to differentiate between the various taxa. At present taxonomy is based upon numerical and dimensional

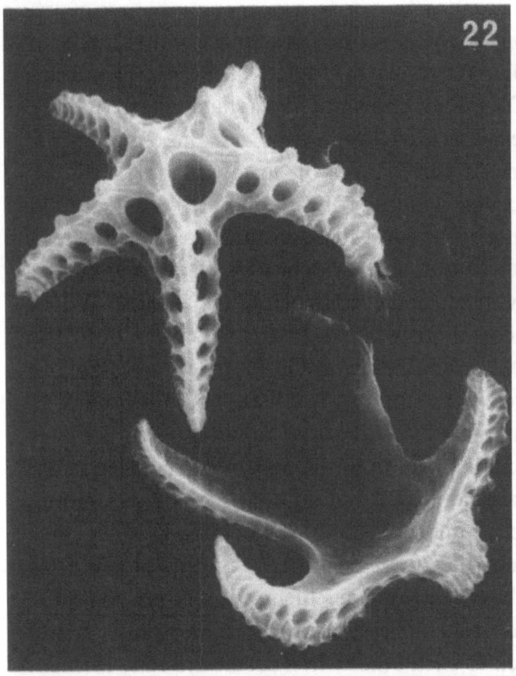

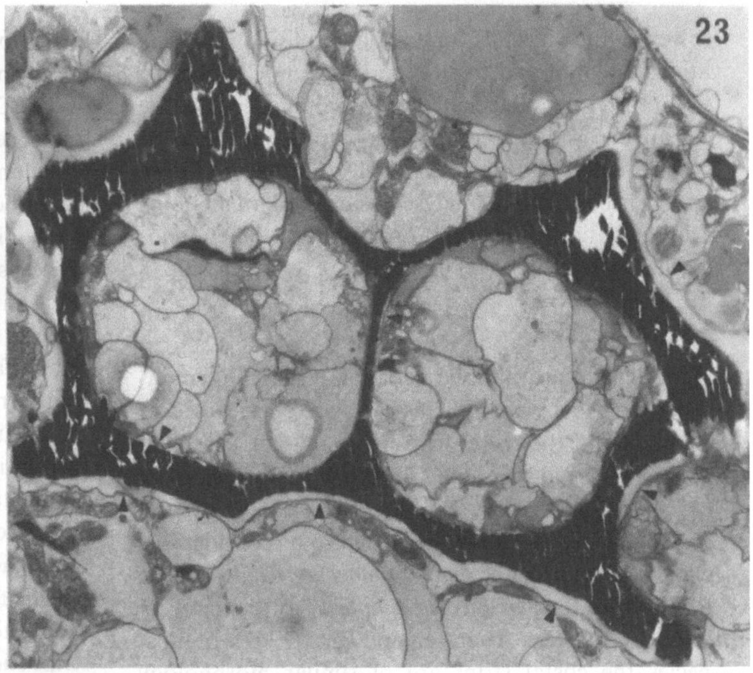

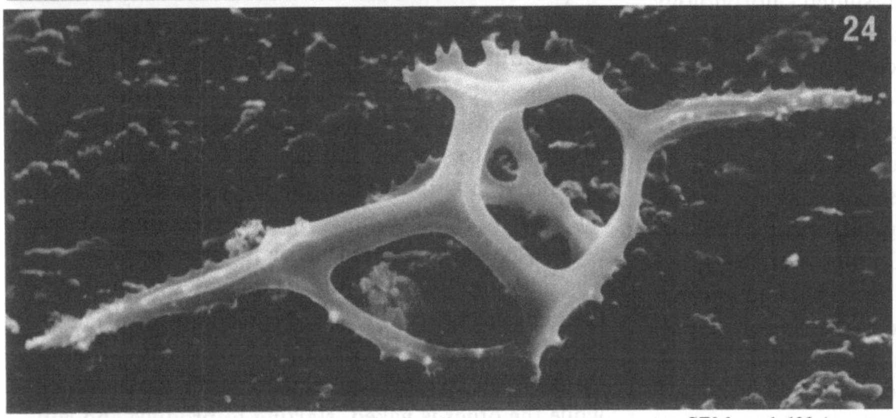

Fig. 22. Internal skeleton of the naked dinoflagellate *Actiniscus pentasterias*. SEM, × 1,600 (courtesy of G. M. Hallegraeff)

Fig. 23. Section through the central part of the skeleton of *Actiniscus pentasterias* inside a cell. The surrounding membrane is marked with arrowheads. TEM, × 13,000 (courtesy of G. Hansen)

Fig. 24. Internal skeleton of the ebriid *Hermesinum adriaticum*. SEM, × 2,000 (courtesy of P. E. Hargraves)

Fig. 25. *Ebria tripartita*. The theca (arrow) of the cell is visible covering the internal skeleton. TEM, × 2,200 (courtesy of H. A. Thomsen)

aspects of the lorica, as well as on the type of connections between component costal strips, costal strip morphology and orientation, and the position of the protoplast in the lorica (for more details, see Thomsen and Buck 1991). Little is known about the variability in lorica construction within these organisms, though in some species, variations (e.g., in the numbers of costae and costal strips within a lorica) may occur in clonal culture. Some costal strips can be arranged to form a pedicel at the posterior end of the lorica in some species, but in "*Stephanoeca pedicellata*" this character proved to be of no taxonomic significance (and *S. pedicellata* had to be subsumed in *S. diplocostata*). It has been shown that pedicels can both be present or lacking in the same clonal culture, depending on whether the cells are free-floating or attached prior to lorica assembly (Leadbeater 1979). In other choanoflagellates, however, pedicels may still serve as valid taxonomic features (Thomsen and Buck 1991). Variations also occur in the silica content of costal strips and in the dimensions of the loricae (Leadbeater 1991). Variation may be minimal in some species (e.g., *Diaphanoeca pedicellata*), but in others (e.g., *Bicosta spinifera*) much more variation may occur (Thomsen et al. 1990). It should also be noted that during silica impoverishment thinner costal strips may be produced in culture. Eventually, after prolonged silica deprivation, completely naked cells may be formed which survive and divide normally even though a lorica is lacking (Leadbeater 1989). Nevertheless, it is beyond doubt that loricae are prime taxonomic characters in acanthoecid choanoflagellates, and future findings concerning lorica variability and development (e.g., incidence of tectiform and nudiform replication) will certainly lead to a better circumscription of taxonomic entities.

Dinoflagellates

A small number of dinoflagellates produce siliceous skeletons. The best known species is the heterotrophic non-armoured *Actiniscus pentasterias* (Figs. 22 and 23), which commonly possesses a pair of five-armed internal skeletal elements (pentasters) capping the nucleus, oriented with their concave sides opposed (Hansen 1993). The degree of silicification may vary considerably, but nothing is known about skeleton morphogenesis. Other species of *Actiniscus* have skeletons of either radial or biradial symmetry. Living *Actiniscus* spp. have been recorded from most oceans as well as from some freshwater localities. Fossil skeletons attributed to this genus are widely distributed in the marine diatomites of Miocene age and later, but are never abundant (Tappan 1980).

Internal skeletons are also known from several other living dinoflagellate genera (e.g., *Achradina, Dicroerisma, Pavillardinium, Plectodinium*; see Tappan 1980), some of which have been placed together with *Actiniscus* in the same family, Actiniscaceae (see Taylor 1987), but the composition of their endoskeletons is uncertain. Silicified cyst walls have been reported primarily from some fossil cysts (e.g., *Peridinites*), but in these it is not certain whether silicification occurred before, during or after fossilization (Tappan 1980). The strongly developed pellicle of *Achradina*, which may be homologous with a cyst wall even though formed in the motile stage, appears to have some silica content (Taylor 1987). Silica granules are also known from the wall of zygotic cysts of a freshwater species of *Ceratium* (Chapman et al. 1982).

Ebriids

This group of cosmopolitan marine heterotrophic flagellates is characterized by a basket-like internal siliceous skeleton (Figs. 24 and 25). Only three (possibly four) species, assigned to two genera (*Ebria* and *Hermesinum*), are known from present time (Taylor 1990), but in the Tertiary the group has been much more diverse (Tappan 1980, Lipps 1993).

Ebriids have often been classified with dinoflagellates (e.g., Loeblich 1982) on the basis of the endoskeleton which resembles that of some dinoflagellates (e.g., *Dicroerisma*), but also because ebridian nuclei resemble dinokaryons in apparently having condensed chromosomes at interphase. This latter feature has not yet been confirmed by electron microscopy. In earlier literature, a relationship between ebriids and the silicoflagellates has also been suggested. Today the ebriids are generally classified as a separate group (Ebriophyceae/Ebriida) with uncertain phylogenetic affinities.

The ebridian skeleton consists of a solid framework of siliceous rods, 3–4 µm in thickness, branching in a regular manner from an initial branching point which may comprise three (*Ebria*) or four (*Hermesinum*) branches. Electron microscopy shows many little spines on all branches of the skeleton (Hargraves and Miller 1974). A special terminology for the different skeletal parts has been established, based on that for sponge spicules (see Tappan 1980). Skeleton morphogenesis has not been studied in detail. Daughter skeletons appear to be formed, although not necessarily completely, before cell division takes place (Hovasse 1932). Occasional errors in the formation process may lead to double or even triple skeletons.

Silicoflagellates (dictyochophytes)

This group of cosmopolitan marine phototrophic flagellates is characterized by a unique form of external siliceous skeleton (Fig. 26). There is probably only one extant genus, *Dictyocha*, with just three extant species (Moestrup and Thomsen 1990, Henriksen et al. 1993). Numerous taxa are known only as fossils from the early Cretaceous onwards, reaching a peak development in the Tertiary (Tappan 1980, Spaulding 1992, Lipps 1993).

In earlier algal literature the silicoflagellates have usually been classified within the Chrysophyceae (e.g., Kristiansen 1982) on the basis of chloroplast structure and pigments in *Dictyocha*, but now they are often placed in a separate class, Dictyochophyceae (Silva 1980). Recently the pedinellids have been suggested to be related to the silicoflagellates, owing to similarities in the flagellar apparatus, and have been included in the same class (Moestrup 1994).

The skeletons of silicoflagellates consist of hollow, tubular, siliceous rods fused together forming a network.

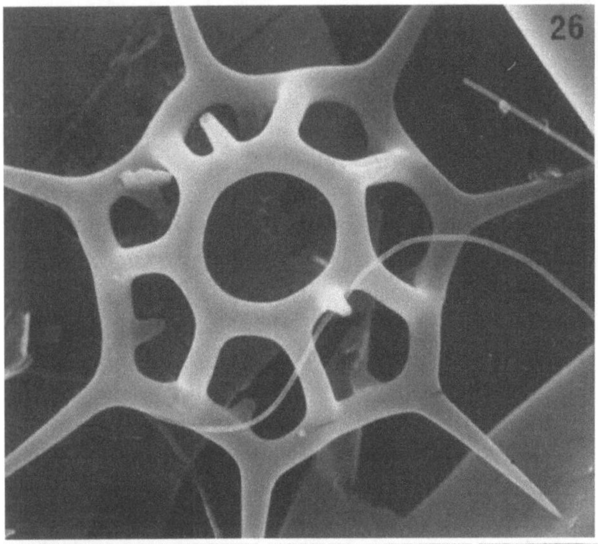

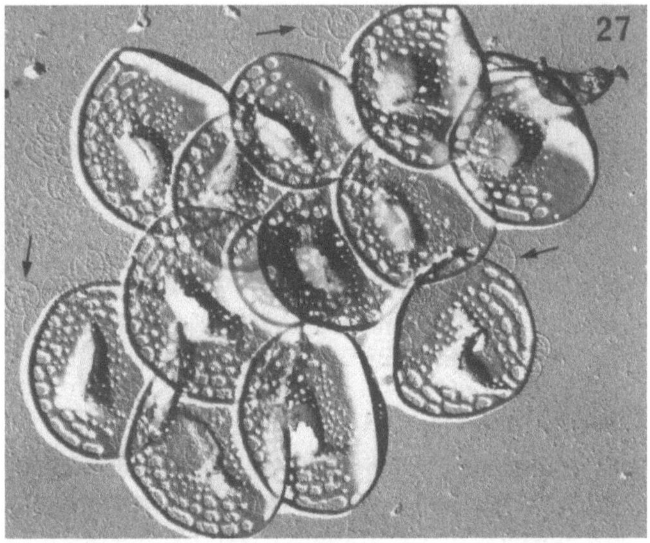

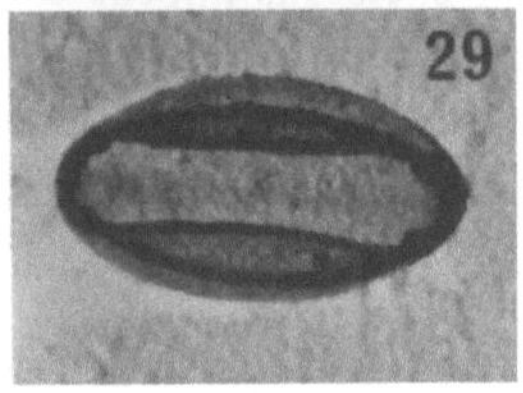

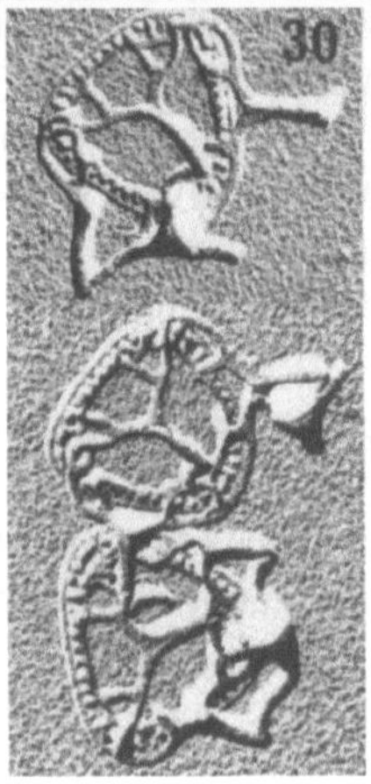

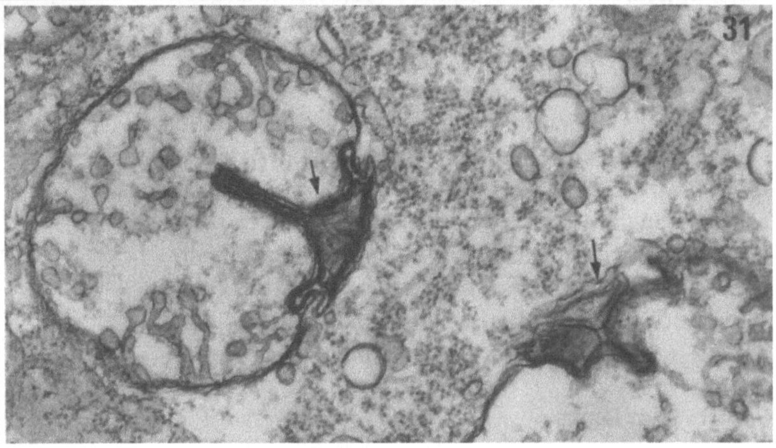

Fig. 26. External skeleton of the silicoflagellate *Dictyocha speculum*. SEM, × 2,500 (courtesy of H. A. Thomsen)

Fig. 27. *Petasaria heterolepis*. Group of large siliceous scales and small unmineralized scales (arrows). TEM, × 10,100 (courtesy of Ø. Moestrup)

Fig. 28. Two scales of the thaumatomastigid *Gyromitus disomatus*, one in top view (top), the other in side view. TEM, × 45,000

Fig. 29. *Gyromitus limax*. Single scale in top view. TEM, × 40,000

Fig. 30. *Thaumatomastix sagittifera*. Group of scales. TEM, × 29,000 (courtesy of Ø. Moestrup)

Fig. 31. *Thaumatomastix sagittifera*. Section showing developing scales (arrows) located in vesicles which project into the lumen of a mitochondrion. TEM, × 31,000 (courtesy of Ø. Moestrup)

Spines of different sizes may radiate from this meshwork, the spine tips sometimes being decorated with spinose conical collars. The cell lies in the cavity of the skeleton and is closely appressed to the skeletal elements. The plasmalemma is internal to the skeleton, though the cell periphery is very irregular, with numerous large indentations, and the skeletal elements are often present in these cavities (Moestrup and Thomsen 1990). In some cells a wall-like structure has been observed surrounding both the cytoplasm and the skeleton (Van Valkenburg 1971 b). It should also be noted that silicoflagellate cells without siliceous skeletons have been observed in both cultures and natural populations (Henriksen et al. 1993).

Detailed ultrastructural studies on the origin of the skeleton are missing, but evidence indicates that the skeleton is formed internally and then extruded by the cell. Silicification appears to occur sequentially rather than the entire skeleton being formed at once. The number of spines and degree of elaboration of the skeleton in *Dictyocha* may depend on the time elapsed in skeletal deposition. Evidence from electron microscopy also suggests that any parts of the skeleton that come to touch during growth will fuse at the juncture, and

that such junctures are of varied age in a single individual (Van Valkenburg 1971 a, 1980).
The taxonomy has been based exclusively on the morphology of the skeleton. The validity of this taxonomy when applied to living species has been questioned because of the plasticity in skeleton configuration found in clonal cultures (Van Valkenburg and Norris (1970).

Thaumatomastigids

Cells of *Gyromitus, Thaumatomastix,* and *Thaumatomonas* are covered by characteristic species-specific scales (Figs. 28–30). These are siliceous at least in *Gyromitus disomatus* (Nicholls 1979), *Thaumatomastix salina* (Beech and Moestrup 1986), and *Thaumatomonas lauterbornii* (Shirkina 1987), but silica is probably present in the scales of all three species of *Gyromitus* (Swale and Belcher 1975), ten species of *Thaumatomastix* (Thomsen et al. 1993), and the two species of *Thaumatomonas* (Karpov and Zhukov 1987, Mylnikov and Karpov 1993) known at present. The phylogenetic affinity of these biflagellate, heterotrophic organisms is not clear, and they cannot be assigned to any well-circumscribed supraordinal group of protists. Shirkina (1987) and Karpov (1990) placed them in a separate order (Thaumatomonadida) while Patterson and Zölffel (1991) included them with four other genera (*Hyaloselene, Protaspis, Rhizaspis, Synoikomonas*) in the family Thaumatomastigaceae. However, for the latter genera it is not known whether they produce scales at all.
The scales of *Gyromitus, Thaumatomastix,* and *Thaumatomonas* are oval or triangular, more or less solid or perforated and in many cases formed by the partial fusion of two almost equally sized plates. In some species of *Thaumatomastix* spine-scales and small flagellar scales also occur. Scale morphogenesis has not yet been studied in detail, but in all species examined so far [*Gyromitus disomatus, G. limax, Thaumatomastix sagittifera* (= *Reckertia sagittifera*), *Thaumatomonas lauterbornii, Thaumatomonas seravinii*] it has been clearly shown that they are formed in SDVs on the outer mitochondrial membrane (Swale and Belcher 1974, 1975, Moestrup 1982, Karpov and Zhukov 1987, Karpov 1993). Each mitochondrion may be covered simultaneously by several scale-containing vesicles that often project far into the lumen of the mitochondrion (Fig. 31).

Petasaria

In this monotypic genus the cells are covered by siliceous, large hat-like scales and unmineralized, small cobweb-like scales (Fig. 27). The organism is uniflagellated and though it appears to be wide-spread in marine waters (Moestrup 1979, pers. comm.), it is so far shown from whole mount preparations only. It shows no clear relationship with any of the known groups of flagellates and nothing is known about the mode of scale formation.

Siliceous structures in other flagellated protists

Chlorophytes

Silicified cell walls are known to occur in some coccoid genera (*Pediastrum, Tetraedron*; see Parker 1969), but silica has also been shown to be a major component of loricae in some flagellate genera such as *Hemitoma* (Krienitz et al. 1990). An unusual occurrence of quartz (silica in micro-crystalline form) has been reported from *Pedinomonas tuberculata,* where this mineral is apparently a constituent (amongst cellulose and calcite) of the characteristic tubercular excrescences on the cell surface (Brandenberger and Frey-Wyssling 1947, Manton and Parke 1960).

Euglenophytes

The loricae of *Trachelomonas* are usually predominantly mineralized by iron and manganese compounds, but in some species silica can also be a major component (Steinberg and Klee 1984). In some species of *Urceolus* (e.g., *U. sabulosus*) extraneous particles of siliceous or other composition can be accreted on the cell surface (see Huber-Pestalozzi 1955).

Haptophytes (prymnesiophytes)

Siliceous structures appear to be largely absent in this algal group, but it is noteworthy that cyst walls of *Prymnesium* have been observed to be composed of layers of scales, with siliceous material being deposited on the distal face of the outermost scales (Pienaar 1980).

Xanthophytes (tribophytes)

The endogeniously produced bipartite cysts produced by members of this algal group (e.g., *Chloromeson, Heterochloris*) have been reported to possess a siliceous wall, but ultrastructural information on cyst morphogenesis is lacking. Silicified cell walls also occur in some coccoid genera (e.g., *Arachnochloris, Chlorallanthus, Goniochloris, Polyedriella*; see Ettl 1978).

Conclusions

The siliceous structures produced by flagellated protists are used as prime tools in the taxonomy and systematics of these organisms. However, relatively little, or in some groups virtually nothing, is known about morphogenesis and silicification of these structures. It is also largely unknown to what extent morphological variations occur and how far these can be influenced by environmental conditions. Such studies combined with molecular and biochemical approaches should now be attempted if we are to make progress in the taxonomy and systematics of these organisms.

Acknowledgements

The provision of micrographs by R. A. Andersen, G. M. Hallegraeff, G. Hansen, P. E. Hargraves, Ø, Moestrup, H. A. Thomsen, and R. Wetherbee is gratefully acknowledged. I am also thankful to A. Zuppiger for assistance with the photographic work.

References

Allison CW (1981) Siliceous microfossils from the Lower Cambrian of northwest Canada: possible source for biogenic cherts. Science 211: 53–55

Andersen RA (1987) Synurophyceae classis nov., a new class of algae. Amer J Bot 74: 337–353

– (1989) The Synurophyceae and their relationship to other golden algae. Beih Nova Hedwigia 95: 1–26

– (1990) The three-dimensional structure of the flagellar apparatus of *Chrysosphaerella brevispina* (Chrysophyceae) as viewed by high voltage electron microscopy stereo pairs. Phycologia 29: 86–97

Anderson OR (1994) Cytoplasmic origin and surface deposition of silicate structures in sarcodine protists. Protoplasma 181: 61–77

Asmund BA, Kristiansen J (1986) The genus *Mallomonas* (Chrysophyceae). Opera Bot 85: 1–128

Beech PL, Moestrup Ø (1986) Light and electron microscopical observations on the heterotrophic protist *Thaumatomastix salina* comb. nov. (syn. *Chrysosphaerella salina*) and its allies. Nord J Bot 6: 865–877

– Wetherbee R, Pickett-Heaps JD (1990) Secretion and deployment of bristles in *Mallomonas splendens* (Synurophyceae). J Phycol 26: 112–122

Booth BC, Marchant HJ (1987) Parmales, a new order of marine chrysophytes, with descriptions of three new genera and seven new species. J Phycol 23: 245–260

Brandenberger E, Frey-Wyssling A (1947) Über die Membransubstanzen von *Chlorochytridion tuberculatum* W. Vischer. Experientia 3: 492–493

Bourrelly P (1963) Loricae and cysts in the Chrysophyceae. Ann NY Acad Sci 108: 421–429

Buck KR (1990) Phylum Zoomastigina, class Choanomastigotes (choanoflagellates). In: Margulis L, Corliss JO, Melkonian M, Chapman DJ (eds) Handbook of Protoctista. Jones and Bartlett, Boston, pp 194–199

Chapman DV, Dodge JD, Heaney SI (1982) Cyst formation in the freshwater dinoflagellate *Ceratium hirundinella* (Dinophyceae). J Phycol 18: 121–129

Cronberg G (1986) Chrysophycean cysts and scales in lake sediments: a review. In: Kristiansen J, Andersen RA (eds) Chrysophytes: aspects and problems. Cambridge University Press, Cambridge, pp 281–315

– (1988) Variability in size and ultrastructure of the statospore of *Mallomonas caudata*. Hydrobiologia 161: 31–39

– Sandgren CD (1986) A proposal for the development of standardized nomenclature and terminology for chrysophycean statospores. In: Kristiansen J, Andersen RA (eds) Chrysophytes: aspects and problems. Cambridge University Press, Cambridge, pp 317–328

De Saedeleer H (1929) Notes de protistologie. II. Craspedomonadines: morphologie et physiologie. Recueil Inst Zool Torley-Rousseau 2: 241–287

Ettl H (1978) Xanthophyceae (I). G Fischer, Stuttgart [Ettl H, Gerloff J, Heynig H (eds) Süßwasserflora von Mitteleuropa, vol 3]

Evered D, O'Connor M (eds) (1986) Silicon biochemistry. Wiley, Chichester (Ciba Foundation Symposium 121)

Gutowski A (1993) Autökologische und taxonomische Studien schuppentragender Chrysophyceen anhand von Untersuchungen im Freiland und in Kulturen. PhD Thesis, Freie Universität Berlin, Berlin, Federal Republic of Germany

Hansen G (1993) Light and electron microscopical observations of the dinoflagellate *Actiniscus pentasterias* (Dinophyceae). J Phycol 29: 486–499

Hargraves PE, Miller BT (1974) The ebridian flagellate *Hermesinum adriaticum* Zach. Arch Protistenk 116: 280–284

Henriksen P, Knipschildt F, Moestrup Ø, Thomsen HA (1993) Autecology, life history and toxicology of the silicoflagellate *Dictyocha speculum* (Silicoflagellata, Dictyochophyceae). Phycologia 32: 29–39

Hibberd DJ (1978) *Bicosoeca accreta* sp. nov., a flagellate accumulating extraneous silica fragments. Br Phycol J 13: 161–166

Hovasse R (1932) Contribution à l'étude des silicoflagellés. Multiplication, variabilité, hérédité, affinités. Bull Biol Fr Belg 66: 447–501

Huber-Pestalozzi G (1955) Das Phytoplankton des Süßwassers, Teil 4: Euglenophyceen. Schweizerbart'sche Verlagsbuchhandlung, Stuttgart [Thienemann A (ed) Die Binnengewässer, vol 16/4]

Karpov SA (1990) Analysis of orders Phalansteriida, Spongomonadida and Thaumatomonadida (Mastigophora). Zool Zh 69 (3): 5–12 (in Russian)

– (1993) The ultrathin structure of the colourless flagellate *Thaumatomonas seravini* Mylnikov et Karpov, 1993 (Thaumatomonadida). Tsitologiya 35: 8–11 (in Russian)

– Zhukov BF (1987) Cytological pecularities of the colourless flagellate *Thaumatomonas lauterborni*. Tsitologiya 29: 1168–1171 (in Russian)

Kosman CA, Thomsen HA, Østergaard JB (1993) Parmales (Chrysophyceae) from Mexican, Californian, Baltic, Arctic and Antarctic waters with description of a new subspecies and several new forms. Phycologia 32: 116–128

Krienitz L, Peschke T, Giering B (1990) Lichtmikroskopische, rasterelektronenmikroskopische und röntgenmikroanalytische Untersuchungen an *Hemitoma maeandrocystis* Skuja (Chlorophyta, Phacotaceae). Arch Protistenk 138: 159–170

Kristiansen J (1982) Chrysophyceae. In: Parker SP (ed) Synopsis and classification of living organisms, vol 1. McGraw-Hill, New York, pp 81–86

– (1986) The ultrastructural bases of chrysophyte systematics and phylogeny. CRC Crit Rev Plant Sci 4: 149–211

– (1990) Phylum Chrysophyta. In: Margulis L, Corliss JO, Melkonian M, Chapman DJ (eds) Handbook of Protoctista. Jones and Bartlett, Boston, pp 438–453

Lavau S, Wetherbee R (1994) Structure and development of the scale case of *Mallomonas adamas* (Synurophyceae). Protoplasma 181: 259–268

Leadbeater BSC (1974) Ultrastructural observations on nanoplankton collected from the coast of Jugoslavia and the Bay of Algiers. J Mar Biol Assoc UK 54: 179–196

– (1979) Developmental studies on the loricate choanoflagellate *Stephanoeca diplocostata* Ellis. I. Ultrastructure of the non-dividing cell and costal strip production. Protoplasma 98: 241–262

– (1986) Silica deposition and lorica assembly in choanoflagellates. In: Leadbeater BSC, Riding R (eds) Biomineralization in lower plants and animals. Clarendon Press, Oxford, pp 345–359 (Systematics Association special volume 30)

– (1989) Lorica production in choanoflagellates – a model system for investigating the mechanics, controls and dynamics of secretion. J Protozool 36: 373–378

– (1990) Ultrastructure and assembly of the scale case in *Synura* (Synurophyceae Andersen). Br Phycol J 25: 117–132

– (1991) Choanoflagellate organization with special reference to loricate taxa. In: Patterson DJ, Larsen J (eds) The biology of free-living heterotrophic flagellates. Clarendon Press, Oxford, pp 241–258 (Systematics Association special volume 45)

– Barker DAN (1994) Biomineralization and scale production in the Chrysophyta. In: Sandgren CD, Smol JP, Kristiansen J (eds) Chrysophyte algae: ecology, phylogeny and development. Cambridge University Press, Cambridge (in press)

Lee RE (1978) Formation of scales in *Paraphysomonas vestita* and the inhibition of growth by germanium dioxide. J Protozool 25: 163–166

Lipps JH (ed) (1993) Fossil prokaryotes and protists. Blackwell, Oxford

Loeblich AR III (1982) Dinophyceae. In: Parker, SP (ed) Synopsis and classification of living organisms, vol 1. McGraw-Hill, New York, pp 101–115

Mann DG, Marchant HJ (1989) The origins of the diatom and its life cycle. In: Green JC, Leadbeater BSC, Diver WL (eds) The chromophyte algae: problems and perspectives. Clarendon Press, Oxford, pp 307–323 (Systematics Association special volume 38)

Mann S, Parker SB, Perry CC, Ross MD, Skarnulis AJ, Williams RJP (1983) Problems in the understanding of biominerals. In: Westbroek P, De Jong EW (eds) Biomineralization and biological metal accumulation. Reidel, Dordrecht, pp 171–183

Manton I, Parke M (1960) Further observations on small green flagellates with special reference to possible relatives of *Chromulina pusilla* Butcher. J Mar Biol Assoc UK 39: 275–298

Mitchell JG, Silver MW (1986) Archaeomonad (Chrysophyta) cysts: ecological and paleoecological significance. BioSystems 19: 289–298

Moestrup Ø (1979) Identification by electron microscopy of marine nanoplankton from New Zealand, including the description of four new species. NZ J Bot 17: 61–95

– (1982) Flagellar structure in algae: a review, with new observations particularly on the Chrysophaceae, Phaeophyceae (Fucophyceae), Euglenophyceae, and *Reckertia*. Phycologia 21: 427–528

– (1994) Current status of chrysophyte "splinter groups": synurophytes, pedinellids, sillicoflagellates. In: Sandgren CD, Smol JP, Kristiansen J (eds) Chrysophyte algae: ecology, phylogeny and development. Cambridge University Press, Cambridge (in press)

– Thomsen HA (1990) *Dictyocha speculum* (Silicoflagellata, Dictyochophyceae), studies on armoured and unarmoured stages. Biol Skrifter 37: 1–57

Mylnikov AP, Karpov SA (1993) A new representative of colourless flagellates, *Thaumatomonas seravini* sp.n. (Thaumatomonadida, Protista). Zool Zh 72(3): 5–10 (in Russian)

Nicholls KH (1979) Is *Hymenomonas prenanti* Lecal (Prymnesiophyceae) really the colourless flagellate *Gyromitus disomatus* Skuja? Phycologia 18: 420–423

Parker BC (1969) Occurrence of silica in brown and green algae. Can J Bot 47: 537–540

Patterson DJ, Zölffel M (1991) Heterotrophic flagellates of uncertain taxonomic position. In: Patterson DJ, Larsen J (eds) The biology of free-living heterotrophic flagellates. Clarendon Press, Oxford, pp 427–476 (Systematics Association special volume 45)

Pienaar RN (1980) Observations on the structure and composition of the cyst of *Prymnesium* (Prymnesiophyceae). Proc Electron Microsc Soc S Afr 10: 73–74

Pipes LD, Leedale GF (1992) Scale formation in *Tessellaria volvocina* (Synurophyceae). Br Phycol J 27: 11–19

Preisig HR (1986) Biomineralization in the Chrysophyceae. In: Leadbeater BSC, Riding R (eds) Biomineralization in lower plants and animals. Clarendon Press, Oxford, pp 327–344 (Systematics Association special volume 30)

– (1994) A modern concept of chrysophycean classification. In: Sandgren CD, Smol JP, Kristiansen J (eds) Chrysophyte algae: ecology, phylogeny and development. Cambridge University Press, Cambridge (in press)

– Hibberd DJ (1982) Ultrastructure and taxonomy of *Paraphysomonas* (Chrysophyceae) and related genera 2. Nord J Bot 2: 601–638

– – (1983) Ultrastructure and taxonomy of *Paraphysomonas* (Chrysophyceae) and related genera 3. Nord J Bot 3: 695–723

– Vørs N, Hällfors G (1991) Diversity of heterotrophic heterokont flagellates. In: Patterson DJ, Larsen J (eds) The biology of free-living heterotrophic flagellates. Clarendon Press, Oxford, pp 361–399 (Systematics Association special volume 45)

Sandgren CD (1983 a) Survival strategies of chrysophycean flagellates: reproduction and the formation of resistant resting cysts. In: Fryxell GA (ed) Survival strategies of the algae. Cambridge University Press, Cambridge, pp 23–48

– (1983 b) Morphological variability in populations of chrysophycean resting cysts. I. Genetic (interclonal) and encystment temperature effects on morphology. J Phycol 19: 64–70

– (1989) SEM investigations of statospore (stomatocyst) development in diverse members of the Chrysophyceae and Synurophyceae. Beih Nova Hedwigia 95: 45–69

– (1991) Chrysophyte reproduction and resting cysts: a paleolimnologist's primer. J Paleolimnol 5: 1–9

– Barlow SB (1989) Siliceous scale production in chrysophyte algae. II. SEM observations regarding the effects of metabolic inhibitors on scale regeneration in laboratory population of scale-free *Synura petersenii* cells. Beih Nova Hedwigia 95: 27–44

Schmid A-MM (1994) Aspects of morphogenesis and function of diatom cell walls with implications for taxonomy. Protoplasma 181: 43–60

Shirkina NI (1987) Morfologija i zhiznennyj cikl *Thaumatomonas lauterborni* De Saedeleer (Mastigophora Diesing). In: Fauna i biologija presnovodnych organizmov. Nauka, Leningrad (St. Petersburg), pp 87–107 (in Russian)

Silva PC (1980) Names of classes and families of living algae. Reg Vegetabile 103: 1–156

Simpson TL, Volcani BE (eds) (1981) Silicon and siliceous structures in biological systems. Springer, New York Berlin Heidelberg

Siver PA (1991) The biology of *Mallomonas*: morphology, taxonomy and ecology. Kluwer, Dordrecht

– Glew JR (1990) The arrangement of scales and bristles on *Mallomonas* (Chrysophyceae): a proposed mechanism for the formation of the cell covering. Can J Bot 68: 374–380

Smol JP (1988) Chrysophycean microfossils in paleolimnological studies. Palaeogeog Palaeoclimatol Palaeoecol 62: 287–297

– (1994) Application of chrysophytes to problems in paleoecology. In: Sandgren CD, Smol JP, Kristiansen J (eds) Chrysophyte algae: ecology, phylogeny and development. Cambridge University Press, Cambridge (in press)

Spaulding S (1992) Bibliography and taxa of silicoflagellates III. INA Newslett (Proceedings International Nannoplankton Association) 14: 42–90

Steinberg C, Klee R (1984) Zur Chemie von *Trachelomonas*-Loricae. Arch Protistenk 128: 283–294

Swale EMF, Belcher JH (1974) *Gyromitus disomatus* Skuja – a free-living colourless flagellate. Arch Protistenk 116: 211–220

– (1975) *Gyromitus limax* nov. sp. – free-living colourless amoeboflagellate. Arch Protistenk 117: 20–26

Tappan H (1980) The paleobiology of plant protists. Freeman, San Francisco

Taylor FJR (ed) (1987) The biology of dinoflagellates. Blackwell, Oxford (Botanical monographs 21)

– (1990) Ebridians. In: Margulis L, Corliss JO, Melkonian M, Chapman DJ (eds) Handbook of Protoctista. Jones and Bartlett, Boston, pp 720–721

Thomsen HA, Buck KR, Coale SL, Garrison DL, Gowing MM (1990) Loricate choanoflagellates (Acanthoecidae, Choanoflagellida) from the Weddell Sea, Antarctica. Zool Scripta 19: 367–387

– Buck KR (1991) Choanoflagellate diversity with particular emphasis on the Acanthoecidae. In: Patterson DJ, Larsen J (eds) The biology of free-living heterotrophic flagellates. Clarendon Press, Oxford, pp 259–284 (Systematics Association special volume 45)

– Hällfors G, Hällfors S, Ikävalko J (1993) New observations on the heterotrophic protist genus *Thaumatomastix* (Thaumatomastigaceae, Protista incertae sedis), with particular emphasis on material from the Baltic Sea. Ann Bot Fenn 30: 87–108

Van Valkenburg SD (1971 a) Observations on the fine structure of *Dictyocha fibula* Ehrenberg. I. The skeleton. J Phycol 7: 113–118

– (1971 b) Observations on the fine structure of *Dictyocha fibula* Ehrenberg. II. The protoplast. J Phycol 7: 118–132

– (1980) Silicoflagellates. In: Cox ER (ed) Phytoflagellates. Elsevier North-Holland, New York, pp 335–350

– Norris RE (1970) The growth and morphology of the silicoflagellate *Dictyocha fibula* Ehrenberg in culture. J Phycol 6: 48–54

Wee JL, Hood CS, Andersen RA (1992) Phylogenetic relationships among the colonial Synurophyceae: an ontogenetic model based on scale biogenesis. J Phycol 28 [Suppl]: 4

Wetherbee R, Koutoulis A, Beech PL (1989) The role of the cytoskeleton during the assembly, secretion and deployment of scales and spines. In: Coleman AW, Goff LJ, Stein-Taylor JR (eds) Algae as experimental systems. AR Liss, New York, pp 93–108

– Koutoulis A, Andersen RA (1992) The microarchitecture of the chrysophycean cytoskeleton. In: Menzel D (ed) The cytoskeleton of the algae. CRC Press, Boca Raton, pp 1–17

– Ludwig M, Koutoulis A (1994) Immunological and ultrastructural studies of scale development and deployment in *Mallomonas* and *Apedinella*. In: Sandgren CD, Smol JP, Kristiansen J (eds) Chrysophyte algae: ecology, phylogeny and development. Cambridge University Press, Cambridge (in press)

Protoplasma (1994) 181: 43–60

PROTOPLASMA

Aspects of morphogenesis and function of diatom cell walls with implications for taxonomy

Anna-Maria M. Schmid*

Institut für Pflanzenphysiologie, Universität Salzburg, Salzburg

Received September 28, 1993
Accepted May 11, 1994

Dedicated to Professor Peter Sitte on the occasion of his 65th birthday

Summary. Aspects of morphogenesis and morphology of diatom cell walls are reviewed to highlight functional correlations between wall structures and three-dimensional cytoplasmic activities during the cell cycle. Morphogenesis of the siliceous valve within the silica deposition vesicle is discussed in the light of the dependency on a precisely orchestrated moulding machinery, involving the cytoskeleton, mitochondria, endoplasmic reticulum, spacer vesicles produced by the Golgi apparatus, and the plasmalemma, in combination with adhesion of the cells to parts of the parental wall and localized "plasmolyses". Sensitivity of morphogenetic events to fluctuations of external factors has implications for taxonomy.

Keywords: Bacillariophyceae; Cell wall; Cytoskeleton; Endoplasmic reticulum; Golgi apparatus; Mitochondria; Morphogenesis.

Abbreviations: CF cleavage furrows; cPL cleavage plasmalemma; GB girdle bands; LP labiate process; LPA labiate process apparatus; MC microtubule center; mLP macro labiate process; MT microtubule; MTOC microtubules organizing center; PL plasmalemma; SDV silica deposition vesicle; SL SDV membrane; SpV spacer vesicles.

Introduction

As eucaryotic algae, diatoms are diploid in their vegetative state, and have a rigid, silicified wall. They are unicellular or colonial, colonies often appearing as individuals of higher organizational order (e.g., *Chaetoceros*; von Stosch et al. 1973). Cell walls display specific appendages for connecting cells or floating, or with apertures for secretion of mucous substances responsible for cell adhesion or motility (Round et al. 1990, Round and Crawford 1990, Pickett-Heaps et al. 1990, Hoagland et al. 1993). Taxonomy is traditionally based on the intricate and symmetrical sculpturing of the siliceous part of the cell wall, the valve, in particular. Primarily because of differences in the valve symmetry, diatoms have been separated into two major groups: the centric diatoms, first appeared in the Cretaceous, with mainly tri- to omniradiate, but also bipolar symmetry; and the pennate diatoms, blooming in the Tertiary onwards, with bipolar and, only as an aberration, tripolar symmetry (Pickett-Heaps et al. 1990 and references therein, Round et al. 1990 and references therein). Currently, about 100,000 species are recognized; scanning electron microscopy (SEM) has revealed the structural heterogeneity of a number of taxa previously united on the basis of light microscopical (LM) observations (Round et al. 1990). Most of the fine detail of wall structures detectable in the SEM and used for identification are understood neither in their function, nor in their formation. This review will focus on some aspects of morphogenesis and morphology of the wall to highlight its function with respect to spatial determinants for intracellular orientation and abutments for transmitting shearing forces throughout the cell cycle as well as implications for taxonomy. More extended coverage on ultrastructural morphology, morphogenesis, and systematics of diatoms can be found in Pickett-Heaps et al. (1990), Round and Crawford (1990), and Round et al. (1990).

* Correspondence and reprints: Institut für Pflanzenphysiologie, Universität Salzburg, Hellbrunnerstrasse 34, A-5020 Salzburg, Austria.

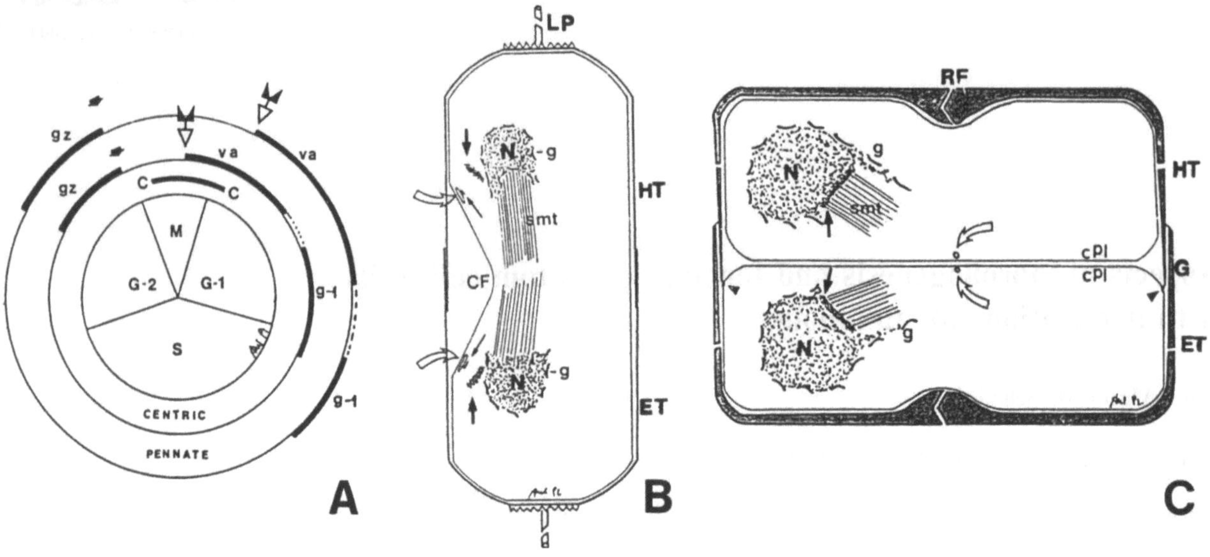

Fig. 1 A–C. Differences between centric and pennate diatoms with respect to valve formation. **A** Correlation of mitotic cycle (inner circle) with cleavage (*C–C*) and formation of cell wall. Release of last GB (*gz*, small arrow) correlated with onset of mitosis. Difference in start of valve formation (large arrows). **B** Centric diatoms (*Ditylum*): start in anaphase/telophase. Based on Li and Volcani (1985 a: figs. 2 and 3), and Pickett-Heaps et al. (1988: fig. 7). **C** Pennate diatoms (*Navicula cuspidata*): start at post cleavage. Based on Edgar and Pickett-Heaps (1984 b: figs. 2 and 3). *g* Golgi bodies; *RF* raphe fissure; *smt* spindle MTs; open arrow, SDV; dark arrows, polar complex; small dark arrows, LPA; arrowheads, connection between protoplast lying in epitheca to last GB of hypotheca

Cell wall morphology and function

The rigid cell wall and consequences for division

A rigid, silicified wall is inextensible; thus to accommodate cell growth, it consists of two overlapping parts, the "hypotheca" as the bottom and the "epitheca" as the top of a box. Each theca is composed of a "valve", and a series of "girdle bands" (GBs) attached to it and less intricately ornamented than the valve. The GBs constitute the overlapping region and protect the protoplast during its expansion through the cell cycle (Fig. 1 A–C). The last GB of the hypotheca, formed immediately prior to cell division (Pickett-Heaps et al. 1990) often differs from the other bands by its characteristic design with fringes (or lamellae in *Thalassiosira eccentrica*), to which the protoplast adheres prior to and throughout mitosis, cleavage and valve formation; and it was proposed to function as a reference

Figs. 2–5. *Biddulphia titiana*; LM, girdle view. Bars: 100 µm

Fig. 2. Interphase cell

Fig. 3. Interphase; contracted protoplast adhered to LPs

Fig. 4. Immuno-fluorescence of cytoplasmic MTs

Fig. 5. Immunolabelled prometaphase cell; no cytoplasmic MTs

Fig. 6–9. *Coscinodiscus wailesii*; LM; Bars: 25 µm

Fig. 6. Interphase cell; strands from nucleus to mLPs (arrows)

Fig. 7. Same cell from 6-o-clock

Fig. 8. Cell in anaphase; nucleus in 12, mLPs in 4 and 8-o-clock

Fig. 9. Same cell from 12-o-clock; strands from nucleus to LPs

Fig. 10. *C. wailesii*; SEM; acid cleaned valve; marginal LPs; single mLP. Bar: 10 µm

Fig. 11. *C. wailesii*; TEM; section through inner part of LP; adhesive material between siliceous tube (*Si*) and PL (arrows) and loose PL beneath the slit; fibrillar structures (? small arrows). *m* Mucilage, *mv* mucous vesicle, *ol* organic layer. Bar: 500 nm

Fig. 12. *Navicula cuspidata*; TEM; TS through inner raphe. Adhesive material between raphe ribs (*Si*) and PL (arrows), and loose PL beneath the raphe fissure. *A* Actin filaments. Bar: 200 nm. (Courtesy of Dr. Lesley Edgar)

line for defining the cell equator for division (Fig. 1 A–C) (Schmid 1984 b, 1986 a, 1987 a, b; and see below). During cell division one parental theca is passed on to each of the progeny cells; these thecae in both siblings become the epithecae, while new hypothecae are intracellularly formed. For most species, this sequence of events results in a decrease in size during successive divisions in some offspring.

Rejuvenation to the original cell size is generally accomplished by sexual reproduction (oogamy in centrics and isogamy in pennates), where the zygote swells into a large auxospore, within which the initial thecae of a new subclone are laid down. Its mode of formation renders it morphologically distinguishable from those thecae, which are created by division. Initial valves are often domed or even hemispherical, and with few exceptions lack external processes (Drebes 1977, Pickett-Heaps et al. 1990, Round et al. 1990, Schmid 1990).

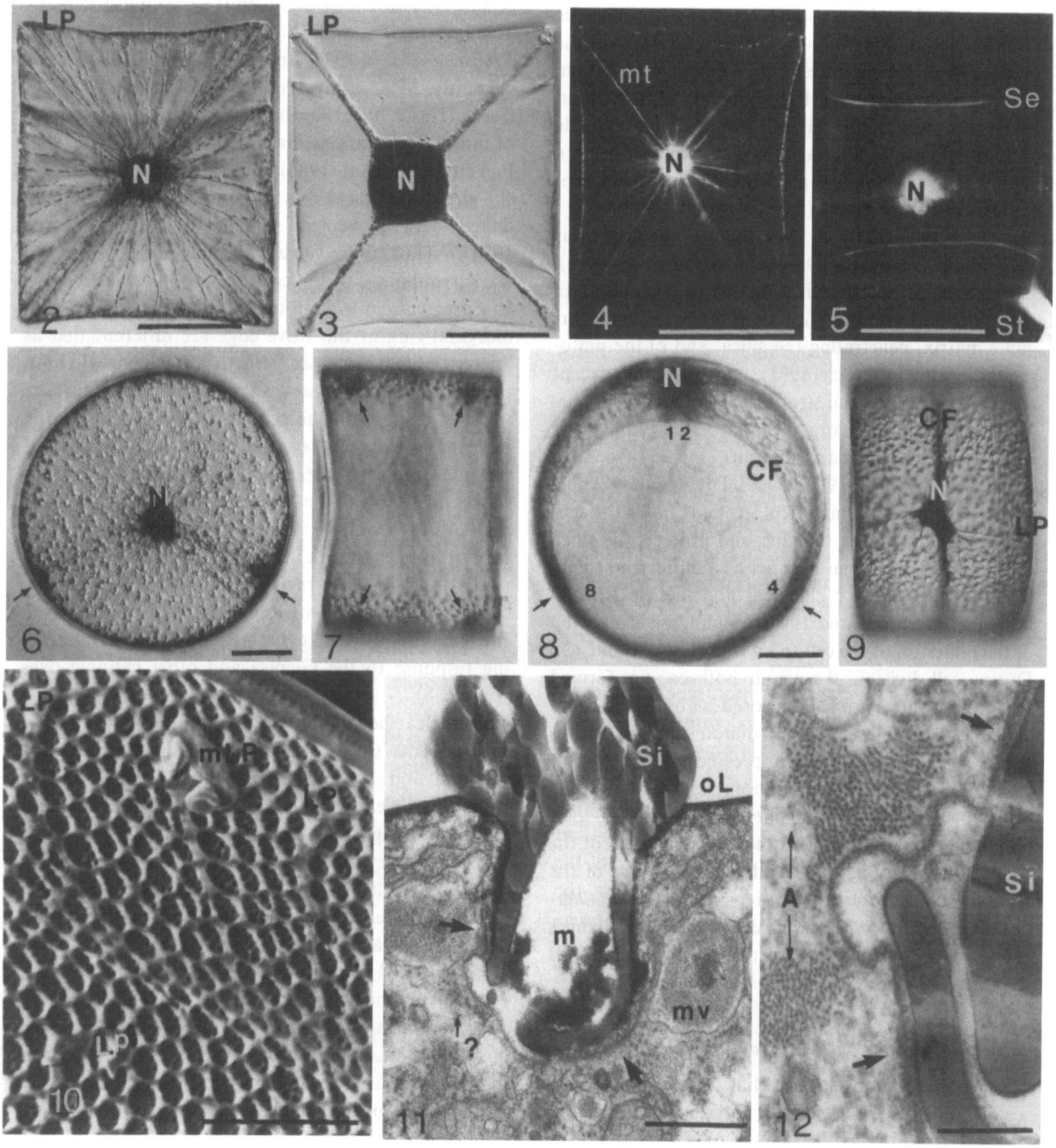

The chemical components of the cell wall and their possible functions

The diatom wall consists of two chemically different components which probably serve different functions.
(1) The rigid part, amorphous silica, shaped into a complex lightweight-construction of rib- or honeycomb-systems, is designed to serve mechanical functions such as maintenance of shape, counterbalancing turgor pressure (Steucek and Schmid 1989, Pickett-Heaps et al. 1990); specific regions, to which the protoplast is anchored, act as attachment for cytoplasmic strands containing tension transducing elements of the cytoskeleton for intracellular guying, orientation, movements and cleavage (Figs. 2–4 and 6–12). Without such connections to the cell wall, diatoms cannot divide (Schmid 1984 a, b, 1986 a, 1987 a, b, and see below). A similar conclusion was also recently drawn for protoplasts of higher plants (Schindler et al. 1989).
(2) The composition of the organic component is known only fragmentarily (refs. in Schmid et al. 1981). It may be differentiated into a) the organic coat, an investment around the siliceous component; b) an organic matrix, penetrating the silica; and c) the diatotepum sensu von Stosch (1981), a distinct, and, except for the labiate (LP) and strutted (SP) processes, raphe, and ocelli, continuous organic layer on the cytoplasmic side of the wall in many diatoms (Fig. 11) (see also Pickett-Heaps et al. 1990: fig. 2, Drebes and Schulz 1990). The organic component not only holds the separate parts of a theca together and protects the silica against dissolution, but may act as a buffer and signal transducing zone (for chemical and tacticle stimuli) between the cell wall and its surroundings; it is involved in the adhesion of the cell to its wall (Figs. 11 and 12). During cell division, the new hypothecae are formed while the siblings are still enclosed within the mother wall. The protoplasts in many diatoms are adhered to it by organic glue, and are appressed to it by turgor. Thus, the already existing diatom wall serves during morphogenesis of the new ones as a mould for their circumferential outline. Morphological features of the mother wall may be imprinted on the surface of the forming hypothecae, e.g., as steps in the valve (*Ellerbeckia*; Crawford 1981); notch deformities and changes in symmetry are perpetuated over many generations (Schmid 1986 a).

Structural features of the wall as spatial cues

Several observations suggest that certain morphological features of the rigid wall serve as spatial determinants which the protoplast might use to order its activities in three dimensions. This interpretation is suggested from the fact that the protoplast is very firmly attached to the wall at structural features that include the SPs, LPs, and raphe area (Schmid 1987 a, b; Pickett-Heaps et al. 1990). A locally thickened organic layer and an electron-dense glycocalyx restricted to this area, connects the protoplast to the raphe (Fig. 12) (Edgar and Pickett-Heaps 1984 a, b) and to LPs (Fig. 11) and SPs. Upon plasmolysis or contraction, these wall parts remain adhesion sites at which the protoplast remains attached to the wall (Figs. 2–4) (Schmid 1984 a, b, 1986 a, 1987 a, b, 1990; Franz and Schmid 1994). Concurrently, SPs are involved in the extrusion of chitin fibrils (Hasle 1972, Herth 1979), LPs in the secretion of mucous substances (Hasle 1973, Medlin et al. 1986, Pickett-Heaps et al. 1986); and the raphe, a pervalvar slit along the apical axis in pennates, is functionally linked with movement (Edgar and Pickett-Heaps 1984 a). Thus, these valve structures are also ports to the surroundings of the cells.
In *Coscinodiscus wailesii* two of the LPs arranged in a marginal ring at the valve edge are differentiated as "macro labiate processes" (Fig. 10), and aligned in both thecae, 120° apart, in a 4- and 8-o-clock position. The nucleus, residing in the center of the epivalve during interphase, connects to these mLPs with thick cytoplasmic strands (Figs. 6 and 7). For mitosis the nucleus moves invariably to the 12-o-clock location in the girdle region (Figs. 8 and 9), and the strands are again prominent during creation of the new valves, where the new mLPs form in close proximity to the old ones. The cytoplasm below the mLPs can be discriminated from the remaining cortical cytoplasm by means of its staining response to the Giemsa and other cationic dyes (Figs. 6–8).
This nuclear behaviour with respect to the position of mLPs seems to be a generic trait in *Coscinodiscus* (Schmid and Volcani 1983, Schmid 1986 b; and in prep.), and clearly indicates some kind of cellular hierarchy among the valve apertures. It may also show an evolutionary trend, culminating in the raphe as proposed by Hasle (1973), with the mLP as intermediate structure. Interestingly, Sims (1989) found in cretaceous and palaeocene *Coscinodiscus* spp. two areas of closely packed LPs within a ring of evenly spaced processes, all identical in structure. The packed LPs occupy the same position where in recent *Coscinodiscus* sensu stricto the mLPs are located, and the valve margin is depressed at these two regions, which appear as notches in the valve outline, indicating tension to the cell in-

terior during valve formation. Unfortunately these species did not survive beyond the Miocene.

Analogous to the mLPs in *Coscinodiscus*, in *Thalassiosira eccentrica*, *T. punctigera*, and others one normal labiate process is present among the marginal ring of SPs, linked to the nucleus. In this case, however, and even in *Planktoniella sol*, bearing two marginal LPs 120° apart (Hargraves and Schmid 1994), the mitotic position of the nucleus seems optional with respect to the location of the LPs, because the LPs of epi- and hypothecae are not aligned as they are in *Coscinodiscus* (Fig. 7), where the polarity of a whole clone is fixed with the formation of the first initial valve of the auxospore. This behaviour is reminiscent of raphid pennate diatoms displaying "cis"-configuration (Mann and Stickle 1988).

Because of such structural features, almost all diatom valves, and thus the cells, are either asymmetric or bilateral, which is perhaps of vital importance for the cytoplasm as reference for orientation.

Variability of the diatom wall and consequences for taxonomy

In many diatoms, the unique division, as a result of the mechanical constraints of a rigid wall, causes a successive decrease in cell size for part of a clone or population. The ratio between the volume of the nucleus, cytoplasm and vacuole changes during diminution, and this may lead to a physiological and, consequently, morphological heterogeneity among the cells of a clone. This renders diatoms quite distinct from other algal groups. As an example, sexual reproduction can be externally triggered only when cells are of a certain size class (Drebes 1977, Round et al. 1990, Schmid 1990). The size decrease is often accompanied by an allometric change in the proportions – shortening of the longitudinal axis occurs faster than that of the transversal axis – and by a loss of structural elements of the cell wall, i.e., taxonomically important characters vary with the age of a clone (Pickett-Heaps et al. 1990, Round et al. 1990, Schmid 1990).

Superimposed on this type of variation are those which are manifested as changes in gross morphology and valve pattern as a response to changes in environmental conditions, culminating in the production of resting spores (Syvertsen 1977, Schmid 1979, French and Hargraves 1985, Round et al. 1990). This variation may be coupled to changes in protein composition and immunochemistry as was demonstrated for *Phaeodactylum tricornutum* (Gutenbrunner et al. 1994). Also, chain forming diatoms frequently produce morphologically different intercalary- and separation-valves. Von Stosch et al. (1973) and Rines and Hargraves (1988) encountered up to seven different valves within a single clone of *Chaetoceros*.

These examples demonstrate the problems lurking in a taxonomical concept based solely on valve structure, or solely on biochemical analyses. The problem arises since the valve producing machinery is highly sensitive to the environment, a feature diatoms share with other silicifying organisms (Preisig 1994 and references therein). This sensitivity, together with the specific division mode, results in a great phenotypical plasticity in many species. Thus, some authorities, estimate the number of bona fide-taxa around 15,000, and not, as has been described, more than 100,000 species (Van Landingham 1967–79, Werner 1977 and references therein).

Events preceding valve morphogenesis

One reason for this sensitivity may be that the processes of mitosis and cytokinesis, both of which under normal circumstances precede valve formation, have a profound impact on the morphology of the prospective valve. Several cytoplasmic organelles change their distribution through the division cycle, and the nucleus invariably assumes its mitotic position at the girdle bands (Figs. 5, 8, 9, 13 B, 19, and 20), moves to the center of the prospective valve in telophase and back to interphase position after valve formation (Pickett-Heaps et al. 1990, Round et al. 1990; Pickett-Heaps 1991 a, b). The cytoskeleton is depolymerized during prometaphase (Figs. 5 and 14) (Wordemann et al. 1986, Franz and Schmid 1994) and has to be reestablished after cleavage and prior to or during valve formation. Hence, valve morphogenesis is susceptible to external influences affecting cell division and the cytoskeleton (Pickett-Heaps et al. 1990), and teratologies are more often encountered in valves than in girdle bands.

In *Biddulphiopsis titiana*, for example, the cytoplasmic strands from the LPs anchoring nucleus and its microtubule center (MC)/spindle-precursor-complex within the cell (Figs. 2–4) become reorganized after formation of the last GB of the hypotheca and prior to nuclear migration, forming a diaphragm in the prospective plane of division. Then they completely retract. Later the cleavage furrow cuts in precisely from this circumferential line where the strands were inserted in the cortical cytoplasm (Franz and Schmid 1994). These strands contain microtubules (MTs) as indicated with

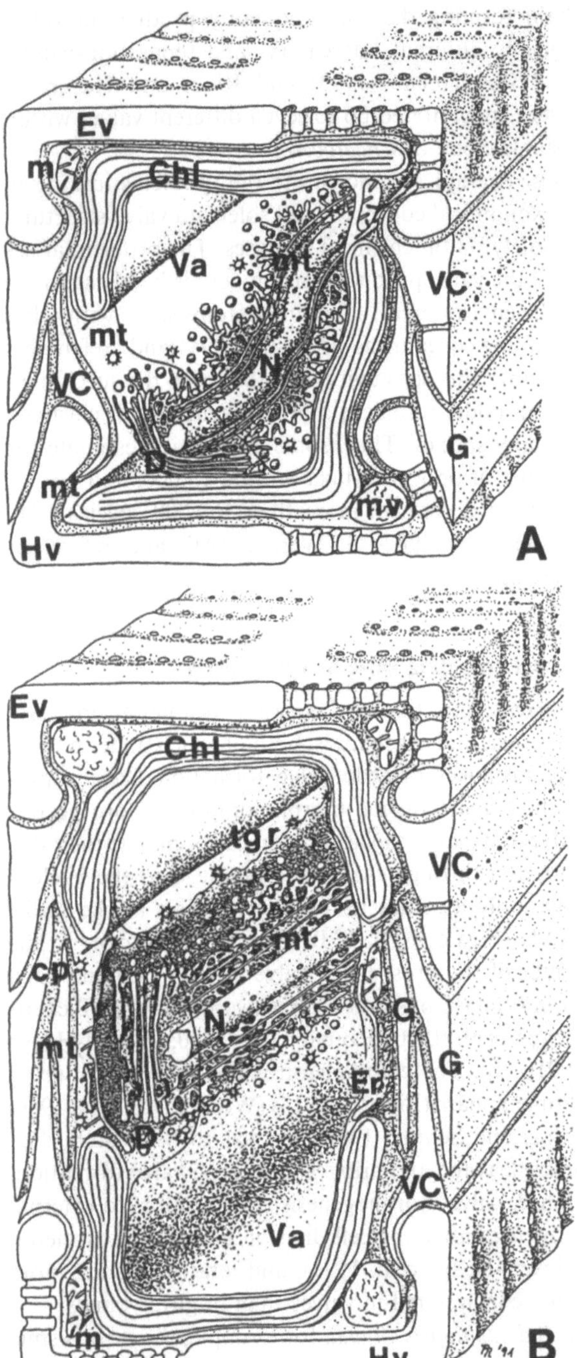

Fig. 13 A, B. Diagrams of *Synedra ulna* **A** Interphase cell; epitheca complete; hypotheca incomplete. Rows of valve pores = striae. Protoplasm in diagonal arrangement. Nuclear extensions (*N*) with series of dictyosomes (*D*), supported by MTs (*mt*); single MTs at the chloroplast (*Chl*) edges. **B** Cell in preprophase; girdle band of HT exocytosed; rotated protoplast; nuclear extensions aligned with the girdle band (*G*); MTs in the equatorial cortex; subjacent ER; coated pits (*cp*) in the equatorial PL; trans-Golgi-reticulum (*tgr*) forms large cisternae. *Ev* Epivalve, *Hv* hypovalve, *m* mitochondria, *Va* vacuole, VC valvocopula

monoclonal antibodies against α-tubulin, and they are depolymerized by pro- to prometaphase (Figs. 4 and 5) and reassembled after cleavage, exactly as described for *Stephanopyxis* (Wordemann et al. 1986, Wordemann 1992).

This MT breakdown has severe consequences for the distribution of organelles. In *Synedra ulna*, a narrow pennate diatom up to 450 μm long, the nucleus extends, supported by MTs, the entire length of the cell. The dictyosomes are serially arranged along the nucleus throughout interphase (Figs. 13 A, B and 18). When the cytoplasmic MTs depolymerize during mitosis, the nuclear extensions retract and the dictyosomes cluster around the central part of the nucleus until some time after cleavage. Nuclear extensions are not sent out prior to the onset of valve formation (Figs. 16 and 17) (Schmid 1989, and in prep.). The situation in *Pinnularia* is similar. The dictyosomes are paired along ER-cisternae (Drum 1966; Pickett-Heaps et al. 1978 a, b, 1979), which are continuations of nuclear tentacles (Fig. 15). Again, they are withdrawn in mitosis and, until after cleavage, the MC is formed close to the cleavage plasmalemma (cPL) (reorganized at/from the spindle pole; Pickett-Heaps et al. 1978 a, b, 1979), sending out cytoplasmic MTs, nuclear tentacles are lacking. The ER and, now single, dictyosomes are randomly oriented around the telophase nucleus (Fig. 14) (Schmid in prep.). Restoration of the cell's interphase symmetry seems to be dependent upon the aster of cytoplasmic MTs focussing in the MC (Fig. 15), similar as described for other cells (Kreis 1990).

Mitosis was studied the last two decades in great detail by Pickett-Heaps and co-workers, whose EM-work made the diatom spindle the best known of all organisms (see review by Pickett-Heaps 1991 b; Pickett-Heaps et al. 1990). In their exquisite research they concentrated also on the two types of microtubule organizing centers (MTOCs) in diatoms, the MC, as the focus of cytoplasmic MTs, and the spindle, and their interdependence and apparent transformation into each other (Pickett-Heaps 1991 b). Two types of MTOCs, however, are obviously not continuous through the whole group. In *Stephanopyxis* the MC seems to be completely absent, and the spindle precursor, that in some species resembles ultrastructurally a striated flagellar root (e.g., *Lithodesmium*, Manton et al. 1969; *C. wailesii*, Pickett-Heaps et al. 1990: fig. 77) (Figs. 21–22), was suggested to serve also the functions of an interphase MC (Wordemann et al. 1986, Jeckel and Kowallik 1990). A similar proposal was put forward for *Ditylum* (Li and Volcani 1985 a), although in the latter an ill-defined and very transient MC was eventually detected in an indentation of the nucleus (Pickett-Heaps et al. 1988). Thus the question is raised, whether MC or spindle precursor alone, as perhaps the functionally and structurally interdependent parts of a "centrosome" sensu Mazia (1987) can serve both functions, similar to the situation in *Ochromonas* (Bouck and Brown 1973). If this is indeed the case, then these genera would truly differ from the other diatoms investigated so far, where the spindle precursor (or primordial spindle) develops in preprophase (Lauterborn 1896, Pickett-Heaps et al. 1990). In *Lithodesmium* (Manton et al. 1969) and in *C. wailesii* (Schmid 1987 b), on the contrary, both entities seem to persist through the entire interphase, with the MC admittedly less defined than in some pennate diatoms (Fig. 21). The relevance for valve formation is given below.

Cytokinesis involves cytochalasin-sensitive microfilaments (Pickett-Heaps et al. 1990). In several *Coscinodiscus* spp., e.g., *C. subbuliensis* (Ikari 1923), *C. biconicus* (Hofker 1928), *C. wailesii* (Figs. 8 and 9) (Schmid and Volcani 1983), and some *Thalassiosiraceae, T. eccentrica, T. punctigera* and *P. sol* (Schmid 1984 a, b, and unpubl.), as well as in *S. turris* (Drebes 1969) and in *Ditylum* (Fig. 1 B) (Li and Volcani 1985 a) the cleavage furrow (CF) was reported to grow unilaterally from the mitotic nuclear region, while in other centrics the CF seems to constrict in a rather symmetric fashion. Various reports note an initiation at the apices in pennates, or its homologue, the narrow pole in *Surirella* (Lauterborn 1896, Pickett-Heaps et al. 1990), while in *S. ulna* the CF appears to develop initially unilaterally, assisted by MTs emanating from the MC lying subjacent to the last GB (Figs. 19 and 20).

There are also different responses to MT-inhibitors: they did not seem to influence cytokinesis in *Stephanopyxis* (Wordemann et al. 1986) and *Ditylum* (Li and Volcani 1988) whereas in a number of other centric and pennate species cleavage can be reversed at a stage when the CF is not yet closed to a ring furrow (e.g., *S. ulna, C. wailesii, T. eccentrica*; Schmid 1980, 1984 a, b; Pickett-Heaps et al. 1990). This response to anti-MT drugs and the situation in *S. ulna* (Figs. 19 and 20), suggest that the MC and some of its MTs are involved in the establishment of the CF in some species.

In addition, the possibility may exist that the region where the last GB of the hypovalve has been exocytosed (Fig. 1 A), acts as a reference region for the cytoskeleton establishing the cleavage furrow, perhaps analogous to higher plant cells in which the MT-preprophase band is thought to define this region. In *S. ulna*, entry into mitosis is signalled with the perfectly straight coalignment of the nuclear extensions with this GB (Fig. 13 B), and in *B. titiana* with the formation of the diaphragm, i.e., cytoplasmic spokes from the nuclear region to the last GB (Franz and Schmid 1994), and reminiscent of the "phragmosome' in vacuolated cells of higher plants (Lloyd et al 1992). These observations coincide with the suggestion of Wordemann et al. (1986), that the plane of division is cortically determined prior to mitosis. Clearly this matter merits further intensive research.

Valve formation

Morphogenetic studies usually focus on the formation of the valve, partly perhaps because valves bear the important characters for species identification, such as diameter, outline (centric, pennate, undulations), surface (hemispherical or flat, elevations, horns), and the pattern (orientation and density of fascicles, striae, areolae, fibulae). Valve formation is tightly bound to mitosis (Geitler 1963, von Stosch et al. 1973) and is strictly controlled by a non-reversible sequence of instructions. Formation of the siliceous part of the cell wall occurs intracellularly in a membrane bound compartment, the silica deposition vesicle (SDV) (Drum and Pankratz 1964, Gordon and Drum 1994), which is presumably conservative in all silicifying organisms because it is fundamental for controlling the chemistry of silica deposition (Mann 1983, 1986; Simkiss 1986). Modelling of this reaction vessel in a species-specific manner creates the species-specific valve, which is exocytosed at maturity. In diatoms, modelling is accomplished by a precisely timed sequential interaction between the plasmalemma, cytoskeleton, endoplasmic reticulum, spacer-vesicles, and other Golgi-derived vesicles and mitochondria. Eventually a protective organic coat and the diatotepum are added to the valve.

The membrane of the SDV (Fig. 26) ultrastructurally resembles a Golgi-membrane, and is thus clearly distinguishable at least from an ER-membrane. Controversial opinions exist as to its origin and enlargement. It was suggested by some workers to be of dictyosomal origin, while others doubt this, paralleling the situation in chrysophytes and thecameba (see discussion in Crawford and Schmid 1986, Pickett-Heaps et al. 1990; and Anderson 1994, Preisig 1994). A single observation in *C. wailesii* points indeed towards a Golgi-derivation of the SDV (Schmid 1986 a, b; Pickett-Heaps et al. 1990: fig. 30 a–c), and also the organic wall component seems Golgi-derived in this species (see below). A similar de-

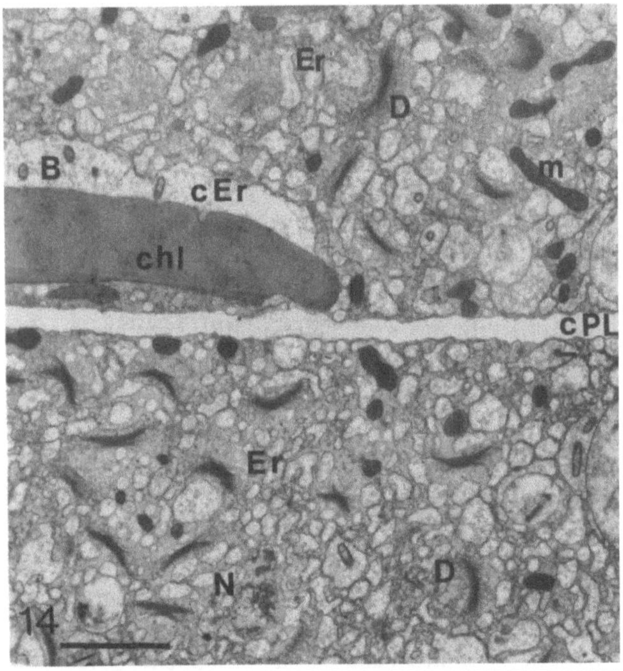
Er
D
B
cEr
m
chl
cPL
Er
N
D
14

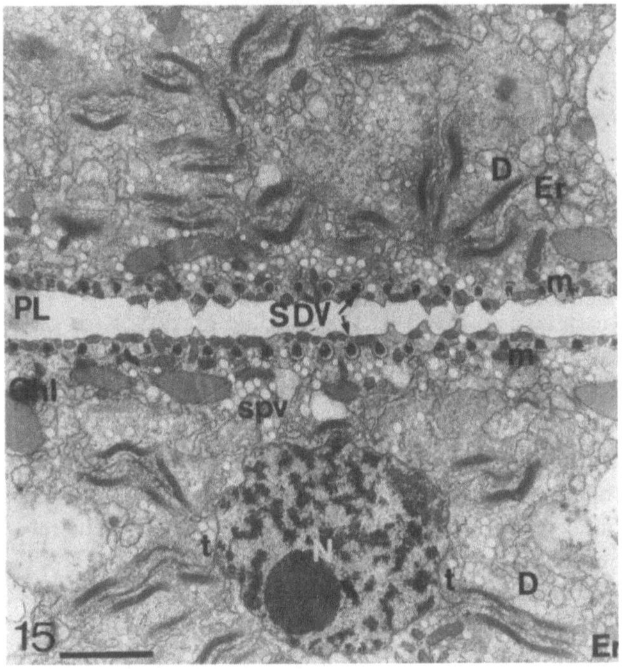
D
Er
PL
SDV
m
m
chl
spv
t
N
t
D
Er
15

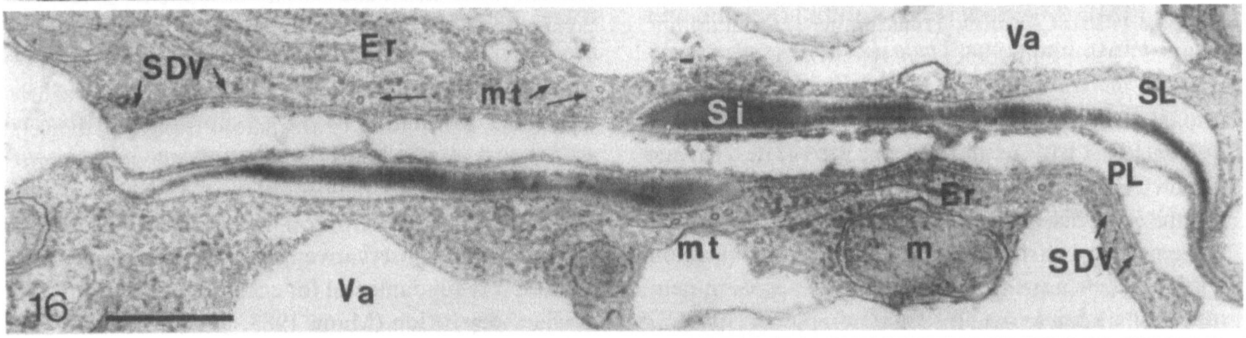
Er
SDV
mt
Va
Si
SL
PL
Er
mt
m
SDV
Va
16

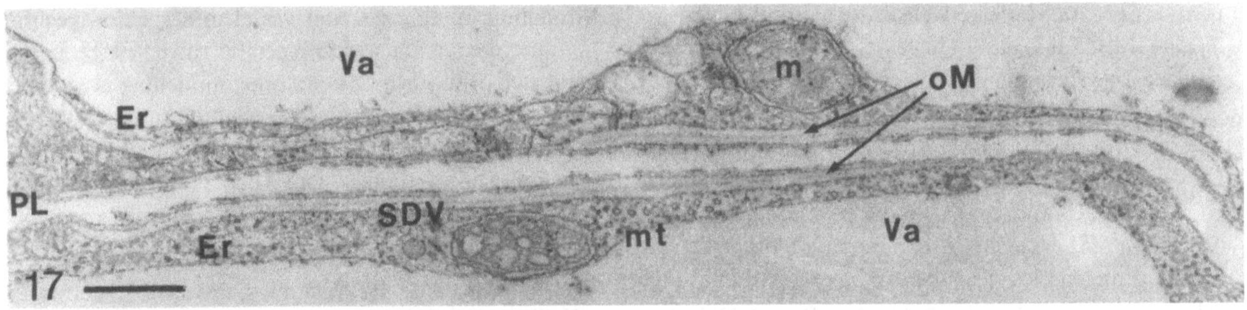
Va
m
oM
Er
PL
SDV
mt
Va
Er
17

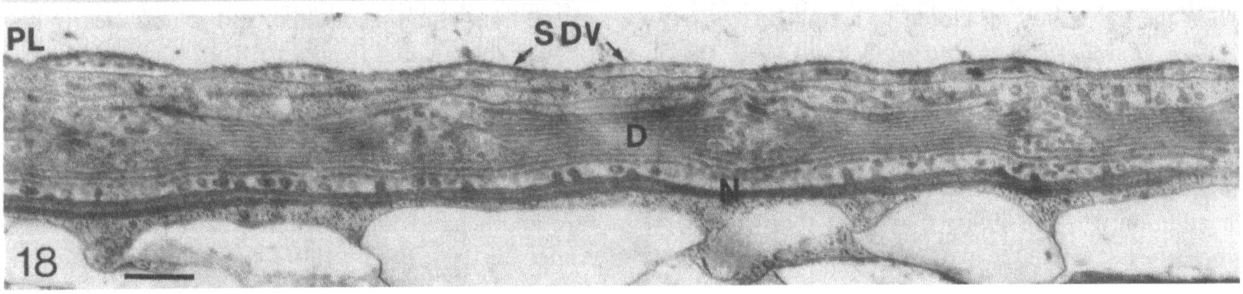
PL
SDV
D
N
18

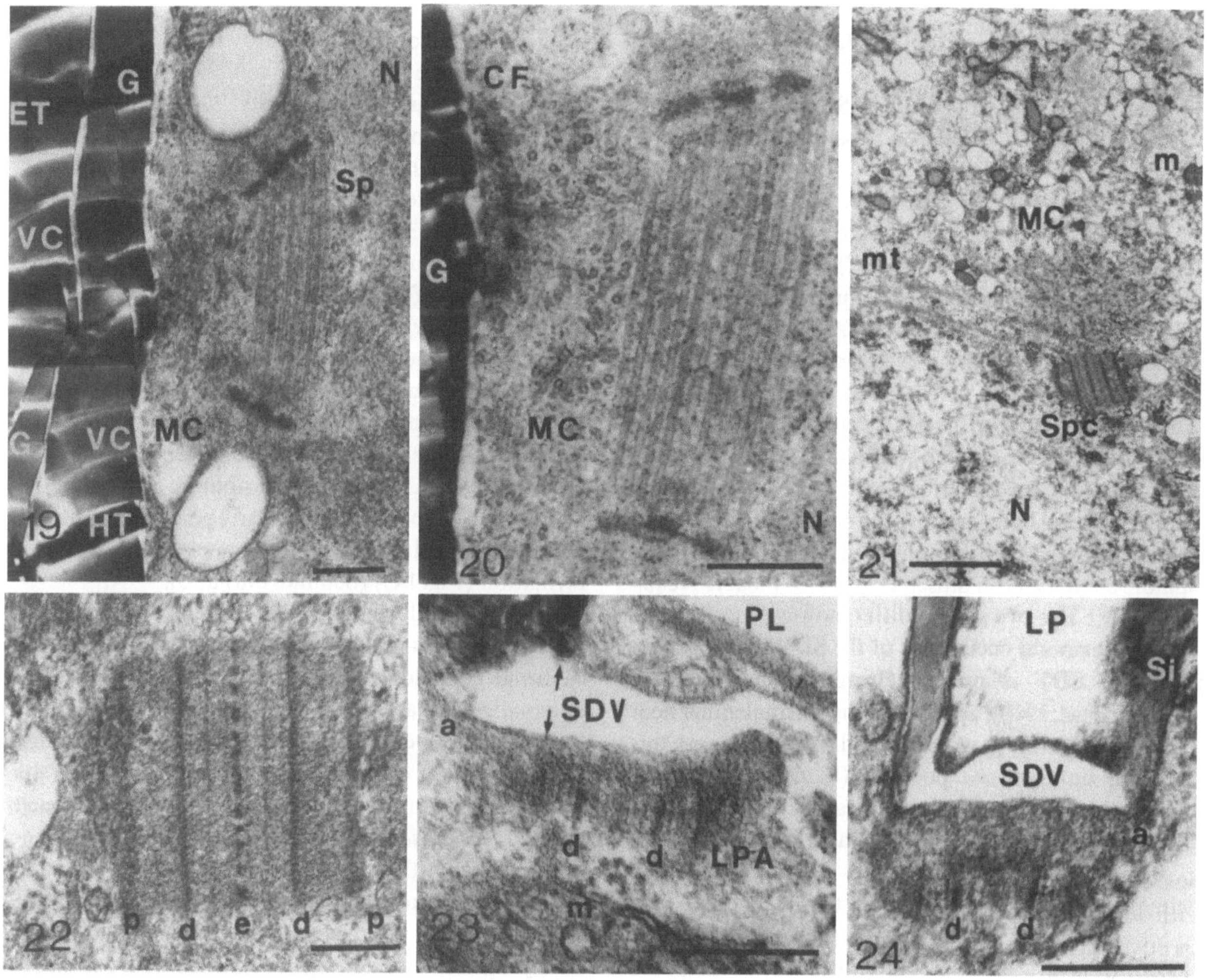

Figs. 19 and **20.** *Synedra ulna.* TEM; TS (pole view) through prophase cell. Bars: 300 nm

Fig. 19. MC between extranuclear spindle (*Sp*) and PL subjacent to the VC – G junction of the hypotheca

Fig. 20. Same cell, a few sections away. Initial stage of cleavage furrow (*CF*) involving MTs

Figs. 21–24. *Coscinodiscus wailesii.* TEM

Fig. 21. Interphase nucleus with MC and spindle precursor (*Spc*); MTs. Bar: 1 µm

Fig. 22. Spindle precursor (slightly oblique); *p* poles; *e* equatorial loops; *d* dense "band". Bar: 200 nm

Fig. 23. Early valve formation. Labiate process apparatus (LPA) at the cytoplasmic side of the SDV: two components, amorphous structure (*a*) adjacent to SDV, striated structure with similar striation as the Spc. Bar: 200 nm

Fig. 24. LPA at later stage of formation of the internal tube (*Si*) of the LP; amorphous and striated components of LPA. Bar: 200 nm

---

Figs. 14 and **15.** *Pinnularia maior.* TEM, crossections in girdle view. Bars: 3 µm

Fig. 14. Sibling cells at post cleavage, prior to MC, MT, and SDV formation. ER and single dictyosomes randomly oriented; *B* endobacteria

Fig. 15. Sibling cells after MC-assembly, at early valve formation. Transapical ribs in SDV; spacer vesicles (*spv*) gave way to mitochondria; paired dictyosomes along nuclear tentacles (*t*) and ER-continuations

Figs. 16–18. *Synedra ulna*; TEM. Bars: 300 nm. Figures 16 and 17, transversal sections, early valve formation; Fig. 18, apical section (girdle view), later stage of valve formation.

Fig. 16. Alternating transapical ribs (*Si*) within the SDV (*SL*); narrow profile of the striae-SDV; MTs present, but no nuclear extensions

Fig. 17. Same cell, different section, extraction of silica with HF: organic matrix (*oM*)

Fig. 18. Later stage of valve formation; nuclear extensions present (*N*), dictyosomes (*D*)

bate exists as to enlargment of the SDV, which was suggested to occur by coalescense of small vesicles in some marine centric diatoms (Schmid 1987 a, b), whereas in others this could not be observed (Pickett-Heaps et al. 1990, Gordon and Drum 1994). In pennates, the situation is also obscure. The first indication of the SDV is a membranous tube running the entire length of the future valve (Fig. 1 C), and this tube then expands laterally in a way not yet understood (Figs. 16 and 17) (Pickett-Heaps et al. 1979, Edgar and Pickett-Heaps 1984 b, Boyle et al. 1984; and discussion in Pickett-Heaps et al. 1990, Gordon and Drum 1994).

The initiation site for the SDV determines the pattern center of the valve, i.e., the "annulus" or a central area in centrics, and an axial rib in pennates (Pickett-Heaps et al. 1990). In both groups, the initiation site is correlated with the position of the nucleus and its MC-spindle complex (von Stosch et al. 1973, Pickett-Heaps et al. 1979) The two groups differ, however, in the spatial and temporal occurrence of the SDV. In most centrics, the SDV is present already at mid-cleavage (Schnepf et al. 1980), apposed to the plasmalemma near the spindle poles. Both SDV and nucleus are at this stage in an excentric position (Pickett-Heaps et al. 1988), whereas in pennates it appears at a region equivalent to the former spindle equator and not before postcleavage (Edgar and Pickett-Heaps 1984 b) and, with few exceptions, already at the site of the prospective valve center (Fig. 1 B, C) (Pickett-Heaps et al. 1990). In centrics, if the comigration of the excentric SDV and nucleus to the future valve center is inhibited while valve formation continues, then the pattern center remains excentric (Schmid 1984 a).

Valve outline and topography

Determining the forces and interactions that regulate valve morphogenesis, we must distinguish between moulding the circumferential outline, the topography, and the pattern. In diatoms the SDV is closely apposed to the plasmalemma created during cleavage (cPL) (Figs. 16–18, 31–34, 36, and 37) and the protoplasts in many diatoms are tightly appressed to the parental wall. Thus the outline of the new valve is predetermined by the outline of the mother cell wall and so the issue of how the circumferential form is ultimately controlled can be ascribed to the stage of the initial valve arising within the auxospores following sexual reproduction. In cells with a circular outline the auxospore cover may be sufficient as a reference for the initial valve, whereas all forms deviating from a circular outline need a moulding "perizonium" (von Stosch 1982). In pennate diatoms, the size of the single complete siliceous band of the perizonium directly controls the cell diameter (Mann 1982). In centric diatoms, the more complex shapes arise with the aid of a sequentially deposited complicated system of siliceous bands and hoops (von Stosch et al. 1973, von Stosch 1982).

The moulding surface for the new valve's topography is the cleavage furrow, which is shaped by a combination of turgor, local contractions or "spontaneous plasmolyses" and tension of the new membrane and cytoplasmic cortex, mediated through adhesion to the cell wall. In many diatoms, the cleavage furrow (CF) appears to adhere tightly to the girdle bands of the parental hypotheca, i.e., both siblings adhere to the mother cell's hypotheca, a necessity perhaps, to prevent accidental opening of the mother wall prior to valve formation (Fig. 1 C) (Schmid 1987 a, b).

Two types of divisions have been distinguished by Mann (1984) that correlate with the shape of the CF. A broad CF allows the sibling cells to develop their hypovalves apparently independent from each other ("non-interactive"-type); thus their gross morphology resembles that of initial valves formed within the auxospore. In many diatoms, the CF is narrow, such as in *T. eccentrica*, a typical member of the "interactive"-type, and the siblings are connected during morphogenesis by adhesive material. The valve surfaces, as they develop, are thus mutually influenced, as beautifully demonstrated in *Cymatopleura*, with the undulations of the valves exactly in register (Pickett-Heaps 1991 a). Transient adhesion to both the sibling cell and the mother wall appear to be important for spine formation in *T. eccentrica*, and it seems very likely that this mode of formation also accounts for interlocking spines, leading to colonies (Schmid 1984 b, c, 1986 a, 1987 a, b). Growth of setae, outside the parental wall, in *Chaetoceros* or *Bacteriastum* follows a different mechanism (Pickett-Heaps et al. 1994).

Generation of the valve pattern

The cleavage plasmalemma and its fibrous lamina

The patterned diatom valve with its functionally different domains, such as ribs, chambers, porefields, LPs, SPs, and raphe, arises as a result of the cell's ability to localize silica and moulding organelles to particular regions of the cPL; the SDV in all investigated species is adhered to the cPL during expansion and shaping. Thus it appears possible that the cPL and a fibrous lamina (fLA) (as present in *Coscinodiscus*; Schmid

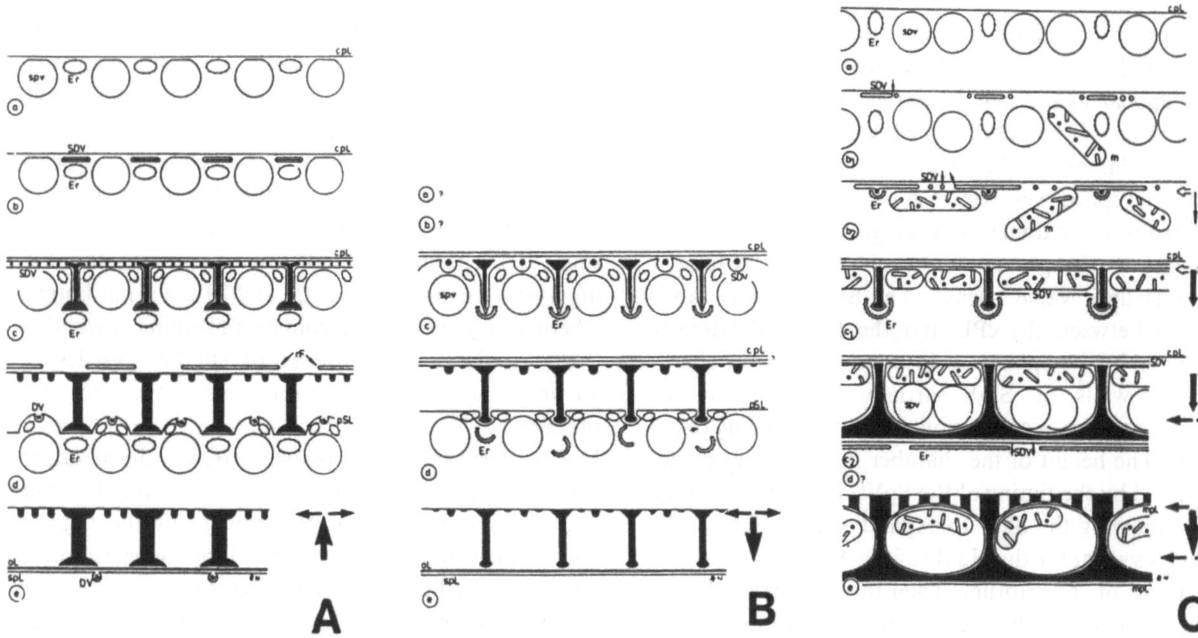

Fig. 25 A–C. Schematic summary of valve formation and exocytosis, and the involvement of spacers. **A** *Coscinodiscus wailesii* and **B** *Achnanthes longipes* with "split walls"; "empty" chambers sealed by organic layer (= diatotepum). Similar moulding events and valve release. Silicification (large arrow) from proximal to distal in *Coscinodiscus* and in reverse direction in *Achnanthes*. **C** *Pinnularia maior*. Chambers (= alveoli) filled with cytoplasm. SpV give way to mitochondria; ER in the same role as in *Coscinodiscus* and *Achnanthes*. Silicification of the chamber side walls from distal to proximal and then the internal plate; chamber roofs are silicified last

1984c, 1986b, 1987a, b) carry morphogenetic information for the new valve. How this information could be transmitted to the SDV and auxiliary moulding organelles is completely unknown, as is the regulation of the precisely timed steps during formation of the three-dimensional architecture. Since pattern formation in diatoms cannot be followed in living cells, information gained with the EM is, by necessity, interpretative, but serial sections of cells at different developmental stages may allow a schematic summary as presented in Fig. 25. Several regions of plasmamembranes (PL) are distinguishable in the EM in diatoms possessing a "split wall" (in which a distinct diatotepum seals the siliceous chambers from the cytoplasm: *C. wailesii* and *Achnanthes longipes*; Figs. 11 and 25A, B), suggesting the possibility that the cPL and its fLA are involved in valve formation. These regions are: a) the PL with its associated fLA and glycocalyx created during cleavage, as a moulding surface for the valve; b) the PL created upon completion of the silicified part of the cell wall by incorporation of the membranes of the "dense vesicles" (see below), involved in the formation of the organic diatotepum; c) the metabolically active PL, ruffled into a patterned array of microplicae and thus sharply demarkated against the smooth membrane domains involved in morphogenesis. This non-random distribution of PL areas appears to be mediated by (or correlated with) adhesion of defined regions of the heteropolar protoplast to defined regions in the heteropolar wall, in a sequential manner (Schmid 1987a, b). Another example of a locally modified PL region was demonstrated in *Pinnularia* where a continuous callosic strip runs along the distal edge of the hypotheca, thought to function as a flexible joint sealing the two thecae during interphase and cell division (Waterkeyn and Bienfait 1987). Callose-synthase complexes are usually located in the PL (Roberts 1990, Kauss 1994). If this is the case also for diatoms, then also in *Pinnularia* the PL region next to the cell wall part formed last, is a specific membrane domain, which travels over the cytoplasmic surface as new girdle bands are exocytosed.

Endoplasmic reticulum, mitochondria, and spacer vesicles

Centric diatoms: The construction units of the *C. wailesii* valve are loculate areolae, hexagonal in outline, a complex cribrum as an outer cover, and a basewall with large circular openings, the foramina, sealed by a diatotepum (schematic summary in Fig. 25 A). ER-cisternae and spacer vesicles (SpV) tightly adhered to the cPL, appear to lay out the pattern in advance of the appearance of the SDV. The SDV grows centrifugally as a tubular membrane system in the distal gaps between the spacer vesicles, with a discontinuous system of fibrous dense domains at the growing tips. The ER attaches to the proximal side of the SDV, where it remains until valve release. The SDV then is sandwiched between the cPL and the ER, and laterally bordered by the SpVs. For differentiation of the chamber side walls, the SDV partially detaches from the cPL, remaining in contact only with the growing distal edges. The height of the chamber side walls seems determined by the timing of the SpVs detaching from the cPL, allowing the SDV to expand again laterally. ER vesicles seen at the distal side of the AVs seem to mould the pores of the cribrum. Then these pores fill with a delicate lattice, with a similar mesh-size to that of the dictyosomal cisternae (Figs. 25 A and 27) (Schmid 1984 c, 1986 b, 1987 a, b). Although in *T. eccentrica* the architecture of the loculate areolae is reversed, with the foramina at the valve outside and the cribrum facing the cytoplasm, the valve is constructed in the same sequence from proximal to distal, but without the involvement of spacer vesicles. They are perhaps not necessary because of the inverted construction (Schmid and Schulz 1979). In both these centric species, silica is deposited into the SDV as it grows, with a small time lag, allowing the observer to distinguish a growing zone and a compacting zone where the silica is already hardened (Schmid and Schulz 1979; reviewed in Pickett-Heaps et al. 1990).

Pennate diatoms: Achnanthes longipes also possesses a split valve, in which evenly spaced, empty chambers with deeply protruding side walls, are transversely oriented to the raphe and closed with a diatotepum at the cytoplasmic side. Involvement of the ER and spacer vesicles in chamber formation, as well as the process of valve release and sealing of the chambers, are comparable to *Coscinodiscus* (see schematic summary in Fig. 25 B). The spacer vesicles are enormous, often filling a whole chamber (Figs. 28 and 29). It is not known whether they are this size from the beginning or whether they are the fusion product of smaller vesicles or whether they grow. As in *Coscinodiscus* their origin is obscure.

In *Pinnularia* and *Caloneis* alveoli are serially arranged transversally to the raphe. They are covered in the mature valve by a perforated roof facing the exterior and an unstructured plate with a large opening to the inside. In the living cell these chambers are occupied by mitochondria which travel to the SDV early in valve formation (Pickett-Heaps et al. 1979, Edgar, 1980) (schematic summary in Fig. 25 C). In post-cleavage stages of *Pinnularia*, prior to the appearance of the SDV, spacer vesicles and ER can be found in close proximity to the cPL, similar to the situation in *C. wailesii* (Figs. 25 C and 30). These spacer vesicles resemble those produced by the medial to distal cisternae of dictyosomes (Fig. 38), and seem aligned also below the forming raphe ribs (Fig. 31). The early stages of SDV formation and the involvement of the cytoskeleton were illustrated by Pickett-Heaps et al. (1979). In the interim stages, not observed before, the SDV is seen as fine tubular projections at the sites of the future partition walls of the chambers; gradual expansion in apical direction may form the SDV for the alveolar

---

Fig. 26. *Pinnularia maior*. TEM; section through plasmalemma (*cPL*), SDV and ER to show the differences in thickness of the membranes. Bar: 200 nm

Fig. 27. *Coscinodiscus wailesii*. TEM; TS through forming valve prior to release showing spacer (= areolae vesicles, *AV*). Bar: 3 µm

Figs. 28 and **29.** *Achmanthes longipes*; TEM. Bars: 1 µm

Fig. 28. Valvar plane section at stage of cribrum formation: elongated spacer vesicles (*SpV*) and mitochondria

Fig. 29. Section in apical plane (girdle view) through siblings shortly prior to valve release

Figs. 30–34. *Pinnularia major*. TEM; section in apical plane

Fig. 30. Post cleavage prior to MC, MT, and SDV formation. Spacer vesicles (*SpV*) and ER below the new cPL. Bar: 1 µm

Fig. 31. SpV below the forming raphe ribs (*RR*) Bar: 1 µm

Fig. 32. SpV between transapical ribs, close to the raphe ribs. Bar: 1 µm

Fig. 33. SpV gave way to mitochondria, attaching to the SDV of the chamber roofs. Transapical ribs (*Si*) capped by ER-cisternae. Bar: 1 µm

Fig. 34. Mitochondria in their final position. Bar: 500 nm

roofs (Fig. 34). Spacer vesicles then leave this area, except in the proximity of the raphe ribs (Fig. 32), while mitochondria invade, apparently adhering to the SDV of the chamber roofs (Figs. 25 C, 33, and 34). The forming side walls, are capped by ER cisternae (Figs. 25 C, 32, and 33) as illustrated by Pickett-Heaps et al. (1979). Growth of the SDV commences from the distal side (i.e., the cPL) inwards to the cell interior, the side walls

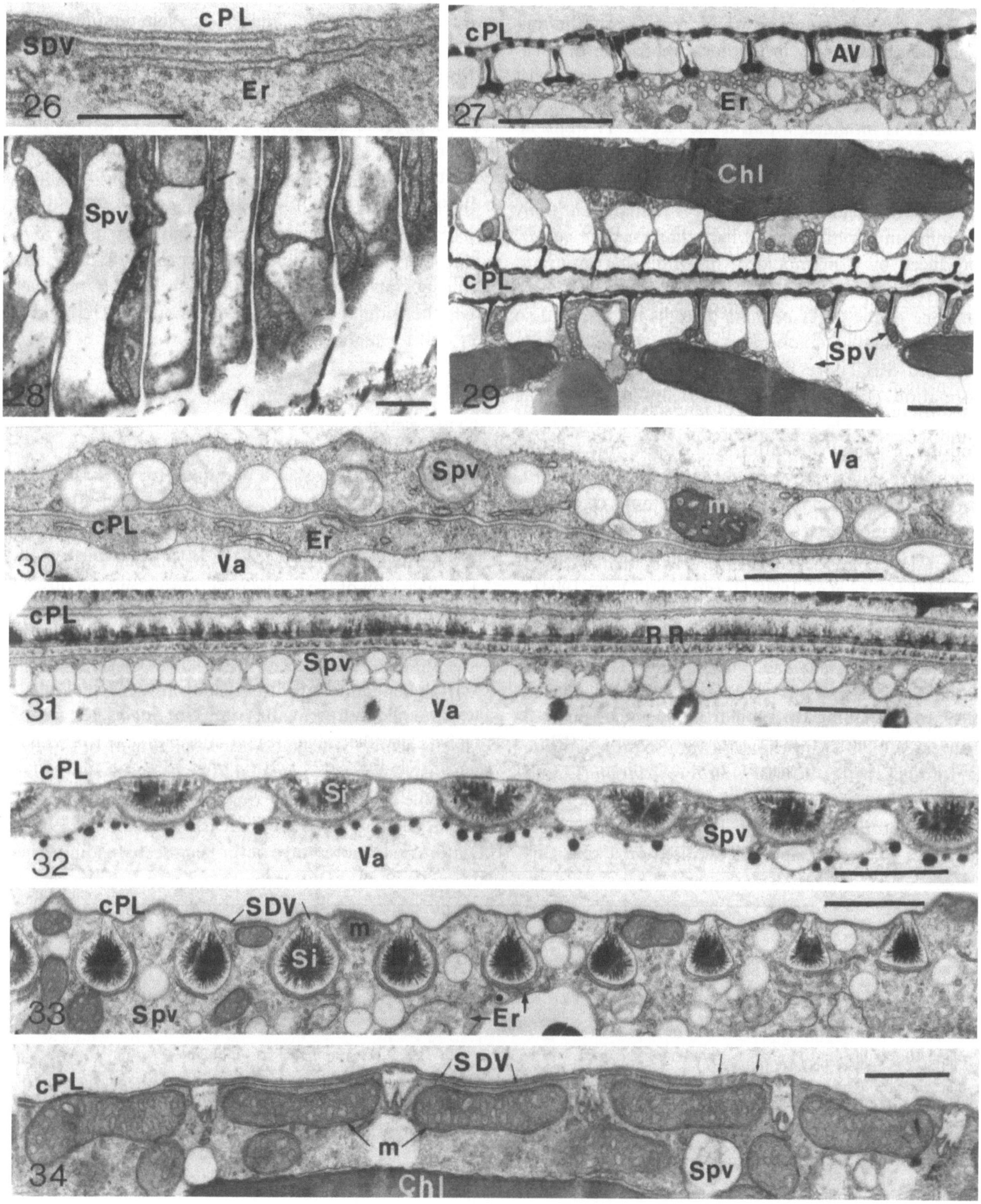

thicken and then the internal, unstructured plate is formed and silicified. Despite the fact that the SDV for the chamber roofs is present from the very early stages, it is moulded and silicified last (Figs. 25 C and 32–34) (Schmid in prep.). This sequence of silicification has been described also for *Caloneis* (Edgar 1980), and in both cases it is unclear which role, if any, the mitochondria may play in this local inhibition of silicification. Two modes of cribrum formation thus exist in diatoms. Here in *Pinnularia* (and some others), the cribrum is moulded within a sac-like SDV, while in *Coscinodiscus*, the siliceous meshwork is tightly enveloped by a reticulate SDV, which does not become a continuous sac until valve release. A similar dichotomy is found in chrysophytes, where the cyst pore can be produced by a premoulded, or within a continuous SDV (Sandgren 1980).

In none of the other pennate diatoms investigated so far has a similar involvement of spacer vesicles been reported. Strial pores involve fibrillar material for their formation (Edgar and Pickett-Heaps 1984 b). The growth of the SDV seems to progress to both the distal and the proximal side, after the layout of the valve is established. Silicification proceeds concurrently and silica is also suggested to be precipitated onto an organic template which serves perhaps as a nucleating surface (Fig. 17) (for more detailed information and references, see Edgar and Pickett-Heaps 1984 b, Simkiss 1986, Pickett-Heaps et al. 1990, Gordon and Drum 1994).

Mitochondria also lie subjacent to the forming raphe in genera characterized by a canal raphe, in Nitzschiaceae, and Surirellaceae (Fig. 39). They appear to be involved in shaping the tubular raphe canal (Pickett-Heaps et al. 1990). Mitochondria are also present inside the forming rostrae in *Chaetoceros rostratrum* (Li and Volcani 1985 b; for the role of mitochondria in other silicifying protists, see Preisig 1994).

In species where the raphe slit is raised on a keel, this region is stabilized by fibulae, i.e., more or less evenly spaced, siliceous ribs holding the raphe ribs together (Fig. 35). In *Nitzschia*, the spacing of these "angle irons" seems achieved by giant vesicles (Fig. 36). The whole cisterna of the trans face of the dictyosomes appears to slough off and act as a spacer vesicle, moving to the SDV (Fig. 39). The ER has again an undefined role in covering the forming fibulae and transapical ribs (Fig. 37) (Schmid in prep.).

The ER may act as a mould for the SDV in some chrysophyceae, thus serving the same function as the chloroplast ER in synurophyceae (refs. in Preisig 1994). Its function in diatoms may be equivalent: in certain diatoms, the ER displays during valve formation a patterned arrangement in close proximity to the SDV, and this suggests a function in exercising local control during valve formation; just how this is achieved, whether similar to the proposed model of Hepler et al. (1990) for higher plants, is unknown.

Cytoskeleton

Biddulphiopsis serves as a paradigm for illustrating the interaction of cell wall and cytoskeleton. Its cytoplasmic strings anchor the nucleus firmly to the labiate processes at the poles (Figs. 2–4). A protoplast would not normally fill the corners of an angular valve, due to the tension of the cytoplasmic cortex and the turgor pressure. The protoplast, therefore, must be anchored. This becomes apparent comparing initial valves of *C. wailesii* (produced within the auxospore each after an acytokinetic mitosis) with vegetative valves: the initial valve is hemispherical and bears one ring of LPs at the valve edge; the valve created after cleavage is angular in profile and bears two rings of LPs, the additional row at the face mantle-corner (Schmid 1987 a, 1990). Diatom auxospores have never been seen dividing. The

Figs. 35–37. *Nitzschia sigmoidea*. Bars: 1 µm

Fig. 35. SEM, view into the raphe canal (RC); *F* fibulae

Figs. 36 and 37. TEM; longitudinal sections through the forming fibulae

Fig. 36. Spacer vesicles between fibulae (*F*)

Fig. 37. ER subjacent to transapical ribs and on both sides of the fibulae-SDV. SDV between the TR is continuous (arrows)

Fig. 38. *Pinnularia maior*. TEM, crossection through a dictyosome producing spacer vesicles. Cisternal membranes in the central region closely appressed; with wider lumen towards the margin; inflated at the periphery. Bar: 300 nm

Fig. 39. *Nitzschia sigmoidea*, TS (pole view) through raphe canal (*RC*) and dictyosome (*D*) producing a spacer vesicle (*Spv*): the whole distal cisterna sloughs off. *F* Fibula; *R* raphe ribs; arrows point to MTs. Bar: 300 nm

acytokinetic mitoses in connection with initial valve formation might be interpreted as the auxospore's incapacity to divide because of lacking anchors (Schmid 1990), a conclusion consistent with that drawn recently for wall free protoplasts of higher plants (Schindler et al. 1989). In the family *Thalassiosira* the SPs serve as these anchors (Schmid 1984 a), while in pennates the raphe ribs serve this purpose (Höfler 1940).

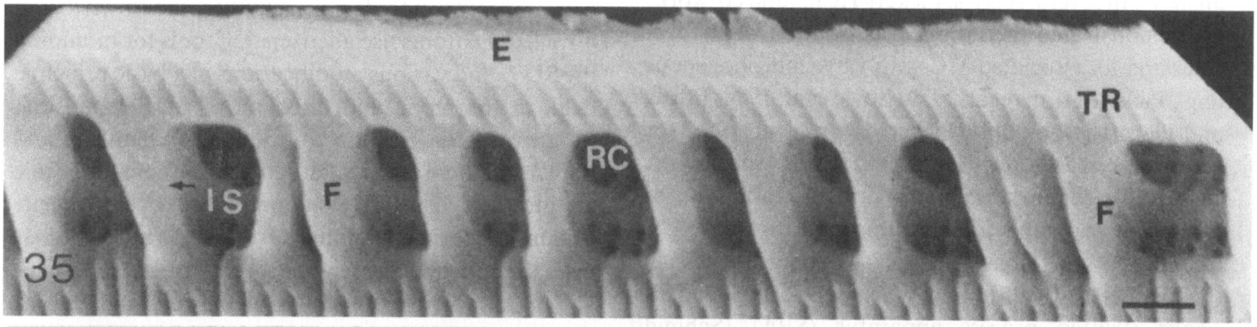

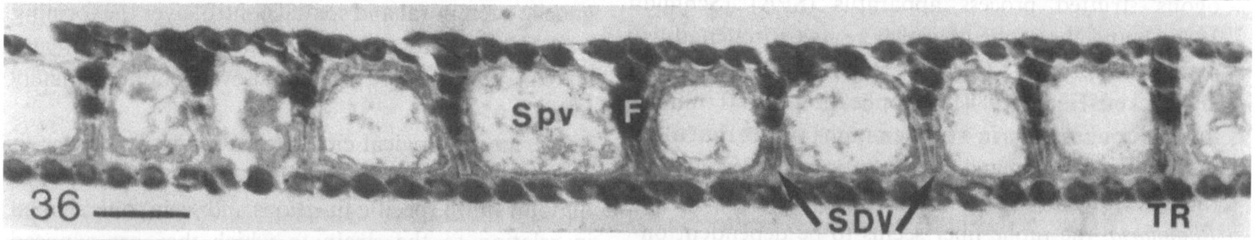

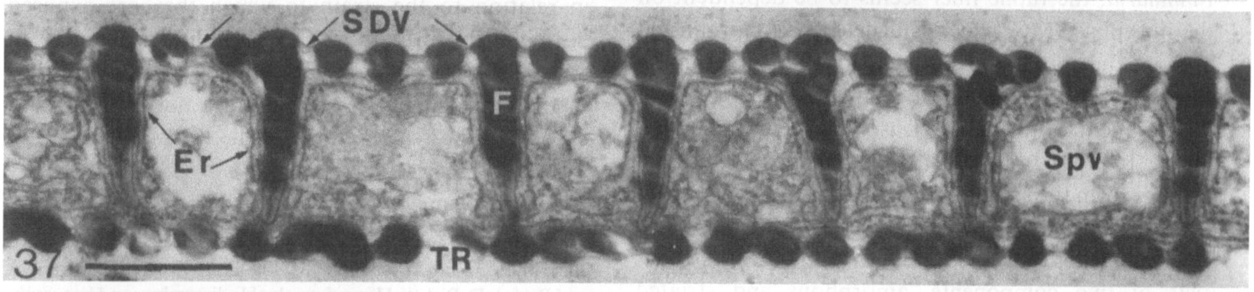

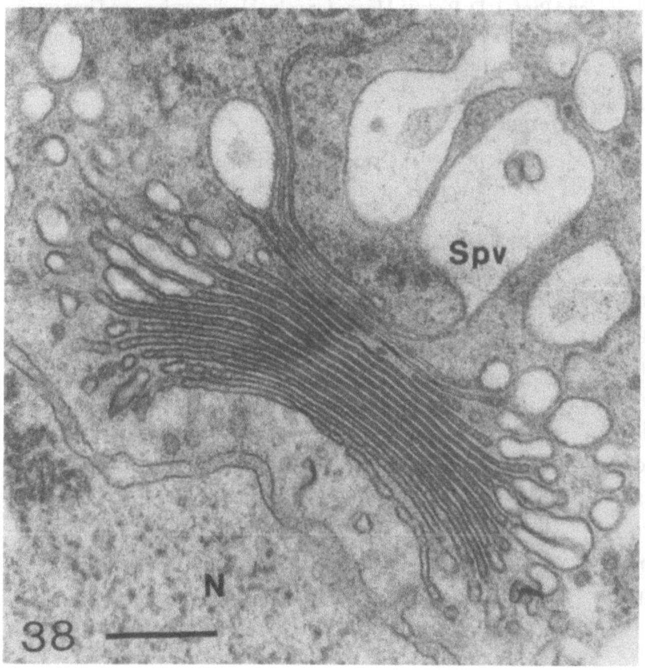

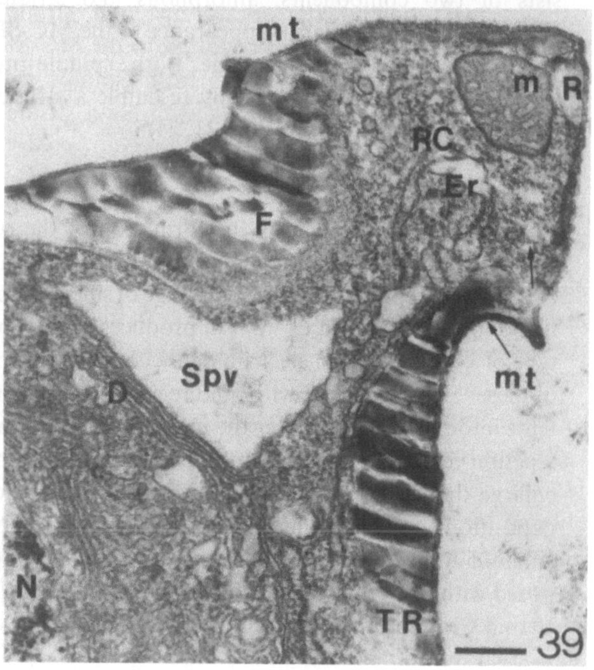

Since the anchorage of the protoplast, and in particular the nucleus-MC-spindle-complex, to the wall seems of vital importance to the cell for various reasons, it is not surprising that certain cytoskeletal elements are involved in the formation and correct location of the anchors. In several species, Pickett-Heaps and co-workers demonstrated the intimate association of the nucleus and its associated MC with valve morphogenesis (e.g., Pickett-Heaps et al. 1979, Boyle et al. 1984, Edgar and Pickett-Heaps 1984, Pickett-Heaps et al. 1988; for exceptions, see Pickett-Heaps et al. 1990). Specific cytoskeletal structures are involved in the formation of LPs, SPs, and raphe, arising in all cases after SDV initiation: the labiate process apparatus (LPA) (Schnepf et al. 1980, Li and Volcani 1985 a); the analogous strutted process apparatus (SPA) (Schmid 1986 a); and the raphe fiber, where its entire length is associated with MTs (Pickett-Heaps et al. 1979, Edgar and Pickett-Heaps 1984 b). Application of MT inhibitors not only results in a broken raphe fissure, but also in a maldistribution and malformation of LPs and SPs (Schmid 1980, 1987 a, b; Pickett-Heaps et al. 1990). In *Pinnularia*, the raphe fiber seems to be dependent on the interphase-MC, which is at telophase reorganized at/from the spindle pole, and is located subjacent to the prospective valve center from the early stages of development until valve release (Pickett-Heaps et al. 1978 a, b, 1979). In *Ditylum*, a dense, ill defined material extends from the spindle poles towards the SDV, where a distinct structure, the LPA, lies beneath the SDV (Pickett-Heaps et al. 1988). The LPA in *C. wailesii* consists of two components, amorphous and striated (Figs. 23 and 24), similar in appearance to the MC-SP-complex, only on a smaller scale. Both, spindle precursor and LPA morphologically resemble a striated flagellar root (rhizoplast) (Figs. 21–24).

Valve release

Valve release is preceded by the formation of the organic coat and followed by the formation of the organic diatotepum, if present. At the final stage of silicification the Golgi apparatus in *C. wailesii* produces a different type of vesicles with an electron dense content. Fusion of these "dense vesicles" with, and release of their contents into the SDV may alter the internal milieu of this reaction vessel and also its membrane composition, and enlarge the surface area, thus perhaps stopping silica deposition and allowing exocytosis of the valve (Figs. 25 A, B and 29). This formation step was proposed with respect to other models by Schmid (1986 a, b) and Crawford and Schmid (1986).

Conclusions

Impressive progress is being made towards an understanding of cellular dynamics and morphogenesis of the diatom cell wall. The various functions of wall features, i.e., anchor surfaces, reference points, etc., with respect to cytoplasmic activities, are slowly being unveiled. Diatoms use an arsenal of tools for moulding the SDV, and various modifications of a basic principle (e.g., in the mode and direction of SDV-expansion, progress of silicification, formation of chambers and pores) and there are great differences in the cytoplasmic architecture. This diversity may have phylogenetic origins. The understanding of the underlying mechanism of morphogenesis, the orchestrated action involving almost all organelles of the cell, and the organelle's temporal and spatial control over the forming wall part is, however, in its infancy. Initiation and stabilization of these sequential events are completely unknown.

Using morphological characters to identify species, it is necessary to bear in mind that cell wall structures in diatoms fulfill specific functions and seem to be shaped in relation to the strain to which they are exposed during their formation. The moulding machinery in many species responds very sensitively to alterations in the environment, and thus does not always work exactly to the scale or wish of the taxonomist.

Acknowledgements

I am grateful to the Austrian FFWF for financial support (P-8727-Bio), to Dr. L. Edgar for supplying Fig. 12, to Prof. P. E. Hargraves and Prof. J. D. Pickett-Heaps for valuable discussions and improving the English, to Dr. H. R. Preisig for scientific cooperation, and to Mag. Roman Auer for excuting part of the line drawings.

References

Anderson OR (1994) Cytoplasmic origin and surface deposition of silicate structures in sarcodine protists. Protoplasma 181: 61–77

Bouck GB, Brown DL (1973) Microtubule biogenesis and cell shape in *Ochromonas*. I. The distribution of cytoplasmic and mitotic microtubules. J Cell Biol 56: 340–359

Boyle JA, Pickett-Heaps JD, Czarnecki D (1984) Valve morphogenesis in the pennate diatom *Achnanthes coarctata*. J Phycol 20: 563–573

Crawford RM (1981) Some considerations of size reduction in diatom cell walls. In: Ross R (ed) Proceedings of the 6th International Diatom Symposium, Budapest 1980. O Koeltz, Koenigstein, pp 253–265

– Schmid AM (1986) Ultrastructure of silica deposition in diatoms. In: Leadbeater BSC, Riding R (eds) Biomineralization in lower plants and animals. Clarendon Press, Oxford, pp 291–314 (Systematics Association special volume 30)

Drebes G (1969) Asexual and sexual reproduction in *Stephanopyxis*, b/w 16 mm cine films. IWF, Göttingen, E 1341–1344

– (1977) Sexuality. In: Werner D (ed) The biology of diatoms. Bot Monogr 13: 250–283
– Schulz D (1990) Taxonomy and morphology of *Fragilaria oblonga* sp. nov., an araphid diatom from the Wadden Sea of the German Bight (North Sea). Beih Nova Hedwigia 100: 3–17
Drum RW (1966) Electron microscopy of paired Golgi structures in the diatom *Pinnularia nobilis*. J Ultrastruct Res 15: 100–107
– Pankratz HS (1964) Post mitiotic fine structure of *Gomphonema parvulum*. J Ultrastruct Res 10: 217–223
Edgar LA (1980) Fine structure of *Caloneis amphisbaena* (Bacillariophycea). J Phycol 16: 62–72
– Pickett-Heaps JD (1984 a) Diatom locomotion. Progr Phycol Res 3: 47–88
– – (1984 b) Valve morphogenesis in the pennate diatom *Navicula cuspidata*. J Phycol 20: 47–61
Franz SM, Schmid AM (1994) Cell-cycle and phenotypes of *Biddulphiopsis titiana*. Diatom Res (in press)
French FW, Hargraves PE (1985) Spore formation in the life cycles of the diatoms *Chaetoceros diadema* and *Leptocylindrus danicus*. J Phycol 21: 447–483
Geitler L (1963) Alle Schalenbildungen der Diatomeen treten als Folge von Zell- oder Kernteilungen auf. Ber Deutsch Bot Ges 75: 393–396
Gordon R, Drum RW (1994) The chemical basis of diatom morphogenesis. Int Rev Cytol (in press)
Gutenbrunner SA, Thalhamer J, Schmid AM (1994) Proteinaceous and immunochemical distinctions between the oval and fusiform morphotype of *Phaeodactylum tricornutum* (Bacillariophyceae). J Phycol 30: 129–136
Hargraves PE, Schmid AM (1994) Morphology, cytology and growth characteristics of the diatom *Planktoniella sol* (Wall) Schütt. In: Kociolek P (ed) Proceedings of the 11th International Diatom Symposium, San Francisco 1990. California Academy of Sciences, San Francisco (in press)
Hasle GR (1972) Two types of valve processes in centric diatoms. Beih Nova Hedwigia 39: 55–78
– (1973) The "mucilage pore" of pennate diatoms. Beih Nova Hedwigia 45: 167–186
Hepler PK, Palevitz BA, Lancelle SA, McCauley MM, Lichtscheidl I (1990) Cortical endoplasmic reticulum in plants. J. Cell Sci 96: 355–373
Herth W (1979) The site of β-chitin fibril formation in centric diatoms. II. The chitin-forming cytoplasmic structures. J Ultrastruct Res 68: 6–15
Hoagland KD, Rosowski JR, Gretz MR, Roemer SC (1993) Diatom extracellular polymeric substances: function, fine structure, chemistry, and physiology. J Phycol 29: 537–566
Hofker J (1928) Die Teilung, Mikrosporen- und Auxosporenbildung von *Coscinodiscus biconicus*. Ann Protistol (Paris) 1: 167–194
Höfler K (1940) Aus der Protoplasmatik der Diatomeen. Ber Deutsch Bot Ges 58: 97–120
Ikari J (1923) On the nuclear and cell division of a plankton diatom *Coscinodiscus subbuliensis* Bot Mag (Tokyo) 37: 96–108
Jeckel B, Kowallik KV (1990) Development of the mitotic spindle in *Stephanopyxis palmeriana* (Grev.) Grunow and the nature of the diatom centrosome. Beih Nova Hedwigia 100: 61–76
Kauss H (1994) Callose synthesis. In: Smallwood M, Knox P, Bowles DJ (eds) Membranes: specialized functions in plant cells. JAI Press, London (in press)
Kreis TE (1990) Role of microtubules in the organisation of the Golgi apparatus. Cell Motil Cytoskeleton 15: 67–70
Lauterborn R (1896) Untersuchungen über Bau, Kernteilung und Bewegung der Diatomeen. Engelmann, Leipzig
Li CW, Volcani BE (1985 a) Studies on the biochemistry and fine structure of silica shell formation in diatoms. VIII. Morphogenesis of the cell wall in a centric diatom, *Ditylum brightwellii* Protoplasma 124: 10–29
– – (1985 b) Studies on the biochemistry and fine structure of silica shell formation in diatoms. IX. Sequential valve formation in a centric diatom, *Chaetoceros rostratum*. Protoplasma 124: 30–41
– – (1988) Aberrant valve formation in a centric diatom, *Ditylum brightwellii*. Protoplasma 145: 1–6
Lloyd CW, Venverloo CJ, Goodbody KC, Shaw PJ (1992) Confocal laser microscopy and three-dimensional reconstruction of nucleus-associated microtubules in the division plane of vacuolated plant cells. J Microsc 166: 99–109
Mann DG (1982) Structure, life history and systematics of *Rhoicosphenia* (Bacillariophyta). II. Auxospore formation and perizonium structure of *Rh. curvata*. J Phycol 18: 264–274
– (1984) An ontogenetic approach to diatom systematics. In: Mann DG (ed) Proceedings of the 7th Internationl Diatom Symposium, Philadelphia 1982. O Koeltz, Koenigstein, pp 113–141
– Stickle AJ (1988) Nuclear movements and frustule symmetry in raphid pennate diatoms. In: Round FE (ed) Proceedings of the 9th International Diatom Symposium, Bristol 1986. Biopress, Bristol, O Koeltz, Koenigstein, pp 281–291
Mann S (1983) Mineralization in biological systems. Struct Bond 54: 125–174
– (1986) Biomineralization in lower plants and animals – chemical perspectives. In: Leadbeater BSC, Riding R (eds) Biomineralization in lower plants and animals. Clarendon Press, Oxford, pp 39–54 (Systematics Association special volume 30)
Manton I, Kowallik K, v. Stosch HA (1969) Observations on the fine structure and development of the spindle at mitosis and meiosis in a marine centric diatom (*Lithodesmium undulatum*). II. The early meiotic stages in male gametogenesis. J Cell Sci 5: 271–298
Mazia D (1987) The chromosome cycle and the centrosome cycle in the mitotic cycle. Int Rev Cytol 100: 49–92
Medlin LK, Crawford RM, Andersen RA (1986) Histochemical and ultrastructural evidence for the function of the labiate process in the movement of centric diatoms. Br Phycol J 21: 297–301
Pickett-Heaps JD (1991 a) Post-mitotic cellular reorganisation in the diatom *Cymatopleura solea*: the role of microtubules and the microtubule centre. Cell Motil Cytoskeleton 18: 279–292
– (1991 b) Cell division in diatoms. Int Rev Cytol 128: 63–108
– Tippit DH, Andreozzi JA (1978 a) Cell division in the pennate diatom *Pinnularia*. I. Early stages in mitosis. Biol Cell 33: 71–78
– – – (1978 b) Cell division in the pennate diatom *Pinnularia*. II. Later stages in mitosis. Biol Cell 33: 79–84
– – – (1979) Cell division in the pennate diatom *Pinnularia*. IV. Valve morphogenesis. Biol Cell 35: 199–206
– Hill D, Wetherbee R (1986) Cellular movement in the centric diatom *Odontella*. J Phycol 22: 334–339
– Wetherbee R, Hill DRA (1988) Cell division and morphogenesis of the labiate process in the centric diatom *Ditylum brightwellii*. Protoplasma 143: 139–149
– Schmid AM, Edgar L (1990) The cell biology of diatom valve formation. Progr. Phycol. Res 7: 1–168
– Carpenter J, Koutoulis A (1994) Valve and seta (spine) mor-

phogenesis in the centric diatom *Chaetoceros peruvianus*. Protoplasma 181: 269–282

Preisig HR (1994) Siliceous structures and silicification in flagellated protists. Protoplasma 181: 29–42

Rines JEB, Hargraves PE (1988) The Chaetoceros Ehrenberg (Bacillariophyceae) flora of Narragansett Bay, Rhode Island, USA. Bibl Phycol 79: 1–196

Roberts K (1990) Structures at the plant cell surface. Curr Opin Cell Biol 2: 920–928

Round FE, Crawford RM (1990) Phylum Bacillariophyta. In: Margulis L, Corliss JO, Melkonian M, Chapman DJ (eds) Handbook of Protoctista. Jones and Bartlett, Boston, 574–596

– – Mann DG (1990) The diatoms. Biology and morphology of the genera. Cambridge University Press, Cambridge

Sandgren CD (1980) Resting cyst formation in selected chrysophyte flagellates: an ultrastructural survey including a proposal for the phylogenetic significance of interspecific variations in the encystment process. Protistologia 16: 289–303

Schindler M, Meiners S, Cheresh DA (1989) RGD-dependent linkage between plant cell wall and plasmamembrane: consequences for growth. J Cell Biol 108: 1955–1965

Schmid AM (1979) Influence of environmental factors on the development of the valve in diatoms. Protoplasma 99: 99–115

– (1980) Valve formation in diatoms: a pattern-related filamentous system in pennates and the effect of APM, colchicine and osmotic pressure. Nova Hedwigia 33: 811–847

– (1984a) Wall morphogenesis in *Thalassiosira eccentrica*: comparison of auxospore formation and the effects of MT-inhibitors. In: Mann DG (ed) Proceedings of the 7th International Diatom Symposium, Philadelphia 1982. O Koeltz, Koenigstein, pp 47–70

– (1984b) Tricornate spines in *Thalassiosira eccentrica* as a result of valve modelling. In: Mann DG (ed) Proceedings of the 7th International Diatom Symposium, Philadelphia 1982. O Koeltz, Koenigstein, pp 71–95

– (1984c) Valve morphogenesis in diatoms. In: Bach K, Burkhardt B (eds) Diatoms I. Shells in nature and techniques. Cramer, Stuttgart, pp 300–317 (Communications of the Institute of Lightweight Structures 28)

– (1986a) Organization and function of cell structures in diatoms and their morphogenesis. In: Ricard M (ed) Proceedings of the 8th International Diatom Symposium, Paris 1984. O Koeltz, Koenigstein, pp 271–292

– (1986b) Wall morphogenesis in *Coscinodiscus wailesii*. II. Cytoplasmic events of valve morphogenesis. In: Ricard M (ed) Proceedings of the 8th International Diatom Symposium, Paris 1984. O Koeltz, Koegstein, pp 293–314

– (1987a) Morphogenetic forces in diatom cell wall formation. In: Bereiter-Hahn J, Anderson OR, Reif WE (eds) Cytomechanics. Springer, Berlin Heidelberg New York Tokyo, pp 183–199

– (1987b) Wall morphogenesis in diatoms. In: Wiessner W, Robinson DG, Starr RC (eds) Molecular and cellular aspects of algal development. Springer, Berlin Heidelberg New York Tokyo, pp 34–41

– (1989) Geitler's "Plattenband" in the diatom *Synedra* cf. *ulna* in the light of TEM investigations. Plant Syst Evol 164: 239–252

– (1990) Intraclonal variation in the valve structure of *Coscinodiscus wailesii*. Beih Nova Hedwigia 100: 101–119

– Schulz D (1979) Wall morphogenesis in diatoms: deposition of silica by cytoplasmic vesicles. Protoplasma 100: 267–288

– Volcani BE (1983) Wall morphogenesis in *Coscinodiscus wailesii*. I. Valve morphology and development of its architecture. J Phycol 19: 387–402

– Borowitzka MA, Volcani BE (1981) Morphogenesis and biochemistry of diatom cell walls. In: Kiermayer O (ed) Cytomorphogenesis in plants. Springer, Wien New York, pp 63–97 [Alfert M et al (eds) Cell biology monographs, vol 8]

Schnepf E, Deichgräber G, Drebes G (1980) Morphogenetic processes in *Attheya decora* (Bacillariophyceae, Biddulphiineae). Plant Syst Evol 135: 265–277

Simkiss K (1986) The process of biomineralization in lower plants and animals – an overview. In: Leadbeader BSC, Riding R (eds) Biomineralization in lower plants and animals. Clarendon Press, Oxford, pp 19–37 (Systematics Association special volume 30)

Sims PA (1989) Some cretaceous and palaeocene species of *Coscinodiscus*: a micromorphological and systematic study. Diatom Res 4: 351–371

Steucek GL, Schmid AM (1989) Light weight architecture of the diatom *Thalassiosira*. In: Reiner R, Wirth H (eds) Natürliche Konstruktionen – Leichtbau in Architektur und Natur. Universität Stuttgart SFB 230, Stuttgart pp 195–203

Syvertsen EE (1977) *Thalassiosira gravida* and *T. rotula*: ecology and morphology. In: Simonsen R (ed) Proceedings of the 4th Symposium on Recent and Fossil Marine Diatoms. Cramer, Vaduz, pp 99–112

Van Landingham SL (1967–79) Catalogue of the fossil and recent genera and species of diatoms and their synonyms, vols I–VII. Cramer, Vaduz

Von Stosch HA (1981) Structural and histochemical observations on the organic layers of the diatom cell wall. In: Ross R (ed) Proceedings of the 6th International Symposium on Recent and Fossil Diatoms, Budapest 1980, pp 231–252

– (1982) On auxospore envelopes in diatoms. Bacillaria 5: 127–156

– Theil G, Kowallik KV (1973) Entwicklungsgeschichtliche Untersuchungen an zentrischen Diatomeen. V. Bau und Lebenszyklus von *Chaetoceros didymum*, mit Beobachtungen über einige andere Arten der Gattung. Helgol Wiss Meeresunters 25: 384–445

Waterkeyn L, Bienfait A (1987) Localisation et rôle des β-1,3-glucanes (callose et chrysolaminarine) dans le genre *Pinnularia* (diatomées). Cellule 74: 199–226

Werner D (1977) Introduction with a note on taxonomy. In: Werner D (ed) The biology of diatoms. Bot Monogr 13: 1–17

Wordemann L, (1992) The cytoskeleton in the diatoms. The mitotic spindle and cell cycle dependent organization. In: Menzel D (ed) The cytoskeleton of algae. CRC Press, Boca Raton, pp 39–57

– McDonald KL, Cande ZW (1986) The distribution of cytoplasmic microtubules throughout the cell cycle of the centric diatom *Stephanopyxis turris*: their role in nuclear migration and positioning the mitotic spindle during cytokinesis. J Cell Biol 102: 1688–1698

Protoplasma (1994) 181: 61–77

PROTOPLASMA

Cytoplasmic origin and surface deposition of siliceous structures in Sarcodina

O. R. Anderson*

Biological Oceanography, Lamont-Doherty Earth Observatory of Columbia University, Palisades, New York

Received August 5, 1993
Accepted November 2, 1993

Summary. Siliceous products, deposited at the cell surface of amoeboid protists, include a wide variety of species-specific structures; i.e., spicules, scales, solid plates, granules, meshworks frustules, and other elaborate geometric forms. A common secretory mechanism has been reported in testate amoebae, heliozoa and heliozoon-like amoebae, and radiolaria. Silica deposition vesicles (SDVs), either situated in the cell cytoplasm (as in testate amoebae and heliozoa and relatives) or within an expanded portion of the peripheral cytoplasm known as a cytokalymma (in radiolaria), are the site of silicification. In some testate amoebae, moreover, Golgi-derived vesicles fuse with the membrane surrounding silica deposition sites. These vesicles possibly contribute additional silica-secreting membrane into the surface of the SDV while increasing the membrane surface area. Silica products of testate amoebae and heliozoa are deposited on the cell surface by exocytosis. The cytokalymma of radiolaria, while containing a silica-secreting vacuolar space, is decidedly different in form and activity from the intracellular secretory spaces of testate amoebae and heliozoa. The cytokalymma is a dynamic structure exhibiting cytoplasmic flowing activity, and in a mold-like manner determines the remarkable species-specific shape of the skeleton. Consequently, the deposited silicate product of radiolaria is an endoskeleton and is not released on the surface by exocytosis. Further research is needed to determine if Golgi-derived vesicles, designated Golgi-fibrillar vesicles (GFV) in some testate amoebae, are also the source of SDV membranes in other silicate secreting sarcodines.

Keywords: Amoeboid protists; Biomineralization; Heliozoa; Radiolaria; Silicification; Testate amoebae.

Introduction

A wide variety of mineral compounds are deposited at the surface of protistan cells including calcium carbonates ($CaCO_3$), gypsum ($CaSO_4$), calcium phosphate ($Ca_3(PO_4)_2$), celestite ($SrSO_4$), silica ($SiO_2\,n\,(H_2O)$), and iron and manganese oxides (Manton and Peterfi 1969; Ogden 1979 b, 1991; Ogden and Hedley 1980; Brook 1981; Lowenstam 1986; Green et al. 1989; Anderson 1981, 1983; Anderson et al. 1990). The form of the deposition varies across taxa from a relatively structureless surface coat to particulates of complex design. Surface silica deposits of diverse shape and size occur among a broad range of protists, especially algal, and heterotrophic flagellated and amoeboid taxa (e.g., Pickett-Heaps et al. 1979; Ogden and Hedley 1980; Bovee 1981, 1991; Crawford 1981; McGrory and Leadbeater 1981; Harrison et al. 1976, 1981; Leadbeater 1981; Volcani 1981; Anderson 1981, 1983; Patterson and Dürrschmidt 1986; Faber and Preisig 1994; Preisig 1994; Schmid 1994). These structures encompass spicules, scales, solid plates, granules, meshworks, frustules, and other elaborate geometric forms. Surface structures are used to define species, and in some cases characterize a taxonomic class, such as the skeletal architecture of radiolaria. This paper reviews current knowledge of the cytoplasmic origin and surface deposition of silicate structures among amoeboid protists. More general summaries can be found in Anderson (1988 b) and Bovee (1991).

Morphology and taxonomy

General morphology

Based on cytoplasmic organization and degree of regional specialization, the major taxa of sarcodines can

* Correspondence and reprints: Biological Oceanography, Lamont-Doherty Earth Observatory of Columbia University, Palisades, NY 10964, U.S.A.

be categorized in three groups (Anderson 1984): (1) diffuse, without major partitioning barriers within or surrounding the cytoplasm (e.g., gymnamoebae; Fig. 1); (2) transitional, with some form of organic or mineralized barrier that encloses the cytoplasm or separates it into specialized regions that are not sharply delimited, but intergrade in organization and function (e.g., the testate amoebae; Fig. 2) and foraminifera extending to heliozoa (Fig. 3); and (3) zonal, with distinct organizational regions, with functional specializations,

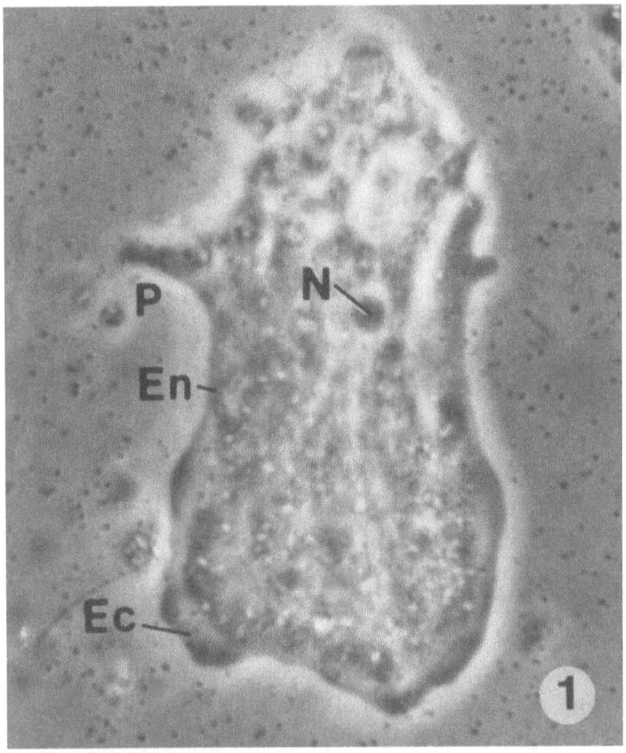

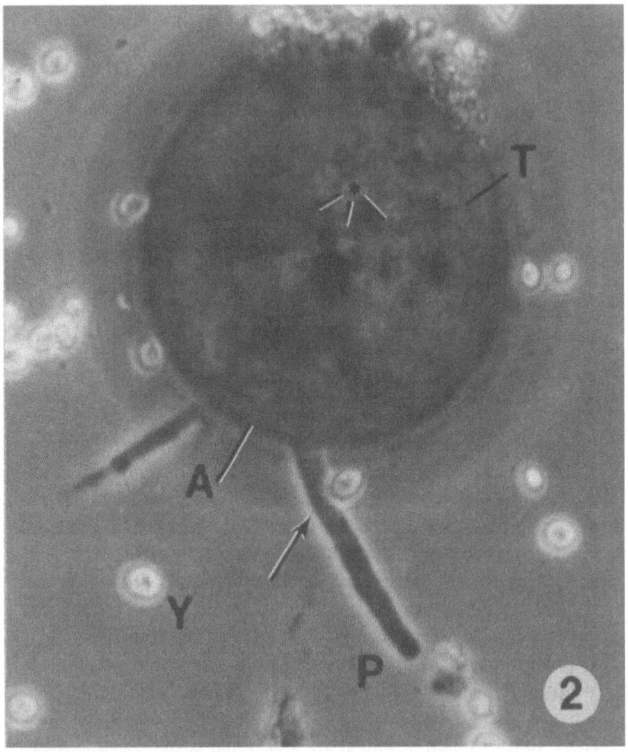

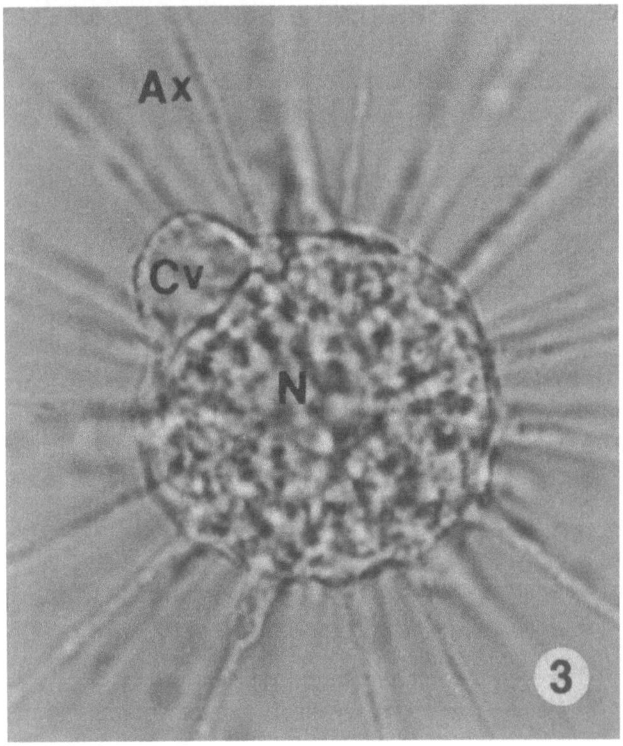

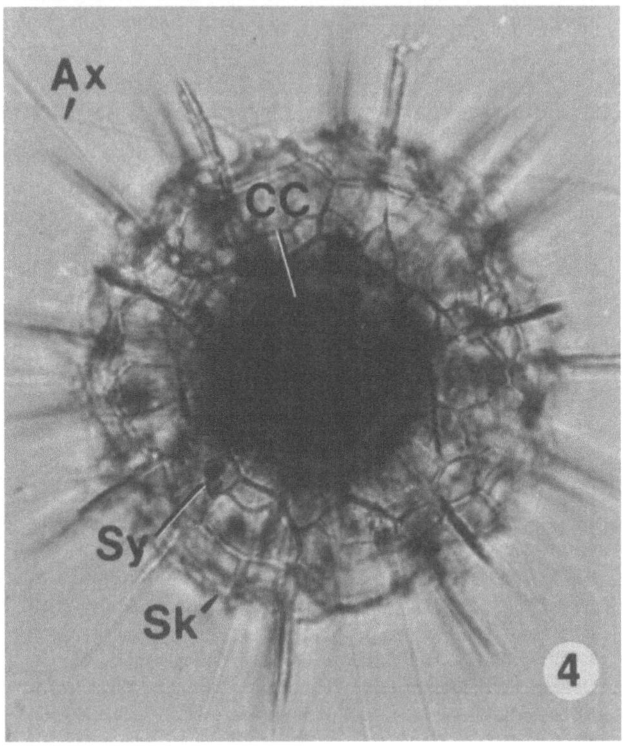

Table 1. Categories of cellular specialization in sarcodines

1 Diffuse	2 Transitional	3 Zonal
Cytoplasm without specialized barriers or clearly delimited major regions serving differentiated functions. Cellular organization is fluid and organelles are not distributed within stable specialized regions (e.g., naked amoebae with endoplasm and ectoplasm, Ec). Surface structures largely organic (glycocalyx, glycostyles, or scales, Sc).	To varying degrees, the cytoplasm is partially delimited by organic and/or mineralized barriers. Cytoplasmic regions intergrade often across pores or apertures within the enclosures (e.g., testate amoebae and foraminifera). Surface structures are mineralized plates, particles, scales, or perforated tests (T) enclosing the cytoplasm (Cy) and nucleus (N).	Cytoplasm is organized in clearly delimited regions with structural and functional specializations, e.g., radiolaria with a central capsule (CC) enclosed by a stable organic wall and surrounded by a vacuolated ectoplasm. Mineralized surface structures are often siliceous and vary from spicules to elaborate, enclosing skeletons (Sk). From Anderson (1984).

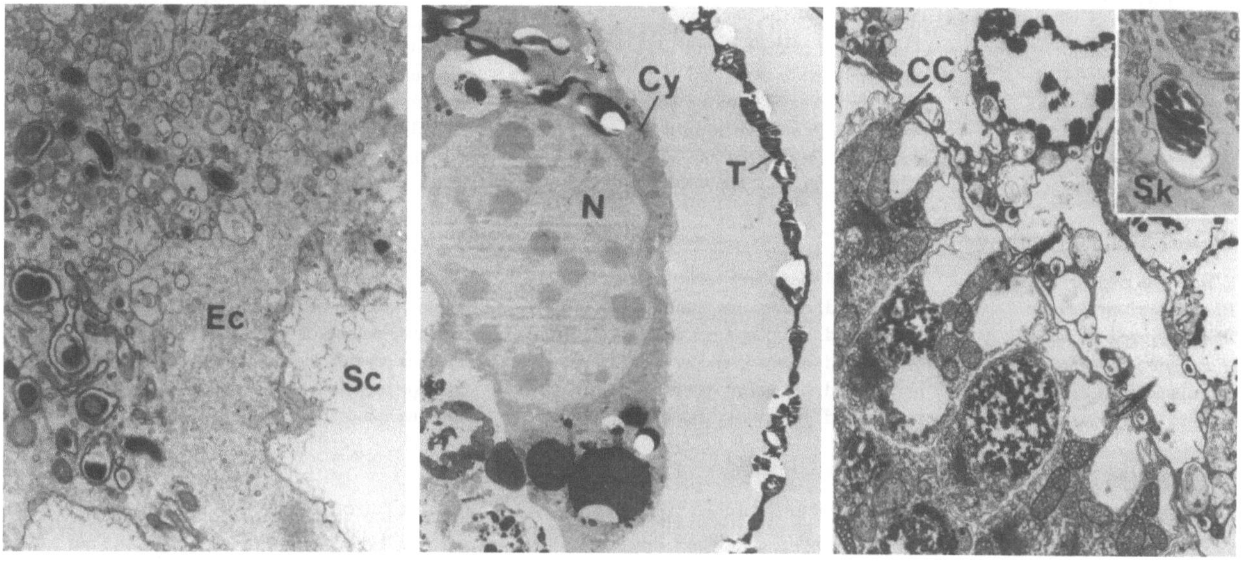

Figs. 1–4. Light micrographs of living sarcodines

Fig. 1. A large, mayorellid-type gymnamoeba from brackish water is categorized in group 1 (Table 1). The fluid cytoplasm is diffusely organized. A small subpseudopodium (*P*) extends from the thin ectoplasmic layer (*Ec*) surrounding the more dense endoplasm (*En*) containing the nucleus (*N*). × 280

Fig. 2. A testate amoeba, *Netzelia tuberculata*, is categorized in group 2 (Table 1), and is enclosed by an organic test (*T*) containing siliceous particles secreted by the cell. Lobose pseudopodia (*P*) protrude from the aperture (*A*) hidden beneath the shell. Yeast food particles (*Y*) are captured by attachment to the pseudopodia (arrow) where they are carried by cytoplasmic streaming into the aperture and engulfed in digestive vacuoles. Particles enclosed by the cytoplasm (asterisk) are visible within the shell. × 700

Fig. 3. A freshwater heliozoan shows the radiating array of axopodia (*Ax*) containing bundles of microtubules. The heliozoa are more clearly delineated in cytoplasmic organization compared to the gymnamoebae. They are categorized as advanced members of group 2. The microtubular bundles of the axoneme are anchored in the central cytoplasm near a microtubular organizing center in the nuclear region (*N*). A contractile vacuole (*Cv*) occurs at the margin of the cell. × 1,000

Fig. 4. A radiolarian with axopodia (*Ax*) protruding from the more dense central capsular cytoplasm (*CC*) is surrounded by a siliceous skeleton (*Sk*) and contains algal symbionts (*Sy*) in the extracapsulum. Radiolaria, with a clear delineation between capsular cytoplasm and extracapsular cytoplasm, are categorized in group 3 (Table 1). The axopodia protrude from specialized pores in the capsular wall surrounding the central cell body containing the nucleus. × 200

Table 2. Some surface-secreting sarcodines categorized according to the groups in Table 1

Species	Morphology	Secretory product
Group 1: Gymnamoebae (Fig. 1)		
Dactylamoeba stella	lobose amoeba (20–120 μm)	organic, porous boat-shaped scales (550 nm)
Vexillifera minutissima	lobose amoeba (5–16 μm) with long-tapered subpseudopodia	organic glycostyles forming hexagonal towers (c. 80 nm diam. and 60 nm high)
Group 2: Testate amoebae (Figs. 2, 5, and 7)		
Trichosphaerium platyxyrum	subspherical, oblong amoeboid (67 μm); flexible, organic test	Calcareous spicules, blade-like with square edges (7 μm)
Lesquereusia spiralis	spheroidal to ovoid test (89–117 μm) compressed laterally	siliceous, curved rods, closely packed (9–20 μm × 3–4 μm)
Quadrulella symmetrica	ovoid or pyriform test (72–103 μm), compressed laterally	quadrangular, siliceous plates (c. 8 μm), regularly arranged
Difflugia lanceolata	elongate and tapered at ends (116–159 μm)	angular, siliceous particles (3–20 μm)
Netzelia tuberculata	spheroidal test (c. 80–100 μm)	siliceous ovoid to peg-shaped (5–8 μm)
Tracheleuglypha dentata	ovoid test, circular in cross sect. (52–61 μm)	circular, siliceous plates (8 μm diam.)
Euglypha rotunda	ovoid-tapered test (34–54 μm)	oval siliceous plates (5–7 μm)
Euglypha strigosa	test, ovoid, tapered flattened (73–89 μm)	siliceous plates (10 μm) with spines (2–23 μm)
Assulina muscorum	test, ovoid, laterally flattened (c. 50 μm)	oval shell plates in rows (c. 3–4 μm)
Group 2: Heliozoa (Figs. 3 and 17)		
Clathrulina elegans	body round, periphery vacuolated (c. 20 μm)	organic porous coat (c. 75 μm diam.)
Heterophrys myriopoda	body round, endoplasm granular (50–80 μm)	organic spicules as needles (10–13 μm)
Pompholyxophrys punicea	body round, ectoplasm not distinct (c. 30 μm)	siliceous, pebble- or pearl-like (1–8 μm)
Rabdiophrys anulifera	body round, eccentric endosome (30 μm)	siliceous needles with plate base (8–12 μm)
Raphidiocystis lemani	body round, ectoplasm distinct (25 μm)	cup-like and radial silic. needles (75 μm)
Acanthocystis longiseta	body round, endoplasm distinct (35–60 μm)	siliceous spines long, oval base (30 μm)
Group 3: Radiolaria (Figs. 4, 19–21)		
Physematium mulleri	large gelatinous spheroidal (3 mm)	siliceous curved spicules (150 μm)
Lampoxanthium pandora	spherical, hyaline with gelatinous coat (3 mm)	spicules, Y-shaped or branched (300 μm)
Spongoplegma antarcticum	spherical cell body enclosed by shell	siliceous, porous shell (200 μm)
Hexacontium laevigatum	spheroidal cell body in spherical shell	siliceous shell (100 μm), six spines (10–65 μm)
Didymocyrtis tetrathalamus	cell body within porous spheroidal inner shell	outer shell hour-glass shaped (> 80 μm).
Hymeniastrum euclidis	cell body within porous discoidal inner shell	spongiose, three arms, Y-shaped (c. 200 μm)
Spongaster tetras	cell body within spongiose inner shell	spongiose, flattened quadrangular (> 200 μm)
Eucyrtidium acuminatum	cell body prolate-spheroidal	shell conical, porous with segments (100–200 μm)
Pterocanium trilobum	cell body prolate-spheroidal	shell conical with three basal "feet" (250 μm)

separated by stable, organic or mineralized barriers (e.g., radiolaria; Fig. 4). Illustrations of foraminifera are presented by Faber and Preisig (1994). Radiolarian cytoplasm is organized into two major domains separated by an organic capsular wall: (1) a central capsule containing the nucleus, numerous mitochondria, and reserve substances, and (2) the extracapsulum composed of axopodia and a frothy cytoplasm where prey is captured and digested. Reproductive swarmers are produced in the central capsulum. Descriptions of each group are summarized in Table 1.

The characteristics of biomineralized surface deposits vary in a consistent pattern across the three groups. Each of the major taxa will be described in greater detail to provide a context for the descriptions of secretory processes presented later. A synopsis of some species, categorized according to Table 1 and listed with surface secretory products, is presented in Table 2.

Gymnamoebae

The gymnamoebae (categorized in group 1) are known to secrete organic structures (glycostyles and scales) as in *Dactylamoeba* spp. (Page 1983, Page and Siemensma 1991). These are typically attached to the plasmalemma. The term "naked amoeba" often used for this group is not quite correct, since many species have some type of surface coat known as a glycocalyx. This may be a very thin filamentous deposit or a rather thick glycoprotein coat. The coat may be sufficiently thickened to produce a wrinkled appearance when the amoeba is in locomotion (e.g., *Thecamoeba* sp. with longitudinal surface folds).

Testate amoebae

The testate amoebae (categorized in group 2) have an organic test, often augmented by mineralized particles. *Trichosphaerium* sp., with morphological features resembling gymnamoebae in the naked stage (*Pontifex maximus*), secretes calcitic spicules (Page 1983). *Trichosphaerium* contains a pliable organic test containing pores through which pseudopodia project and thus is included with other testate amoebae in group 2. The silicate particles incorporated in the organic walls of testate amoebae vary markedly across species (e.g., Ogden and Hedley 1980; Harrison et al. 1976, 1981; Anderson 1986, 1988 a), and include approximately spheroidal, pebble-like structures (c. 5 to 20 μm diam.), straight or curved rod-shaped structures (c. 10 to 20 μm long), and scales or plates varying in shape from circular or oval to elongate, square, or polyhedral (c. 10 to 20 μm in major dimension) (Figs. 5–15).

Surface enclosures include totally organic tests (e.g., *Arcella* and *Centropyxis* spp.), smooth tests composed of amorphous calcium phosphate (*Cryptodifflugia oviformis*), and granular tests composed of siliceous particles (e.g., *Euglypha rotunda, Trinema lineare, Assulina muscorum,* and *Netzelia tuberculata*). Some species are agglutinate (e.g., *Difflugia urceolata* and *Phryganella nidulus*), with tests composed of mineral matter collected from the environment. Test ultrastructure and deposition have been investigated in several major genera including *Arcella* (Netzel 1975 a, b), *Centropyxis* (Netzel 1975 c, 1976 a), *Euglypha* (Hedley and Ogden 1973, Ogden 1979 a), *Trinema* (Hedley and Ogden 1974), *Netzelia* (Anderson 1987, 1988 a), and *Difflugia* (Netzel 1972, 1976 b, 1977; Ogden 1989).

Heliozoa

The heliozoa, such as *Heterophrys marina* with a microtubular organizing center in the centroplast (Bardele 1975), are categorized as advanced members of group 2. The trophic stages secrete plate-scales, lacking major ornamentation, and spine-scales (Fig. 16), with prominent spicule-shaped or branching projections attached to a base plate (c. 2 to 5 μm). Cysts (Figs. 17 and 18) contain walls composed of spheroidal particles or plate-like surface structures (5 to 10 μm). Some species of nucleariid filose amoebae (amoebae with filose pseudopodia and mitochondria with disc-like cristae) produce siliceous artifacts in the form of spherical or ovoid "perls", or plate-scales and spines (Patterson 1979, 1983, 1984, 1985; Patterson and Dürrschmidt 1986).

The axonemal bundle of microtubules (when present), forming a rod in the center of the axopodium, has a characteristic lattice-like pattern of organization in cross-section. This pattern is used in conjunction with cytoplasmic features and skeletal components to classify genera. *Actinophrys* (e.g., Fig. 3) and *Actinosphaerium* are common floating freshwater species lacking mineralized secretions. Numerous axopodia radiate from the round central body. *Clathrulina, Hedriocystis,* and *Monomastigocystis* are sessile, stalked species with an organic porous coat surrounding the cell. *Acanthocystis*, a member of the centrohelidian group with a massive microtubular organizing center in the cell body, secretes long siliceous spines and oval overlapping scales. Other biomineralizing genera include: *Raphidiocystis* (cup-shaped and radiate tubular spicules), *Raphidiophrys* (curved, spindle-shaped spicules), *Lithocolla* (pebble-like particles), *Rabdiophrys* (needle-like spicules with swollen tips), *Pinaciocystis* (biconvex, circular plates), *Pinaciophora* (round, biconvex, perforated scales), and *Pompholyxophrys* (minute spherules).

Radiolaria

The radiolaria (categorized in group 3) include three major groups: (1) Spumellarida with a generally spheroidal body plan and skeletal structures generally built on a spherical design; (2) Nassellarida with a prolate-shaped cell body and skeletal structures organized around a tripodal design; and (3) Phaeodaria containing a massive central capsule and three major cytoplasmic protrusions with organo-siliceous skeletons (e.g., Anderson 1983). The radiolarian skeleton (when present) varies from scattered, simple, tapered rods (c. 200 μm long) to ornate, latticed or geodesic-shaped shells (20 μm to 1000 μm) enclosing the main cytoplasmic mass of the cell (Figs. 4 and 19;124). *Thalassicolla*, a large species (1–3 mm diam.), lacks siliceous products, and is surrounded by a frothy bubble-like layer of cytoplasm. *Physematium* has a very large central capsule (several millimeters in diam.) with curved, siliceous spicules on the surface. Other large gelatinous species contain siliceous spicules varying in shape from rod-like to branched, and with a texture that can be

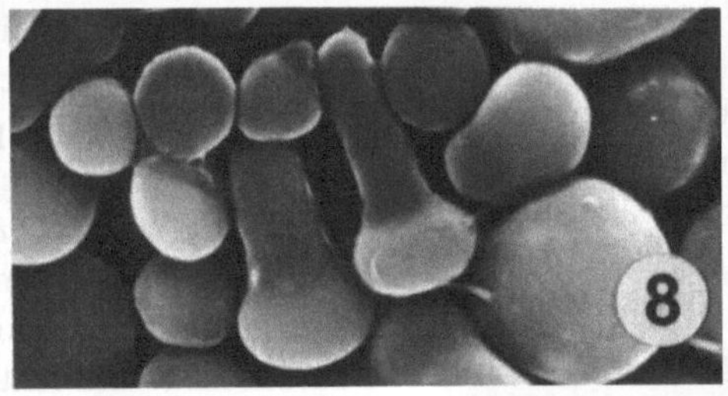

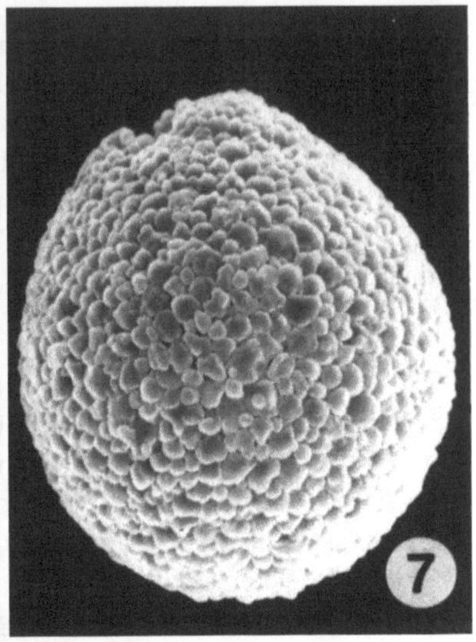

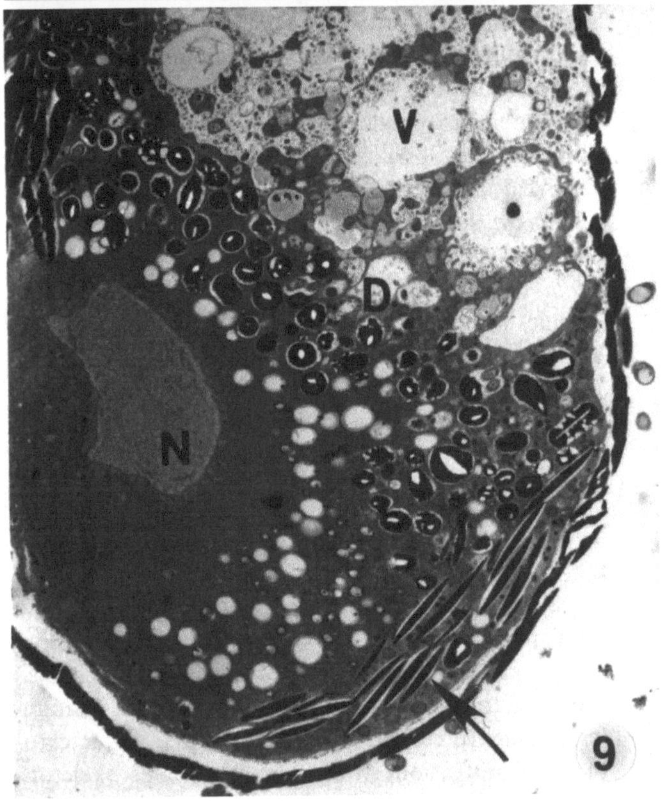

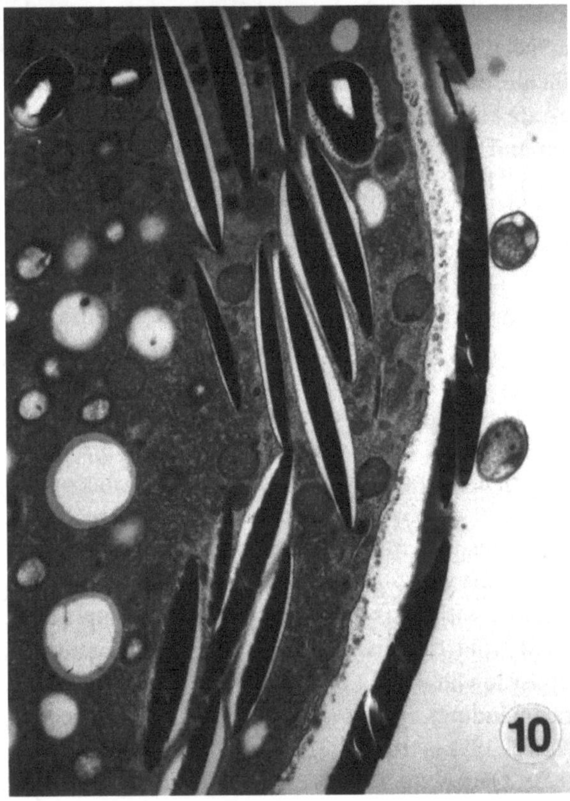

Figs. 5 and **6.** Scanning electron microscopic views of *Assulina muscorum*. Overview of the shell (Fig. 5). × 1,600. Imbricated siliceous plates are bound into an organic matrix of the shell (Fig. 6). × 4,400

Figs. 7 and **8.** Scanning electron microscopic views of *Netzelia tuberculata*. Overview of shell (Fig. 7). × 500. Siliceous particles within the shell are irregular or peg-shaped (Fig. 8). × 4,800

Figs. 9 and **10.** Transmission electron microscopic views of cytoplasm within the test of *A. muscorum*

Fig. 9. The nucleus (*N*), small hyaline vacuoles, and silica deposition vesicles containing shell plates (arrow) occur in the posterior of the cell; while vacuoles filled with electron dense matter (*D*), possibly adhesive granules, and digestive vacuoles (*V*) occur near the anterior of the cell. × 4,500

Fig. 10. A higher magnification view shows the siliceous plates still within the SDVs stacked in nearly parallel organization. × 8,970

smooth, or embellished with fine bristles (Fig. 19). Spongiose skeletons as in *Spongaster* sp. (e.g., Fig. 21) are commonly found among some genera of Spumellarida. The siliceous skeletons of Nassellarida can vary substantially. Some consist only of a simple tripodal shaped spicule, others contain a helmet-shaped enclosure surrounding the central capsule, and some such as *Callimitra* sp. (Fig. 20) produce very elaborate latticed wings at the periphery. Phaeodaria produce siliceous structures varying from porous clam-shaped enclosures to very ornate latticed or multiply branched skeletons varying in shape from polyhedra to spongiose stellate forms with long arm-like extensions.

Silica secretion

Current knowledge of silica secretion in three major groups will be presented: (1) testate amoebae, (2) heliozoa and related taxa, and (3) radiolaria.

Testate amoebae

The testate amoebae are convenient organisms to use in studying silica secretion and deposition. Many species are relatively easily maintained in laboratory culture and the individuals are sufficiently small (approximately 20 to several hundred micrometers) to fully examine a single cell in an ultrathin section using transmission electron microscopy. Illustrative events during silica secretion will be summarized for two species: *Assulina muscorum* and *Netzelia tuberculata*. *A. muscorum* is a small euglyphid-type species with a flattened, elongate shell of approximately 50 µm length (Fig. 5). The test is composed of imbricated siliceous plates (Fig. 6) bound to an organic matrix. *N. tuberculata* is a relatively large, spheroidal species with a shell diameter of about 80 to 100 µm (Fig. 7). It secretes pebble-like siliceous particles (Fig. 8) embedded in an organic matrix. Under some conditions, the shells also contain peg-shaped particles. In both species, the siliceous particles are secreted within silica deposition vesicles (SDVs) typically located near the posterior part of the cell (opposite the shell aperture).

In *A. muscorum*, SDVs containing siliceous plates in various stages of development occur near the posterolateral portions of the cytoplasm (Fig. 9). Golgi bodies occur scattered throughout this region and are thought to be the source of the SDVs. The earliest formed plates are somewhat granular or sintered in appearance. Mature plates are more solid and occur near the anterior portion of the cytoplasm. They are arranged in stacks with the long axes oriented parallel to the long axis of the cell (Fig. 10). This pattern of storage plates is also observed in other euglyphid testate amoebae (e.g., Ogden and Coûteaux 1987).

A clear developmental pattern of scale maturation progresses from the posterior to the anterior of the cytoplasm. Immature plates are largely located in the posterior, while more mature plates occur mid to anterior. Cellular digestion and scale storage occur largely in the anterior part of the cytoplasm. A more frothy, vacuolated cytoplasm containing food in various stages of digestion fills the most anterior part of the shell. Vacuoles, enclosing an electron opaque substance (possibly adhesive granules), are scattered throughout the cytoplasm near the frothy layer. The adhesive material produces the organic wall of the daughter cell during binary fission. The siliceous scales are deposited at the surface by exocytosis.

In *N. tuberculata*, silica is deposited either de novo within secretory vesicles (Figs. 11 and 12), or upon the surface of ingested particles within larger vesicles (Fig. 13). The particles deposited de novo are known as idiosomes, while those formed by surface deposition on ingested particles are known as coated xenosomes (Anderson 1987, Ogden 1991). It is not known how the primary SDVs originate, but there is good evidence that additional silica-secreting membrane is added to the SDVs by fusion with small Golgi-derived vesicles. There are two types of Golgi-derived vesicles (Anderson 1988 a, 1990): (1) plaque-containing vesicles (GPV) that produce the organic plaques cemented in the test among the siliceous granules at the time of test construction and (2) fibrillar vesicles (GFV) that contribute additional membrane to the silica depositional vesicles (SDVs). In addition, there are frequently small Golgi-secreted vesicles containing adhesive substances that contribute to the organic cement in the wall.

The GFVs contain a fine fibrillar deposit. They occur near the periphery of the Golgi apparatus, are scattered throughout the nearby cytoplasm (Fig. 14), and fuse with existing silica deposition vesicles (Fig. 15). GFVs have been observed fusing with both SDVs containing idiosomes and those containing xenosomes. The fine fibrillar substance is expelled into the larger SDV space at the time of fusion. Fine deposits of polymerized silica, forming chain-like beads, are observed on the inner surface of the membrane near the site of fusion and at other locations in the SDV. An hydrofluoric acid resistant matrix is present in the SDVs that produce idiosomes. This matrix appears to be the initial site for silica polymerization. As additional silica is deposited within the SDV, the siliceous product en-

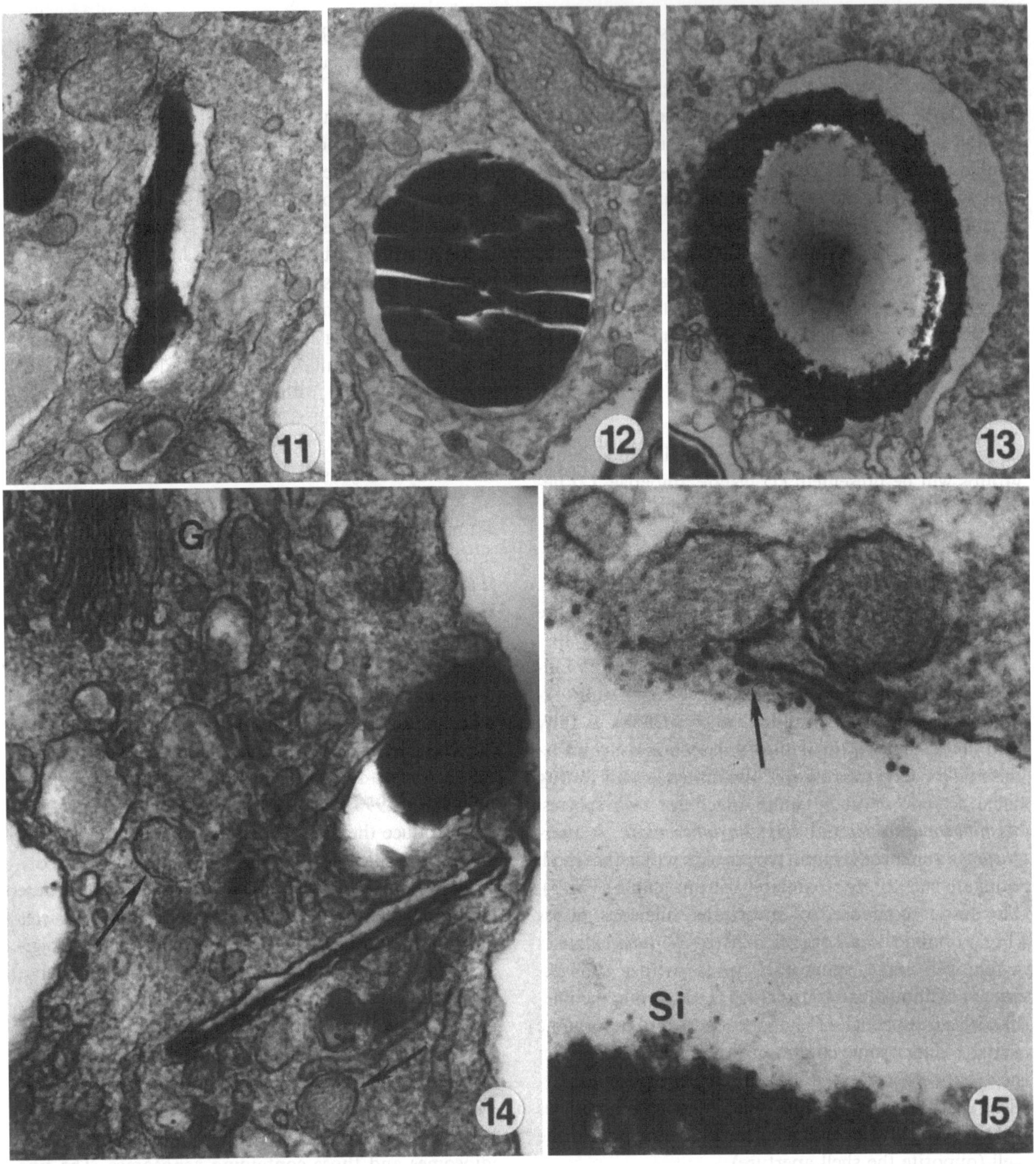

Figs. 11–15. Transmission electron microscopic views of the cytoplasm of *N. tuberculata* showing deposition of siliceous particles within SDVs

Fig. 11. An early stage of idiosome deposition shows the granular quality of the silica and the enclosing membrane of the SDV. × 36,000

Fig. 12. A more mature idiosome is electron dense and "glassy" in appearance with a smoother margin. × 30,200

Fig. 13. An ingested particle is coated by a layer of somewhat granular silica during formation of a coated xenosome within an SDV. × 12,980

Fig. 14. A Golgi apparatus (*G*) and several Golgi fibrillar vesicles (arrows) occur in the vicinity of SDVs with electron dense deposits of silica in various stages of idiosome formation. × 83,000

Fig. 15. Fusion of GFV (arrow) with membrane of a vesicle containing a rather granular deposit of silica (*Si*) in an early stage of deposition. × 128,300

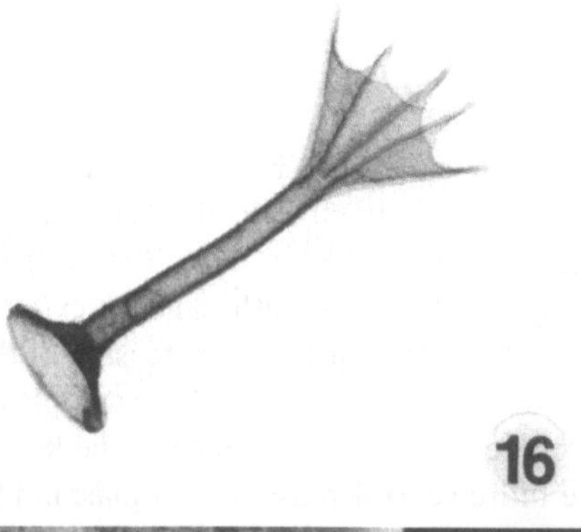

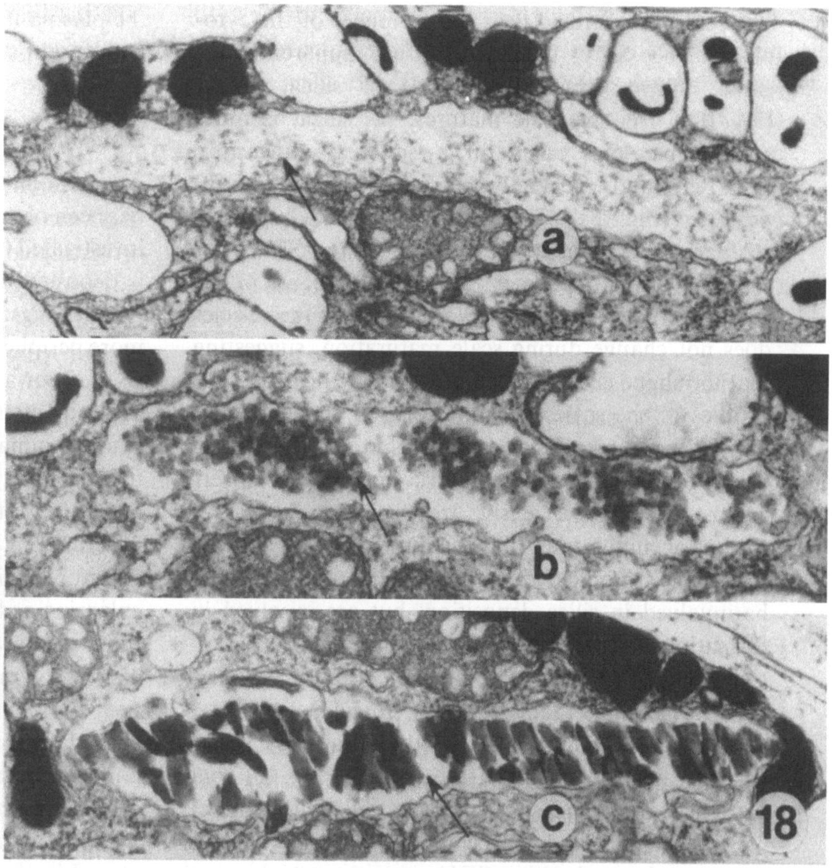

Figs. 16–18. Silica secretion in heliozoa. Adapted from Patterson and Dürrschmidt (1986) with permission from Dr. D. J. Patterson

Fig. 16. Scanning electron microscopic view of a forked and webbed spine-scale from *Acanthocystis pectinata.* × 20,000

Fig. 17. Cysts of *Actinophrys sol* surrounded by overlapping siliceous plates. × 1,750

Fig. 18. Silica plate deposition within silica deposition vesicles of *Actinophrys sol* as viewed by transmission electron microscopy showing an early stage (**a**) with beads of silica (arrow), a more advanced stage (**b**) with a sintered appearance (arrow), and a more mature stage (**c**) with a glassy deposit (arrow) in the SDV. × 30,000

larges. It appears somewhat granular in earlier stages (e.g., Figs. 13 and 15). At later stages, it is more homogeneous and electron opaque (Fig. 12). The fully formed silica particles remain enclosed within the SDVs and occur scattered throughout the cytoplasm. These are reserve particles used to form the shell of the daughter cell during binary fission (Anderson 1989). Xenosomes, engulfed in vacuoles, are also coated with silica. GFVs fuse with the membrane surrounding the engulfed xenosome and may contribute additional silica-secreting membrane to the vacuole. A variety of engulfed particles are coated with silica including starch grains from algal prey, and yeast cells (Figs. 13) offered as food in laboratory culture. A similar process of silica deposition on xenosomes has been observed in *Heleopera petricola* (Ogden 1991). Small, Golgi-derived vesicles were shown fusing with the SDV membrane. It is not clear at present if these Golgi vesicles are equivalent in function to the GFV observed in *N. tuberculata.*

Heliozoa

Silica secretion in heliozoa and heliozoon-like protists (Figs. 16–18) has been concisely summarized by Patterson and Dürrschmidt (1986). The actinophryid heliozoa produce siliceous structures as irregular plates or spheres (Fig. 17) that are deposited in one or many layers of the cyst wall (e.g., Patterson and Thompson 1981). Deposition of siliceous structures was investigated more thoroughly in *Actinophrys sol* (Patterson 1979, Mignot 1980). Scales are deposited within SDVs of unknown origin in very early stages of encystment.

The scales are formed near the periphery of the cytoplasm. Thes SDVs contain a diffuse, apparently organic, matrix where fine beads of silica appear (Fig. 18 a). As the scale matures, the silica becomes more dense and takes on a sintered appearance (Fig. 18 b) and eventually becomes "glassy" in appearance (Fig. 18 c). There are no microtubules or microfilaments visible in the vicinity of the SDVs, although a layer of filamentous material occurs in the vicinity of newly formed SDVs. The size of the vesicles does not change during scale maturation, suggesting that the shape is determined prior to silicification. The profiles of the scales take the form of series of concavities, perhaps produced by pressure on SDVs from nearby structures. Mitochondria may press closely against the SDVs. Extensive arrays of endoplasmic reticulum accumulate around the SDVs, especially during later stages of scale formation. They do not appear to be involved in silica deposition, but are involved in formation of other layers of the cyst.

Centrohelid heliozoa (e.g., *Acanthocystis, Cienkowskya, Heterophrys, Raphidiophrys, Raphidocystis, Oxnerella,* and *Chlamydaster*) produce flattened scales (plate scales) or spicules (spine-scales) (Fig. 16) that form a layer (periplast) around the cell (Dürrschmidt 1985, Patterson and Dürrschmidt 1986). Developing scales are located within SDVs near the periphery of the cell, beneath the plasma membrane. There is no evidence that the SDVs arise from any other membranous system in the trophic cells. The initial deposits of silica are spongy in appearance and during maturation become more electron dense ("glassy"), presumably due to additional deposition of silica. The earliest spongy matrix may contain an organic template, an interpretation that is supported by the presence of an hydrofluoric-acid-resistant component of mature scales. Mature scales are deposited by exocytosis on the plasma membrane and some may be endocytosed with food particles and recycled to the surface.

Radiolaria

A common cytoplasmic mechanism for silica secretion has been observed in all Spumellarida and Nassellarida investigated (e.g., Anderson 1983: p. 126). The skeleton is deposited in a cytoplasmic sheath called a cytokalymma (Figs. 22–24). This sheath is an extension of the more peripheral web-like components of the cytoplasm (Fig. 22) or an elaboration of the surface of the lobate portions of the more central mass of cytoplasm. The cytokalymma varies in thickness from c. 0.2 to several micrometers, depending on the location within a cell and perhaps the physiological state. Light microscopic observations indicate the cytokalymma is a dynamic structure exhibiting cytoplasmic streaming and actively enlarging as the internal siliceous skeleton grows. The inner membrane of the cytokalymma presumably acts as a silicalemma (Fig. 24). Initial deposits of silica appear granular (c. 80 nm diam.) and are more widely spaced at growing surfaces compared to an increasingly compact organization in more mature locations. The early stages appear to have a loose aggregated organization similar to the "sintered" state (Fig. 15) observed in early stages of silica deposition in SDVs of testate amoebae (Anderson 1988 a, 1990) and heliozoa (Patterson and Dürrschmidt 1988). As the siliceous product matures, the silica becomes more electron opaque, and the surface of the skeletal structures becomes smoother. There is a thin, final deposition of silica known as a veneer (Anderson 1983). The endoskeleton is deposited directly within the cytokalymma, hence there is no exocytosis as with testate amoebae and heliozoa. The endoskeleton is exposed, however, when the cytokalymma or surrounding cytoplasmic

Figs. 19–21. Scanning electron microscopic views of siliceous products of radiolaria. A spicule from a colonial radiolarian (Fig. 19). × 2,500. *Callimitra* sp. produces a very elaborate skeleton with a fine latticed meshwork forming three wings surrounding the more heavily silicified framework at the center (Fig. 20). × 750. *Spongaster tetras tetras* produces a quadrangular, spongiose skeleton (Fig. 21). × 200

Figs. 22–24. Views of cytokalymma (the silica-secreting cytoplasmic envelope of radiolaria)

Fig. 22. A scanning electron microscopic view of a radiolarian spine enclosed by a thin cytokalymma (*Cy*) with cytoplasmic projections (arrow) on the upper surface. × 21,500

Fig. 23. An ultrathin section through the skeleton of a radiolarian showing the very thin cytokalymma with a characteristic bulge of cytoplasm on one side, which in this case contains a small mitochondrion with tubular cristae. × 80,000

Fig. 24. A view of the cytokalymma surrounding a segment of the skeleton (*Sk*) of *S. tetras tetras* showing the cytoplasmic bulge with small fibrillar vesicles (*Ve*) possibly equivalent to GFV, and the inner membrane of the cytokalymma that serves as a silicalemma (arrow) during silica deposition. × 80,000. Note that the fibrillar vesicles in Fig. 14 and Fig. 24 are about equivalent in size (0.15 μm)

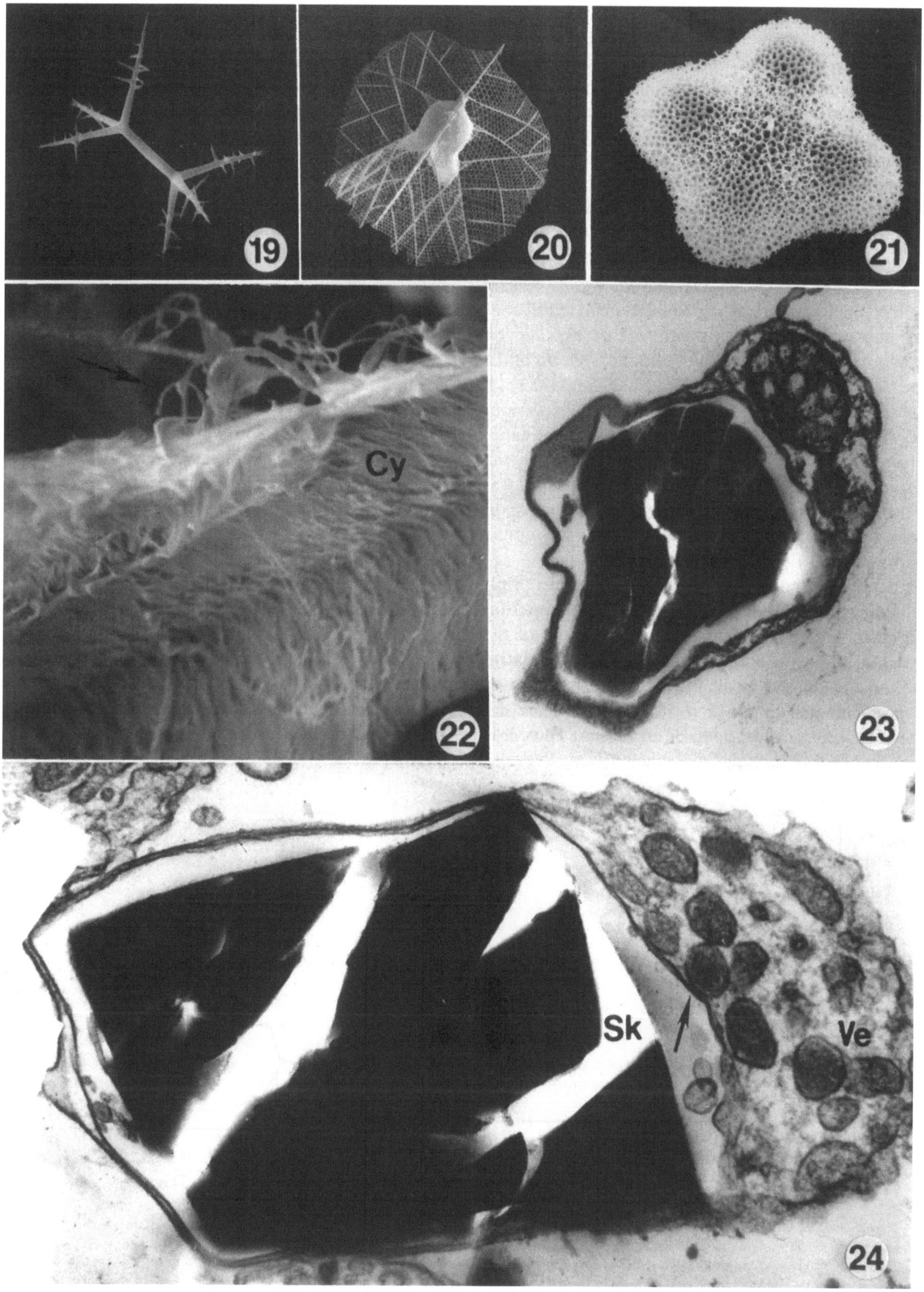
19
20
21
Cy
22
23
Sk
Ve
24

network is withdrawn from the surface by cytoplasmic streaming or contraction.

The origin of the silicalemma membrane during the expansive growth of the enclosed skeleton is not known. However, light microscopic observations of spine growth indicate that small, hyaline vesicles stream outward in the cytoplasmic sheath surrounding the growing spine and become lodged at specific sites of silicification, such as surface spines (Anderson 1983). These may be equivalent to the Golgi-derived fibrillar vesicles (GFV) in testate amoebae. Similar small, fibril-containing vesicles have been observed in transmission electron microscopic images of ultrathin sections of the cytokalymma surrounding delicate radiolarian skeletal structures (Fig. 24). These are observed only occasionally, and none have been seen fusing with the inner membrane of the cytokalymma. However, given the large expanse of the cytokalymma, the probability is low that an ultrathin section will pass through a region where a GFV had just fused with the inner membrane at the time of fixation.

It is difficult to determine the origin of these fibril-containing vesicles in radiolaria. Skeletal deposition frequently occurs in the more peripheral parts of the cytoplasm and the Golgi bodies occur within the central capsular cytoplasm. In the larger spumellarida, the distances between the Golgi regions and the silica depositional sites can be up to 80 to 100 µm. Thus, it is not easy to convincingly link Golgi associated vesicles in the central capsule with similar appearing ones at the periphery. Future research may be more profitable if molecular markers can be developed for the fibrillar deposit within GFVs, thus providing more convincing evidence for the equivalence of vesicles in different regions of the cytoplasm.

It is not immediately clear how energy is supplied to sustain silica secretory activity. Mitochondria are occasionally observed in the thicker portions of the cytoplasmic sheath (Fig. 23), but in the most cases the layer of cytoplasm is so thin as to preclude mitochondria. It is possible that ATP from more distantly located mitochondria diffuses toward, or is transported to, sites of silica deposition. This will require further biochemical clarification.

The rate of silica deposition in laboratory culture has been computed for one radiolarian species (*Spongaster tetras*). This species has a quadrangular skeleton (Fig. 21) with a mature size of approximately 280 µm as measured along the diagonal. The equation relating mass (W) in µg to length (L) in µm is:

$W = (3.338 \times 10^{-6})\ L^{2.205}$.

This equation was derived by regression analysis. The mean weight of fully cleaned skeletons in each size class was obtained using a Cahn electrobalance. Size was based on the length of the diagonal of the quadrangular skeleton. The weights were entered in the regression analysis as a dependent variable and the sizes of the diagonal as an independent variable (Anderson et al.

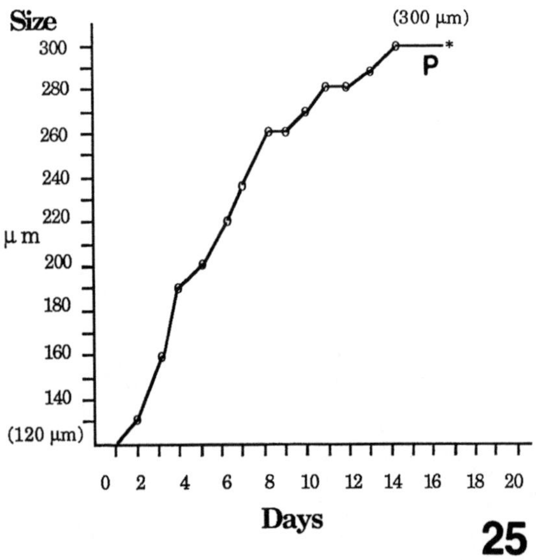

25

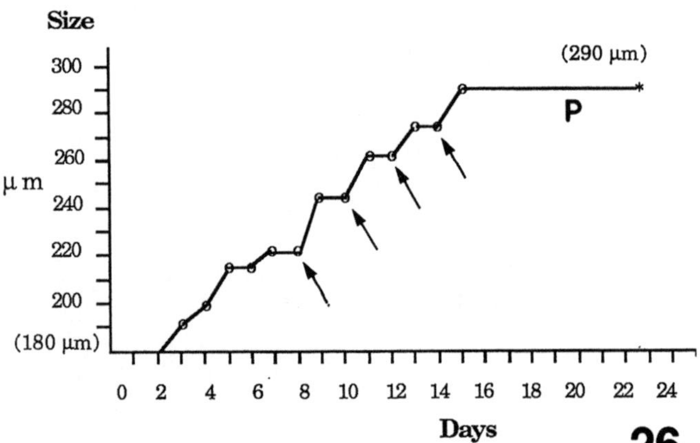

26

Figs. 25 and **26.** Growth curves for *Spongaster tetras tetras* in laboratory culture showing a pattern of rather continuous growth (Fig. 25) and a pattern with intermittent phases as denoted by arrows (Fig. 26). The maximum diagonal size achieved in 20 to 24 days of laboratory growth was 300 and 290 µm, respectively, thereafter a plateau phase (*P*) occurred until death

1989). The equation can be used to predict the weight of a living individual by measuring the length of the diagonal observed by light microscopy. The weights are used to plot growth curves and calculate rate of silica deposition as explained below.

The daily deposition of silica in laboratory cultured *S. tetras* was determined using light microscopic observations of skeletal size (Anderson et al. 1989). Due to its relative opacity, the perimeter of the skeleton is readily visible within the cytoplasm. The amount of silica deposited varied from 2.6 ng/day for small individuals (100–120 µm) to 18.7 ng/day for mature individuals (241–280 µm). The mean deposition rate for all sizes was 7.8 ng/day. Observations of shell growth in laboratory culture indicate a wide variability of growth patterns across individuals of *S. tetras* (Figs. 25 and 26). In some cases, growth is fairly continuous with only occasional plateaus (Fig. 25). In other cases, growth is sporadic with plateaus alternating at nearly daily intervals with growth increments (Fig. 26).

Conclusions

Cytoplasmic organization and form of siliceous products

Based on the organization and compartmentalization of cytoplasm, the Sarcodines can be categorized into three main groups: diffuse, transitional, and zonal. Differentiation of cytoplasmic structures increases across the three major groups including a greater tendency toward increased compartmentalization of the cytoplasm and greater specialization of function. This is also paralleled by increasingly complex secretory products that span simple pebble-shaped or scale-like secretions (testate amoebae and heliozoa), to highly complex geometric forms with geodesic design and complex latticed extensions (radiolaria).

The function of many of these biomineralized surface structures is not fully known. Secretion of some products at the surface could be an adaptive mechanism to expel excessive concentrations of ions that accumulate in the cytoplasm. This may be a plausible explanation for the phylogenetic origins of biomineralization, especially calcium and silica deposits. Calcium concentrations are at near saturation levels in ocean water (c. 0.4 mg/ml), and silica concentrations in the geological past were undoubtedly much higher than present (c. 20 mg/m^3 at the surface). Thus, calcium and silica secretion may have arisen initially during evolution as a mechanism to reduce intracellular concentrations. However, given the low concentrations of silica in modern oceanic surface water, it is not likely that sufficient silica would accumulate in the cytoplasm to account for the massive and sometimes elaborate skeletons secreted by some marine protists. Observational evidence indicates that radiolarian skeletons, for example, are protective, and also provide a scaffolding to anchor cytoplasmic structures used to capture prey (Anderson 1983). In some cases the prey are actively swimming metazoa several times larger than the body of the radiolarian. If this is the case, the metabolic cost of secreting these massive structures could be more than offset by the nutritional advantages gained in capture of larger prey. The frustules, moreover, of some algal protists are protective and provide a wall to contain the cytoplasm under high turgor pressure.

The role of smaller surface structures such as spicules, or delicate, ornamented scales and plates, requires further clarification. It is fairly clear from a functional morphological perspective that the siliceous plates and granules embedded in the organic matrix of the wall of testate amoebae can provide increased strength and protection (e.g., Anderson 1990, Ogden 1991). It is not known why they often have such elaborate geometries.

Silicate deposition in SDVs

Current evidence indicates that silica deposition in testate amoebae and heliozoa occurs within intracytoplasmic silica depositional vesicles (SDVs). An hydrofluoric-acid-resistant matrix (possibly organic in heliozoa) is present within the SDVs of both groups and apparently is a binding site for initial deposits of silica. The earliest stages of silicification appear loosely granular or sintered in ultrathin sections viewed by transmission electron microscopy. As silicification progresses, the sintered stage becomes more "glassy" in appearance, and electron opaque. Scanning electron microscopic views of the surfaces of the mature silicified product indicate it is typically quite smooth. This may be attributed to biological activity during the addition of the final phases of silica, and/or to the physical chemical properties of siliceous particles within closed aqueous spaces (e.g., Iler 1979). Surface geometry of siliceous particles tends to approach an equilibrium at the lowest free energy, with minimum surface perturbations. This is due to the tendency for projections from the surface to have a high silica solubility, whereas depressions tend to have lower silica solubility. Consequently, there is a net movement of silica from convex regions to concave. At equilibrium, a smooth surface is produced. It is interesting to note, however, *N. tub-*

erculata grown in low silicate medium deposits pitted and deformed granules in the test (Anderson 1992). This may be attributed to early deposition of partially formed particles. Test particles are fewer and more scattered compared to *N. tuberculata* grown in silica-enriched medium. If silica particles are ejected from the SDV prematurely, there may not be sufficient time within the closed space of the SDV for the surface to reach an equilibrium, and a smooth texture. Moreover, cytoplasmic concentrations of silica may be insufficient for completion of biological deposition, thus further contributing to the irregular surface texture.

Origin, intracytoplasmic distribution, and morphology of SDVs

There is good evidence that the Golgi apparatus in some testate amoebae secretes fibrillar vesicles (GFV) that migrate to and fuse with the SDV membrane, thus increasing its area and possibly contributing additional silica-secreting sites into the membrane. Early stages of silica polymerization occur in the near vicinity of sites where the GFV have fused with the SDV membrane. It is not known how the exact shape of the secreted product is determined. Present electron microscopic evidence indicates that there are no microtubules or microfilaments in the vicinity of the SDVs. However, further investigations using fixatives that especially preserve cytoskeletal elements are needed to clarify this issue.

Clearly, the testate amoebae must possess some mechanism for determining the peculiar geometry of the siliceous plates that are species specific. For example, *Quadrulella symmetrica* produces quadrangular plates that decrease in size from the posterior to the anterior region of the test (Ogden and Hedley 1980). Peg-shaped particles are frequently observed among the more irregular particles in the test of *N. tuberculata* (Anderson 1988 a). Other species, such as *Lesquerusia spiralis*, produce curved to vermiform rods of rather constant diameter (Harrison et al. 1976). Other species produce oval plates of regular dimension and often with characteristic species-specific decorations on the margin. This does not appear to be solely the result of localized pressures or constraints produced by occasional nearby cytoplasmic structures.

In heliozoa, silica products are continuously secreted and exocytosed on the surface (Patterson and Dürrschmidt 1986). This may account for the peripheral location of the SDVs. In testate amoebae, silica products are stored in the cytoplasm as reserve plates or particles in advance of cell division, and are deposited on the surface of the daughter cell cytoplasm during binary fission. Thus, surface deposition is cyclical and occurs only at the time of daughter cell production. A clear progression of plate development is observed from the posterior to the anterior region. Early deposition occurs in SDVs near the broad base of the cell where the nucleus and Golgi bodies are typically located. As the siliceous products mature, they occur increasingly more abundant toward the anterior region of the cytoplasm. It is not known how this sequence is organized within the cell. Since the Golgi bodies are located at the posterior end, and are likely the site of the initial SDVs, it is not surprising that silica deposition begins in this region. The displacement of maturing SDVs toward the anterior of the cell suggests additional translational mechanisms possibly involving microtubules and/or microfilaments (e.g., Bereiter-Hahn 1987, Weiss et al. 1987). Further research is needed to clarify how the mature plates are moved.

The cytokalymma deposits endoskeletons in radiolaria

Secretion of large skeletal structures in radiolaria differs markedly from plate production in testate amoebae and heliozoa. The siliceous product is deposited within a thin cytoplasmic sheath, known as a cytokalymma. It is produced by cytoplasmic extension of the cell surface. The origin of the silicalemma membrane lining the surface of the cytokalymma is unknown. Vesicles have been observed in the vicinity of the silica-secreting membrane, but it remains to be determined if these vesicles arise from the Golgi and if they indeed do fuse with the inner membrane of the cytokalymma thus increasing its area. Light microscopic observations of developing skeletal spines surrounded by a cytokalymma indicate the cytoplasmic layer is highly dynamic exhibiting cytoplasmic flowing along the surface and extending beyond the tip of the growing spine. The cytokalymma appears to dictate the shape of the developing spine by serving as a "living mold" for the silica deposition. Hence, it is possible to explain the species-specific design of the siliceous skeleton. The molding action of the cytokalymma may be genetically determined, although we do not know how the genetic information is translated at the molecular level into dynamic activity of the cytokalymma. The cytokalymma is not surrounded by other massive cytoplasmic structures especially during distal spine growth. Therefore, it is not possible to explain the particular shape of the siliceous product by constraints imposed by closely surrounding cytoplasmic elements. In some cases, the cytokalymma is connected by cytoplasmic threads to the surrounding cytoplasm. Further research is needed to determine the dynamic and structural relationships between the cytokalymma and these nearby cytoplasmic structures.

Control of vesicular traffic and silicate deposition

There is clearly a complex traffic of vesicle movement and fusion in most eukaryotic cells (e.g., Farquhar 1985, Mellman et al. 1987), and the control mechanisms that regulate this orderly flow need further clarification. The issue becomes more intriguing for testate amoebae that produce idiosomes and coated xenosomes. Where do the idiosome vesicles originate? It is clear that some of the coated xenosomes come from ingested food. For example *N. tuberculata* segregates starch grains from ingested algal cytoplasm and encloses them in silica-secreting vesicles (Anderson 1988 a). Also, yeast cells are engulfed in phagocytic vacuoles and some are segregated into vacuoles where silica is deposited on the surface of the yeast (Anderson 1989). Xenosomes are also isolated into silica-secreting vacuoles in *H. petricola* (Ogden 1991). We need to clarify more fully what determines the fate of ingested particles, how they become segregated into silica-secreting vacuoles, and what environmental conditions and cellular mechanisms regulate these functions.

One of the major issues to be resolved is the origin of the silicalemma membrane in the various groups of amoeboid protists. An appealing hypothesis is that the silicalemma originates as Golgi vesicles. This would provide a consistent explanation across the major groups. Fibrillar vesicles occur in the vicinity of the silicalemma in testate amoebae and radiolaria (Figs. 15 and 24) and have been observed to fuse with the silicalemma in *N. tuberculata.* The origin of the initial SDV is unknown. Hypothetically, the Golgi vesicle may give rise directly to the SDV without further modification. Or, fusion with additional Golgi vesicles could increase the size of the SDV while contributing additional membrane to the silicalemma. It is also possible that the initial vesicle containing the HF-resistant matrix could arise directly from the endoplasmic reticulum and become activated as an SDV by fusion with Golgi vesicles (GFVs). Likewise, SDVs that enclose and coat xenosomes, as occurs in some testate amoebae, could arise from Golgi vesicles that fuse with the enclosing vacuole, thus increasing its surface area and contributing silica-secreting membrane to the SDV. This has been suggested for *N. tuberculata* (Anderson 1988 a). Similarly, the inner membrane of the cytokalymma of radiolaria may be enlarged by fusion with Golgi-derived vesicles. As appealing as this unifying model is, we need much additional research to test it, and to more fully characterize the origin of silica-secreting membranes.

Recent research has begun to elucidate how certain signalling molecules in vesicular membranes determine binding specificity and fusion. Some of these signalling molecules, e.g., "NEM-sensitive fusion protein" (NSF) and "soluble NSF attachment proteins" (SNAP) or equivalent molecules, occur broadly in eukaryotic cells from yeast to mammalian neurons (e.g., Schekman 1992, Bennett and Scheller 1993, Söllner et al. 1993). NSF and SNAP occur in target membranes (e.g., plasma membrane) and may bind to synaptobrevin (VAMP) in vesicular membranes to initiate fusion. This knowledge base may provide insights for research studies that clarify vesicular traffic in silicate secreting protists.

The intriguing question also remains, how does the cell control the sometimes complex and species-specific shape of the siliceous products? Are some processes common to all groups? What processes, if any, are peculiar to a given group? The silicified products provide a tangible record of cellular activity. They can be used as a convenient source of visible evidence in research on the effects of modifications in the physical and/or chemical environment on secretory activity of protists (Anderson 1992). A variety of metabolic inhibitors, including colchicine, cytochalasin B, nocadazole, tunicamycin, and caffeine, have been used to experimentally assess intracellular events during silica scale production in chrysomonads (Sandgren and Barlow 1989). Control of silica secretion in amoeboid protists may be profitably investigated using similar techniques. In addition to characterizing the specific cellular mechanisms of secretion in these protists, an analysis of secretory structures related to experimental manipulations of cellular activity, may help to elucidate more general principles of eukaryotic cellular secretory processes.

Acknowledgements

Research at our laboratories was supported at various times by grants from the Biological Oceanography of the National Science Foundation. My thanks to Paul Bennett, Martha Bryan-Yarinsky, and Dee Breger for technical assistance. Dr. David J. Patterson of the University of Sydney, Australia, kindly supplied figures for heliozoan silica deposition. This is Lamont-Doherty Earth Observatory Contribution no. 5225.

References

Anderson OR (1981) Radiolarian fine structure and silica deposition. In: Simpson TL, Volcani BE (eds) Silicon and siliceous structures in biological systems. Springer, New York Berlin Heidelberg, pp 347–379

– (1983) Radiolaria. Springer, New York Berlin Heidelberg

– (1984) Cellular specialization and reproduction in planktonic foraminifera and radiolaria. In: Steidinger KA, Walker LM (eds) Marine plankton life cycle strategies. CRC Press, Boca Raton, pp 35–66

– (1986) Silicification in radiolaria – deposition and ontogenetic origins of form. In: Leadbeater BSC, Riding R (eds) Biomineralization in lower plants and animals. Clarendon Press, Oxford, pp 375–391 (Systematics Association special volume 30)

– (1987) Fine structure of a silica-biomineralizing testate amoeba, *Netzelia tuberculata*. J Protozool 34: 302–309

– (1988 a) Fine structure of silica deposition and the origin of shell components in a testate amoeba *Netzelia tuberculata*. J Protozool 35: 204–211

– (1988 b) Comparative protozoology. Springer, Berlin Heidelberg New York Tokyo

– (1989) Some observations of feeding behavior, growth, and test particle morphology of a silica-secreting testate amoeba *Netzelia tuberculata* (Wallich) (Rhizopoda, Testacea) grown in laboratory culture. Arch Protistenk 137: 211–221

– (1990) Effects of silicate deficiency on test morphology, cytoplasmic fine structure, and growth of the testate amoeba *Netzelia tuberculata* (Wallich) Netzel (Rhizopoda, Testacea) grown in laboratory culture. Arch Protistenk 138: 17–27

– (1992) The effects of silicate depletion and subsequent replenishment on the cytoplasmic fine structure of the silica-secreting testate amoeba *Netzelia tuberculata* in laboratory culture. J Morphol 211: 285–293

– Bennett P, Bryan M (1989) Experimental and observational studies of radiolarian physiological ecology: 1. growth, abundance and opal productivity of the spongiose radiolarian *Spongaster tetras tetras*. Mar Micropaleontol 14: 257–265

– Perry CC, Hughes NP (1990) Transmission and scanning electron microscopic evidence for cytoplasmic deposition of strontium sulphate crystals in colonial radiolaria. Philos Trans R Soc Lond [Biol] 329: 81–86

Bardele CF (1975) The fine structure of the centrohelidian heliozoan *Heterophrys marina*. Cell Tissue Res 161: 85–102

Bennett MK, Scheller RH (1993) The molecular machinery for secretion is conserved from yeast to neurons. Proc Natl Acad Sci USA 90: 2559–2563

Bereiter-Hahn J (1987) Mechanical principles of architecture of eukaryotic cells. In: Bereiter-Hahn J, Anderson OR, Reif W (eds) Cytomechanics. Springer, Berlin Heidelberg New York Tokyo, pp 3–30

Bovee EC (1981) Distribution and forms of siliceous structures among protozoa. In: Simpson TL, Volcani BE (eds) Silicon and siliceous structures in biological systems. Springer, New York Berlin Heidelberg, pp 275–279

– (1991) Sarcodina. In: Harrison FW, Corliss JO (eds) Microscopic anatomy of invertebrates, vol 1, protozoa. Wiley-Liss, New York, pp 161–259

Brook AJ (1981) Calcium sulphate inclusions in the desmids *Bambusina* and *Gonatozygon*. Br Phycol J 16: 267–272

Crawford RM (1981) The siliceous components of the diatom cell wall and their morphological variation. In: Simpson TL, Volcani BE (eds) Silicon and siliceous structures in biological systems. Springer, New York Berlin Heidelberg, pp 129–156

Dürrschmidt M (1985) Electron microscopic observations on scales of species of the genus Acanthocystis (Centrohelida, Heliozoa) from Chile, I. Arch Protistenk 129: 55–87

Faber WW Jr, Preisig HR (1994) Calcified structures and calcification in protists. Protoplasma 181: 78–105

Farquhar MG (1985) Progress in unraveling pathways of Golgi traffic. Annu Rev Cell Biol 1: 447–488

Green JC, Perch-Nielsen K, Westbroek P (1989) Phylum Prymnesiophyta. In: Margulis L, Corliss JO, Melkonian M, Chapman DJ (eds) Handbook of Protoctista. Jones and Bartlett, Boston, pp 293–333

Harrison FW, Dunkelberger D, Watabe N, Stump AB (1976) The cytology of the testaceous rhizopod *Lesquereusia spiralis* (Ehrenberg) Penard. II. Ultrastructure and shell formation. J Morphol 150: 343–358

– – – – (1981) Ultrastructure and deposition of silica in rhizopod amebae. In: Simpson TL, Volcani BE (eds) Silicon and siliceous structures in biological systems. Springer, New York Berlin Heidelberg, pp 281–294

Hedley RH, Ogden CG (1973) Biology and fine structure of *Euglypha rotunda* (Testacea: Protozoa). Bull Br Mus Nat Hist (Zool) 26: 119–137

– – (1974) Observations on *Trinema lineare* Penard (Testacea: Protozoa). Bull Br Mus Nat Hist (Zool) 26: 187–199

Iler R (1979) Chemistry of silica. Wiley, New York

Leadbeater BSC (1981) Ultrastructure and deposition of silica in loricate choanoflagellates. In: Simpson TL, Volcani BE (eds) Silicon and siliceous structures in biological systems. Springer, New York Berlin Heidelberg, pp 295–322

Lowenstam HA (1986) Mineralization processes in monerans and protoctists. In: Leadbeater BSC, Riding R (eds) Biomineralization in lower plants and animals. Clarendon Press, Oxford, pp 1–17 (Systematics Association special volume 30)

Manton I, Peterfi LS (1969) Observations on the fine structure of coccoliths, scales and the protoplast of a freshwater coccolithophorid, *Hymenomonas roseola* Stein, with supplementary observations on the protoplast of *Cricosphaera carterae*. Proc R Soc Lond [Biol] 172: 1–15

Mellman I, Howe C, Helenius A (1987) The control of membrane traffic on the endocytotic pathway. Curr Top Membr Transport 29: 255–288

McGrory CB, Leadbeater BSC (1981) Ultrastructure and deposition of silica in the Chrysophyceae. In: Simpson TL, Volcani BE (eds) Silicon and siliceous structures in biological systems. Springer, New York Berlin Heidelberg, pp 201–230

Mignot J-P (1980) Etude ultrastructurale de la pédogamie chez *Actinophrys sol* (Héliozaire) III. gamétogenèse, fécondation, enkystement. Protistologica 16: 533–547

Netzel H (1972) Die Schalenbildung bei *Difflugia oviformis* (Rhizopoda, Testacea). Z Zellforsch 135: 55–61

– (1975 a) Struktur und Ultrastruktur von *Arcella vulgaris* var. *multinucleata* (Rhizopoda, Testacea). Arch Protistenk 117: 219–245

– (1975 b) Die Entstehung der hexagonalen Schalenstruktur bei der Thekamöbe *Arcella vulgaris* var. *multinucleata* (Rhizopoda, Testacea). Arch Protistenk 117: 321–357

– (1975 c) Morphologie und Ultrastruktur von *Centropyxis discoides* (Rhizopoda, Testacea). Arch Protistenk 117: 369–392

– (1976 a) Die Abscheidung der Gehäusewand bei *Centropyxis discoides* (Rhizopoda, Testacea). Arch Protistenk 118: 53–91
– (1976 b) Die Ultrastruktur der Schale von *Difflugia oviformis* (Rhizopoda, Testacea). Arch Protistenk 118: 321–339
– (1977) Die Bildung des Gehäuses bei *Difflugia oviformis* (Rhizopoda, Testacea). Arch Protistenk 119: 1–30
Ogden CG (1979 a) An ultrastructural study of division in *Euglypha* (Protozoa: Rhizopoda). Protistologia 15: 541–556
– (1979 b) Siliceous structures secreted by members of the subclass Lobosia (Rhizopoda: Protozoa). Bull Br Mus Nat Hist (Zool) 36: 203–207
– (1989) Morphology of the organic shell matrix of *Difflugia* (Rhizopoda) in culture, including modification by the addition of agglutinate particles. Arch Protistenk 136: 365–376
– (1991) The ultrastructure of *Heleopera petricola* an agglutinate soil amoeba; with comments of feeding and silica deposition. Eur J Protistol 27: 238–248
– Coûteaux M-M (1987) The biology and ultrastructure of the soil testate amoeba *Tracheleuglypha dentata* (Rhizopoda: Euglyphidae). Eur J Protistol 23: 28–42
– Hedley RH (1980) An atlas of freshwater testate amoebae. British Museum (Natural History), London, Oxford University Press, Oxford
Page FC (1983) Marine gymnamoebae. Institute of Terrestrial Ecology, Cambridge
– Siemensma FJ (1991) Nackte Rhizopoda und Heliozoa. G Fischer, Stuttgart
Patterson DJ (1979) On the organization and classification of the protozoon, *Actinophrys sol* Ehrenberg, 1830. Microbios 26: 165–208
– (1983) On the organization of the naked filose amoeba, *Nuclearia moebiusi* Frenzel 1897 (Sarcodina, Filosea) and its implications. J Protozool 30: 301–307
– (1984) The genus Nuclearia (Sarcodina, filosea): species composition and characteristics of the taxa. Arch Protistenk 128: 127–139
– (1985) On the organization and affinities of the amoeba *Pompholyxophrys punicea* Archer, based on ultrastructural examination of individual cells from wild material. J Protozool 32: 241–246
– Dürrschmidt M (1986) Siliceous structures formed by heliozoa and heliozoon-like amoebae. In: Leadbeater BSC, Riding R (eds) Biomineralization in lower plants and animals. Clarendon Press, Oxford, pp 361–374 (Systematics Association special volume 30)
– Thompson DW (1981) Structure and elemental composition of the cyst wall of *Echinosphaerium nucleofilum* Barrett (Heliozoea, Actinophryida). J Protozool 28: 188–192
Pickett-Heaps JD, Tippit DH, Andreozzi JA (1979) Cell division in the pennate diatom *Pinnularia*. IV. Valve morphogenesis. Biol Cell (Paris) 35: 199–203.
Preisig H (1994) Siliceous structures and silicification in flagellated protists. Protoplasma 181: 29–42
Sandgren CD, Barlow SB (1989) Siliceous scale production in chrysophyte algae. II. SEM observations regarding the effects of metabolic inhibitors on scale regeneration in laboratory population of scale-free *Synura petersenii* cells. Nova Hedwigia 95: 27–44
Schekman R (1992) Genetic and biochemical analysis of vesicular traffic in yeast. Curr Opin Cell Biol 4: 587–592
Söllner T, Whiteheart SW, Brunner M, Erdjument-Bromage H, Geromanos S, Tempst P, Rothman JE (1993) SNAP receptors implicated in vesicle targeting and fusion. Nature 362: 318–324
Volcani BE (1981) Cell wall formation in diatoms: morphogenesis and biochemistry. In: Simpson TL, Volcani BE (eds) Silicon and siliceous structures in biological systems. Springer, New York Berlin Heidelberg, pp 157–200
Weiss DG, Langford GM, Allen RD (1987) Implications of microtubules in cytomechanics. In: Bereiter-Hahn J, Anderson OR, Reif W (eds) Cytomechanics, Springer, Berlin Heidelberg New York Tokyo, pp 100–113

Protoplasma (1994) 181: 78 – 105

PROTOPLASMA

Calcified structures and calcification in protists

W. W. Faber Jr.[1,]* and **H. R. Preisig**[2]

[1] Drinking Water Quality Control, Bureau of Water Supply and Wastewater Collection, New York City Department of Environmental Protection, Corona, New York and [2] Institute of Systematic Botany, University of Zurich, Zurich

Received September 20, 1993
Accepted April 19, 1994

Summary. The diversity of calcified structures found in protists, the mechanisms utilized to form these structures, and the role these structures play in the taxonomy and systematics of the protists are presented. The two most frequently studied orders of protists which produce calcified structures, the coccolithophorids and foraminifera, are featured. However, consideration is given to the less known and least studied organisms.

Keywords: Calcification; Coccolithophorids; Coccolithogenesis; Foraminifera; Protists; Taxonomy.

Introduction

Calcified structures which are produced by protists are often found in limnological and oceanographic sediments. Nearly 50% of pelagic sediments are composed of the skeletal remains of calcium carbonate secreting organisms, such as coccolithophorids and foraminifera (Sverdrup et al. 1942, Saito and Bé 1967), and testacea are often found in lake core samples (Schoenborn 1990). The abundance of these organisms and their long evolutionary life histories have allowed scientists to utilize these skeletal remains extensively in biostratigraphic and paleoecological research.

The biomineralization process of calcification is seen in a wide range of unrelated taxa within the protists (Pennak 1978, Ogden and Hedley 1980, Lee et al. 1985, Green 1986, Anderson 1988, Loeblich and Tappan 1988). Depending on which systematic scheme is utilized, these protists are in at least 10 orders (Table 1). The resultant calcified structures produced are used as basic morphologic tools in the classification of these protists at several taxonomic levels. Also since morphology is considered an adaptive characteristic, the modifications which are seen in these structures are believed to have arose as a means to exploit a particular environment and resources (Hallock et al. 1991). However, the central question to understanding the relationships in the phylogeny of these protists is which mechanism is used for the secretion of the calcified structures.

As research has increased, most notably in the coccolithophorids and foraminifera, this central question has expanded to include to what extent is the calcification due to the cell (i.e., genetic control) and to the environment (i.e., ecological control). For example in the foraminifera, the shape and the size of the calcite crystals may be genetically controlled whereas the morphology of the calcite shell may be influenced by the environment (Towe and Cifelli 1967).

This review will present the diversity of calcified structures found in protists, the mechanisms utilized to form these structures, and the role these structures play in the taxonomy and systematics of the protists. Macrophytic algae (rhodophytes, phaeophytes, chlorophytes) are not dealt with here (for a recent review, see Pentecost 1991 a).

Coccolithophorids

The coccolithophorids comprise a distinct assemblage of prymnesiophyte (= haptophyte) algae which bear variously ornamented calcified plates (coccoliths) of

* Correspondence and reprints: Drinking Water Quality Control, Bureau of Water Supply and Wastewater Collection, NYC Department of Environmental Protection, 59-17 Junction Boulevard, Corona, NY 11368, U.S.A.

Table 1. A partial list of protists that produce calcified structures, based on the taxonomic scheme of Lee et al. (1985)

Phylum	Subphylum	Class	Order	Example
Sarcomastigophora	Mastigophora	Phytomastigophorea	Dinoflagellida	Thoracosphaera
			Chrysomonadida	Anthophysa
			Prymnesiida (Coccolithophorales)	Coccolithophorids
			Volvocida	Phacotus
	Sarcodina	Granuloreticulosea	Foraminiferida	Foraminifera
		Mycetozoea	Physarida	Physarum
		Lobosea	Trichosida	Trichosphaerium
			Arcellinida	Cryptodifflugia
Ciliophora	Postciliodesmatophora	Spirotrichea	Choreotrichida	Codonella, Codonaria
	Rhabdophora	Protomatea	Prorodontida	Plagiopogon, Coleps Nolandia

micron-size (ca. 1–25 µm) on the cell surface. Coccolithophorids are especially common in marine waters, but a few freshwater species also exist. Both biflagellate and non-motile species are known; some of the biflagellate species also have a third flagellum-like appendage, a haptonema. Some 60–70 genera and almost 200 species are known from present time (Chrétiennot-Dinet 1990, Jordan and Kleijne 1994), but many more have been described from fossils (Loeblich and Tappan 1966, Tappan 1980). The fossil records go back to the Carboniferous (approximately 300 million years ago), but great abundances of coccoliths have only been recorded from the Early Jurassic on (ca. 190–170 Ma). The taxonomy of coccolithophorids is based mainly on the morphology of the species-specific coccoliths. Cell shape, coccosphere arrangement and the presence of different types of coccoliths on a single cell are also taxonomic criteria. In the most recent schemes of classification all extant coccolithophorids were placed in a single order (Coccolithophorales = Coccosphaerales) and 13–14 different families were distinguished (Chrétiennot-Dinet 1990, Chrétiennot-Dinet et al. 1993, Jordan and Kleijne 1994). Separation at the family level is based mainly on the proximal disc construction of the main coccolith type, whereas at the generic level characters like central area structures, flagellar position, and additional coccolith types are important. Species generally differ on small-scale variations on the generic theme. Paleontologists traditionally used schemes of classification based entirely on coccolith morphology. In the scheme of Tappan (1980) the coccolithophorids are distributed among 7 orders and 42 families.

In this review, we primarily concentrate on aspects of coccolith morphology and development as well as on aspects of systematics and taxonomy. Further information on coccolithophorids can be found in Green et al. (1990), Winter and Siesser (1993), and in the references therein.

Coccolith morphology and crystal structure

Many coccolithophorids have organic scales in addition to their coccoliths (Fig. 4). Organic matrix material is present in all coccoliths and base-plate scales often form the proximal part of the coccosphere. In a few genera (such as *Balaniger, Calciarcus, Quaternariella, Trigonaspis, Turrisphaera,* or *Wigwamma*), there are no heavily calcified coccoliths but only lightly calcified scales, or there is only a incomplete covering of variously shaped calcite crystals on primarily organic scales (see Chrétiennot-Dinet 1990). All of these genera have only recently been described and require more extensive investigations on their structure and life cycles. Recently, it has been shown that coccoliths typical for some of these genera may also be produced by different genera typically producing more heavily calcified types of coccoliths (e.g., *Turrisphaera*-like coccoliths may also occur in species of *Papposphaera*; see below and Thomsen et al. 1991).

Typical coccoliths are constructed of rhombohedral calcite crystals in a radial arrangement, a few have hexagonal prisms, and some have a combination of these two forms in a single crystal (Tappan 1980). On the basis of crystallization two main types of coccolith are distinguished: (1) holococcoliths, (2) heterococco-

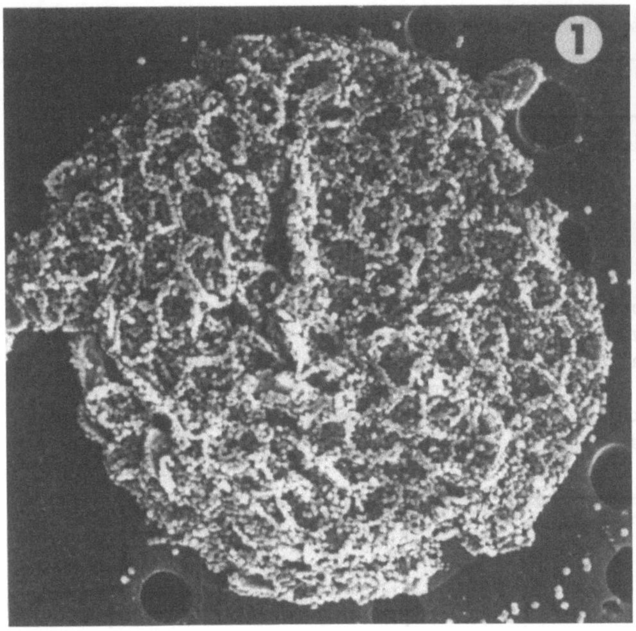

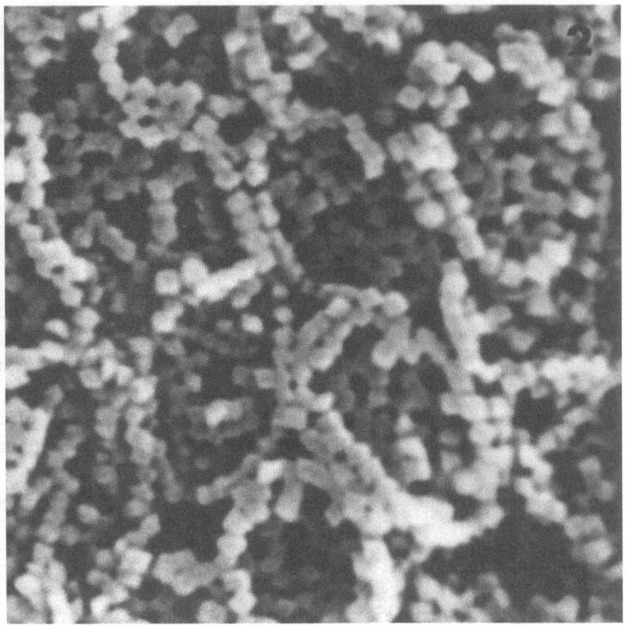

Figs. 1 and **2.** Scanning electron micrographs (SEMs) of *Crystallolithus hyalinus*; collapsed cell showing basic outline and many crystalloliths (holococcoliths) (Fig. 1), × 3,900. Crystalloliths at higher magnification (Fig. 2), × 14,600

liths. The basis of this subdivision is the varying degree of crystal growth.

Holococcoliths are built up of numerous simple calcite

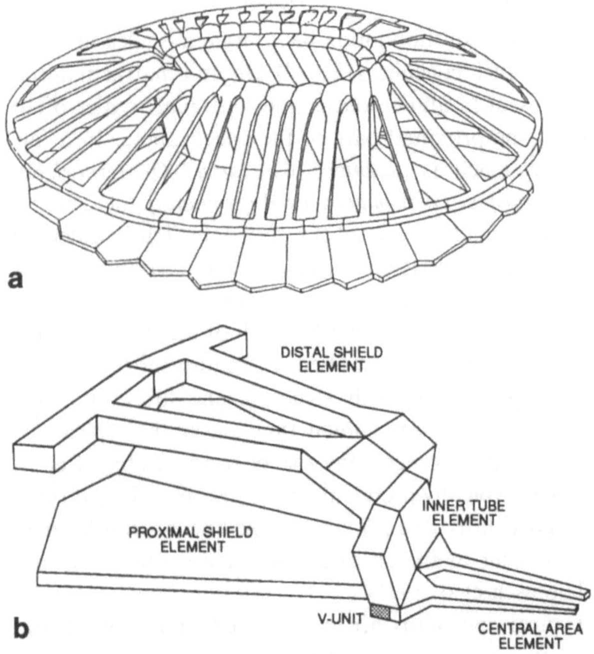

Fig. 3. Model of a heterococcolith (placolith) of *Emiliania huxleyi* showing the radial array of crystal units and their complex multi-element construction (see also Figs. 8 and 16). **a** Entire coccolith, × 32,000; **b** oblique view of two single unit elements, × 64,000 [reproduced from "Nature" (Young et al. 1992), with permission]

crystallites of uniform size, held together by an organic matrix. The shape of the holococcolith itself is quite variable and several different types have been given special names, such as areoliths, calyptroliths, crystalloliths, flosculoliths, gliscoliths, helladoliths, laminoliths and zygoliths (see Norris 1985, Chrétiennot-Dinet 1990, Kleijne 1991). Crystalloliths, for example, are plate- or saucer-like coccoliths, with simple uncrowded rhombohedral crystallites being scattered over the surface of the organic base-plate but not completely covering it (Figs. 1 and 2).

Heterococcoliths are composed of distinguishable subelements, which are a result of the diverse morphology of the calcite structures including rim, shield, tube or plate elements, walls, lamellae, rods and various other elements (Fig. 3). Many different types of heterococcoliths have been described in the literature (e.g., caneoliths, ceratoliths, cribriliths, cricoliths, cyrtoliths, lepidoliths, lopadoliths, osteoliths, pappoliths, pentaliths, placoliths, scapholiths, tremaliths; see Chrétiennot-Dinet 1990).

An example for a relatively simple type of heterococcolith is the cricolith (Figs. 4–6) which consists of a base-plate scale comparable to the unmineralized scales, with organic matrix material and calcite crystallites being stacked peripherally to form a simple tube (Outka and Williams 1971). Such coccoliths occur for example in *Pleurochrysis carterae* (Braarud and Fagerland) Christensen [= *Hymenomonas carterae* (Braa-

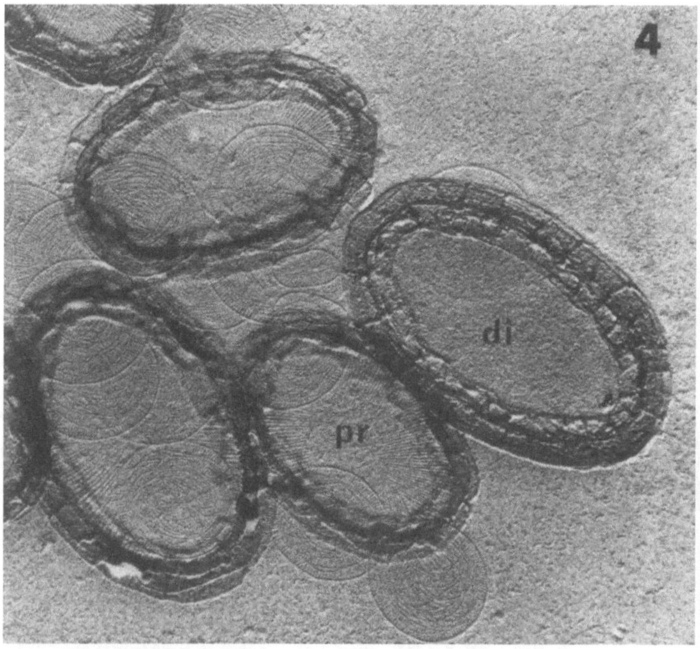

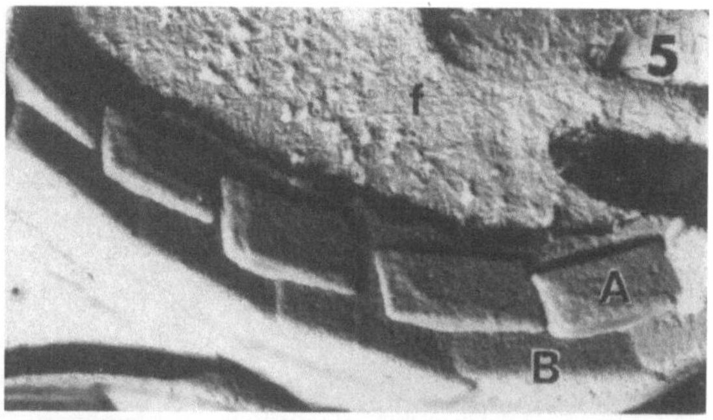

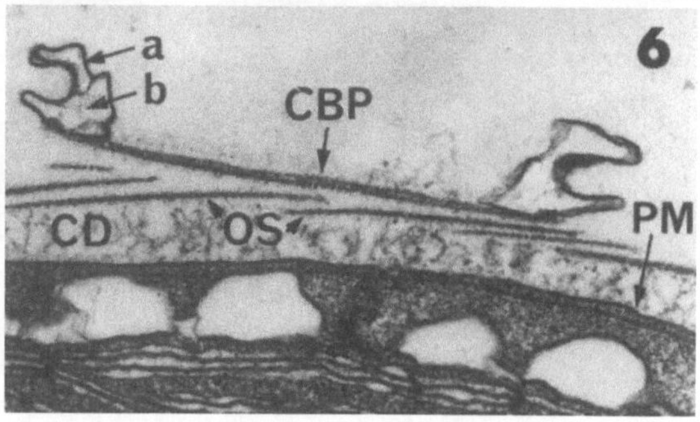

Figs. 4–6. *Pleurochrysis carterae*

Fig. 4. Shadow-cast wholemount preparation of organic scales and partially decalcified cricoliths. The microfibrillar substructure is visible in the organic scales and on the proximal (*pr*) surface of the cricoliths. This substructure is largely obscured by amorphous matter on the distal (*di*) side. Transmission electron micrograph (TEM), × 25,000

Fig. 5. Freeze-fracture (deep-etch) preparation showing the two types of calcite crystals (*A*, *B*) which make up the calcified part of the coccolith. A fibrillar pattern (*f*) is apparent on the distal surface of the coccolith base-plate, × 54,000

Fig. 6. Transverse section through a decalcified cricolith and the organic scales (*OS*) on the cell surface. Note the fusion of the organic matrix elements (*a*, *b*) to each other and to the cricolith base-plate (*CBP*). *CD* Columnar deposit between plasma membrane (*PM*) and scales; × 54,000

rud and Fagerland) Braarud; = *Cricosphaera carterae* (Braarud and Fagerland) Braarud]. In this species, each coccolith carries about 30 units of crystalline $CaCO_3$ on the distal side of the rim (Figs. 4–6). Other heterococcoliths have a more complicated morphology (Figs. 3 and 7–15). Placoliths, for instance, have an upper and a lower shield composed of radial elements interconnected by a central tube. This type of coccolith is found in some of the most common species, e.g., in *Coccolithus pelagicus* (Wallich) Schiller (Fig. 7) and in *Emiliania huxleyi* (Lohmann) Hay and Mohler [= *Coccolithus huxleyi* (Lohmann) Kamptner] (Fig. 8). Coccolith structure of *E. huxleyi* has been analyzed in most detail (Mann and Sparks 1988, Westbroek et al. 1989, Young 1989, Young et al. 1992) and been shown to consist of a radial array of unit elements, each of which is composed of several superficially discrete sub-elements (Fig. 3). Each unit element is a single crystal, as shown by electron diffraction, high-resolution transmission electron microscopy and crystal overgrowth studies.

Electron diffraction patterns and crystal forms have indicated that in nature coccoliths are usually composed of low-magnesium calcite, but from a few coccolithophorids aragonite has also been reported, e.g., from *Polycrater galapagensis* Manton and Oates (1980). Aragonite (orthorhombic crystalline form of $CaCO_3$) differs from rhombohedral calcite in structure, hardness, specific gravity and solubility. It has also been reported that $CaCO_3$-forms other than calcite (such as aragonite or vaterite) may occur in coccoliths of *E. huxleyi*, when grown in a nitrogen-deficient culture medium (Wilbur and Watabe 1963), but this has been questioned by Young et al. (1991) after their attempts to reproduce the same results have not been successful.

Coccosphere arrangement

The arrangement of coccoliths within the coccosphere is very precise and has three general forms. Coccolithophorids with only one type of coccolith are termed monomorphic. Coccospheres of *Pleurochrysis carterae*,

Fig. 7. *Coccolithus pelagicus* (SEM). Whole coccosphere of interlocking placoliths, × 6,500

Fig. 8. *Emiliania huxleyi* (SEM). Three complete coccospheres in the same sample showing variations in placolith morphology, × 7,200

for instance, consist of a single layer of 100–200 uniform coccoliths, whereas in *E. huxleyi* there are usually some 15 uniform coccoliths arranged in a single layer (Fig. 8), but much higher numbers of coccoliths arranged in multiple layers can also be produced (Linschooten et al. 1991). Dimorphic species have two different types of coccoliths on their cell surface. One of the two coccolith types is often restricted to the anterior or both poles of the coccosphere. An example for this type is *Acanthoica quattrospina* Lohmann (Fig. 14) which has simple basket-like heterococcoliths with a short central process (cyrtoliths) on most of the cell body and styliform cyrtoliths (rhabdoliths) at each pole. The third type of arrangement is the dithecate coccosphere, which is comprised of two complete layers of coccoliths, the endotheca and the exotheca. An example for this type is *Syracosphaera nodosa* Kamptner (Figs. 11 and 12) which has an outer layer of circular

---

Fig. 9. *Discosphaera tubifera* (SEM). Complete coccosphere consisting of salpingiform cyrtoliths, × 3,900

Fig. 10. *Pontosphaera syracusana* (SEM). Complete coccosphere consisting of cribriliths, × 2,900

Figs. 11 and **12.** *Syracosphaera nodosa* (SEMs). Fig. 11, complete inner layer of oval incomplete caneoliths (some at the anterior pole bearing rods), × 11,400. Fig. 12, dislodged circular cyrtoliths forming an outer layer, × 7,700

Fig. 13. *Braarudosphaera bigelowii* (SEM). Complete coccosphere showing six of twelve coccolith (pentalith) faces, × 8,800

Fig. 14. *Acanthoica quattrospina* (SEM). Flattened coccosphere consisting of simple cyrtoliths and styliform cyrtoliths (rhabdoliths) at the two poles, × 6,200

Fig. 15. *Calciosolenia murrayi* (SEM). Flattened coccosphere consisting of scapholiths and with only one spine still attached at the anterior pole, × 4,200

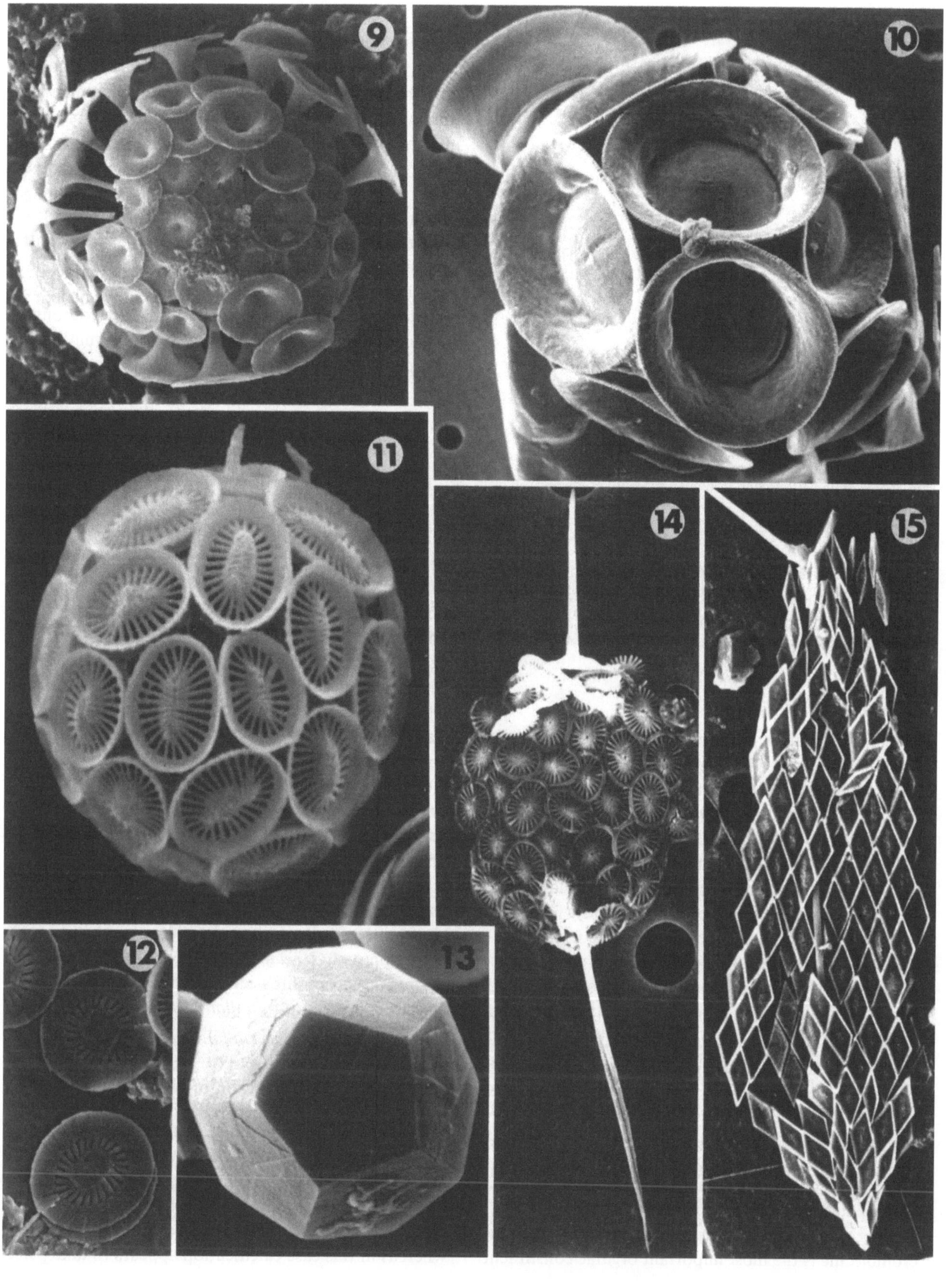
9
10
11
12
13
14
15

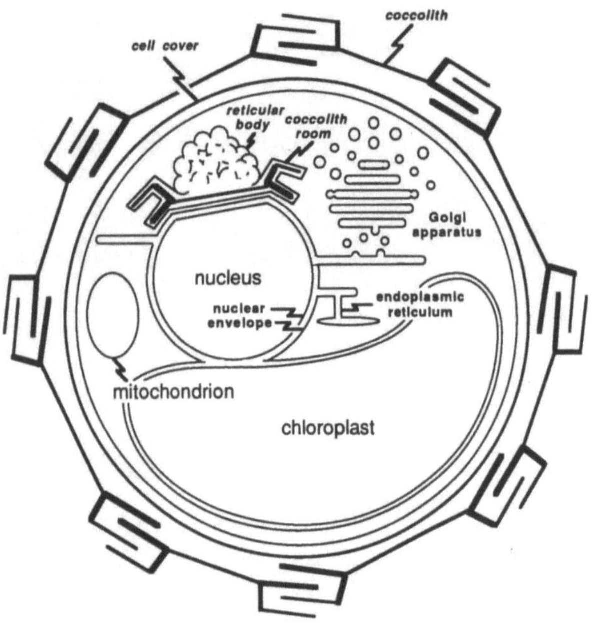

Fig. 16. Diagrammatic section through a coccolith producing cell of *Emiliania huxleyi*, × 13,000

cyrtoliths and an inner layer of oval incomplete caneoliths (elliptical discoid heterococcoliths with petal-shaped upper and lower rims and a central area filled with slatlike elements).

Coccolithogenesis

Coccolith formation has been studied in most detail in *Emiliania huxleyi* and *Pleurochrysis carterae* (Green 1986; de Vrind-de Jong et al. 1986; Westbroek et al. 1986, 1989; Green et al. 1990).

In *E. huxleyi* (Fig. 16), coccoliths are formed in a specialized organelle consisting of a coccolith vesicle (which surrounds the growing coccolith) and a joined system of anastomosing tubes, the reticular body. The coccolith vesicle is closely associated with the nuclear surface and appears to evolve by the fusion of small Golgi-derived vesicles, which contain polysaccharide and probably also calcium and carbonate. At the onset of coccolith formation a thin organic base-plate is formed in the coccolith vesicle and it appears that this plate serves as a substrate for calcite crystal nucleation. Calcification starts around the rim of the plate and from there crystals grow to form single-crystalline elements which constitute the completed coccolith. The formation of coccoliths is regulated, at least in part, by a complicated acidic polysaccharide containing Ca^{2+}-binding carboxyl and ester-sulfate groups. This macromolecule is the most important organic constituent of the coccoliths (Borman et al. 1987) and probably induces nucleation of calcite crystals on stereochemically analogous faces, and subsequent coccolith growth. When finally the coccolith has reached its ultimate shape, the coccolith vesicle is detached from the nucleus, the reticular body disappears and the coccolith is extruded to the periphery of the cell where it interlocks with other coccoliths to form the coccosphere. Production of coccoliths usually ceases when the cells have reached the stationary phase, but new coccolith production can be induced by removal of extracellular coccoliths (Linschooten et al. 1991). In nutrient-depleted media, cells can continue to produce coccoliths in the stationary phase resulting in the formation of multiple layers of coccoliths.

In *P. carterae*, there is no coccolith vesicle-reticular body system associated with the nucleus, but coccolithogenesis takes place inside cisternae of the Golgi apparatus (Westbroek et al. 1986). First the base-plate scale is formed, then the cisternae fuse with special vesicles containing granular particles (the so-called coccolithosomes). These vesicles, derived from the Golgi apparatus, pass through the cytoplasm before they fuse with the scale-containing cisternae. It seems that the coccolithosomes carry both calcium and matrix material to the sites of calcification and then they disintegrate. The further development of the coccoliths and their extrusion onto the cell surface resembles the situation in *E. huxleyi* (Westbroek et al. 1986, van der Wal et al. 1987).

Coccolithogenesis has also been studied in some detail in *Coccolithus pelagicus* (Parke and Adams 1960, Manton and Leedale 1969), which has a placolith-bearing heterococcolithophorid stage (Fig. 7), but also an alternate motile stage bearing holococcoliths (Figs. 1 and 2). This latter stage was originally described as a separate species, *Crystallolithus hyalinus* Gaarder and Markali (1957). Rowson et al. (1986) also showed that some clones of *C. pelagicus* can give rise to motile cells bearing holococcoliths identical to those of another described species, *Crystallolithus braarudii* Gaarder (1962).

Formation and calcification of heterococcoliths in *C. pelagicus* is an intracellular process. They are produced in a T-shaped vesicle, the part corresponding to the stem of the T being situated in the Golgi stack, and the remainder containing the growing coccolith, in an adjacent position perpendicular to the overall cisternal orientation. The holococcolith base-plates are also formed intracellularly in Golgi-associated cisternae, whereas in this case calcification appears to take place within an outer investment, after the base-plates have been discharged outside the cell (Green 1986, Rowson et al. 1986). Although calcification is an extracellular process in this type of coccolith, crystal arrangement is very regular (Fig. 2) and appears to be nevertheless biologically controlled.

Non-coccolith-associated calcification

Coccolithophorids may also produce calcified structures which are not directly associated with coccolith formation. Examples of this are the small irregularly shaped crystals, probably of calcium carbonate (Westbroek et al. 1986), that are produced within a vacuole in S-cells of *E. huxleyi* (for cell types in this species, see below). Calcareous cysts can also be formed by some coccolithophorids (Kamptner 1937). Brief mention should also be given to the coccoid cells of *Ochrosphaera* forming a thick stratified matrix, into which calcareous elements may be deposited (Parke 1970). These elements, resembling nannofossils of uncertain affinity included in the genus *Tetralithus*, seem to be produced by extracellular means as the result of modifications of the environment by biological activity. Similar structures are also formed by the non-coccolithophorid prymnesiophyte *Chrysotila lamellosa* Anand (see below and Green and Course 1983).

Coccolithophorid taxonomy

Much taxonomic confusion has occurred due to the assignment of different growth phases in the life cycles of some species to different taxa. These phases have often been given separate names and sometimes these have even been placed in different orders or families. As mentioned above, *Coccolithus pelagicus* (Fig. 7) is known to have a non-motile heterococcolithophorid (*C. pelagicus*) phase that can alternate with a motile holococcolithophorid (*Crystallolithus hyalinus*/*Crystallolithus braarudii*) phase (Figs. 1 and 2). In *Calcidiscus leptoporus* (Murray and Blackman) Loeblich and Tappan a non-motile heterococcolithophorid phase alternates with a motile holococcolithophorid phase which was originally described as *Crystallolithus rigidus* Gaarder (Kleijne 1991). Heteromorphic life cycles blurring the distinction between the hetero-/holococcolithophorid genera *Papposphaera*/*Turrisphaera*, *Pappomonas*/*Trigonaspis* and *Wigwamma*/*Calciarcus* have also been reported (Thomsen et al. 1991).
Heteromorphy also exists in *Emiliania huxleyi*, in which at least three different cell types exist: (1) coccolith-bearing cells (C-cells), (2) naked cells without conspicuous extracellular scales or coccoliths (N-cells), and (3) motile cells producing non-calcified organic scales (S-cells; see Klaveness 1972). Amoeboid cells also rarely occur. Each of the three main cell types is capable of indefinite vegetative reproduction and each is interconvertible with the other cell types. However, it is not known what processes trigger the different cell types. There is also some evidence that a species of *Ochrosphaera* may have an alternate form that bears *Pleurochrysis*-type coccoliths. Several species of *Pleurochrysis* are known to have a filamentous *Apistonema* stage, which can produce scale-bearing or naked swarmers or motile coccolith-bearing cells (Leadbeater 1971, Johansen et al. 1988, Green et al. 1990, Fresnel and Billard 1991). So far only a very small fraction of coccolithophorid species has been investigated in culture and it is thus conceivable that future life history studies will lead to a drastic reduction of existing coccolithophorid taxa.
Cultures are also useful tools for studying the intraspecific variation in coccolith morphology (e.g., Young and Westbroek 1991) and this is of taxonomic significance in cases where it is demonstrated that previously described taxa are merely polymorphic forms of a single taxon (e.g., Rowson et al. 1986). The puzzling phenomenon that coccoliths considered to be characteristic for distinctly different taxa (e.g., *Emiliania huxleyi*/*Gephyrocapsa oceanica* Kamptner or *Calcidiscus leptoporus*/*Coccolithus pelagicus*) may occasionally occur on the same cell (Clocchiatti 1971, Gard 1987) should also be further investigated by use of cultures.
Studies on coccolith formation and crystal assembly also appear to provide useful information for taxonomy (e.g., Young 1989, Young et al. 1992), but this cannot be clearly assessed at present due to the fact that so few species of coccolithophorids have been investigated in this respect. If one considers the extreme diversity of coccolith morphology (Chrétiennot-Dinet 1990) studies on coccolithogenesis in genera with unusual coccoliths (e.g., in dimorphic or dithecate genera or in *Braarudosphaera*, *Pappomonas*/*Papposphaera*, *Wigwamma*, etc.) could be of greatest value to our understanding of both coccolithogenesis and taxonomy.
Flagellar apparatus, internal cell ultrastructure and biochemical characteristics (e.g., lipids) have also been demonstrated to be important taxonomic criteria in coccolithophorids (Marlowe et al. 1984, Inouye and Pienaar 1988, Preisig 1989, Green et al. 1990, Roberts and Mills 1992). Although only few genera have been investigated in this respect, there is evidence that several

genera (e.g., *Emiliania, Pleurochrysis, Syracosphaera*) bear a closer resemblance to non-coccolithophorid members of the orders Isochrysidales and Prymnesiales than to Coccolithophorales (Coccosphaerales).

Given the above considerations, it is clear that coccolithophorid taxonomy cannot be based exclusively on coccolith morphology, though it is beyond doubt that coccoliths will always remain important characters for species identification. Information received from culture studies on cell cycles, variability, coccolithogenesis, flagellar apparatus and internal cell ultrastructure must certainly be taken into account in species definition and higher level classification. Furthermore, it is probable that other criteria, which have not yet been studied in any detail in most species of coccolithophorids (such as physiological, biochemical and molecular biological characteristics), will become increasingly important for coccolithophorid taxonomy in the future.

Other microalgae

Prymnesiophytes/prymnesiids (except coccolithophorids)

Calcification also occurs in some non-coccolithophorid prymnesiophytes, notably in *Chrysotila lamellosa* (Green and Course 1983, Green 1986), in which clusters of variably shaped calcareous bodies can be formed within a thick layer of mucilage surrounding the cells. The mechanism of calcification observed in this species is not known, but it is conceivable that the mucilage may well provide sites suitable for $CaCO_3$ crystal nucleation. Once crystallization has been established, it may continue by inorganic processes. The mucilage may also act as a diffusion barrier leading to a localized alkalinization due to the long diffusion path for CO_2, resulting in precipitation of calcium carbonate. Similar depositions within mucilage surrounding cells are also known from other algae and cyanobacteria (see below and Borowitzka 1989, Pentecost 1991 a). Calcified bodies of this sort thus cannot be taken as diagnostic of any particular microbial group and seem to be of limited taxonomic significance.

Chrysophytes/Chrysomonads

For this algal group siliceous structure are characteristic (Preisig 1994) but a few members producing calcareous structures are also known (Preisig 1986). There are observations on calcified structures associated with mucilage coatings (similar to those described above for *Chrysotila lamellosa*) in the chrysophyte genera *Celloniella, Chrysonebula, Sarcinochrysis,* and *Tetrasporopsis* (Lund 1953, Parke 1970, Tschermak-Woess 1980). Calcium is also sometimes deposited on extracellular stalks, loricae and pseudocysts. Stalks of the apochlorotic chrysophyte *Anthophysa vegetans* (O. F. Müller) Stein were shown to be composed primarily of spherules of calcium phosphate, but depending on the elements in the culture medium, these spherules may be replaced by granules of iron and manganese compounds (Lee and Kugrens 1989). Nothing is known about the mode of formation of these structures.

Diatoms

Calcification has rarely been reported from diatoms, but some species may contribute to stromatolite formation by depositing and accumulating calcite particles in extracellular gel, stalk or sheath material (Winsborough and Golubic 1987).

Tribophytes (xantophytes)

No calcification is known from flagellate or rhizopodial members of this algal class, but species of the siphonous genus *Botrydium* frequently develop a superficial coating of calcium carbonate on their cell walls (Ettl 1978).

Dinoflagellates

Calcareous structures also occur in some dinoflagellates, notably in the vegetative coccoid cells of *Thoracosphaera*, in which calcium carbonate is deposited between two shell membranes surrounding the cell (Tangen et al. 1982, Inouye and Pienaar 1983). The process of calcification has not been studied in detail, but the latter authors found $CaCO_3$-like intracellular crystals which appear to be produced in the Golgi apparatus and which probably play a role in extracellular calcification. In young cells, just commencing calcite deposition, crystals are irregularly grouped over the surface of the cell and these are probably nucleation sites from where calcium carbonate deposition proceeds, eventually resulting in a shell of uniform thickness. In cyst walls of a number of peridinioid (calciodinellid) taxa an outer layer consisting of numerous calcitic, rod-like crystals has also been reported (Montresor et al. 1993). Calcareous dinoflagellate cysts are also known from the fossil record (Keupp 1991).

Chlorophytes

Calcification commonly occurs in the macrophytic chlorophytes belonging to the Dasycladales and Caulerpales, as well as in charophytes (which are often included in the chlorophytes). For some filamentous and crustose chlorophytes growing epilithic in freshwater (e.g., *Cladophora, Gongrosira*) calcification is also well-known (for references, see Pentecost 1991 a).

Considerable calcareous deposits may also be formed by the colonial desmid *Oocardium* (Pentecost 1991 b). Intracellular calcium carbonate deposition is rare in chlorophytes, but unique cross-shaped aragonite crystals may occur within the cell in *Spirogyra* (Mann et al. 1987).

In unicellular coccoid and flagellated chlorophytes, calcification has only rarely been reported. In the coccoid genus *Gloeotaenium*, characteristic band-shaped deposits of calcium carbonate are formed in the mucilage surrounding the single cells or between groups of 2–4 cells. Deposition takes place along the lines of division of the cells, being transverse in two-celled but more or less cruciately arranged in four-celled colonies. The deposits are mainly calcium carbonate, but strontium, magnesium and barium, probably also in the form of carbonates, may also be present (Devi Prasad and Chowdary 1982).

Calcification in flagellated chlorophytes (Volvocida) is mainly known from genera such as *Phacotus* which possess loricae (Müller and Oti 1981, Krienitz et al. 1993). In this case, the lorica appears to be initially organic but calcite crystals may be gradually incorporated and cemented together to form a thick, perforated armor (Giering et al. 1990). Another type of calcification is found in *Chlamydomonas*, where calcification may occasionally occur in mucilage surrounding the cells (Green 1986) or in cell walls (Lewin 1990), but in this latter case unusual calcium phosphate mineralization has been reported. Calcium phosphate is a mineral characteristic of vertebrate bone, but almost unknown among plants.

Another unusual type of mineralization is known from a pedinophyte, *Pedinomonas tuberculata*, in which the cell membrane is covered by deposits of cellulose, calcite and quartz (Manton and Parke 1960).

Foraminifera

There are over 34,000 identified species of foraminifera, with greater than 12% of these being alive today (Lee et al. 1985). Recent foraminifera are ubiquitous and cosmopolitan, although most species are tropical or subtropical. As a result of their abundance and distribution, the foraminifera are important constituents of most marine environments (Lee et al. 1985).

In the last quarter century, research on the biology of the foraminifera has increased, with many of the investigations being conducted on the planktonic and the larger foraminifera (for reviews, see Hemleben et al. 1989, Lee and Anderson 1991). The planktonic foraminifera are classified together in one superfamily, with 3 families and 37 species (Bé 1967, 1977; Loeblich and Tappan 1988), whereas "larger" foraminifera has no taxonomic meaning. The term is descriptive of the relatively large size which certain species attain, although these species seem to share common ecological and physiological features (Ross 1974). Modern species of larger foraminifera belong to two different suborders and 8 different families (Ross 1974).

There are three types of tests which foraminifera can produce: (1) agglutinated tests, (2) tests composed of calcite needles, and (3) calcium carbonate secreted tests with "glassy" appearance. Members of the suborder Textularia produce agglutinated tests (Fig. 17) and those of suborder Carterinina produce tests composed of calcite needles or spicules. For example, species of *Rhumberinella* possess a sheet of uncemented small calcite spicules (Loeblich and Tappan 1964). In another species with calcareous spicules, *Zaninettia manaarensis* (Figs. 18 and 19), recent ultrastructural work suggests that the spicules are secreted in its protoplasm, then positioned in the test (Lee 1993). The majority of foraminifera produce calcium carbonate shells (tests). In recent foraminifera, the shells are usually calcite, although those species in the family Robertinacea produce aragonite shells (Lynts and Pfister 1967). The 12 suborders of foraminifera are grouped by the composition and structure of the foraminiferal test, including crystalline arrangement and lamellarity (Loeblich and Tappan 1988, Hallock et al. 1991). Foraminifera in the suborders that produce calcified shells are further classified by the characteristics of these shells: chamber shape and arrangement, the coiling mode, the aperture, and other accessory and internal structures (Loeblich and Tappan 1988).

Observations on the growth of foraminifera in laboratory cultures and in situ have provided insights into the functional morphology of these protists (for review, see Hallock et al. 1991), and into the possible mechanisms of calcification used by the foraminifera (for review, see Kuile 1991). Much evidence exists (and has been extensively reviewed) that the shells of those foraminifera which possess algal endosymbionts seem well adapted for endosymbioses (Haynes 1965, Ross 1974, Lee 1983, Lee and McEnery 1983, Hallock 1985, Lee and Hallock 1987). Some morphological characteristics, which seem to reinforce this adaptation to endosymbioses, include overall final size, test thickness, and complex internal morphologies (Hallock et al. 1991), which may reflect a concession between hydrodynamic factors and endosymbiotic requirements (Haynes 1965).

While individual species of foraminifera exhibit a wide variability in their morphology (Thompson 1961), which is partially based on changes in environment (Boltovskoy and Wright 1976, Boltovskoy et al. 1991), several key morphological features are relatively constant within a taxonomic grouping. Cushman (1969), in his classic text on the foraminifera, describes the major morphological characteristics of those foraminifera which produce calcite shells, and the possible variations. These included the chamber arrangement and test shape (flabelliform, stellate, rectilinear, biserial, triserial, arborescent or fistulose), the sutures, the chamber wall (including spines and retal processes), ornamentation (raised bosses, peripheral keels, acicular spines, tuberculate marginal spines, and test textures which are reticulate, cancellated, costate, punctate or hispid), and apertures (simple, dendritic, dentate) (Cushman 1969).

The introduction of scanning and transmission electron microscopy allowed scientists to observe morphological features better, which elucidated minute details of shell microstructure, crystal types and growth patterns (Bé et al. 1969). This review will discuss several of these morphological features which are utilized as criteria in the taxonomy of foraminifera including ornamentation, pores, canal systems and spines. These structures may be better characteristics for distinguishing species than other gross morphological features such as chamber arrangement.

Ornamentation

Ornamentation can be a consistent character utilized to separate species. For example, *Calcarina gaudichaudii* (Fig. 20) has a smooth test surface with smooth wart-like elevations, whereas *C. spengleri* (Figs. 21–23) has a rough surface with pointed warts (Röttger et al. 1990). Ornamentation on the chamber wall of species of *Peneroplis*, a genus considered extremely variable in its morphology (Thompson 1961), can be used to distinguish between the different species: *P. pertusus* (Figs. 24–26) has elongated grooves with multiple rows of fused pits; *P. planatus* (Figs. 27 and 28) has slightly depressed furrows with single to triple rows of pits; *P. acicularis* (Figs. 29–31) has evenly dispersed pits in single rows; *P. antillarum* has evenly dispersed pits in single, double or triple rows; and *P. arietinus* (Fig. 32) has deep grooves covered with multiple rows of pits (Gudmundsson 1990). These pit structures may enhance the exchange of gases between the ambient seawater and the foraminiferal cytoplasm, through the shell wall (Reiss and Hottinger 1985). Recently, Pawlowski and Lee (1991) utilized shell microstructure to separate five genera of small Rotaliinas (*Schaferina, Rotaliella, Glabratellina, Microglabratella,* and *Selenita*), based on the surface texture, perforations and ornamentation of the chamber walls.

In *Sorites orbiculus* and *Marginopora kudakajimensis*, calcareous ridges are seen in some specimens. Possibly the ridges add strength to the test (Kloos 1981). Species of the tiny foraminiferal genus *Rotaliella* possess radial grooves on the umbilical and/or spiral side (Pawlowski and Lee 1991).

Some foraminifera exhibit sensitivity to the ecological conditions under which they live. Such sensitivity can alter the development of ornamentation, such as in *Amphistegina* (Todd 1976). If salinity decreases, the degree of ornamentation may also decrease (Boltovskoy and Wright 1976).

Pores

Pore size, shape, density (porosity) and arrangement are relatively constant within a species (Hofker 1951 a), and should be considered as important in the taxonomic considerations of the foraminifera (Bé et al. 1969). Some species have no pores (the so-called imperforate foraminifera) and those that do may not have them

Fig. 17. *Textularia* sp. (SEM). Test composed of agglutinated foreign material, × 5,000

Figs. 18 and **19.** *Zaninettia manaarensis* (SEMs). Fig. 18, close-up of calcareous spicules, × 1,000. Fig. 19, spiral view, × 446

Fig. 20. *Calcarina gaudichaudii* (SEM). Spiral view showing smooth, wart-like elevations, × 33

Figs. 21–23. *Calcarina spengleri* (SEMs). Fig. 21, spiral view showing rough pointed elevations, × 40. Fig. 22, pore opening, × 1,250. Fig. 23, close-up of pointed elevations, × 925

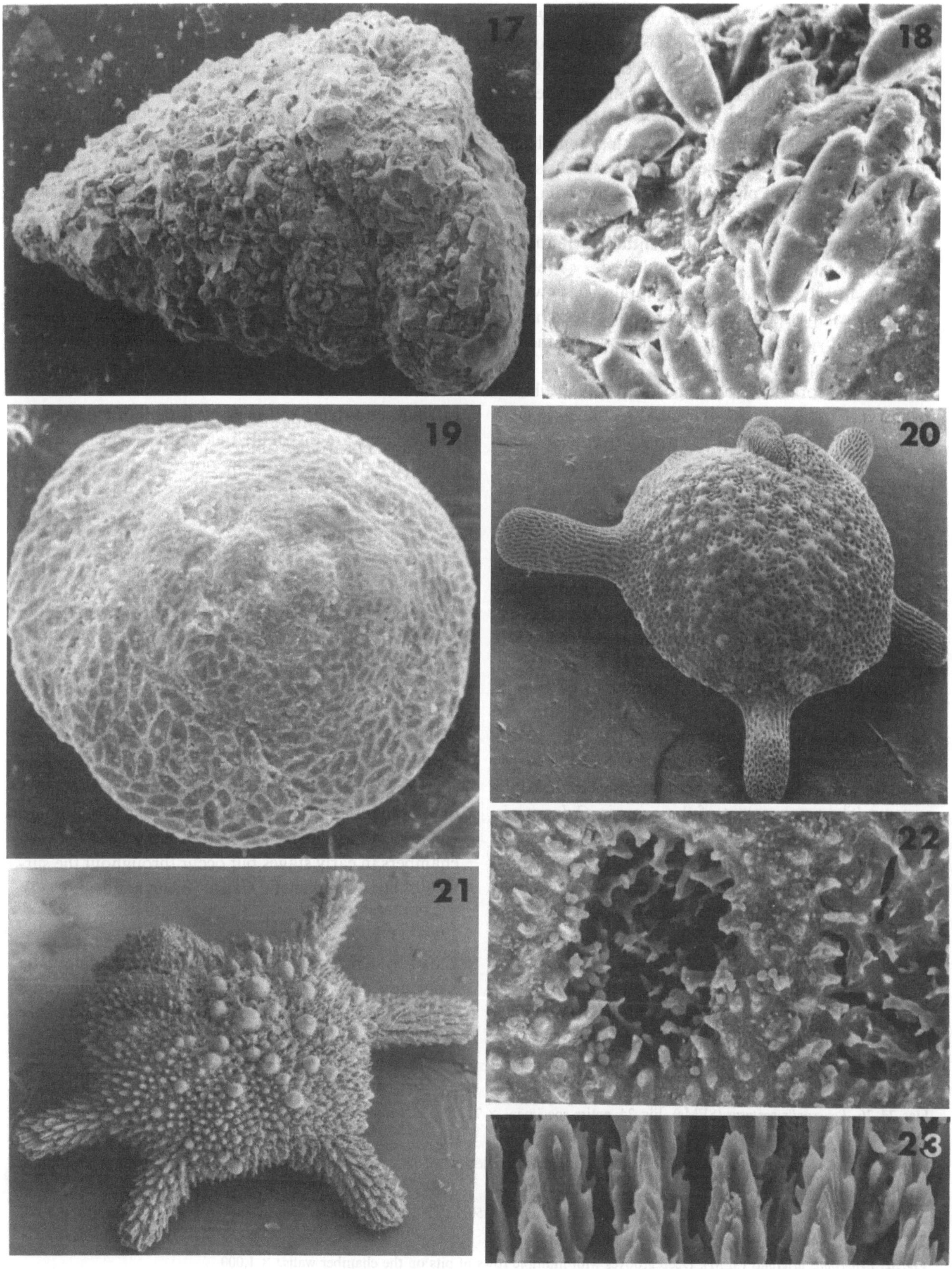
17
18
19
20
21
22
23

throughout the entire test. In *Microglabratella tabaensis*, only the spiral side has coarse pores, whereas the umbilical side is imperforate (Pawlowski and Lee 1991). In planktonic foraminifera, the pore walls are lined by organic membranes which form pore plates. In non-spinose species, these pore plates are almost completely sealed, whereas in the spinose species, numerous small micropores within the plate allow fine rhizopodia, and materials to pass through (Hemleben et al. 1989). Pores in a chamber wall of perforate benthic foraminifera function as passageways for pseudopodia (Loeblich and Tappan 1964). Hence, in those foraminifera where cytoplasm can pass through the pores, sites of exchange between the cytoplasm of the foraminifera and the environment occur (Bé et al. 1980).

In planktonic foraminifera, the pores are either tubular (cylindrical or barrel-shaped) in cross-section as in *Globorotalia* or funnel-shaped as in *Globigerinoides* and *Globoquadrina* (Fig. 42) (Bé 1968, Hemleben and Spindler 1983). Test porosity is used as an indicator of surface water temperature since it varies with latitude (Frerichs et al. 1972, Bé et al. 1973). Tropical and subtropical species build high porosity tests, and polar and subpolar species construct low-porosity tests (Bé 1968). *Orbulina universa* has pores of two different size classes: large pores of 8 to 14 µm and more numerous small pores of 3 to 9 µm (Bé et al. 1976). The differences in test porosity with latitudinal variation can be seen in the density of the small pores of *O. universa*, whereas the large pores are relatively uniform (Bé et al. 1976).

In those species with canal systems, such as the Nummulitidae, pores usually represent openings to the canal systems. The arrangement of the pores in *Calcarina* is another characteristic utilized to distinguish species: *C. gaudichaudii* has irregularly distributed canal openings always located between two neighboring elevations, and *C. spengleri* (Fig. 22) has one series of canal openings between two neighboring elevations (Röttger et al. 1990).

In the Elphidiidae, the pores are small septal pits (fossettes) with tuberculate borders (tooth-lined) and connect to a complex canal system (Figs. 33–36) (Lee and Hallock 1987, Lee et al. 1988, Lee and Lee 1990). This anatomy may be an adaptation for the elphidids to husband chloroplasts (Lopez 1979, Lee and Hallock 1987, Lee et al. 1988, Lee and Lee 1990). In benthic foraminifera, digestion is often extracameral, a trait which may be basic to many foraminifera, and may predispose the foraminifera to endosymbioses (Koestler et al. 1985, Lee et al. 1991 b). In elphidids, pseudopodia emerge through the pores. The external digestion of diatoms, coupled with the small size of the pore opening and the sieve-like arrangement of tuberculate borders, may permit the chloroplasts from the diatoms to enter the foraminiferal cytoplasm by endocytosis whereas the diatom frustule is left outside of the foraminiferal cell (Lee et al. 1988).

In some endosymbiont-bearing foraminifera, the endosymbionts are distributed along the cortical surface of the chambers beneath the pores. In the Nummulitidae, the endosymbionts usually lie below the lateral wall of the chambers against the pore depressions (Hottinger and Dreher 1974). Arranged along the shell wall in *Amphistegina lobifera* are fine pores (Figs. 39 and 40), which are associated with short canals (Fig. 37). The endosymbionts lie at the end of these canals in the small cup-like depressions in the shell wall (Fig. 40).

Pores often vary with the ontogeny of the foraminifera. In *Bolivina* cf. *doniezi*, juvenile stages have pores only near the basal margins of the chambers (Pawlowski and Lee 1991). However, in the adult, the pores are located throughout the surface of the last chambers (Pawlowski and Lee 1991). A similar pattern of pore development is observed in planktonic foraminifera where pores are restricted to the suture area between chambers in early stages but occur throughout the surface in later ontogenetic stages (Brummer et al. 1987).

Canal systems

Canal systems permit the extrusion of pseudopodia from any point of the marginal cord through canal openings (pores) on the surface of the test, and hence

---

Figs. 24–26. *Peneroplis pertusus* (SEMs). Fig. 24, entire specimen, × 71. Fig. 25, ornamentation of elongated grooves with multiple rows of fused pits, × 1,000. Fig. 26, aperture, × 200

Figs. 27 and **28.** *Peneroplis planatus* (SEMs). Fig. 27, single to triple row of pits on chamber walls, × 500. Fig. 28, multiple aperture on planispiral shell, × 125

Figs. 29–31. *Peneroplis acicularis* (SEMs). Fig. 29, entire specimen, × 50. Fig. 30, evenly dispersed pits in single file on chamber walls, × 1,750. Fig. 31, aperture, × 190

Fig. 32. *Peneroplis arietinus* (SEM). Deep grooves with multiple rows of pits on the chamber walls, × 1,000

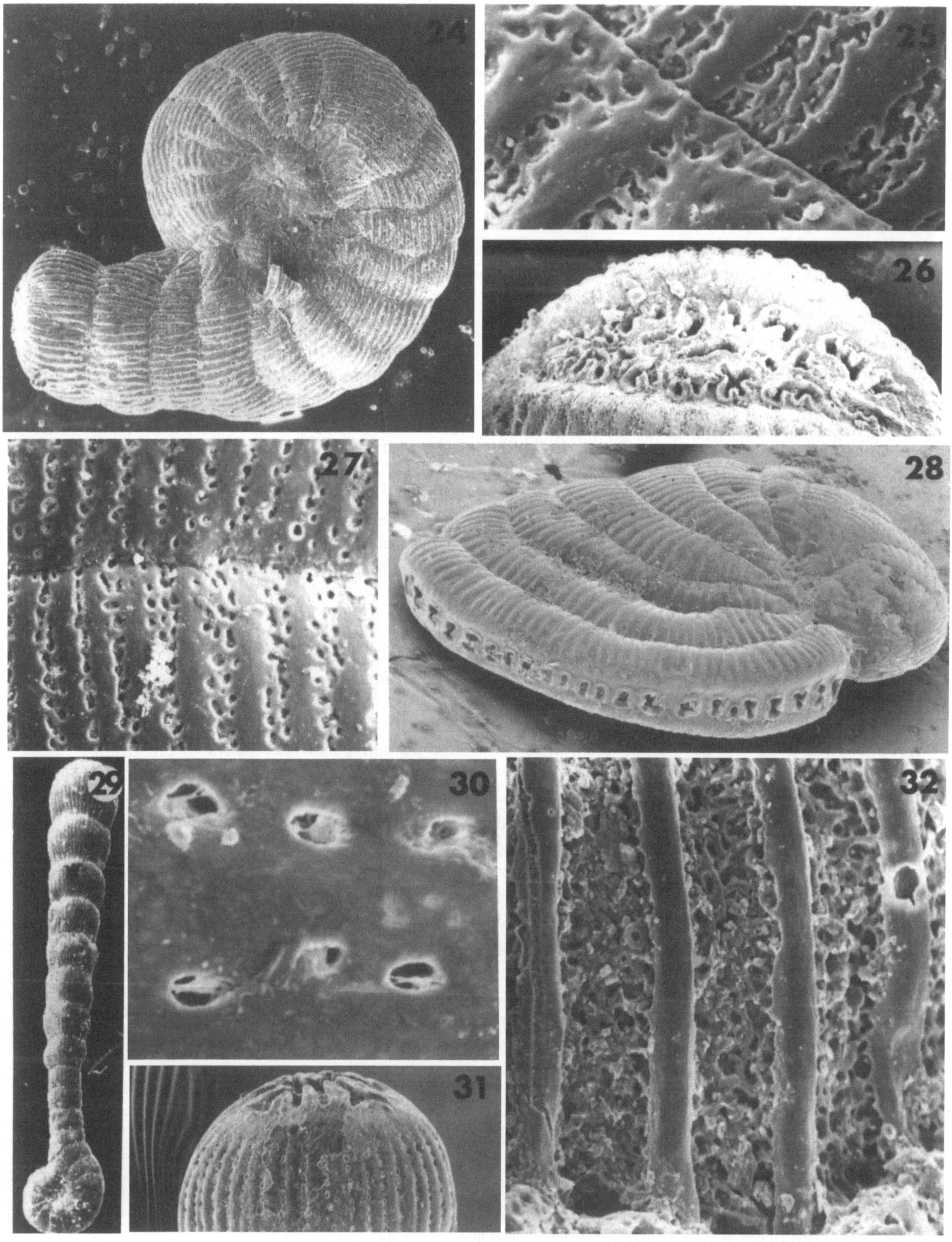
24
25
26
27
28
29
30
31
32

are important in all the principle biological functions of the foraminiferal cytoplasm such as locomotion, growth, excretion, metabolic exchange, substrate attachment, reproduction and protection (Röttger et al. 1984). The families Nummulitidae, Calcarinidae and Elphidiidae are characterized by canal systems (Lee et al. 1988, Röttger and Krüger 1990). In *Operculina ammonoides*, the canal system allows communication between individual chambers and the surfaces of the test (Röttger and Krüger 1990). There are two major types of canals: marginal canals and septal canals (Fig. 35) (Spindler 1978). Marginal canals lie within the keel along the test, and on top of each other, or side by side, where they branch and rejoin (Glaessner 1947, Smout 1954, Loeblich and Tappan 1964, Hofker 1971, Hottinger and Dreher 1974, Hottinger 1978, Spindler 1978). Septal canals are the old marginal canals along a keel which has become part of the septum dividing one chamber from another (Loeblich and Tappan 1964, Lutze et al. 1971, Spindler 1978). Septal canals connect marginal canals from different whorls, and secondary septal canals connect chambers and chamberlets (Spindler 1978). In *Calcarina* the marginal canals open at spine ends (Fig. 38) (Röttger and Krüger 1990).

Spines

The presence or absence of spines is utilized as a major criterion distinguishing the families of planktonic foraminifera (Bé 1967). Spineless forms in the sediments can also be used to distinguish specimens which have reproduced and those that have simply died, since the spinose planktonic foraminifera shed their spines leaving characteristic spine bases during the process of gametogenesis (Bé et al. 1983). These spines are elongated calcitic crystals, which usually lack internal organic matrices, and are implanted in the chamber wall, supported by a calcite collar (Hemleben 1975, Hemleben et al. 1989, 1991). The spines are flexible (Fig. 43) and are either circular (Fig. 44), triangular or tri-radiate (Fig. 45) in cross-section, and might change from circular to triangular, or triangular to tri-radiate from the base of the spine to the tip (Saito and Bé 1976; Hemleben et al. 1989, 1991).

The spines of the Calcarinidae are extensions of the marginal canals (Fig. 38) (Röttger and Krüger 1990). There is great morphological variation in the size and number of these spines within the various species. However, agamonts (microspheric forms) have more numerous, broader spines which branch at the ends as compared to the megalospheric forms (gamonts and schizonts) (Hofker 1970, Röttger et al. 1990).

Chamber formation and calcification

Many researchers have studied chamber formation in foraminifera in the laboratory and in the natural environment (for example, Jepps 1942; Angell 1967, 1979, 1980; Sliter 1970, Spindler and Röttger 1973; Berthold 1976; Röttger and Spindler 1976; Hemleben et al. 1977; Bé et al. 1979). These studies have demonstrated that chamber formation involves a series of events, beginning with the emergence of a cytoplasmic mass (Bé et al. 1979). In those species which form sediment fringes or food mats near their apertures (such as *Sorites orbiculus* and *Peneroplis planatus*) the fringe or mat is pushed outward away from the living foraminifer (Kloos 1981). The rhizopoda of the cytoplasmic bulge forms a primary organic membrane (POM), followed by the formation of an outer protective envelope (Fig. 47). The POM is composed of protein, which is the so-called "active matrix", and a "passive matrix" which is made up of polysaccharides (Towe and Cifelli 1967, Bé and Hemleben 1970). The passive matrix is believed to provide form to the shape of the chambers and the active matrix is the site of calcification (Towe and Cifelli 1967, Hemleben et la. 1989). In *S. orbiculus*, the septula appear after the POM (Kloos 1981). Small calcareous plaques appear over the surface of the POM, and begin to increase in number and size (Bé et al. 1979). Finally, these plaques coalesce to form the chamber wall (Bé et al. 1979).

In transmission electron microscopy, a dense electronic opaque layer is seen within the chamber wall (Hemleben et al. 1977). This layer, which is called the inner organic layer (IOL) (Fig. 48), was first thought to be the site of calcification (Hemleben et al. 1977), but is currently believed to be formed after calcification, as

Figs. 33–35. *Elphidium crispum* (SEMs). Fig. 33, entire specimen, × 90. Fig. 34, fossettes with tuberculate borders, × 380. Fig. 35, internal cast showing canal, chamber and pore arrangement, × 350

Fig. 36. *Elphidium translucens* (SEM). Denticle-lined fossae seen along the septae of the chambers, × 230

Fig. 37. SEM of an internal cast of *Amphistegina lobifera*, showing the pores and associated canals, × 540

Fig. 38. SEM of internal cast of *Calcarina* with the marginal canals in a spine, × 125

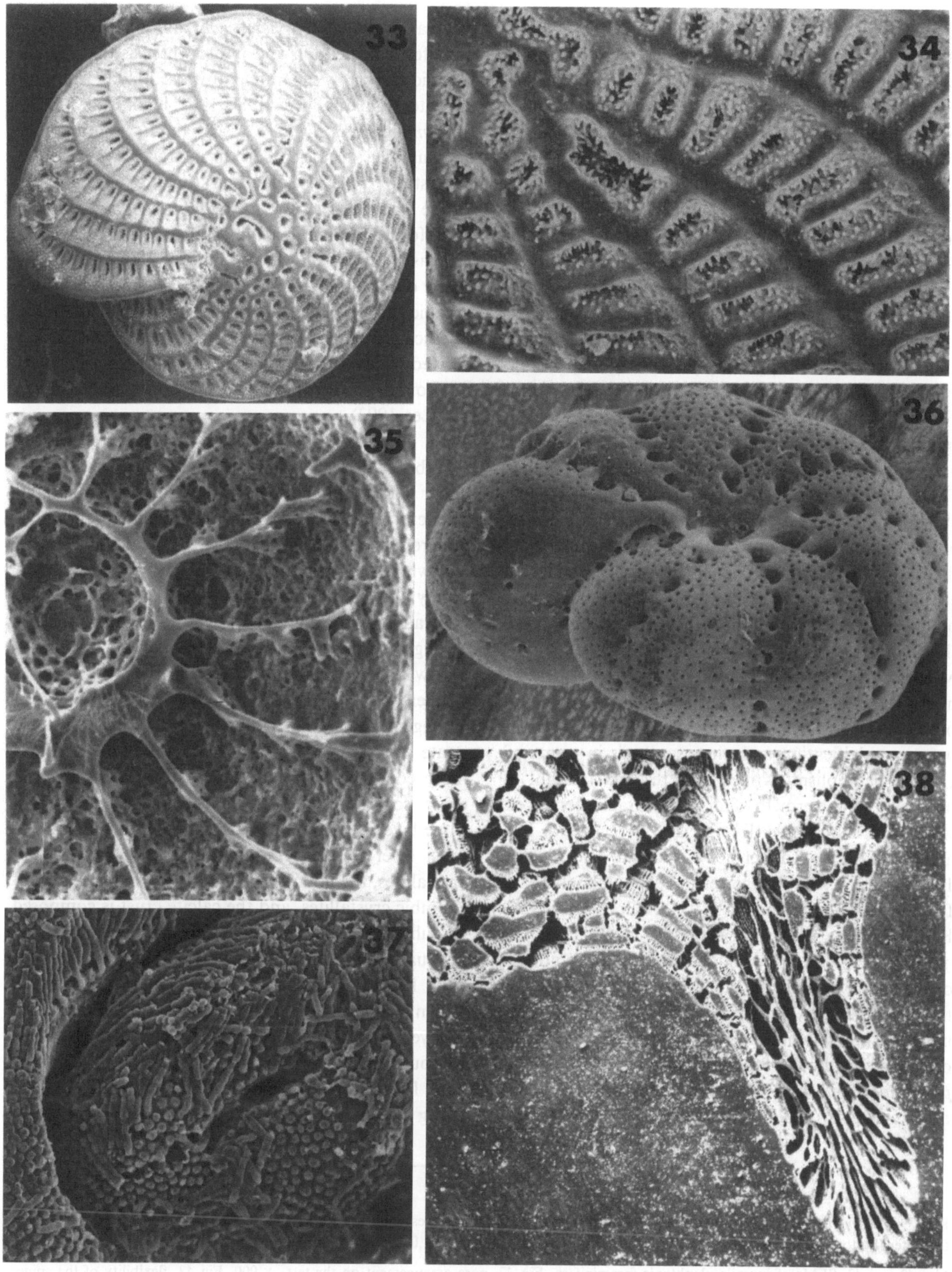
33
34
35
36
37
38

a protective lining (Hemleben et al. 1977, Bé et al. 1979). The mechanisms used for calcification seem different for perforate and imperforate foraminifera (for review, see Kuile 1991). Perforate foraminifera, which include the planktonic species, possess internal inorganic pools of calcium (Anderson and Faber 1984, Kuile and Erez 1988, Kuile et al. 1989 a), and calcium carbonate, which is derived from these pools, is deposited directly on the organic matrix, (POM) (Hemleben et al. 1977, 1986, 1989; Anderson and Bé 1978; Bé et al. 1979; Röttger et al. 1984). Calcification seems to be rate limited by an unidentified enzymatic step (Kuile 1991). The rate of calcification using ^{45}Ca as a tracer has been estimated to be 53 to 87 × 10^{-2} μg calcite/chamber/h depending on the size for *Globigerinoides sacculifer* (Anderson and Faber 1984).

In imperforate foraminifera, calcification depends less on energy and is diffusion limited (Kuile et al. 1989 a). These foraminifera possess no internal pools (Angell 1979; Kuile and Erez 1987, 1988). In vesicles, needles of calcium carbonate are formed and then moved to the sites of calcification (Angell 1979, 1980; Hemleben et al. 1986). These needles may form in the vesicles when other ions, which inhibit calcification, are removed (Kuile 1991).

Much of this research focused on species of foraminifera which possessed endosymbiotic algae. A major problem exists in trying to understand the relationship between calcification and photosynthesis in terms of carbon metabolism (Kuile and Erez 1987). Light is necessary for the growth of all larger foraminifera even when food is available (Röttger and Berger 1972, Lee and Bock 1976, Röttger 1976, Duguay and Taylor 1978, Muller 1978, Röttger et al. 1980, Hallock 1981, Duguay 1983, Kuile and Erez 1984, Hallock et al. 1986, Faber and Lee 1991, Lee et al. 1991 c). This has led many investigators to conclude that the photosynthesis of the endosymbionts of the larger foraminifera is intimately linked to the calcification of the host foraminifera. Possibly light has some other biogenic effect on the calcification of foraminifera (Erez 1978). In *Amphistegina lobifera*, a perforate foraminifer, calcification competes with photosynthesis for inorganic carbon whereas in *Amphisorus hemprichii*, an imperforate foraminifer, calcification and photosynthesis take up inorganic carbon by separate pathways directly from the seawater (Kuile and Erez 1987, 1991; Kuile et al. 1989 a, b). *Amphistegina lobifera* also exhibits considerable dark uptake of labeled carbon in tracer experiments, suggesting recycling of respiratory carbon dioxide (Kuile and Erez 1987). *Peneroplis planatus*, another imperforate foraminifer, may utilize a similar mechanism as *A. hemprichii* (Faber and Lee 1992). However, based on ^{14}C tracer experiments, the amounts of inorganic carbon taken up directly from the medium for calcification may need to be supplemented by carbon derived from translocated photosynthates, ingested food or recycled from respired carbon dioxide (Faber and Lee 1992).

Towe and Cifelli (1967) discussed the three possible chamber wall types which are based on observations under polarized light microscopy: (1) radial where the calcite crystals are elongate and regularly arranged, preferentially on the hexagonal face of calcite; (2) granular where the crystals are irregularly shaped and randomly arranged, preferentially on the rhombohedron face; and (3) porcelaneous where the crystals are disarranged. In planktonic foraminifera, which have granular walls, epitaxy seems to best describe calcification (Anderson and Bé 1978). In this theory the crystal arrangement is dictated by the structure of the substrate. The situation of chamber wall type becomes more complicated as our means of analysis become more sophisticated. In *Globorotalia truncatinoides*, three layers are apparent in cross section. The innermost layer is microgranular, consisting of anhedral crystallites, the intermediate layer is composed of subrhombic crystallites, and the calcite crust is composed of euhedral crystallites which are the product of the gradual enlargement of the subrhombic crystallites (Bé et al. 1975).

When forming a new chamber, some foraminifera form a layer over the exterior of the previous chambers (Smout 1954; Reiss 1957, 1958, 1963; Orr 1967; Hansen et al. 1969). In planktonic foraminifera, this shell thickening is seen as either an adaptive response to a deeper

---

Figs. 39–41. *Amphistegina lobifera* (SEMs) Fig. 39, fine pores seen along entires shell, × 100. Fig. 40, broken test showing the pores along the outer (bottom of picture) and inner (top of picture) sides of the shell, × 1,000. Fig. 41, aperture, × 175

Figs. 42 and **43.** SEMs of juvenile planktonic foraminifera. Fig. 42, pore arrangement on the test, × 900. Fig. 43, flexibility of the spines, × 1,000

Fig. 44. SEM of spines with ovoid to round cross sections found in some planktonic foraminifera, × 1,800

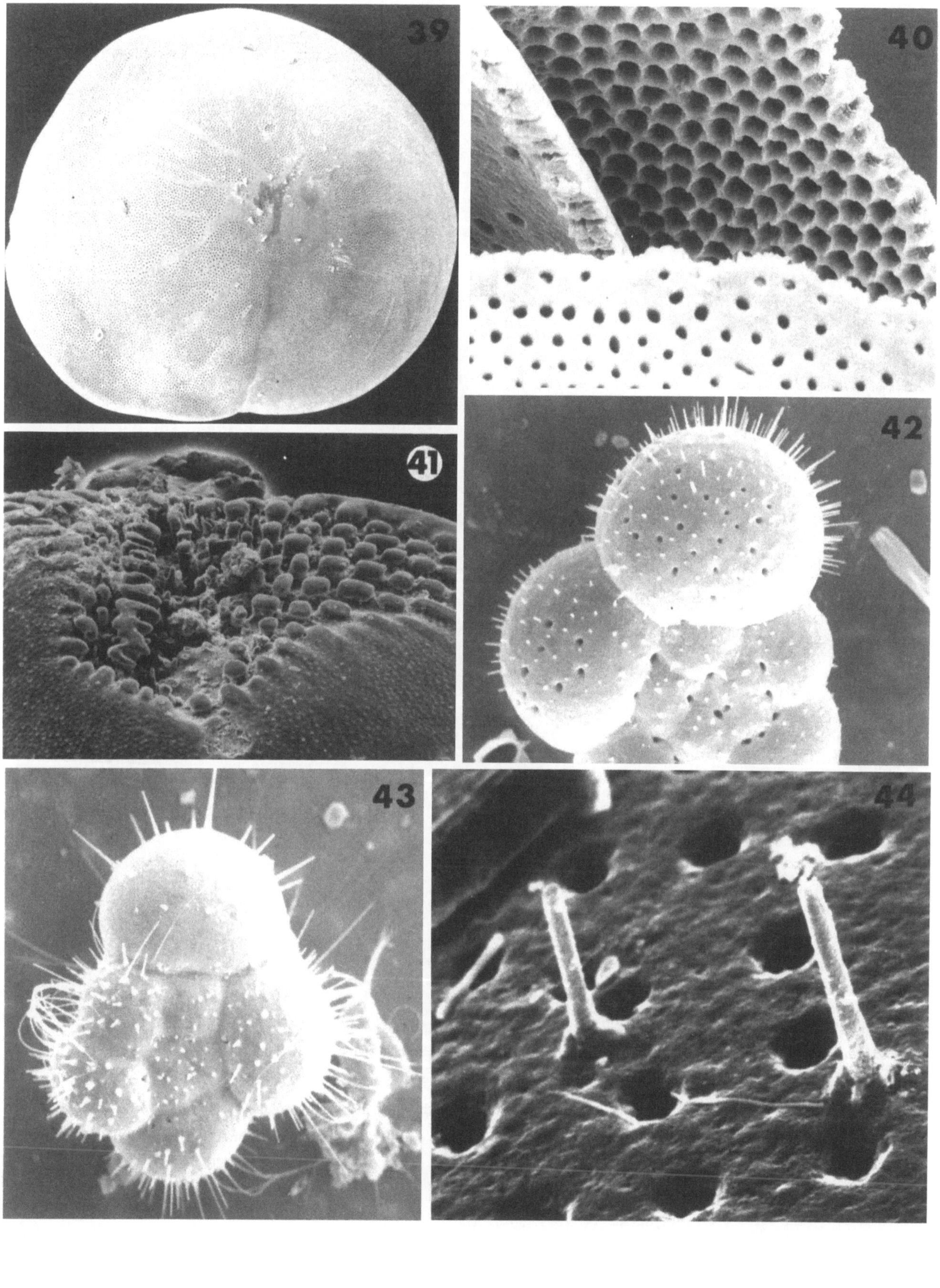
39
40
41
42
43
44

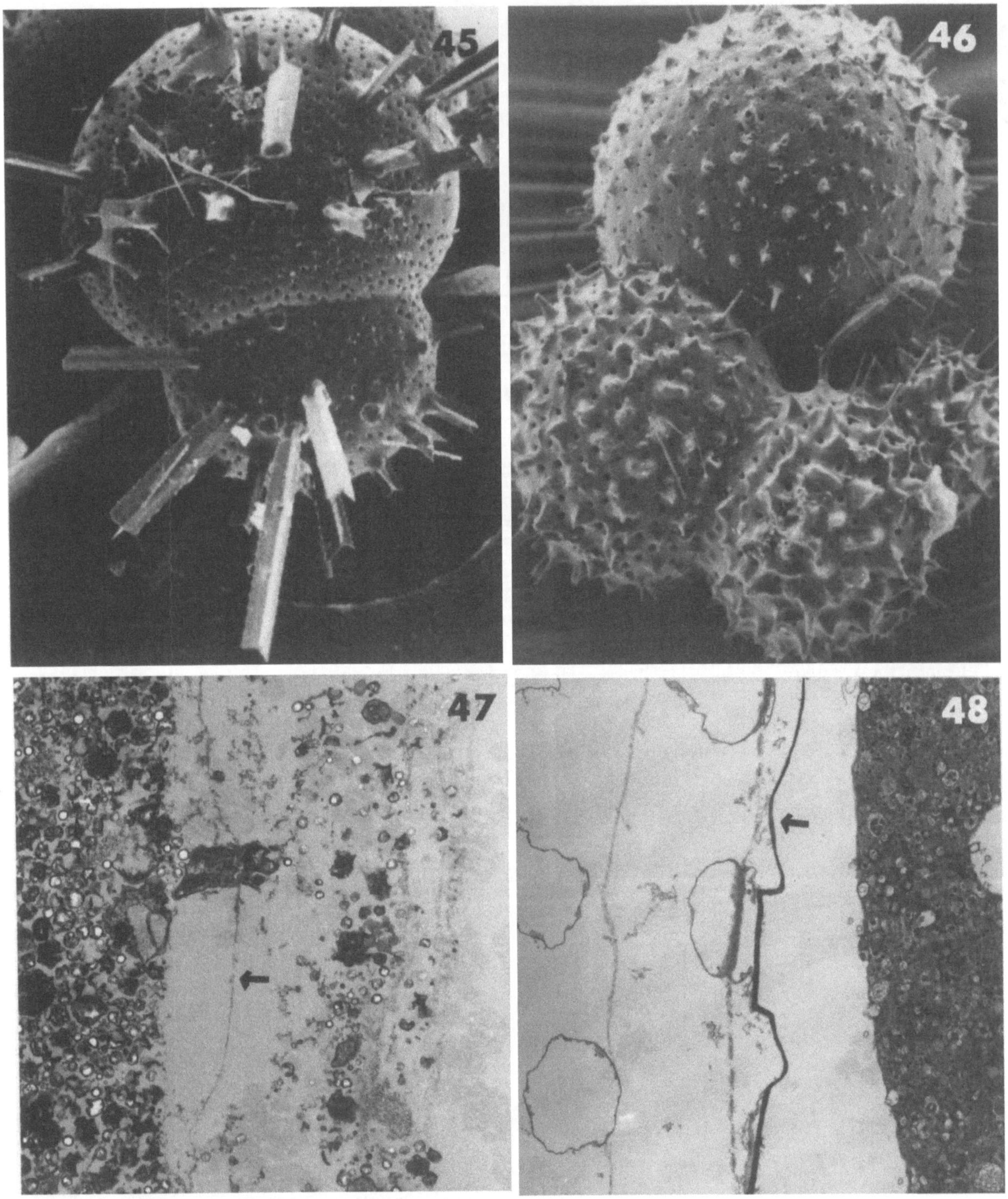

Fig. 45. *Hastergerina pelagica* (SEM). Early whorl chambers with triradiate spines visible, × 180

Fig. 46. *Globigerinoides sacculifer* (SEM). Spinose planktonic foraminifera, × 150

Fig. 47. TEM of early stage of calcification in *Globorotalia truncatinoides*, showing the primary organic membrane (arrow), × 8,500

Fig. 48. TEM of mature wall in *Globorotalia menardii* showing the primary and the inner organic membranes (arrow), × 10,000

habitat, such as with *Globorotalia menardii* (Bé 1960, Bé and Ericson 1963), or as a late ontogenic calcite crust deposited prior to gametogenesis, as in *Globigerinoides sacculifer* (Bé and Hemleben 1970, Bé 1980, Bé et al. 1983).

Kummerform and aberrant chambers

A foraminifer grows by the addition of chambers. With favorable growth conditions, each new chamber is larger in volume than the previous chamber. At times, a kummerform chamber forms which is diminutive. These kummerform chambers are relatively common among the planktonic foraminifera, and may be indicative of some environmental stress (Berger 1969). In *Globigerinoides ruber*, a kummerform chamber records colder isotopic temperatures than normal forms (Hecht and Savin 1972). Possibly, the occurrence of kummerform chambers is a response to a growth limit as has been suggested for *Globigerina pachyderma* (Olsson 1973).

Abnormalities and aberrant growth may be caused by rapid changes in various environmental parameters such as temperature, salinity, pH, or light (Boltovskoy et al. 1991). For example, aberrant chamber growth may be a response to a change in the quality of the light reaching a foraminifer (Hallock 1979) or to food shortage (Murray 1963). Also, damage to the shell of a foraminifer may lead to unusual chamber growth and irregularly shaped chambers (Rhumbler 1911, Hofker 1951 b, Bé and Spero 1981).

In laboratory cultures of *Peneroplis planatus*, aberrant growth was common even though temperature, salinity, pH and light were kept constant, and no physical damage to the shells of the foraminifera occurred (Faber and Lee 1991). Foraminiferal growth involves the episodic expansion of pseudopodia and chamber formation as previously described. Each chamber is formed separately from the rest of the organism (Thompson 1961). Obstructions, which are on the substrates where the foraminifer is living, have been shown to create abnormalities in the chamber shape of several foraminifera (Arnold 1954, Kloos 1981). The pseudopods simply extend around the obstruction, forming an irregular chamber. Hence the pseudopodial mass may be the organizer of the growth axis and responsible for the shape of the subsequent chamber which is formed.

Foraminiferal taxonomy

Foraminifera are divided into 12 suborders, 65 superfamilies, 296 families, 300 subfamilies, and 2446 genera (Loeblich and Tappan 1988, Haynes 1990). The structure of the test is the most basic distinction among the suborders (Lee et al. 1985). Carpenter et al. (1862) separated the shelled foraminifera into two suborders the Perforata (with pores) and the Imperforata (without pores). The more recent separations of the suborders include the crystallography and lamellarity (Towe and Cifelli 1967; Loeblich and Tappan 1964, 1982, 1988) (Table 2).

Table 2. The suborders of foraminifera, based on Loeblich and Tappan (1988)

Suborder	Wall structure and material	Crystallography	Lamellarity
Allogromiina	membranous, organic wall	–	–
Textulariina	agglutinated	–	–
Fusulinina	calcitic	microgranular	–
Involutinina	aragonitic, non-septate	–	–
Spirillinina	calcitic, non-septate	single crystal	–
Carterinina	spicular, calcite crystals	–	–
Miliolina	calcitic, imperforate	porcelaneous	bilamellar
Silicoloculinina	opaline silica, imperforate	–	–
Lagenina	calcitic	hyaline-radial	monolamellar
Robertinina	aragonitic, septate, finely perforate	hyaline-radial	bilamellar
Globigerinina	calcitic, perforate	hyaline radial	bilamellar
Rotaliina	calcitic, perforate	hyaline-radial or oblique	bilamellar

Superfamilial distinctions are based on the chamber characteristics of the test, perforations, canal structures and apertural features (Haynes 1990). These are the same characteristics which Cushman (1969) utilized to distinguish the foraminifera at the family level. There is also some disagreement whether the current superfamily distinctions are natural units since the ancestry and evolution of the groups cannot be distinguished (Fleury and Fourcade 1990). The current family level classification is based on the presence of internal subdivisions, surface texture, coiling mode and the free or attached habit (Haynes 1990). Loeblich and Tappan (1988) distinguish different genera by the test architecture, coiling mode, aperture form, accessory structures and internal structures. Of all the morphological characteristics in the shells of foraminifera, the aperture form seems the most important (Loeblich and Tappan 1988, Haynes 1990).

One major difficulty in the taxonomy of the foraminifera is in the morphological variations seen within one particular species. The very characteristics that are utilized to distinguish the species can take on several different expressions, often due to environmental stress (Boltovskoy et al. 1991). Sometimes variations are seen without any apparent environmental parameter as the cause of the physical change, with the change being simply an alternative expression of a genetic trait. In planktonic foraminifera, the mode of coiling is emphasized as a major characteristic (Bolli et al. 1957, Banner and Blow 1959, Brummer and Kroon 1988). Damage to a shell of a planktonic foraminifer can cause the foraminifer to change direction in its coiling (Bé and Spero 1981). However, in *Globorotalia truncatunlinoides*, despite being considered a left coiling species, a consistent number of specimens collected throughout the year, that show no apparent damage or evidence of environmental stress, are right coiling (Bé 1960).

Another obstacle encountered in the taxonomy of the foraminifera is in the differences between microspheric and megalospheric specimens in those foraminifera which possess dimorphic, trimorphic or paratrimorphic life cycles (for review, see Lee et al. 1991 a). Despite thousands of species being identified, the actual life cycles have been documented for only a handful of these species. Conceivably, the morphologies of gamonts, agamonts and schizonts could be different, and their responses of morphological changes to environmental parameters may differ.

Research on living specimens, especially juvenile forms, will be necessary to investigate whether particular morphological characteristics are valid as taxonomic criteria. As previously mentioned, the species of *Peneroplis* can be distinguished based on the ornamentation of the chamber wall. Classically, the shape of the aperture was utilized to distinguish the species (Hofker 1951 b). The dissection of shells of *Peneroplis* and the microscopic investigation of the apertures of each chamber, revealed that the shape of the aperture changes during the ontogeny of the organism (Gudmundsson 1990). Even Cushman (1969) realized that the aperture shape could change from a juvenile to an adult, and he based the differences between species on adult apertures. Laboratory cultures of clonal populations of *P. planatus* and *P. pertusus* demonstrated that sometimes organisms can grow apertures in adult specimens which are not characteristic of the particular species.

The standard morphological approach, still utilized by taxonomists to classify the foraminifera, may not be adequate. Robbins and Healy-Williams (1991) have demonstrated that protein analyses and other biochemical data may provide a better foundation to examine the genotypic relationships of foraminifera. These analyses combined with continued experimentation on the functional aspects of morphological features may help in understanding the evolution of these protists.

Other Sarcodina

Myxamoebae

One order of slime molds, Physarida, is divided into two families, the Didymiidae and the Physaridae, based primarily on the location of calcium carbonate deposits (Copeland 1956, Martin and Alexopoulos 1969, Hutner and Olive 1985). In the Didymiidae, sporangia have calcareous deposits in the wall of their peridium and often in the stalks, but the capillitium is usually free of mineral deposits. Some species of *Didymium* do have crystalline bodies and nodes held by threads in their capillitium (Keller and Schoknecht 1989). In the Physaridae, the capillitium is calcareous and the entire sporangium contains considerable non-crystalline calcareous deposits. Calcium fluxes and the role of calcium in contraction of the actomyosin fibrils within members of the genus *Physarum* have been researched extensively, with *P. polycephalum* being the most commonly studied (e.g., Aldrich and Daniel 1982, Wohlfarth-Bottermann et al. 1983, Häder 1985, Hutner and Olive 1985, Beylina and Belyavskii 1986, Anderson 1988).

Testate amoebae

This group of Sarcodina are divided into two classes, Lobosea and Filosea, which are separated by differences in their pseudopodia, whereas the species are distinguished by the shell shapes and compositions (Anderson 1988). The organisms are found in a variety of habitats from the marshy areas of lakes or sphagnum bogs to shallow marine waters (Page 1966, Bovee 1985, Schoenborn 1990). Many possess organic tests, agglutinated mineral tests, or siliceous deposits in the test secreted by the amoeba, but at least one, *Cryptodifflugia oviformis*, deposits a calcium phosphate test and others as explained below secrete calcite (Ogden and Hedley 1980).

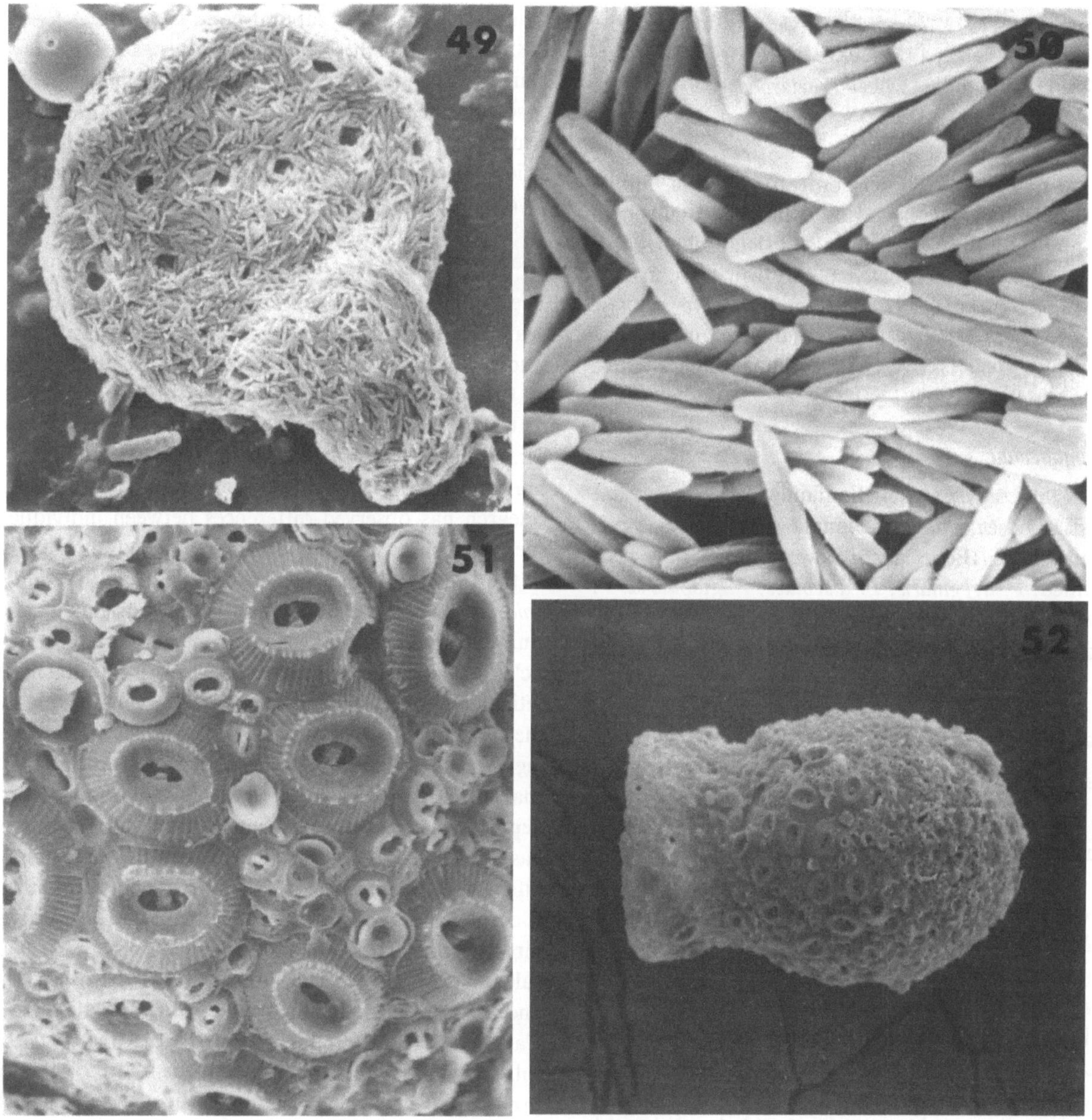

Figs. 49 and **50.** *Trichosphaerium sieboldi* (SEM). Fig. 49, schizont phase, × 1,000. Fig. 50, close-up of solid spicules, × 10,000

Figs. 51 and **52.** Tintinnid (SEMs). Fig. 51, agglutinated coccoliths on the surface of a tintinnid, × 2,000. Fig. 52, entire tintinnid, × 500

Trichosphaerium

The members of this genus, which is the only genus in the order Trichosida (class Lobosea), possess fibrous sheaths with calcite spicules during the schizont phase of its life cycle (Sheehan and Banner 1973, Bovee 1985). Three species have been identified based on the spicule structure and orientation of the schizont phase, as well as size of the organism (Angell 1975): *T. sieboldi* has solid, 3-sided, pointed spicules, which are perpendicular to the organic sheath (Sheehan and Banner 1973) (Fig. 49); *T. micrum* has hollow, 3-sided blunt-ended spicules which lie tangential to the body of the cell (Angell 1975); and *T. platyxyrum* has thin, blade-shaped, beveled or square-ended spicules which are perpendicular to the sheath (Angell 1976). Sheehan and Banner (1973) also observed one culture of *Trichosphaerium* with large spicules which were hexagonal in cross-section. The spicules are formed in vacuoles within the cell (Angell 1975). Possibly, the spicules are formed by the accretion of calcium carbonate onto organic material (Sheehan and Banner 1973). Sheehan and Banner (1973) suggest that the sheath of calcite spicules may act as a protective barrier to salinity variations in the habitats where *Trichosphaerium* is found.

Cryptodifflugia

This genus (class Lobosea, order Arcellinida, family Hyalospheniidae) possesses tests which are usually hyaline and smooth (Grospietsch 1964, Page 1966). Most protists which deposit calcium into their tests, usually do so in the form of calcium carbonate. However, these organisms possess a thick layer of amorphous calcium phosphate, which is covered by a thinner outer organic layer (Anderson 1988).

Paraquadrula

This is the only genus of the family Paraquadrulidae (class Lobosea, order Arcellinida) which secretes calcareous plates (Bovee 1985). Several species have been identified: *P. irregularis, P. discoides, P. globulosa, P. penardi,* and *P. pachylepsis.* The calcareous plates, which are composed of cuboidal crystals, are embedded and arranged along the chitonoid test (DeCloitre 1961).

Ciliophora

Colepidae

The ciliates of this family are usually barreled-shaped with regularly arranged calcium carbonate plates (Pennak 1978, Small and Lynn 1985, Anderson 1988). The three major genera are *Plagiopogon, Coleps*, and *Nolandia* (Small and Lynn 1985). *Plagiopogon* lacks a spiny body but has distinctive characteristic plates (Czapik and Jordan 1976), whereas *Coleps* and *Nolandia* have plates which are a different shape and bodies that are spiny (Small and Lynn 1985). Some species have single layers of plates, such as *C. hirtus*, whereas others, like *N. nolandia*, show a double-layered architecture (Huttenlauch 1985). The genus *Coleps* is separated from *Nolandia* by the position of the brosse files; in *Coleps*, the brosse files are perpendicular to the body kinetics (i.e., dikinetids) whereas the brosse files of *Nolandia* are parallel to the body kinetics (i.e., monokinetids) (Wilbert and Schmall 1976, Small and Lynn 1985).

Tintinnids

Members of the family Codonellidae construct their loricae by agglutinating primarily coccoliths. The two genera, *Codonella* and *Codonaria*, differ in their collar structure: the collar is slightly flared or with ridged rings in *Codonella*; and the collar in *Codonaria* is around a clear inner suboral cone (Small and Lynn 1985). Although these organisms do not secrete any structures composed of calcium, their tests are composed of calcite, albeit vicariously secreted. Other tintinnid genera also utilize coccoliths in their agglutinated loricae (Figs. 51 and 52), such as *Poroecus* (family Ptychocyclidae) (Small and Lynn 1985).

Conclusion

The fundamental basis of calcification in protists appears the same in those organisms which have been studied; calcium carbonate crystals are deposited on an organic matrix. In coccolithophorids, the organic matrix is the base plate of the coccolith, whereas in foraminifera, the matrix is the primary organic membrane (POM). Although these processes seem similar, research has observed differences between individual species, such as with the location of coccolithogenesis in the three well-studied coccolithophorids. In *Emiliania huxleyi*, base plate formation and subsequent calcium carbonate deposition occur in the coccolith vesicle, a nucleus-associated organelle with an attached reticular body system, whereas these events occur in the Golgi in *Pleurochrysis carterae*. In *Coccolithus pelagicus*, the holococcolith base plates are formed in T-shaped vesicles, associated with the Golgi but calcite deposition occurs extracellularly after discharge of the base plate. Such contrasts can be seen in the foraminifera also. In *Amphistegina lobifera* calcium carbonate is derived from internal inorganic pools and deposited directly on the POM, whereas in *Amphisorus hemprichii* calcium carbonate needles are formed first in vesicles and then are deposited at the sites of calcification.

Calcified structures found in protists are extremely useful in their systematics. However, in many of these protists, other biochemical and genetic markers, which need further investigation, may prove to be more useful in creating phylogenetic linkages, and eliminating artificially created groups. Also, calcification needs to be investigated in the less studied groups. As Boltovskoy et al. (1991) stated in regard to the environmental responses of foraminifera, there are "many individual trends ... but few broad ones". This is true for calcification in protists. Researchers should be cautious in making generalizations. In addition, the question as to what extent the calcification is due to ecological control or genetic control remains to be answered.

Acknowledgements

We gratefully acknowledge provision of unpublished micrographs by Ms. Kerryn M. Callaghan, Melbourne (Figs. 1–2 and 7–15), Dr.

Peter L. Beech, Melbourne (Figs. 4–6), and Dr. O. Roger Anderson, New York (Figs. 44–48). We thank Dr. Jeremy R. Young, London, and Dr. Peter R. van Emburg, Leiden, for the permission to reproduce their diagrams (Figs. 3 and 16, respectively). The assistance with the photographic work by Mr. A. Zuppiger, Zurich, is kindly acknowledged. And also we thank Dr. John J. Lee, New York, for permission to utilize his darkroom, expertise, and micrographs (Figs. 17–43 and 49–52).

References

Aldrich HC, Daniel JW (1982) Cell biology of *Physarum* and *Didymium*, vols I, II. Academic Press, New York

Anderson OR (1988) Comparative protozoology: ecology, physiology, life history. Springer, New York Berlin Heidelberg Tokyo

– Bé AWH (1978) Recent advances in foraminiferal fine structure research. In: Hedley RH, Adams CG (eds) Foraminifera, vol 3. Academic Press, London, pp 121–202

– Faber WW Jr (1984) An estimation of calcium carbonate deposition rate in a planktonic foraminifer *Globigerinoides sacculifer* using ^{45}Ca as a tracer: a recommended procedure for improved accuracy. J Foram Res 14: 303–308

Angell RW (1967) The process of chamber formation in the foraminifer *Rosalina floridana* (Cushman). J Protozool 14: 556–574

– (1975) Structure of *Trichosphaerium micrum* sp.n. J Protozool 22: 18–22

– (1976) Observations on *Trichosphaerium platyxyrum* sp.n. J Protozool 23: 357–364

– (1979) Calcification during chamber development in *Rosalina floridana*. J Foram Res 89: 341–353

– (1980) Test morphogenesis (chamber formation) in the foraminifera *Spiroloculina hyalina* Schultze. J Foram Res 10: 89–101

Arnold ZM (1954) *Discorinopsis aguayoi* (Bermudez) and *Discorinopsis radescens* Cushman and Bronnimann: a study of variation in cultures of living foraminifera. Cushman Found Foram Res Contrib 5: 4–13

Banner FT, Blow WH (1959) The classification and stratigraphic distribution of the Globigerinaceae. Palaeontology 2: 1–27

Bé AWH (1960) Ecology of recent planktonic foraminifera: part 2. Bathymetric and seasonal distributions in the Sargasso Sea off Bermuda. Micropaleontology 6: 372–392

– (1967) Foraminifera families: Globigerinidae and Globorotaliidae. Fiche no. 108. In: Fiches d'identification du zooplankton. Conseil Internationale Exploration Mer, Charlottenlund, Denmark

– (1968) Shell porosity of recent foraminifera as a climatic index. Science 161: 881–884

– (1977) A taxonomic and zoogeographic review of recent planktonic foraminifera. In: Ramsey AT (ed) Oceanic micropaleontology. Academic Press, London, pp 1–100

– (1980) Gametogenic calcification in a spinose planktonic foraminifer, *Globigerinoides sacculifer* (Brady). Mar Micropaleontol 5: 117–128

– Ericson DB (1963) Aspects of calcification in planktonic foraminifera (Sarcodina). NY Acad Sci Ann 109: 65–81

– Hemleben C (1970) Calcification in a living planktonic foraminifer, *Globigerinoides sacculifer* (Brady). N Jb Ged Paläontol Abh 134: 221–234

– Spero HJ (1981) Shell regeneration and biological recovery of planktonic foraminifera after physical injury induced in laboratory culture. Micropaleontology 27: 305–316

– Jongebloed WL, McIntyre L (1969) X-ray microscopy of the recent planktonic foraminifera. J Paleontol 43: 1384–1396

– Harrison SM, Lott L (1973) *Orbulina universa* d'Orbigny in the Indian Ocean. Micropaleontology 19: 150–192

– Morse JW, Harrison SM (1975) Progressive dissolution and ultrastructural breakdown of planktonic foraminifera. In: Sliter WV, Bé AWH, Berger WH (eds) Dissolution of deep-sea carbonates. Cushman Found Spec Publ 12: 27–55

– Harrison SM, Frerichs WE, Heiman ME (1976) Variability in test porosity of *Orbulina universa* d'Orbigny at two Indian Ocean localities. In: Takayanagi Y, Saito T (eds) Progress in micropaleontology: selected papers in honor of Prof. Kiyoshi Asano. Micropaleontology Press, New York, pp 1–9

– Hemleben C, Anderson OR, Spindler M (1979) Chamber formation in planktonic foraminifera. Micropaleontology 25: 294–307

– – – – (1980) Pore structures in planktonic foraminifera. J Foram Res 10: 117–128

– Anderson OR, Faber WW Jr, Caron DA (1983) Sequence of morphological and cytoplasmic changes during gametogenesis in the planktonic foraminifer *Globigerinoides sacculifer* (Brady). Micropaleontology 29: 310–325

Berger WH (1969) Planktonic foraminifera: basic morphology and ecologic implications. J Paleontol 32: 1369–1383

Berthold W-U (1976) Test morphology and morphogenesis in *Patellina corrugata* Williamson, Foraminifera. J Foram Res 6: 167–185

Beylina SI, Belyavskii MA (1986) Calcium exchange in *Physarum polycephalum* plasmodium at different stages of the cell cycle. Stud Biophys 116: 195–203

Bolli H, Loeblich AR, Tappan H (1957) Planktonic foraminiferal families Hantkeninidae, Orbulinidae, Globorotaliidae and Globotruncanidae. US Nat Mus Bull 215: 3–50

Boltovskoy E, Wright R (1976) Recent foraminifera. W Junk, The Hague

– Scott DB, Medioli FS (1991) Morphological variations of benthic foraminiferal tests in response to changes in ecological parameters: a review. J Paleontol 65: 175–185

Borman AH, de Jong EW, Thierry R, Westbroek P, Bosch L, Gruter M, Kamerling JP (1987) Coccolith-associated polysaccharides from cells of *Emiliania huxleyi* (Haptophyceae). J Phycol 23: 118–123

Borowitzka MA (1989) Carbonate calcification in algae: initiation and control. In: Mann S, Webb J, Williams RJP (eds) Biomineralization: chemical and biochemical perspectives. VCH, Weinheim, pp 63–94

Bovee EC (1985) Class Lobosea Carpenter 1861. In: Lee JJ, Hutner SH, Bovee EC (eds) An illustrated guide to the protozoa. Society of Protozoologists, Lawrence, KS, pp 158–211

Brummer G-JA, Kroon D (1988) Genetically controlled planktonic foraminiferal coiling ratios as tracers of past ocean dynamics. In: Brummer G-JA, Kroon D (eds) Planktonic foraminifera as tracers of ocean-climate history. Free University Press, Amsterdam, pp 293–297

– Hemleben C, Spindler M (1987) Ontogeny of extant globigerinid planktonic foraminifera, a concept exemplified by *Globigerinoides sacculifer* (Brady) and *G. ruber* (d'Orbigny). Mar Micropaleontol 12: 357–381

Carpenter WB, Parker WK, Jones TR (1862) Introduction to the study of foraminifera. R Hardwicke, London

Chrétiennot-Dinet M-J (1990) Classe des Prymnesiophycées. In: Sournia A (ed) Atlas du phytoplancton marin, vol 3. Éditions du Centre National de la Recherche Scientifique, Paris, pp 75–106
– Sournia A, Ricard M, Billard C (1993) A classification of the marine phytoplankton of the world from class to genus. Phycologia 32: 159–179
Clocchiatti M (1971) Sur l'existence de coccosphères portant des coccolithes de *Gephyrocapsa oceanica* et de *Emiliania huxleyi* (Coccolithophoridés). C R Acad Sci Paris 273: 318–321
Copeland HF (1956) The classification of lower organisms. Pacific Books, Palo Alto, CA
Cushman JA (1969) Foraminifera: their classification and economic use, 4th edn. Harvard University Press, Cambridge, MA
Czapik A, Jordan A (1976) Les ciliés psammophiles de la mer Baltique aux environs de Gdánsk. Acta Protozool 15: 423–445
DeCloitre L (1961) Le genre *Paraquadrula* (Thekamoebina). Int Rev Ges Hydrobiol 46: 321–330
Devi Prasad PV, Chowdary YBK (1982) Nature and composition of the mineral deposition in the freshwater alga *Gloeotaenium loitlesbergerianum* Hansgirg (Chlorophyta, Chlorococcales). Phycologia 21: 323–326
de Vrind-de Jong EW, Borman AH, Thierry R, Westbroek P, Grüter M, Kamerling JP (1986) Calcification in the coccolithophorids *Emiliania huxleyi* and *Pleurochrysis carterae* II. Biochemical aspects. In: Leadbeater BSC, Riding R (eds) Biomineralization in lower plants and animals. Clarendon Press, Oxford, pp 205–217 (Systematics Association special volume 30)
Duguay LE (1983) Comparative laboratory and field studies on calcification and carbon fixation in foraminiferal-algal association. J Foram Res 13: 252–261
– Taylor DL (1978) Primary production and calcification by the soritid foraminifera *Archaias angulatus* (Fichtel & Moll). J Protozool 25: 356–361
Erez J (1978) Vital effect on stable isotope composition seen in foraminifera and coral skeletons. Nature 273: 199–202
Ettl H (1978) Xanthophyceae (I). In: Ettl H, Gerloff J, Heynig, H. (eds) Süßwasserflora von Mitteleuropa, vol 3. G Fischer, Stuttgart, pp 1–530
Faber WW Jr, Lee JJ (1991) Feeding and growth of the foraminifer *Peneroplis plantus* (Fichtel & Moll) Montfort. Symbiosis 10: 63–82
– – (1992) Pathways of carbon in the *Peneroplis planatus* (Foraminifer)-*Porphyridium purpureum* (Rhodophyte) endosymbiosis. Symbiosis 14: 439–463
Fleury J-J, Fourcade E (1990) La super-lamille Alveolinacea (Foraminifères): systématique et essai d'interprétation phylogénétique. Rev Micropaléontol 33: 241–268
Frerichs WE, Heiman ME, Borgman LE, Bé AWH (1972) Latitudinal variations in planktonic foraminiferal test porosity: part I. Optical studies. J Foram Res 2: 6–13
Fresnel J, Billard C (1991) *Pleurochrysis placolithoides* sp. nov. (Prymnesiophyceae), a new marine coccolithophorid with remarks on the status of cricolith-bearing species. Br Phycol J 26: 67–80
Gaarder KR (1962) Electron microscope studies on holococcolithophorids. Nytt Mag Bot 10: 35–51
– Markali J (1957) On the coccolithophorid *Crystallolithus hyalinus* n.gen., n.sp. Nytt Mag Bot 5: 1–5
Gard G (1987) Observation of a dimorphic coccosphere. Abh Geol Bundesanst Wien 39: 85–87
Giering B, Krienitz L, Casper SJ, Peschke T, Raidt H (1990) LM and SEM observations on the asexual reproduction and lorica formation of *Phacotus lendneri* Chodat (Chlamydophyceae, Phacotaceae). Arch Protistenk 138: 75–88
Glaessner MF (1947) Principles of micropaleontology. Wiley, New York
Green JC (1986) Biomineralization in the algal class Prymnesiophyceae. In: Leadbeater BSC, Riding R (eds) Biomineralization in lower plants and animals. Clarendon Press, Oxford, pp 173–188 (Systematics Association special volume 30)
– Course PA (1983) Extracellular calcification in *Chrysotila lamellosa* (Prymnesiophyceae). Br Phycol J 18: 367–382
– Perch-Nielsen K, Westbroek P (1990) Phylum Prymnesiophyta. In: Margulis L, Corliss JO, Melkonian M, Chapman DJ (eds) Handbook of Protoctista. Jones and Bartlett, Boston, pp 293–317
Grospietsch T (1964) Die Gattungen *Cryptodifflugia* und *Difflugiella* (Rhizopoda, Testacea). Zool Anz 172: 243–257
Gudmundsson G (1990) Systematics of recent species of the superfamily Soritidae Ehrenberg 1839 with notes on systematic principles. PhD thesis, City University of New York, New York, NY
Häder D-P (1985) Role of calcium in phototaxis of *Physarum polycephalum*. Plant Cell Physiol 26: 1411–1417
Hallock P (1979) Trends in test shape with depth in large symbiont-bearing foraminifera. J Foram Res 9: 61–69
– (1981) Light dependence in *Amphistegina*. J Foram Res 11: 40–46
– (1985) Why are larger foraminifera large? Paleobiology 11: 195–208
– Forward LB, Hansen HJ (1986) Influence of environment on the test shape of *Amphistegina*. J Foram Res 16: 224–231
– Röttger R, Wetmore K (1991) Hypotheses on form and function in foraminifera. In: Lee JJ, Anderson OR (eds) Biology of the foraminifera. Academic Press, London, pp 41–72
Hansen HJ, Reiss Z, Schneidermann N (1969) Ultrastructure of bilamellar walls in foraminiferida. Rev Esp Micropaleontol 1: 293–316
Haynes J (1965) Symbiosis, wall structure and habitat in foraminifera. Cushman Found Foram Res Contrib 16: 40–43
– (1990) The classification of the foraminifera – a review of historical and philosophical perspectives. Palaeontology 33: 503–528
Hecht AD, Savin SM (1972) Phenotypic variation and oxygen isotope ratios in recent planktonic foraminifera. J Foram Res 2: 55–67
Hemleben C (1975) Spine and pustule relationships in some recent planktonic foraminifera. Micropaleontology 21: 334–341
– Spindler M (1983) Recent advances in research on living planktonic foraminifera. Utrecht Micropaleontol Bull 30: 141–170
– Bé AWH, Anderson OR, Tuntivate S (1977) Test morphology, organic layers and chamber formation of the planktonic foraminifer *Globorotalia menardii* (d'Orbigny). J Foram Res 7: 1–25
– Anderson OR, Berthold W, Spindler M (1986) Calcification and chamber formation in foraminifera: a brief overview. In: Leadbeater BSC, Riding R (eds) Biomineralization in lower plants and animals. Clarendon Press, Oxford, pp 237–249 (Systematics Association special volume 30)
– Spindler M, Anderson OR (1989) Modern planktonic foraminifera. Springer, New York Berlin Heidelberg Tokyo
– Mühlen D, Olsson RK, Berggren WA (1991) Surface texture and

the first occurrence of spines in planktonic foraminifera from the early Tertiary. Geol Jahrb A 128: 117–146

Hofker J (1951 a) Pores of foraminifera. Micropaleontology 5: 38

– (1951 b) Recent Peneroplidae, part 2. J Microsc Soc 71: 342–356

– (1970) Studies of foraminifera, part II. Publ Naturhist Genootschap Limburg 20: 1–98

– (1971) The foraminifera of Piscadera Bay, Curacao. Stud Fauna Cur Carib Isl 35: 1–57

Hottinger L (1978) Comparative anatomy of elementary shell structures in selected larger foraminifera. In: Hedley RH, Adams CG (eds) Foraminifera, vol 3. Academic Press, London, pp 203–266

– Dreher D (1974) Differentiation of protoplasm in Nummulitidae (foraminifera) from Elat, Red Sea. Mar Biol 25: 41–61

Hutner SH, Olive LS (1985) Class Mycetozoea de Bary 1859. In: Lee JJ, Hutner SH, Bovee EC (eds) An illustrated guide to the protozoa. Society of Protozoologists, Lawrence, KS, pp 214–227

Huttenlauch I (1985) Scanning electron microscopy study of the skeletal plates of *Coleps nolandia*. Protistologia 21: 499–503

Inouye I, Pienaar RN (1983) Observations on the life cycle and microanatomy of *Thoracosphaera heimii* (Dinophyceae) with special reference to its systematic position. S Afr J Bot 2: 63–75

– – (1988) Light and electron microscope observations of the type species of *Syracosphaera*, *S. pulchra* (Prymnesiophyceae). Br Phycol J 23: 205–217

Jepps MW (1942) Studies on *Polystomella* Lamarck. J Mar Biol Assoc UK 25: 607–666

Johansen JR, Doucette GJ, Barclay WR, Bull JD (1988) The morphology and ecology of *Pleurochrysis carterae* var. *dentata* var.nov. (Prymnesiophyceae), a new coccolithophorid from an inland saline pond in New Mexico. U.S.A. Phycologia 27: 78–88

Jordan RW, Kleijne A (1994) A classification system for living coccolithophores. In: Winter A, Siesser WG (eds) Coccolithophores. Cambridge University Press, Cambridge, pp 83–105

Kamptner E (1937) Über Dauersporen bei marinen Coccolithineen. Akad Wiss Wien Sitzungsber Math Naturwiss Kl Abt 1 146: 67–76

Keller HW, Schoknecht JD (1989) Life cycle of a new annulate-spored species of *Didymium*. Mycologia 81: 248–265

Keupp H (1991) Fossil calcareous dinoflagellate cysts. In: Riding R (ed) Calcareous algae and stromatolites. Springer, Berlin Heidelberg New York Tokyo, pp 267–286

Klaveness D (1972) *Coccolithus huxleyi* (Lohm.) Kamptn. II. The flagellate cell, aberrant cell types, vegetative propagation and life cycles. Br Phycol J 7: 309–318

Kleijne A (1991) Holococcolithophorids from the Indian Ocean, Red Sea, Mediterranean Sea and North Atlantic Ocean. Mar Micropaleontol 17: 1–76

Kloos DP (1981) Growth and embryogenesis of the foraminifer *Sorites orbiculus*. Proc Konink Nederl Akad Wet 84 B: 145–159

Koestler RJ, Lee JJ, Reidy J, Sheryll RP, Xenophontos X (1985) Cytological investigation of digestion and re-establishment of symbiosis in the larger benthic foraminifera *Amphistegina lessonii*. Endocytol Cell Res 2: 21–54

Krienitz L, Koschel R, Giering B, Casper SJ, Hepperle D (1993) Phenomenology of organismic calcite precipitation by *Phacotus* in hardwater lakes and ponds of northeastern Germany. Verh Int Verein Limnol 25: 170–174

Kuile B ter (1991) Mechanisms for calcification and carbon cycling in algal symbiont-bearing foraminifera. In: Lee JJ, Anderson OR (eds) Biology of the foraminifera. Academic Press, London, pp 73–89

– Erez J (1984) In situ growth rate experiments on the symbiont-bearing foraminifera *Amphistegina lobifera* and *Amphisorus hemprichii*. J Foram Res 14: 262–276

– – (1987) Uptake of inorganic carbon and internal carbon cycling in benthonic symbiont-bearing foraminifera. Mar Biol 94: 499–510

– – (1988) The size and function of the internal inorganic carbon pool of the foraminifer *Amphistegina lobifera*. Mar Biol 99: 481–487

– – (1991) Carbon budgets for two species of benthonic symbiont-bearing foraminifera. Biol Bull 180: 489–495

– – Padan E (1989 a) Mechanisms for the uptake of inorganic carbon by two species of symbiont-bearing foraminifera. Mar Biol 103: 241–251

– – – (1989 b) Competition for inorganic carbon between the processes of photosynthesis and calcification in the symbiont-bearing foraminifer *Amphistegina lobifera*. Mar Biol 103: 253–259

Leadbeater BSC (1971) Observations on the life history of the haptophycean alga *Pleurochrysis scherffelii* with special reference to the microanatomy of the different types of motile cell. Ann Bot 35: 429–439

Lee JJ (1983) Perspective on algal endosymbionts in larger foraminifera. Int Rev Cytol [Suppl] 14: 49–77

– (1993) "On a piece of chalk" – updated. J Eukar Microbiol 40: 395–410

– Anderson OR (eds) (1991) Biology of the foraminifera. Academic Press, London

– Bock WD (1976) The importance of feeding in two species of soritid foraminifera with algal symbionts. Bull Mar Sci 26: 530–537

– Hallock P (1987) Algal symbiosis as the driving force in the evolution of larger foraminifera. Ann NY Acad Sci 503: 330–347

– Lee RE (1990) Chloroplast retention in elphidids (foraminifera). In: Nardon P, Gianinazzi-Pearson V, Grenier AM, Margulis L, Smith DC (eds) Endocytobiology IV. Institut National de la Recherche Agronomique, Paris, pp 215–220

– McEnery ME (1983) Symbiosis in foraminifera. In: Goff LJ (ed) Algal symbiosis: a continuum of interaction strategies. Cambridge University Press, Cambridge, pp 105–128

– Hutner SH, Bovee EC (eds) (1985) An illustrated guide to the protozoa. Society of Protozoologists, Lawrence, KS

– Lanners E, Kuile B ter (1988) The retention of chloroplast by the foraminifer *Elphidium crispum*. Symbiosis 5: 45–60

– Faber WW Jr, Anderson OR, Pawlowski J (1991 a) Life cycles of foraminifera. In: Lee JJ, Anderson OR (eds) The biology of foraminifera. Academic Press, London, pp 285–334

– – Lee RE (1991 b) Granuloreticulopodial digestion – a possible preadaptation to benthic foraminiferal symbiosis? Symbiosis 10: 47–61

– Sang K, Kuile B ter, Strauss E, Lee PJ, Faber WW Jr (1991 c) Nutritional and related experiments on laboratory maintenance of three species of symbiont-bearing larger foraminifera. Mar Biol 109: 417–425

Lee RE, Kugrens P (1989) Biomineralization of the stalks of *Anthophysa vegetans* (Chrysophyceae). J Phycol 25: 591–596

Lewin RA (1990) Calcification of cell walls of *Chlamydomonas* (Volvocales, Chlorophyta) on agar media. Phycologia 29: 536–540

Linschooten C, Bleijswijk JDL, van Emburg PR, de Vrind JPM, Kempers ES, Westbroek P, de Vrind-de Jong EW (1991) Role of the light-dark cycle and medium composition on the production of coccoliths by *Emiliania huxleyi* (Haptophyceae). J Phycol 27: 82–86

Loeblich AR Jr, Tappan H (1964) Sarcodina, chiefly "Thecamoebians' and Foraminiferida. In: Moor RC (ed) Treatise on invertebrate paleontology, Protista 2, part C. University of Kansas Press, Lawrence, pp 1–900

– – (1966) Annotated index and bibliography of the calcareous nannoplankton. Phycologia 5: 81–216

– – (1982) Classification of the Foraminiferida. In: Broadhead TW (ed) Foraminifera, notes for a short course. Department of Geology, University Tennessee, Knoxville, pp 22–36 (Stud Geol 6)

– – (1988) Foraminiferal genera and their classification, vols 1, 2. Van Nostrand Reinhold, New York

Lopez R (1979) Algal chloroplasts in the protoplasm of three species of benthic foraminifera: taxonomic affinity, viability and persistence. Mar Biol 53: 201–211

Lund JWG (1953) New or rare British Chrysophyceae. II. *Hyalobryon polymorphum* n.sp. and *Chrysonebula holmesii* n.gen., n.sp. New Phytol 52: 114–123

Lutze GF, Grabert B, Seibold E (1971) Lebendbeobachtungen an Großforaminiferen (*Heterostegina*) aus dem Persischen Golf. Meteorol Forsch Erg C 6: 21–40

Lynts GW, Pfister RM (1967) Surface ultrastructure of some tests of recent foraminiferida from the Dry Tortugas, Florida. J Protozool 14: 387–399

Mann H, Mann S, Fyfe WS (1987) Aragonite crystals in *Spirogyra* sp. (Chlorophyta). J Phycol 23: 506–509

Mann S, Sparks NHC (1988) Single crystalline nature of coccolith elements of the marine alga *Emiliania huxleyi* as determined by electron diffraction and high-resolution transmission electron microscopy. Proc R Soc Ser B 234: 441–453

Manton I, Leedale G (1969) Observations on the microanatomy of *Coccolithus pelagicus* and *Cricosphaera carterae*, with special reference to the origin and nature of coccoliths and scales. J Mar Biol Assoc UK 49: 1–16

– Oates K (1980) *Polycrater galapagensis* gen. et sp.nov., a putative coccolithophorid from the Galapagos Islands with an unusual aragonitic periplast. Br Phycol J 15: 95–103

– Parke M (1960) Further observations on small green flagellates with special reference to possible relatives of *Chromulina pusilla* Butcher. J Mar Biol Assoc UK 39: 275–298

Marlowe IT, Green JC, Neal AC, Brassell SC, Eglinton G, Course PA (1984) Long chain (n-C_{37}-C_{39}) alkenones in the Prymnesiophyceae. Distribution of alkenones and other lipids and their taxonomic significance. Br Phycol J 19: 203–216

Martin GW, Alexopoulos CJ (1969) The myxomycetes. University of Iowa Press, Iowa City, IA

Montresor M, Zingone A, Marino D (1993) The calcareous resting cyst of *Pentapharsodinium tyrrhenicum* comb.nov. (Dinophyceae). J Phycol 29: 223–230

Müller G, Oti M (1981) The occurrence of calcified planktonic green algae in freshwater carbonates. Sedimentology 28: 897–902

Muller PH (1978) ^{14}Carbon fixation and loss in a foraminiferal-algal symbiont system. J Foram Res 8: 35–41

Murray JW (1963) Ecological experiments on foraminiferids. J Mar Biol Assoc UK 43: 621–642

Norris RE (1985) Indian ocean nannoplankton. II. Holococcolithophorids (Calyptrosphaeraceae, Prymnesiophyceae) with a review of extant genera. J Phycol 21: 619–641

Ogden CG, Hedley RH (1980) An atlas of freshwater testate amoebae. Oxford University Press, Oxford

Olsson RK (1973) What is a kummerform planktonic foraminifera? J Paleontol 47: 327–329

Orr WN (1967) Secondary calcification in the foraminiferal genus *Globorotalia*. Science 157: 1554–1555

Outka DE, Williams DC (1971) Sequential coccolith-morphogenesis in *Hymenomonas carterae*. J Protozool 18: 285–297

Page FC (1966) *Cryptodifflugia operculata* n.sp. (Rhizopodea: Arcellinda, Cryptodifflugiidae) and the status of the genus *Cryptodifflugia*. Trans Amer Microsc Soc 85: 506–515

Parke M (1970) The production of calcareous elements by benthic algae belonging to the class Haptophyceae (Chrysophyta). In: Farinacci A (ed) Proceedings of the 2nd Planktonic Conference. Edizioni Technoscienza, Rome, pp 929–937

– Adams I (1960) The motile (*Crystallolithus hyalinus* (Gaarder & Markali) and non-motile phases in the life-history of *Coccolithus pelagicus* (Wallich) Schiller. J Mar Biol Assoc UK 39: 263–274

Pawlowski J, Lee JJ (1991) Taxonomic notes on tiny, shallow water foraminifera from the northern Gulf of Elat (Red Sea). Micropaleontology 37: 149–162

Pennak RW (1978) Freshwater invertebrates of the United States, 2nd edn. Wiley, New York

Pentecost A (1991 a) Calcification processes in algae and cyanobacteria. In: Riding R (ed) Calcareous algae and stromatolites. Springer, Berlin Heidelberg New York Tokyo, pp 3–20

– (1991 b) A new and interesting site for the calcite-encrusted desmid *Oocardium stratum* Naeg. in the British Isles. Br Phycol J 26: 297–301

Preisig HR (1986) Biomineralization in the Chrysophyceae. In: Leadbeater BSC, Riding R (eds) Biomineralization in lower plants and animals. Clarendon Press, Oxford, pp 327–344 (Systematics Association special volume 30)

– (1989) The flagellar base ultrastructure and phylogeny of chromophytes. In: Green JC, Leadbeater BSC, Diver WL (eds) The chromophyte algae: problems and perspectives. Clarendon Press, Oxford, pp 167–187 (Systematics Association special volume 38)

– (1994) Siliceous structures and silicification in flagellated protists. Protoplasma 181: 29–42

Reiss Z (1957) The Bilamellida, nov. superfam., and remarks on Cretaceous globorotaliids. Cushman Found Foram Res Contrib 8: 127–145

– (1958) Classification of lamellar foraminifera. Micropaleontol 4: 51–70

– (1963) Comments on wall structure of foraminifera. Micropaleontol 9: 50–52

– Hottinger L (1984) The Gulf of Aqaba. Springer, Berlin Heidelberg New York Tokyo [Billings WD et al (eds) Ecological studie vol 50]

Rhumbler L (1911) Die Foraminiferen (Thalamophoren) der Plankton Expedition. Erg Plank Exped Humboldt Stift 3 L Č 1: 1–331

Robbins L, Healy-Williams N (1991) Toward a classification of planktonic foraminifera based on biochemical, geochemical and morphological criteria. J Foram Res 21: 159–167

Roberts KR, Mills JT (1992) The flagellar apparatus of *Hymenomonas coronata* (Prymnesiophyceae). J Phycol 28: 635–642

Ross CA (1974) Evolutionary and ecological significance of large calcareous Foraminiferida (Protozoa), Great Barrier Reef. In:

Proceedings of the 2nd International Symposium Coral Reefs, vol 1, pp 327–333

Röttger R (1976) Ecological observations of *Heterostegina depressa* (Foraminifera, Nummulitidae) in the laboratory and in its natural habitat. Mar Sed Spec Publl 1: 75–80

– Berger WH (1972) Benthic foraminifera: morphology and growth in clone cultures of *Heterostegina depressa*. Mar Biol 15: 89–94

– Krüger R (1990) Observations on the biology of Calcarinidae (Foraminiferida). Mar Biol 106: 419–425

– Spindler M (1976) Development of *Heterostegina depressa* individuals (Foraminifera, Nummulitidae) in laboratory culture. Mar Sed Spec Publ 1: 81–87

– Irwan A, Schmaljohann R, Franzisket L (1980) Growth of the symbiont-bearing foraminifera *Amphistegina lobifera* d'Orbigny and *Heterostegina depressa* d'Orbigny (Protozoa). In: Schwemmler W, Schenk HEA (eds) Endocytobiology I. W de Gruyter, Berlin, pp 125–132

– Spindler M, Schmaljohann R, Richwein M, Fladung M (1984) Functions of the canal systems in the rotaliid foraminifer (*Heterostegina depressa*. Nature 309: 789–791

– Krüger R, Rijk S de (1990) Larger foraminifera: variation in outer morphology and prolocular size in *Calcarina gaudichaudii*. J Foram Res 20: 170–174

Rowson JD, Leadbeater BSC, Green JC (1986) Calcium carbonate deposition in the motile (*Crystallolithus*) phase of *Coccolithus pelagicus* (Prymnesiophyceae). Br Phycol J 21: 359–370

Saito T, Bé AWH (1976) Paleontology of deep-sea deposits. In: Runcorn SK (ed) International dictionary of geophysics, vol 2. Pergamon Press, Oxford, pp 1143–1156

Schoenborn W (1990) Analysis of subfossil shelled protozoa in the sediments of a small acid forest lake East Germany. Limnologia 21: 137–145

Sheehan R, Banner FT (1973) *Trichosphaerium* – an extraordinary testàte rhizopod from coastal waters. Estuarine Coast Mar Sci 1: 245–260

Sliter WV (1970) *Bolivina doniezi* Cushman and Wickenden in clone culture. Cushman Found Foram Res Contrib 21: 87–99

Small EB, Lynn DH (1985) Phylum Ciliophora Doflein 1901. In: Lee JJ, Hutner SH, Bovee EC (eds) An illustrated guide to the protozoa. Society of Protozoologists, Lawrence, KS, pp 393–575

Smout AH (1954) Lower Tertiary foraminifera of the Qatar peninsula. British Museum of Natural History, London

Spindler M (1978) Anatomy of the test of *Heterostegina depressa* (Foraminiferida). J Foram Res 8: 319–331

– Röttger R (1973) Der Kammerbauvorgang der Großforaminifere *Heterostegina depressa* (Nummulitidae). Mar Biol 18: 146–159

Sverdrup HU, Johnson MW, Fleming RH (1942) The oceans. Prentice-Hall, Englewood Cliffs

Tangen K, Brand LE, Blackwelder PL, Guillard RRL (1982) *Thoracosphaera heimii* (Lohmann) Kamptner is a dinophyte: Observations on its morphology and life cycle. Mar Micropaleontol 7: 193–212

Tappan H (1980) The paleobiology of plant protists. Freeman, San Francisco

Thompson DW (1961) On growth and form, abridged edn. Cambridge University Press, New York

Thomsen HA, Østergaard JB, Hansen LE (1991) Heteromorphic life histories in arctic coccolithophorids. J Phycol 27: 634–642

Todd R (1976) Some observations about *Amphistegina* (foraminifera). In: Takayanagi Y, Saito T (eds) Progress in micropaleontology: selected papers in honor of Prof. Kiyoshi Asano. Micropalaeontology Press, New York, pp 382–394

Towe KM, Cifelli R (1967) Wall structure in the calcareous foraminifera: crystallographic aspects and a model for calcification. J Paleontol 41: 742–762, pls 87–99

Tschermak-Woess E (1980) Zur Kenntnis von *Tetrasporopsis fuscescens*. Plant Syst Evol 133: 121–133

Van der Wal P, de Vrind JPM, de Vrind-de Jong EW, Borman AH (1987) Incompleteness of the coccosphere as a possible stimulus for coccolith formation in *Pleurochrysis carterae* (Prymnesiophyceae). J Phycol 23: 218–221

Westbroek P, van der Wal P, van Emburg PR, de Vrind-de Jong EW, de Bruijn WC (1986) Calcification in the coccolithophorids *Emiliania huxleyi* and *Pleurochrysis carterae*. I. Ultrastructural aspects. In: Leadbeater BSC, Riding R (eds) Biomineralization in lower plants and animals. Clarendon Press, Oxford, pp 189–203 (Systematics Association special volume 30)

– Young JR, Linschooten K (1989) Coccolith production (biomineralization) in the marine alga *Emiliania huxleyi*. J Protozool 36: 368–373

Wilbert N, Schmall G (1976) Morphologie und Infraciliatur von *Coleps nolandia* Kahl 1930. Protistologica 12: 193–197

Wilbur KM, Watabe N (1963) Experimental studies on calcification in molluscs and the alga *Coccolithus huxleyi*. Ann NY Acad Sci 109: 82–112

Winsborough BM, Golubic S (1987) The role of diatoms in stromatolite growth: two examples from modern freshwater settings. J Phycol 23: 195–201

Winter A, Siesser WG (eds) (1994) Coccolithophores. Cambridge University Press, Cambridge

Wohlfarth-Bottermann KE, Shraideh Z, Baranowski Z (1983) Contractile and structural reactions to impediments of calcium ion homeostasis in *Physarum polycephalum*. Cell Struct Funct 8: 255–265

Young JR (1989) Observations on heterococcolith rim structure and its relationship to developmental processes. In: Crux J, van Hell SE (eds) Nannofossils and their applications. Ellis Horwood, Chichester, pp 1–20

– Westbroek P (1991) Genotypic variation in the coccolithophorid species *Emiliania huxleyi*. Mar Micropaleontol 18: 5–23

– Didymus JM, Mann S (1991) On the reported presence of vaterite and aragonite in coccoliths of *Emiliania huxleyi*. Bot Mar 34: 589–591

– – Bown PR, Prins B, Mann S (1992) Crystal assembly and phylogenetic evolution in heterococcoliths. Nature 356: 516–518

Protoplasma (1994) 181: 106–122

PROTOPLASMA

Structure and development of the cryptomonad periplast: a review

S. J. Brett**, L. Perasso+, and R. Wetherbee*

School of Botany, University of Melbourne, Parkville, Victoria

Received March 30, 1994
Accepted April 30, 1994

Summary. The structure and development of the complex periplast, or cell covering, of cryptomonads is reviewed. The periplast consists of the plasma membrane (PM) plus an associated surface periplast component (SPC) and cytoplasmic or inner periplast component (IPC). The structure of the SPC and IPC, and their association with the PM, varies considerably between genera. This review, which concentrates on cryptomonads with an IPC of discrete plates, discusses relationships between periplast components and examines the development of this unique cell covering. Formation and growth of inner plates occurs throughout the cell cycle from specialized regions termed anamorphic zones. Crystalline surface plates, which comprise the SPC in many cryptomonad species, appear to form by self-assembly of disorganized subunits. In *Komma caudata* the subunits are composed of a high molecular weight glycoprotein that is produced within the endomembrane system and deposited onto the cell surface within anamorphic zones. The self-assembly of subunits into highly ordered surface plates appears closely associated with developmental changes in the underlying IPC and PM.

Keywords: Cryptophyceae; Cell wall; Periplast; Self-assembly; Freeze fracture-freeze etch.

Introduction

The cryptomonads are a unique group of phytoflagellates, easily distinguished from other microalgae by their asymmetric shape. The distinctive appearance of these cells can be attributed, in part, to the possession of a subapical depression (termed the vestibulum) which may extend internally to form a gullet (Santore 1987; Hill and Wetherbee 1990; Hill 1991 a, b) or progress along the ventral surface into a furrow (Munawar and Bistricki 1979, Klaveness 1981, 1985, Kugrens et al. 1986). The vestibulum/furrow/gullet complex forms an important surface feature of all cryptomonad cells, and three major categories may be recognized within the group (Kugrens and Lee 1991) (Fig. 1).

The characteristic shape of the cryptomonads is maintained by a complex structure termed the periplast, which is present across most peripheral regions of the cell but never extends into the vestibulum, furrow or

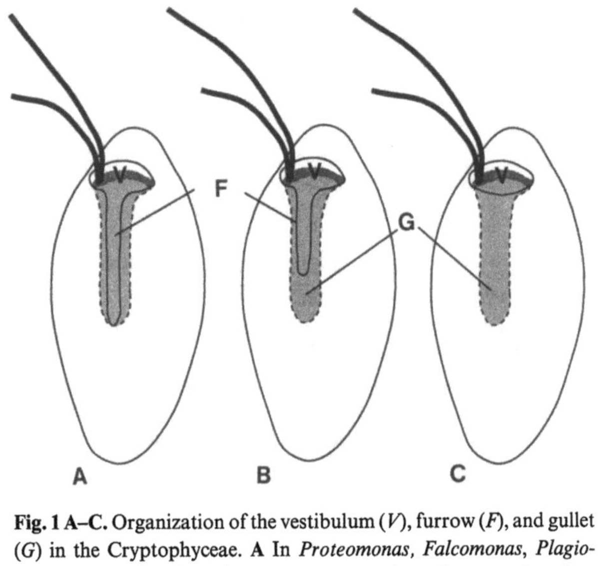

Fig. 1 A–C. Organization of the vestibulum (*V*), furrow (*F*), and gullet (*G*) in the Cryptophyceae. **A** In *Proteomonas, Falcomonas, Plagioselmis,* and *Teleaulax* a furrow progresses along the ventral surface from the vestibulum. **B** *Rhodomonas, Cryptomonas, Capylomonas,* and *Geminigera* possess both furrow and gullet. **C** The vestibulum extends internally to form a gullet in *Komma, Chroomonas, Rhinomonas, Hemiselmis, Guillardia,* and *Storeatula*

* Correspondence and reprints: School of Botany, University of Melbourne, Parkville, Vic 3052, Australia.

** Present address: Bigelow Laboratory for Ocean Sciences, W. Boothbay Harbor, Maine, U.S.A.

+ Present address: Botanisches Institut, Universität zu Köln, Federal Republic of Germany.

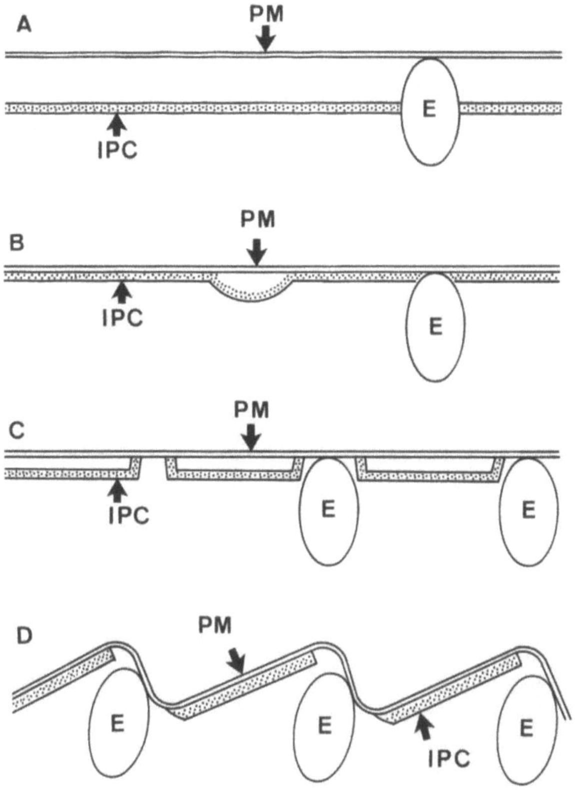

Fig. 2 A–D. Variations in inner periplast component (ICP) morphology throughout the Cryptophyceae. **A** ICP comprising a continuous sheet of material which is never closely associated with the PM. Ejectisomes (*E*) pass through pores in this sheet to contact the PM. **B** IPC comprising continuous sheet of material, closely appressed to the PM. This sheet may occasionally appear separated from the PM. **C** IPC consisting of discrete plates which are strongly attached to the PM at their edges. **D** IPC of discrete anteriorly stepped plates closely appressed to the PM. Ejectisome vesicles associate with PM adjacent to the anterior corners of plates

gullet. The morphology of this organelle differs markedly from the cell wall or cytoskeletal structures of other phytoflagellates. The general organization is similar in all cryptomonads, with the periplast consisting of an inner periplast component (IPC) and an additional surface periplast component (SPC) that support cells via a close association with the plasma membrane (PM) (Dodge 1969, Lucas 1970, Gantt 1971, Hibberd et al. 1971, Faust 1974, Santore 1977). Digestion of both the IPC and SPC using trypsin in several early EM studies (Gantt 1971, Faust 1974) led to the suggestion that the periplast components may be proteinaceous, while a more recent investigation of the composition of the SPC in a single species revealed a high molecular weight glycoprotein (Perasso et al. in prep.).

Examination of cryptomonad cells using scanning electron microscopy (SEM), thin sections and, more recently, freeze-fracture and freeze-etch has enabled accurate determination of periplast morphology in a wide range of genera. Studies incorporating these techniques reveal considerable variation in structure and organization of the IPC and SPC throughout the Cryptophyceae.

The IPC, which appears to form the primary structural elements of the periplast in all cryptomonads, may consist of a continuous sheet of material situated beneath the PM (Grim and Stahelin 1984; Brett and Wetherbee 1986; Hill and Wetherbee 1986; Wetherbee et al. 1986, 1987) or comprise a highly ordered system of discrete internal plates (Gantt 1971; Faust 1974; Hausmann and Walz 1979; Hill and Wetherbee 1986, 1988, 1989; Kugrens and Lee 1987). These differences, and further variations in the relationship between IPC and PM, enable recognition of four major IPC types within the group (Fig. 2).

Morphology of the SPC also varies markedly between taxa. Although early workers recognized some differences in surface organization from thin sectioned material (e.g., Santore 1977), detailed examination of the SPC was not possible until the incorporation of freeze-etch into the methodology. Use of this technique has allowed investigation of surface microarchitecture, revealing that the SPC may range from dense mats of fibrillar material (Fig. 3) to complex rosulate scales (Fig. 4) or highly ordered surface plates (Fig. 5) (Brett and Wetherbee 1986; Wetherbee et al. 1986, 1987; Hill and Wetherbee 1986, 1988, 1989, 1990).

In addition to enabling investigation of the gross details of periplast morphology, use of a combination of EM techniques (in particular freeze-fracture and freeze-etch) has revealed an extremely close relationship between the IPC, PM and SPC in many cryptomonads. Although the association between these periplast components has now been investigated in a wide range of genera (e.g., Hausmann and Walz 1979; Wetherbee et al. 1986, 1987; Kugrens and Lee 1987; Hill 1991 a, b), this review will concentrate on cells with an IPC composed of discrete internal plates. These cryptomonads, which include members of eight distinct genera (Table 1), provide model systems for examination of the interaction between IPC, PM and SPC and have been used extensively to investigate the mode of development of this unique and complex organelle (Brett and Wetherbee in prep.; Perasso et al. in prep.).

The inner periplast component and the plasma membrane

The plates that comprise the IPC in many cryptomonads often display some variation in shape and organization between genera. Details of the IPC, examined directly using thin sections and indirectly in freeze-fracture images, show two major types of inner plate within the group, and reveal differences within each of these types (Table 1). In most genera the inner plates are arranged into offset longitudinal rows which pass from the vestibular margins toward the cell posterior (Gantt 1971; Hill and Wetherbee 1986, 1988, 1989; Hill 1991 b). These plates are closely appressed to the underside of the PM and are anteriorly stepped, giving the periplast a serrated appearance in thin-sections (Fig. 6). Membrane-bound extrusive organelles (termed ejectisomes) are normally positioned adjacent to the anterior corners of each plate, and associate with the

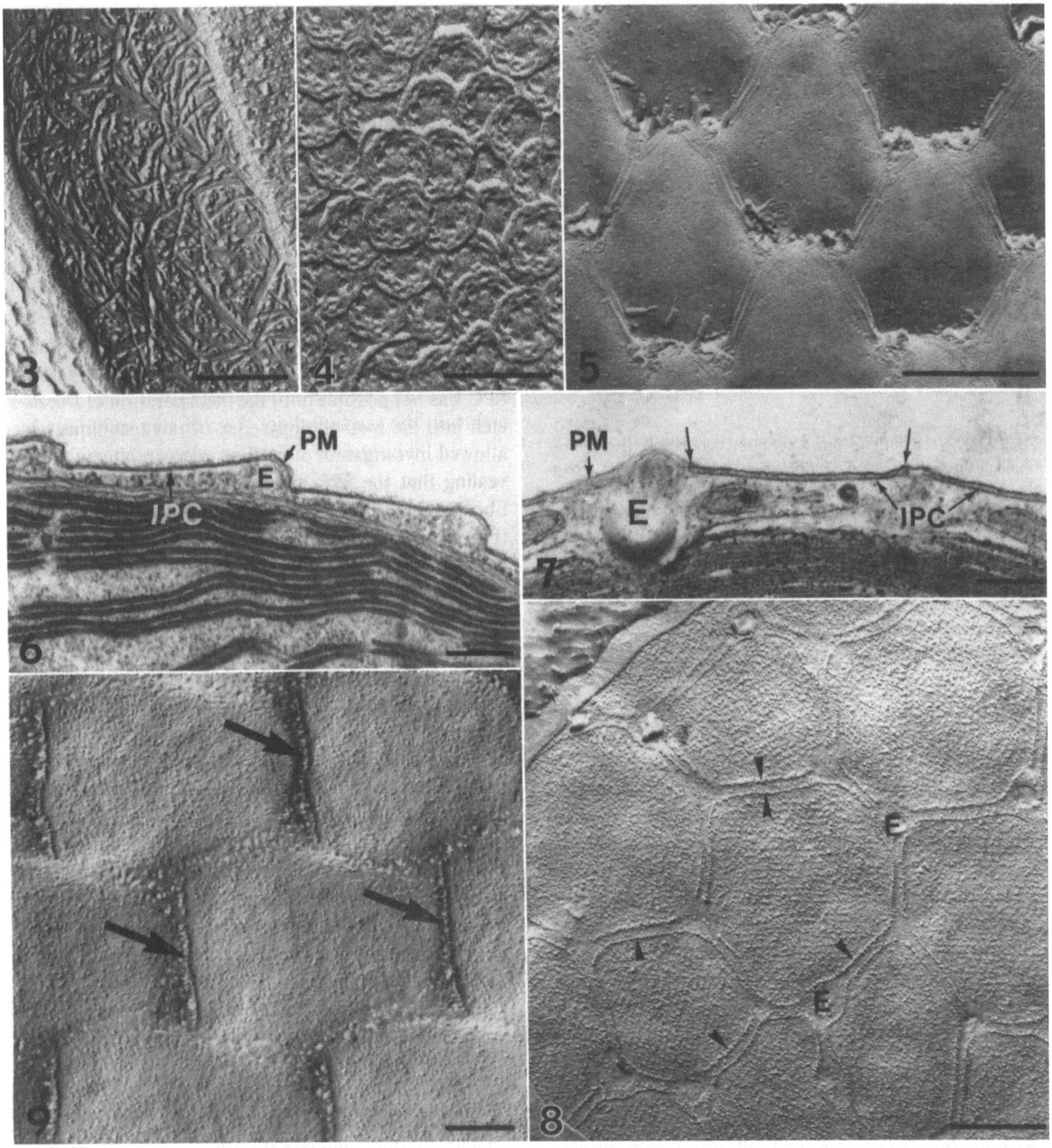

PM in these regions (Antia et al. 1973; Santore 1977, 1982, 1986, 1987; Meyer and Pienaar 1984; Erata and Chihara 1989).

A markedly different inner plate organization is characteristic of the genus *Cryptomonas*. In thin sections the periplast has a flattened appearance, and the inner plates appear most intimately associated with the PM at their edges (Fig. 7). The inner plates and PM commonly appear separate from one another in *Cryptomonas* although this features may result from shrinkage during fixation for electron microscopy (Kugrens and Lee 1987; Kugrens pers. comm., Hill pers. comm.). In contrast to other genera, the inner plates of *Cryptomonas* are not aligned into rows, but instead exhibit a polygonal (generally hexagonal) arrangement (Hibberd et al. 1971; Faust 1974; Santore 1977, 1984, 1985; Brett and Wetherbee 1986; Kugrens and Lee 1987).

Despite the variation in shape and morphology of inner plates, an extremely close relationship between the IPC and PM is evident in all genera. Freeze-fractures reveal that the PM is organized into discrete regions called domains, which are situated directly above inner plates (Figs. 8 and 9). The domains are densely packed with intra-membrane particles (IMPs), and may possess specialized rows of IMPs in regions where the IPC and PM are strongly attached (Hausmannn and Walz 1979; Brett and Wetherbee 1986; Hill and Wetherbee 1986, 1988; Kugrens and Lee 1987). The location and arrangement of attachment sites varies between genera. In *Cryptomonas*, particles are evident around the entire perimeter of each domain (Fig. 8), while in other genera ordered rows of IMPs are commonly observed along the posterior margins (Fig. 9).

In contrast to the ordered domains, regions of PM above the gaps between inner plates contain fewer, less ordered IMPs. These differences in the IMP organization suggest that the inner plates may act as a cytoskeletal template which directly influences the arrangement of IMPs within the PM.

The cell surface

The use of freeze-fracture/-etch preparation has also enabled examination of cell surface features and their relationship to underlying periplast components. In *Cryptomonas* and *Rhodomonas*, the SPC generally consists of elongate fibrils (or scales in *Rhodomonas stigmatica*) which form a dense mat across most of the cell surface (Brett and Wetherbee 1986, Hill and Wetherbee 1989). The arrangement of the SPC in these cryptomonads does not appear closely linked to the organization of the underlying PM and IPC. In all other cryptomonads discussed in this review, however, an intimate relationship is evident between the SPC, PM and IPC. The SPC is composed of discrete plates situated directly above ordered domains in the PM (e.g., Fig. 10).

The details of surface plates vary markedly between genera, and often provide important criteria for the separation of taxa (Table 1). In *Plagioselmis*, the surface plates are composed of particulate material and granular scales (Fig. 11), those of *Rhinomonas* (Hill and Wetherbee 1988) are composed of rod-like subunits with a triangular arrangement (Figs. 12 and 13), while *Komma, Falcomonas, Chroomonas* (Hill 1991 a), and *Proteomonas sulcata* (haplomorph) (Hill and Wether-

Figs. 3–5. Freeze-etch images of cryptomonad cell surfaces. Bars: Figs. 3 and 5, 0.5 μm; Fig. 4, 0.2 μm

Fig. 3. The SPC of *Proteomonas sulcata* (diplomorph), showing a dense mat of elongate fibrils

Fig. 4. Heptagonal scales form the SPC of *Geminigera cryophila*

Fig. 5. Ordered crystalline plates on the cell surface of *Proteomonas sulcata* (haplomorph)

Figs. 6 and **7.** Thin sections through periplast. Bars: Fig. 6, 0.2 μm; Fig. 7, 0.5 μm

Fig. 6. The IPC of *Proteomonas sulcata* consists of discrete anteriorly stepped plates. Ejectisome vesicles (*E*) are located adjacent to the anterior corners of each inner plate

Fig. 7. IPC of *Cryptomonas ovata* has a flattened appearance in T.S. The discrete inner plates (*IPC*) appear intimately associated with the PM at their edges (arrows). Ejectisome vesicles (*E*) associate with the PM in the gaps between plates

Figs. 8 and **9.** Freeze-fracture images of PM. Bars: Fig. 8, 0.5 μm; Fig. 9, 0.2 μm

Fig. 8. Polygonal arrangement of PM domains in *Cryptomonas ovata*. PM domains are densely packed with IMPs, and surrounded by a highly ordered row of particles (arrowheads). Ejectisome vesicles (*E*) are present adjacent to the corners of domains

Fig. 9. A distinct row of IMPs (arrows) defines the posterior margins of PM domains in *Proteomonas sulcata* (haplomorph). Regions of PM between adjacent domains contain fewer IMPs and have a pitted appearance

Table 1. Cryptomonads possessing an IPC of discrete plates. Variations in periplast morphology and arrangement are evident between genera

Organism	Vestibulum: furrow or/ and gullet	IPC type[a]	General shape of plate areas	Attachment sites between IPC and PM	Arrangement of plate areas	Modifications in mid-ventral region	SPC type[b]	Component of SPC	Ordered plate borders	Specialised features of cell posterior
Chroomonas	gullet	A	± rectangular	posterior of PM domain	longitudinal rows	nil	A	hexagonally packed subunit	present	raphe
Komma	gullet	A	± hexagonal	posterior of PM domain	long. rows	nil	A		present	tail-plate
Rhinomonas	gullet	A	± hexagonal	posterior of PM domain	long. rows	plate areas reduced in size	A	triangular packing, rod-like subunits	present	raphe-like structure
Proteomonas sulcata (haplomorph)	furrow	A	± hexagonal (rounded anterior)	posterior of PM domain	long. rows	no plate areas	A	hexag. packed subunits	present	MVB
Falcomonas	furrow	A	± hexagonal	nil	long. rows	no plate areas	A	hexag. packed subunits	absent	MVB
Plagioselmis	furrow	A	± hexagonal	nil	long. rows	no plate areas	A	granules and scales	present	MVB
Rhodomonas	furrow and gullet	A	± rectangular	posterior of PM domain	long. rows	plate areas reduced i.s.	B	fibrils (scales in *R. stigm.*)	absent	plate areas often absent
Cryptomonas	furrow and gullet	B	polygonal (rounded corners)	edges of PM domain	polygonal	plate areas reduced i.s.	B	fibrils	absent	plate areas small and irregular in shape

Variations in the shape of plate areas may occur during preparation of cells for EM (particularly SEM). Santore (1977) cautioned against sole use of this feature in critical taxonomic determinations

[a] *A* Discrete plates closely appressed to PM; *B* discrete plates attached to PM at edges

[b] *A* Organised surface plates; *B* no surface plates

bee 1986) possess highly ordered crystalline plates (Figs. 14–17). Crystalline plates are composed of minute subunits precisely arranged into ordered lattices. Regions of the cell surface between adjacent plates have a markedly different appearance; they contain particulate material that may be randomly arranged (Fig. 14), or aligned into distinct raised borders (Figs. 10 and 14–17). Close examination of these areas reveals that the particulate material is similar in size to (and may display continuity with) the lattice subunits (Figs. 16 and 17).

Recent immunological studies on *Komma caudata* suggest that subunits which comprise the lattices and borders of the SPC in this organism are high molecular weight glycoproteins (Perasso et al. 1994). Monoclonal antibodies selected for surface labelling recognize both the crystalline lattices and plate borders, with most intense labelling occurring in the border regions (e.g., Figs. 50–53). Similar observations have also been reported for the Volvocalean alga *Lobomonas piriformis*, which possesses a layered cell wall overlain by crystalline surface plates surrounded by amorphous boundaries (Roberts et al. 1981, Shaw and Hills 1982). Studies on *L. piriformis* indicate that subunits at plate edges are the same as those within the crystalline lattice, and suggest that conformational changes or biochemical modification of edge subunits is necessary for incorporation into lattices. The intense labelling observed at the plate borders in both *L. periformis* and *K. caudata* suggests that the epitope recognized by the surface antibody is most accessible when subunits are in the border conformation (reviewed in Perasso et al. in prep.).

In the cryptomonads, variations in the arrangement of the border and lattice subunits appear closely associated with the organization of the underlying IPC and PM. Lattices are always located directly above ordered domains and inner plates, whereas the borders are positioned above gaps in the IPC (Fig. 10). These features suggest that alignment of the subunits into the crystalline lattices may be directly influenced by underlying IMPs within PM domains. The domains (and IPC) may act to bind or induce conformational changes in these molecules that will facilitate their consolidation into the ordered lattice arrangement.

Arrangement of plate areas across cryptomonad cells

In all the cryptomonads examined in this review, the inner plates, PM domains and associated surface structures from discrete regions termed plate areas (Gantt 1971). The plate areas are present in most regions of the cell periphery but do not extend into the vestibulum, furrow or gullet. These regions are bounded only by PM, which contains few IMPs and has a disordered appearance in freeze-fracture images (Figs. 18 and 19). Elsewhere, however, plate areas form a highly ordered system.

Studies using SEM and freeze-fracture/-etch have allowed detailed investigation of periplast arrangement and reveal considerable variation in the size (and to some extent shape) of these plate areas across cryptomonad cells (Brett and Wetherbee in prep.). In all cryptomonads examined using these techniques, plate areas appear largest in the equatorial regions, and are generally reduced in size as cells taper toward the anterior and posterior (Figs. 18–21). In addition to these variations in size, the arrangement of plate areas across the ventral surface is often highly specialized. The periplast is commonly separated along the mid-ventral line, with plate areas adjacent to this region reduced in size and modified in shape (Figs. 18, 22, and 23). Although this unusual arrangement is most conspicuous in cryptomonads which possess a furrow, separation of the periplast along the entire mid-ventral line is evident in all genera discussed here except *Komma* and *Chroomonas*. Further modifications in the arrangement of plate areas are also observed in the posterior of cryptomonad cells. The periplast may be absent from this region, or alternately may terminate adjacent to specialized features including the raphe (Gantt 1971, Hill 1991 a), tail plate (Hill 1991 a), or mid-ventral band (Hill and Wetherbee 1986, 1989; Hill 1991 a).

Changes in periplast arrangement throughout the cell cycle

Recent studies have directed attention to changes in periplast arrangement and organization throughout the cell cycle (Brett and Wetherbee in prep.). These reports reveal that substantial development of the periplast occurs during cytokinesis, and also following division as cells enlarge and mature. Cell division in the cryptomonads (Perasso et al. 1992, 1993) involves a unique process termed pole reversal. During cytokinesis the tail regions of daughter cells develop from the anterior of the parental cell, necessitating complete realignment of the periplast. Although slight variations occur between cryptomonad genera, the major features of this unusual mode of cell division arc similar throughout the group, and are described here for *K. caudata*.

At the onset of division, flagella replication occurs in the vestibulum, and lateral expansion of cells is apparent (Figs. 24 and 25). The periplast separates along the mid-ventral line progressing from the vestibulum toward the cell posterior. The edges of the periplast exposed during this process form specialized regions, termed anamorphic zones, which undergo developmental changes both during and after cytokinesis. As the ventral surface divides, these regions, as well as additional anamorphic zones located around the vestibular margins, expand and realign (Figs. 26–28). Distinct pointed structures form where the anamorphic zones converge, and eventually develop into the tail regions of daughter cells. The dorsal surface then divides, the forming tail regions migrate laterally (Figs. 29–31) and rotation of cell halves relative to one

---

Fig. 10. Freeze-fracture/-etch of *Proteomonas sulcata* (haplomorph). The cell surface consists of ordered crystalline plates, surrounded by distinct raised borders (*b*). The plates are located directly above PM domains, while borders are located above less ordered regions of PM (arrows); 0.5 μm

Figs. 11–17. Freeze-etch images of cryptomonad cell surfaces. Bars: Figs. 11 and 13–15, 0.5 μm; Fig. 12, 0.2 μm; Figs. 16 and 17, 0.1 μm

Fig. 11. The surface plates of *Plagioselmis prolonga* are covered in particulate material and granular scales (*s*). Plates are surrounded by raised borders (*b*)

Fig. 12. Surface plates of *Rhinomonas pauca* consisting of rod-like subunits with a triangular arrangement. Border regions (*b*) are composed of elongate subunits

Fig. 13. Arrangement of surface plates in *Rhinomonas pauca*

Fig. 14. SPC of *Falcomonas daucoides*. The crystalline plates are composed of minute subunits. Border regions (*b*) contain disordered particulate material

Fig. 15. The crystalline plates of *Proteomonas sulcata* (haplomorph) are surrounded by distinct raised borders (*b*). Scales are also evident on plate surfaces (*s*)

Fig. 16. Detail of crystalline plate of *Proteomonas sulcata* (haplomorph) showing ordered arrangement of subunits. Border (*b*) subunits may occasionally appear continuous with those of the lattice (arrow)

Fig. 17. SPC of *Komma caudata* showing continuity between lattice and border (*b*) subunits

Figs. 18–21. Freeze-fracture images (PF) of *Proteomonas sulcata* (haplomorph). Bars: Figs. 18 and 21, 0.5 μm; Fig. 19, 0.2 μm; Fig. 20, 1 μm

Fig. 18. Ventral surface of cell, showing reduced PM domains adjacent to the vestibulum (*V*) and furrow (*Fu*). A mid-ventral band (*MV*) extends posteriorly from the furrow

Fig. 19. Small PM domains (arrows) at the margins of the vestibulum (*V*)

Fig. 20. The MV passes along the mid-ventral line to the cell posterior. Regions of PM adjacent to this structure contain few, randomly arranged IMPs (stars). PM domains are smallest at the edges of the periplast (arrow)

Fig. 21. Small PM domains (arrows) at the margins of the periplast adjacent to the mid-ventral line in the cell posterior

Figs. 22 and 23. Freeze-fracture images (PF) of *Rhinomonas pauca*. Bars: 0.5 μm

Fig. 22. Ventral surface showing reduced size of PM domains (arrows) adjacent to the mid-ventral line. This specialization passes from the vestibulum (*V*) to cell posterior

Fig. 23. Small PM domains along mid-ventral line (line) appear less rigidly organized than adjacent domains

Figs. 24–32. SEM images of *Komma caudata* during cell division. Bars: 2 μm

Figs. 24 and 25. Periplast separates along the mid-ventral line at onset of division. Lateral expansion of cells occurs

Figs. 26–28. Mid-ventral (*mva*) and vestibular (*va*) anamorphic zones expand and realign. Pointed tail regions form where the anamorphic zones converge (large arrows). Periplast develops a flattened appearance

Figs. 29–31. Forming tail regions separate laterally and dorsal surface divides. Anamorphic zones align along the mid-ventral line of forming daughter cells (arrowheads)

Fig. 32. Cell halves rotate to facilitate separation

Figs. 33–35. SEM images of *Komma caudata* following cell division. Bars: 2 μm

Fig. 33. Periplast of daughter cells is highly ordered but divided along the mid-ventral line. Small plate areas are evident adjacent to this region

Fig. 34. As cells enlarge the periplast becomes continuous across the ventral surface

Fig. 35. Mature cell showing well defined vestibulum, and increase in number of plate areas comprising the periplast

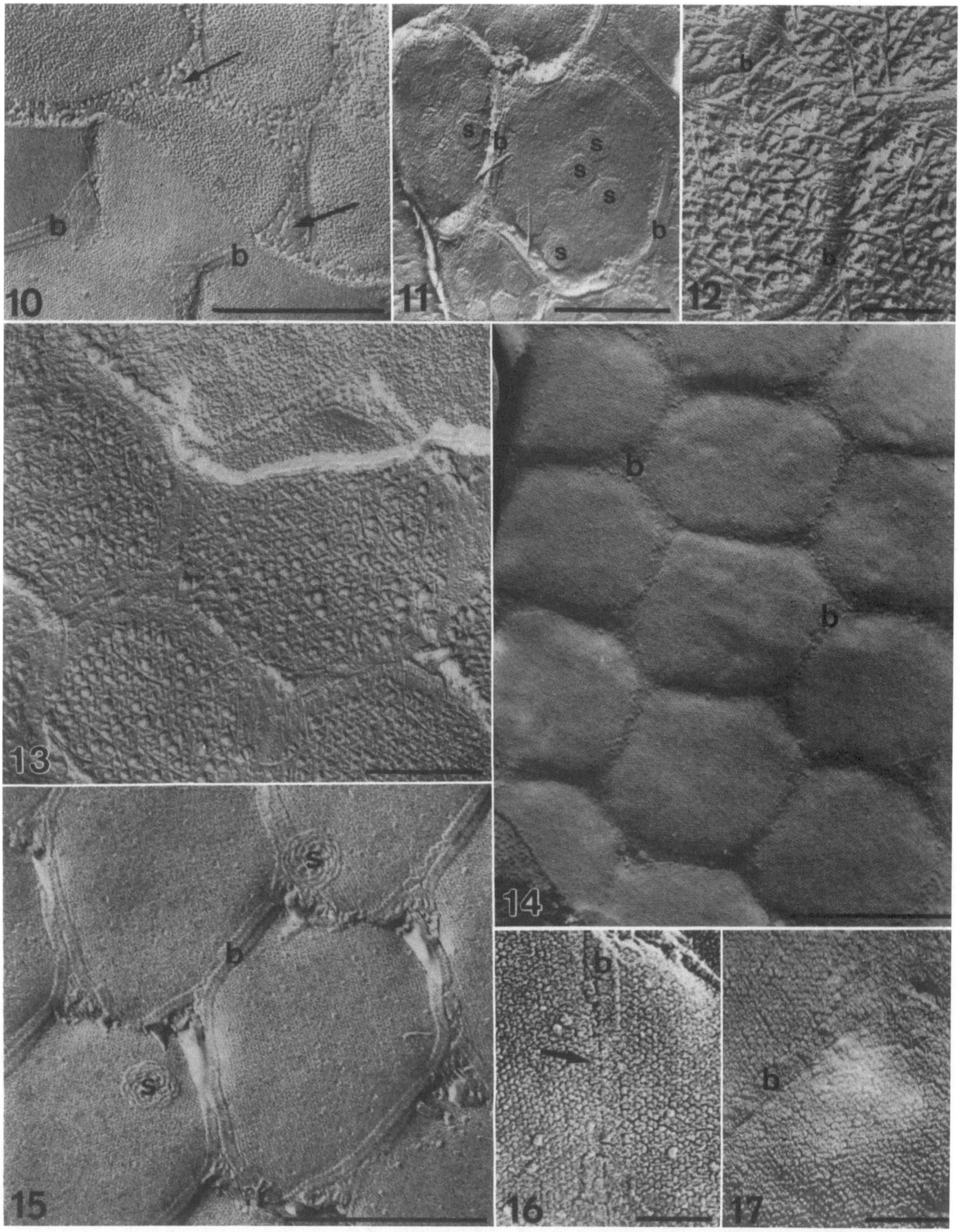
10
b
b
11
s
b
s
s
s
s
b
12
b
b
13
14
b
b
15
s
s
b
16
b
17
b

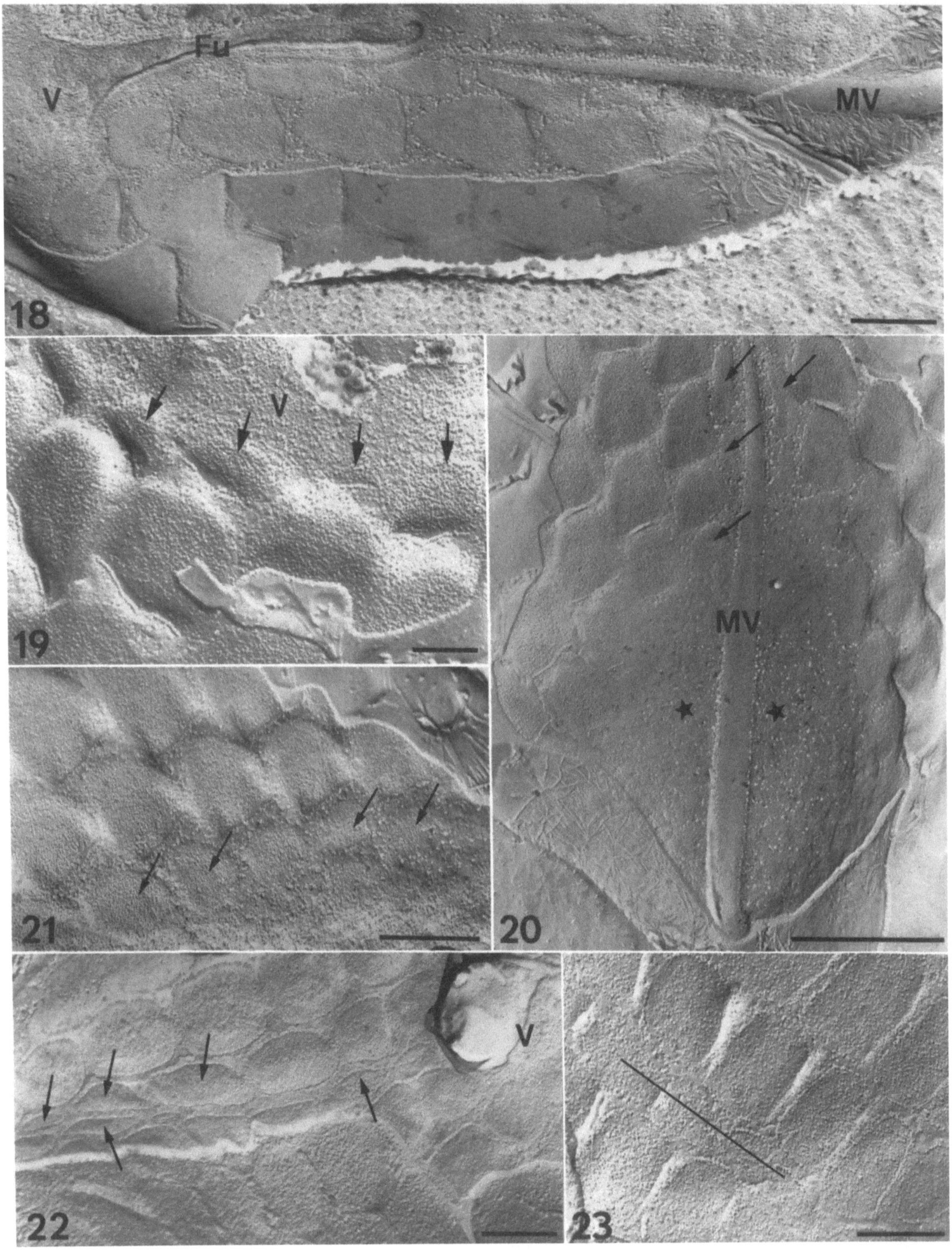
18
Fu
V
MV
19
V
20
MV
21
22
V
23

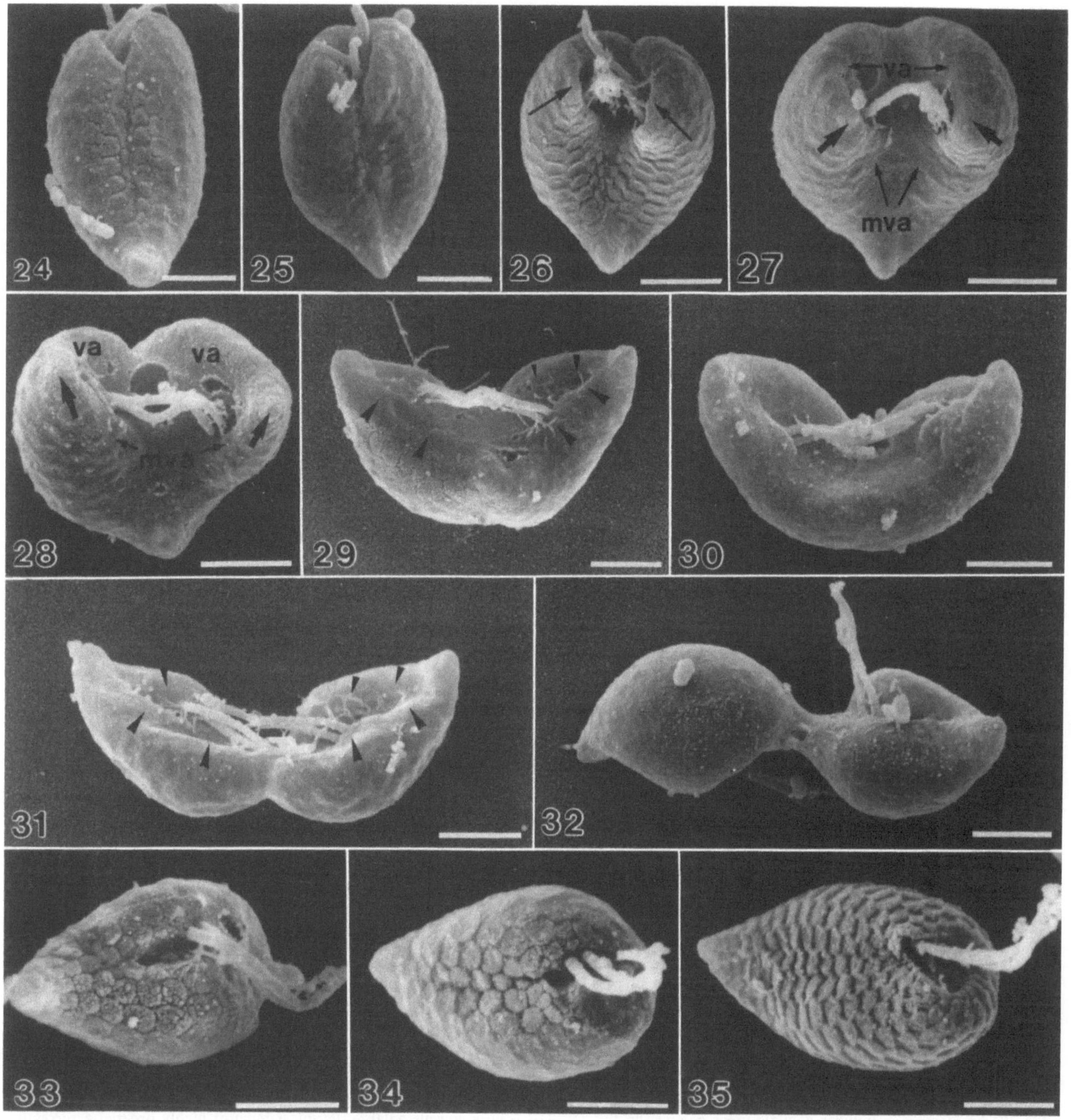
24
25
26
27
va
mva
28
va
va
mva
29
30
31
32
33
34
35

another facilitates separation (Fig. 32). Further realignment of the anamorphic zones occurs, and these eventually position on either side of the mid-ventral line of developing daughter cells (Figs. 32 and 33).

The unusual mode of cell division in the cryptomonads, involving reorientation of daughter cells relative to the parental cell, necessitates major changes in the periplast in many cryptomonads (Brett and Wetherbee 1994 a). In cells with anteriorly stepped plate areas, the periplast persists, but develops a flattened appearance during cell division (Figs. 27–32). The plate areas remain precisely aligned throughout this process, and after cytokinesis adopt a polarity consistent with the newly formed daughter cell. Although the process of periplast reorientation is not fully understood, detailed examination of cells using thin sections and freeze-fracture/-etch suggests that realignment may be facilitated by relocation of the attachment sites between inner plates and the PM (Brett and Wetherbee in prep.).

Following cell division, daughter cells are substantially smaller than their parental counterparts, the vestibular margins are indistinct and the periplast is always divided along the mid-ventral line (Fig. 33). As cells enlarge and mature the vestibulum/furrow/gullet complex becomes more clearly defined, and changes in periplast organization occur. In *Komma* (Figs. 34 and 35) and *Chroomonas* the periplast becomes continuous across the ventral surface as the gullet develops, although in all other organisms the periplast remains separated along the mid-ventral line throughout the cell cycle (Figs. 36–39).

The elongation and lateral expansion of cells that occurs after division is associated with an increase in the number of plate areas comprising the periplast (Figs. 33–39). The periplast remains precisely aligned throughout this process, suggesting that growth occurs in an orderly manner by addition of new plate areas to anamorphic zones (Brett and Wetherbee in prep.). In most genera elongation results from addition of new plate areas to anamorphic zones around the vestibulum and in the posterior, while lateral expansion is facilitated by growth and addition from the mid-ventral line (Figs. 36–39). Growth from these anamorphic zones appears to enable highly ordered expansion of cryptomonads throughout the entire cell cycle. The absence of mid-ventral specializations in *Komma* (Figs. 34 and 35) suggests, however, that lateral expansion in this genus may occur by the enlargement of existing plates in the ventral regions of the cell.

Formation and development of the inner plates

The general changes in the arrangement of plate areas revealed in the SEM studies provided a basis for more detailed investigation of periplast development. Subsequent work (which indirectly examined the formation and growth of the inner periplast plates using freeze fracture images) revealed variations in the size, organization and alignment of PM domains within anamorphic zones (Brett and Wetherbee in prep.). Despite differences in the morphology of the inner plates between genera, similar features were observed in all cryptomonads examined.

Figs. 36–39. SEM images of *Proteomonas sulcata* (haplomorph) following cell division. Bars: 1 µm

Fig. 36. Immediately after cell division vestibular margins and furrow are not clearly defined. Periplast is highly ordered but has a flattened appearance

Fig. 37. As cells enlarge the anteriorly stepped periplast polarity becomes evident

Fig. 38. An increase in number of plate areas comprising the periplast occurs as cells enlarge. Periplast remains highly ordered as growth occurs

Fig. 39. The periplast remains divided along the mid-ventral line throughout the cell cycle. Growth of the periplast occurs by addition of new plate areas from anamorphic zones around vestibulum, mid-ventral line and cell posterior (arrowheads)

Figs. 40 and 41. Freeze-fracture images (PF) of *Cryptomonas ovata.* Bars: 0.5 µm

Fig. 40. PM domains around anamorphic zones are smaller and less rigidly ordered than in other regions of the cell

Fig. 41. Gradation in the size of PM domains adjacent to vestibulum (*V*)

Figs. 42 and 43. Freeze-fracture images (EF) of *Rhodomonas baltica.* Bars: 0.2 µm

Fig. 42. Vestibular anamorphic zone at onset of division. PM domains show variations in size and irregular alignment

Fig. 43. Mid-ventral line at onset of division. Periplast is distinctly divided in this region and small PM domains are located at edges of the anamorphic zone (arrows)

The PM domains within anamorphic zones range over a continuum from minute accumulations or circular aggregates of IMPs to larger, more rigidly ordered regions densely packed with IMPs and possessing specialized attachment sites. Although most conspicuous in dividing and newly formed cells (Figs. 40–43), similar variations are evident adjacent to anamorphic zones throughout the cell cycle (Figs. 18–23). The distinct

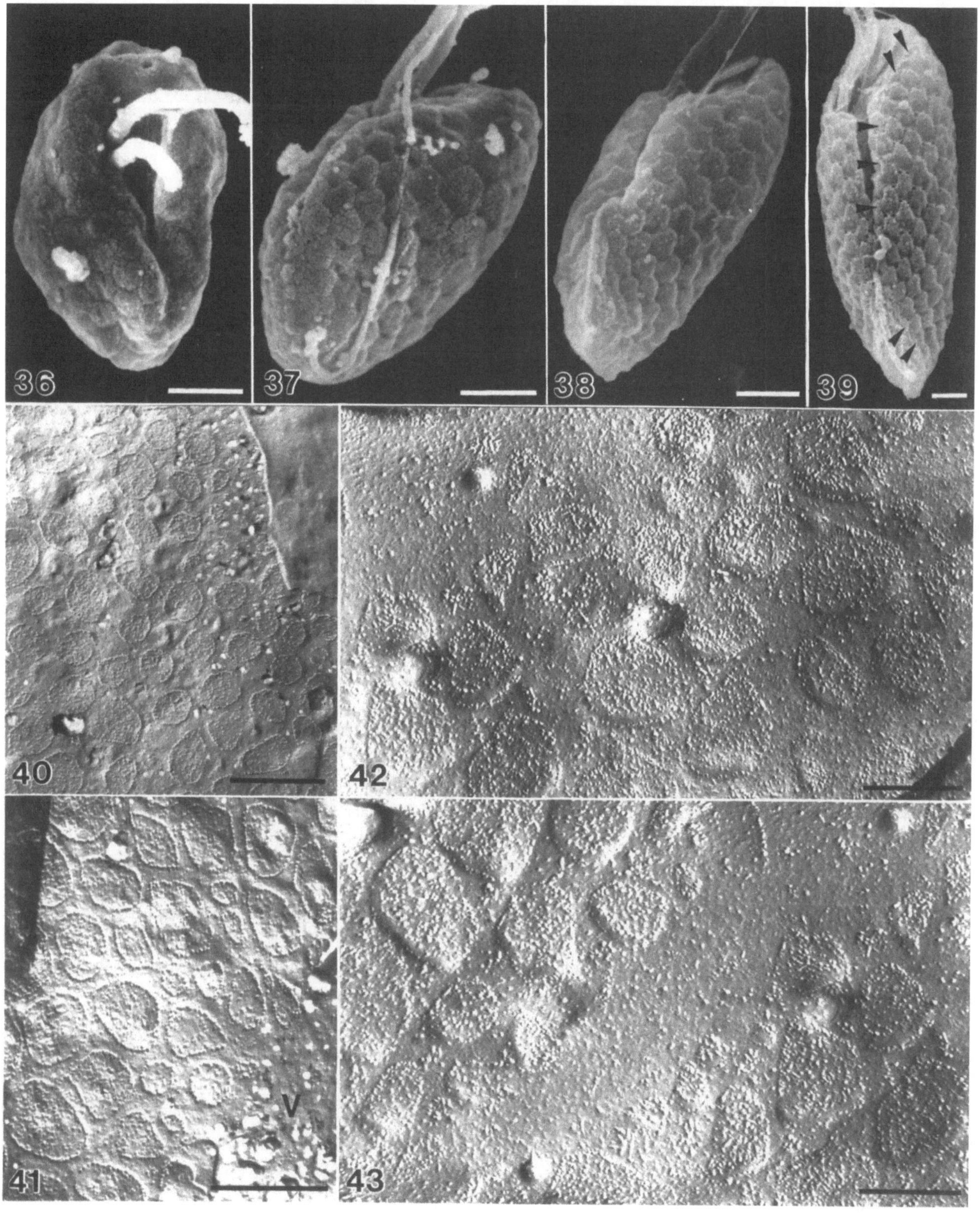

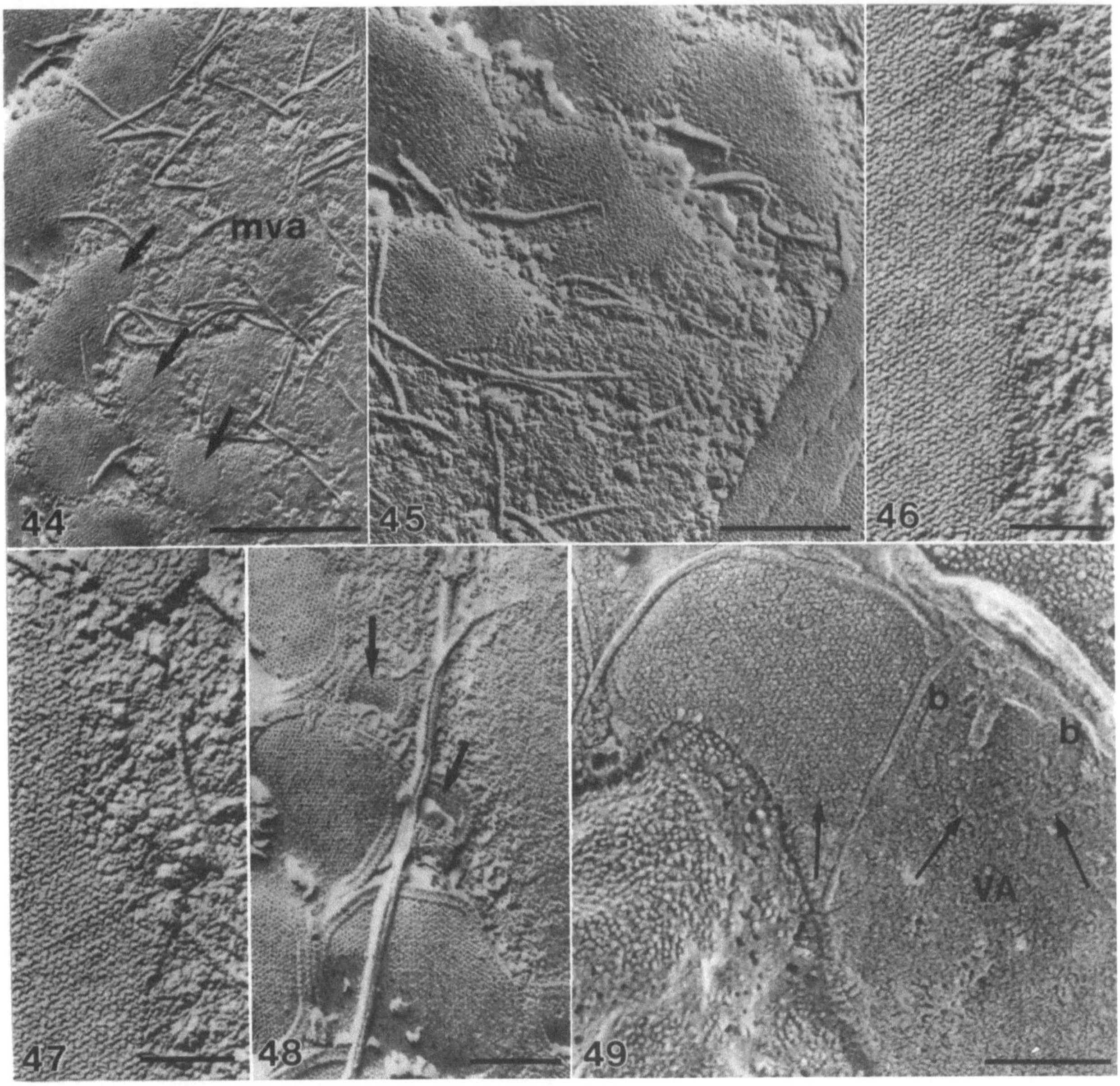

Figs. 44–47. Freeze-etch images around anamorphic zones in *Falcomonas daucoides*. Bars: Fig. 44, 0.5 µm; Fig. 45, 0.2 µm; Figs. 46 and 47, 0.1 µm

Fig. 44. Surface plates adjacent to the mid-ventral anamorphic zone (*mva*) are reduced in size (arrows). Surrounding regions of cell surface contain disordered particles and elongate fibrils

Fig. 45. Particulate material adjacent to the periplast may align into loose aggregates. The plate edges are often indistinct, and closely associated with this material

Figs. 46 and **47.** Surrounding surface particles are similar in size, and often appear continuous with subunits of the crystalline surface lattices

Figs. 48 and **49.** Freeze-etch images around anamorphic zones in *Proteomonas sulcata* (haplomorph). Bars: 0.2 µm

Fig. 48. Vestibular anamorphic zone. Surface plates are reduced in size (arrows), but are surrounded by distinct raised borders

Fig. 49. Vestibular anamorphic zone (*VA*) in dividing cell. The lattice subunits, borders subunits (*b*) and less ordered particulate material may occasionally appear continuous during rapid periplast development (arrows)

gradation in the size of domains within these regions indicates that growth of the periplast does not occur by addition of large, fully formed inner plates at the edges of the IPC. The features observed suggest, instead, that the inner plates form de novo and undergo considerable enlargement within these specialized anamorphic zones. Variations in the spacing and arrangement of PM domains in developing cells (Figs. 40, 42, and 43) imply that the inner plates develop independently, and are capable of some movement relative to one another as they align within the highly ordered periplast.

Development of crystalline surface plates

The intimate relationship between IPC, PM and SPC in the cryptomonads suggests that the development of crystalline surface plates may be closely linked to the formation and growth of the inner plates within anamorphic zones. The unique SPC of *Falcomonas daucoides*, comprising crystalline surface plates which lack distinct borders (Hill 1991 a), provides a model system for investigation of features associated with lattice development (Brett and Wetherbee in prep.). Examination of this organism reveals that surface plates within anamorphic zones are commonly reduced in size, and surrounding areas of the cell surface are covered with disordered particulate material (Figs. 44 and 45). The surface particles are similar in size to lattice subunits, and may be aligned into loose aggregates or closely associated with the edges of periplast plates (Figs. 45–47). Plate margins are often indistinct and surface particles adjacent to these regions may appear continuous with, and are often indistinguishable from, lattice subunits.

The close relationship between the particulate material and lattice subunits suggests that development of the crystalline surface may occur by accumulation and incorporation of less ordered precursors from surrounding regions of the cell surface. After formation of these crystalline plates, enlargement and growth appears facilitated by addition of subunits to the lattice edges. Detailed studies of bacteria (for reviews, see Sleytr and Messner 1983, Messner and Sleytr 1992) and *L. piriformis* (Roberts et al. 1981, Shaw and Hills 1982) describe a similar mode of lattice development and suggest that conformational changes occur to precursor molecules during addition. As stated above, the close spatial relationship between the organization of the IPC, PM and SPC in the cryptomonads suggests that the formation and enlargement of surface plates may be intimately associated with developmental changes in the IPC and PM. Growth of the inner plates and related expansion of PM domains, may induce changes in the precursor molecules which enable incorporation into the highly ordered lattice.

Although the relatively simple arrangement of the SPC in *F. daucoides* provides an ideal system for examination of lattice assembly, the crystalline surface plates in other genera are generally surrounded by distinct raised borders (Figs. 48 and 49) (Hill and Wetherbee 1986, Hill 1991 a). Consequently, growth of these lattices during self-assembly does not appear to occur by direct addition of disordered precursors from surrounding regions of the cell surface. The continuity between borders and lattices in these organisms (Figs. 16, 17, and 49) suggests, instead, that subunits may be rearranged. For example, subunits previously within a border may subsequently become incorporated into the highly ordered surface plates during development (Brett and Wetherbee in prep.). The close relationship between the organization of the SPC and underlying periplast components indicates that incorporation of border subunits into the lattices during self-assembly may be closely associated with the growth of inner plates and PM domains.

The immunological work of Perasso et al. (in prep.) also suggests similarities between the components comprising the lattices, borders and less ordered regions of the cell surface. These studies indicate that the subunits which form the SPC are produced in the Golgi apparatus and secreted through the endomembrane system for deployment at the edges of the periplast (Figs. 50–53). Examination of *K. caudata* through the cell cycle reveals that deposition of these macromolecules occurs predominantly within anamorphic zones, where the majority of lattice formation occurs.

Conclusions

The complexity of the cryptomonad periplast has necessitated use of a wide range of techniques to accurately determine morphology of the IPC, PM and SPC. Although general features of periplast structure and arrangement may be determined using thin sections and SEM, the use of freeze-fracture/-etch has greatly enhanced the understanding of this unique organelle. Studies incorporating these techniques reveal considerable diversity throughout the Cryptophyceae and provide important characters for taxonomic distinctions within the group. The detailed information about PM and surface microarchitecture available from these

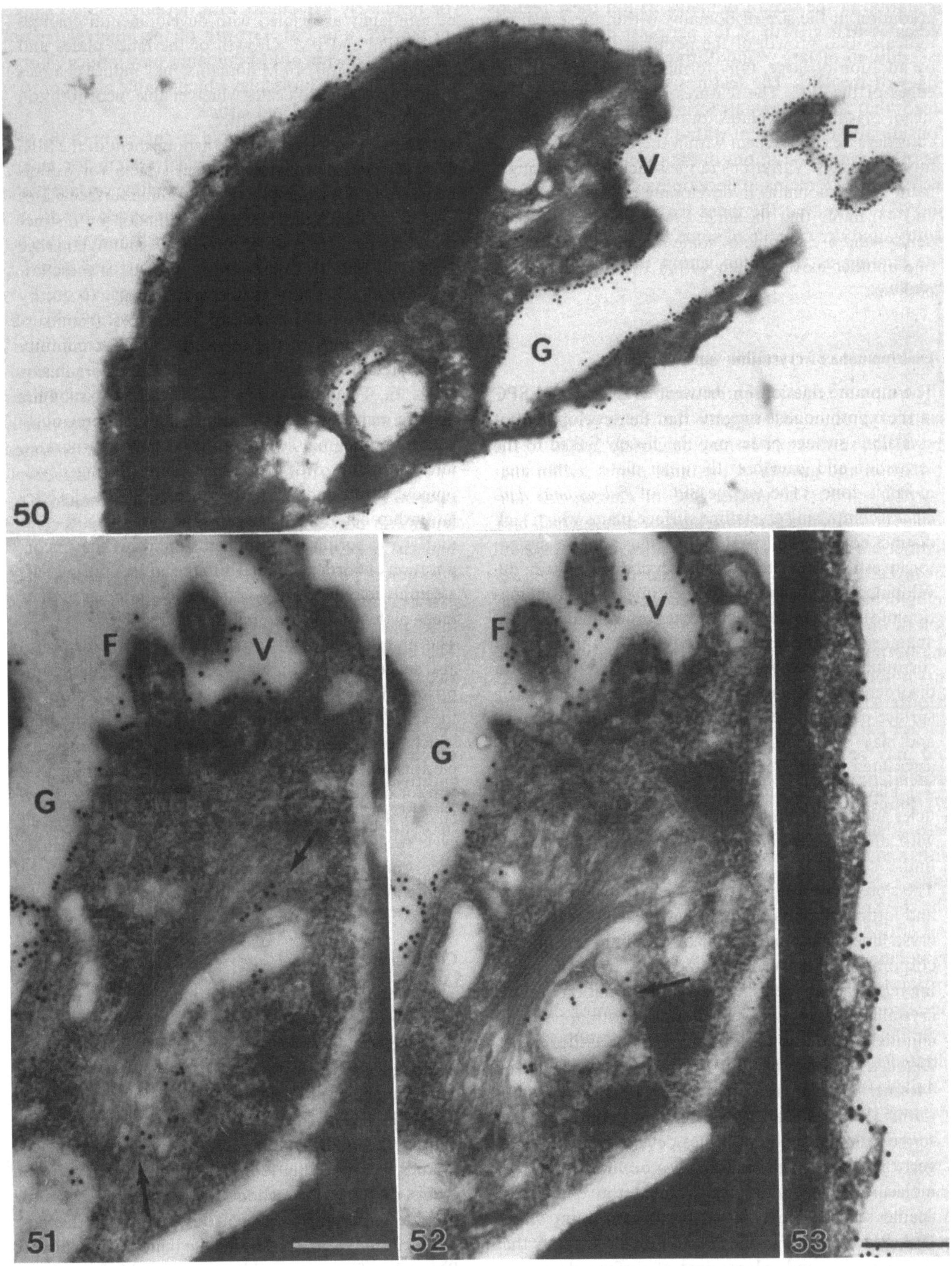
50
V
F
G
51
F
V
G
52
F
V
G
53

reports has also enabled examination of the complex relationships between components comprising the periplast.

The cells discussed in this review all possess an IPC of discrete plates, and have proven ideal systems for investigation of the association between IPC, PM, and SPC. In all these organisms the PM is ordered into discrete domains which are situated directly above the inner periplast plates. In contrast, regions of the PM not supported by the IPC generally contain fewer IMPs and appear less ordered. The IPC-PM association suggests that the inner plates are cytoskeletal, and may influence the arrangement of IMPs within the PM. Organization of the crystalline surface plates found in many cryptomonad genera also appears closely associated with that of the underlying components. The location of these plates directly above ordered PM domains suggests that the PM, and ultimately the IPC, may influence the alignment of subunits within surface lattices.

Development of both the IPC and SPC occurs from specialized anamorphic zones throughout the cell cycle. Unlike the wall components of many microalgal groups, fully formed inner and surface plates are not pre-packaged within cytoplasmic vesicles for later addition to the periplast. Instead the inner periplast plates appear to form de novo and undergo considerable enlargement within anamorphic zones. Growth of the crystalline surface lattices, which occurs by addition and incorporation of less ordered precursors from surrounding surface areas, may be closely associated with this process. The intimate relationship between IPC, PM, and SPC suggests that growth of inner plates and PM domains may result in modifications which enable incorporation of precursor molecules into the highly organized surface lattices. The precise mode of formation may be more fully understood as further details of periplast chemistry are revealed.

Acknowledgements

RW thanks the Australian Research Council for financial assistance. SJB and LP were supported by Commonwealth Postgraduate Scholarships, the University of Melbourne. We thank Dr. Paul Kugrens for Fig. 1 and Dr. David R. A. Hill for valuable discussions and for reviewing the manuscript.

References

Antia NJ, Kalley JP, McDonald J, Bisalputra T (1973) Ultrastructure of the marine cryptomonad *Chroomonas salina* cultured under conditions of photoautotrophy and glycerol-heterotrophy. J Protozool 20: 377–385

Brett SJ, Wetherbee R (1986) A comparative study of periplast structure in *Cryptomonas cryophila* and *C. ovata* (Cryptophyceae). Protoplasma 131: 23–31

Dodge JD (1969) The ultrastructure of *Chroomonas mesostigmatica* Butcher (Cryptophyceae). Arch Microbiol 69: 266–280

Erata M, Chihara M (1989) Re-examination of *Pyrenomonas* and *Rhodomonas* (Class Cryptophyceae) through ultrastructural survey of red pigmented cryptomonads. Bot Mag Tokyo 102: 429–443

Faust MA (1974) Structure of the periplast of *Cryptomonas ovata* var. *palustris*. J Phycol 10: 121–124

Gantt E (1971) Micromorphology of the periplast of *Chroomonas* sp. (Cryptophyceae). J Phycol 7: 177–184

Grim JN, Staehelin LA (1984) The ejectisomes of the flagellate *Chilomonas paramecium*: visualization by freeze-fracture and isolation techniques. J Protozool 31: 259–267

Hausmann K, Walz B (1979) Periplaststruktur und Organisation der Plasmamembrane von *Rhodomonas* spec. (Cryptophyceae). Protoplasma 101: 349–354

Hibberd DJ, Greenwood AD, Griffiths HB (1971) Observations on the ultrastructure of the flagella and periplast in the Cryptophyceae. Br Phycol J 6: 61–72

Hill DRA (1991 a) *Chroomonas* and other blue-green cryptomonads. J Phycol 27: 133–145

– (1991 b) A revised circumscription of *Cryptomonas* (Cryptophyceae) based on examination of Australian strains. Phycologia 30: 179–188

– Wetherbee R (1986) *Proteomonas sulcata* gen. et sp.nov. (Cryptophyceae), a cryptomonad with two morphologically distinct and alternating forms. Phycologia 25: 521–543

– – (1988) The structure and taxonomy of *Rhinomonas pauca* gen. et sp.nov. (Cryptophyceae). Phycologia 27: 355–365

– – (1989) A reappraisal of the genus *Rhodomonas* (Cryptophyceae). Can J Bot 28: 143–158

– – (1990) *Guillardia theta* gen. et sep.nov. (Cryptophyceae). Can J Bot 68: 1873–1876

Klaveness D (1981) *Rhodomonas lacustris* (Pascher & Ruttner) Javornicky (Cryptomonadida): ultrastructure of the vegetative cell. J Protozool 28: 83–90

Figs. 50–53. MAb immunogold labelling of *Komma caudata*. Bars: Figs. 50–52, 1 µm; Fig. 53, 0.5 µm

Fig. 50. Labelling of an interphase cell, labelling is observed over the entire surface, most obvious at the surface plate edges, and also over the vestibular (*V*), gullet (*G*), and flagellar (*F*) surfaces

Figs. 51 and **52.** Labelling of non-consecutive serial sections through a prophase cell, showing labelling about the GA and vesicles associated with the GA (arrows)

Fig. 53. High magnification of labelling of the periplast

Klaveness D (1985) Classical and modern criteria for determining species of Cryptophyceae. Bull Plankton Soc Jap 32: 111–123

Kugrens P, Lee RE, Andersen RA (1986) Cell form and surface patterns in *Chroomonas* and *Cryptomonas* cells (Cryptophyta) as revealed by scanning electron microscopy. J Phycol 22: 512–522

Kugrens P, Lee RE (1987) An ultrastructural survey of cryptomonad periplast using quick-freezing freeze fracture techniques. J Phycol 23: 365–376

– – (1991) Organization of cryptomonads. In: Patterson DJ, Larsen J (eds) The biology of free-living heterotrophic flagellates. Clarendon Press, Oxford, pp 219–233

Lucas IAN (1970) Observations on the fine structure of the Cryptophyceae. 1. The genus *Cryptomonas*. J Phycol 6: 30–38

Messner P, Sleytr UB (1992) Crystalline bacterial cell-surface layers. Adv Microbiol Physiol 33: 213–275

Meyer SR, Pienaar RN (1984) The microanatomy of *Chroomonas africana* sp.nov. (Cryptophyceae). S Afr Tydskr Plank 3: 306–319

Mnawar M, Bistricki T (1979) Scanning electron microscopy of some nanoplankton cryptomonads. Scanning Electron Microsc 3: 247–252

Perasso L, Hill DRA, Wetherbee R (1992) Transformation and development of the flagellar apparatus of *Cryptomonas ovata* (Cryptophyceae) during cell division. Protoplasma 170: 53–67

– Brett SJ, Wetherbee R (1993) Pole reversal during cytokinesis in the Cryptophyceae. Protoplasma 174: 19–24

Roberts K, Shaw PJ, Hills GJ (1981) High-resolution electron microscopy of glycoproteins: the crystalline cell wall of *Lobomonas*. J Cell Sci 51: 295–321

Santore UJ (1977) Scanning electron microscopy and comparative micromorphology of the periplast of *Hemiselmis rufescens*, *Chroomonas* sp., *Chroomonas salina* and members of the genus *Cryptomonas* (Cryptophyceae). Br Phycol J 12: 255–270

– (1982) Comparative ultrastructure of two members of the Cryptophyceae assigned to the genus *Chroomonas* – with comments on their taxonomy. Arch Protistenk 125: 5–29

– (1984) Some aspects of taxonomy in the Cryptophyceae. New Phytol 98: 627–646

– (1985) A cytological survey of the genus *Cryptomonas* (Cryptophyceae) with comments on its taxonomy. Arch Protistenk 130: 1–52

– (1986) The ultrastructure of *Phyrenomonas heteromorpha* comb. nov. (Cryptophyceae). Bot Mar 29: 75–82

– (1987) A cytological survey of the genus *Chroomonas* – with comments on the taxonomy of this natural group of the Cryptophyceae. Arch Protistenk 134: 83–114

Shaw PJ, Hills GJ (1982) Three-dimensional structure of a cell wall glycoprotein. J Mol Biol 162: 459–471

Sleytr UB, Messner P (1983) Crystalline surface layers on bacteria. Annu Rev Microbiol 37: 311–339

Wetherbee R, Hill DRA, McFadden GI (1986) Periplast structure of the cryptomonad flagellate *Hemiselmis brunnescens*. Protoplasma 131: 11–22

– – Brett SJ (1987) The structure of the periplast components and their association with plasma membrane in a cryptomonad flagellate. Can J Bot 65: 1019–1026

Protoplasma (1994) 181: 123–141

PROTOPLASMA

Production and modifications of extracellular structures during development of chytridiomycetes

Martha J. Powell*

Department of Botany, Miami University, Oxford, Ohio

Received August 5, 1993
Accepted March 15, 1994

Summary. In development of the primitive fungi, chytridiomycetes, unwalled zoospores bearing single, posterior flagella are transformed into walled, round-cells which elaborate the thallus. Production, structural modification, or release of extracellular material are involved with each transition of developmental stage. This article reviews the variety and developmental changes of extracellular materials found at the cell surface of chytridiomycetes. A cell coat, produced from Golgi-derived vesicles during zoosporogenesis, is visible around free swimming zoospores of some chytridiomycetes. How the zoospore surface receives and transduces signals is not widely explored, but it is known that fenestrated cisternae and simple cisternae, which are integrated into the microbody-lipid globule complex, are spatially and structurally associated with the plasma membrane and flagellar apparatus. This spatial association, as well as the cytochemical localization of calcium in fenestrated cisternae, suggest a mechanism for signal transduction and for regulation of zoospore motility. Zoospores become encased in a new layer of extracellular material as the zoospore encysts. Among some chytrids the source of this material is preexisting vesicles which fuse with the plasma membrane. Among other zoospores, a readily identifiable population of encystment vesicles is not apparent, demonstrating that there is no single pattern or mechanism for zoospore encystment in chytridiomycetes. Encysted zoospores developing into thalli, typically produce cell walls with a microfibrillar substructure. Ultrastructural analysis of walls reveals distinctive architecture and remarkable sculpturing which have been used in systematics of some members of chytridiomycetes. Nothing is known as to underlying controls of cytoskeletal elements and plasma membrane enzyme complexes in wall biogenesis. Many changes in cell surface structures accompany thallus maturation. Septa, many traversed with plasmodesmata, are produced in most chytrid thallus types. As sporangia and resting spores prepare for the production and release of zoospores, additional extracellular layers of material are frequently produced. Polarized deposits of extracellular material become discharge plugs, discharge vesicles, or endoopercula. Interstitial material is also released into cleavage furrows. Circumscissile or localized digestion of walls produce operculate or inoperculate exit ports for zoospore release. Cryofixation preserves more extensive extracellular material than does conventional chemical fixation, and broader application of cryofixation may radically alter our current view of cell surface structure. Thus chytridiomycetes exhibit a range in patterns for the occurrence and subsequent modifications of extracellular materials, even for members within the same order. The most universally recognized role for these extracellular materials is protection. Although there is a reasonable view of the types of extracellular material involved in chytridiomycete development, we have only limited understandings of their biogenesis or roles in regulation and communication, areas awaiting more investigations.

Keywords: Carbohydrates; Chytridiomycetes; Extracellular material; Membranes; Ultrastructure; Zoospores.

Abbreviations: DIC Nomarski-differential contrast optics; TEM transmission electron microscopy.

Introduction

Chytridiomycetes are primitive fungi (Bowman et al. 1992, Sparrow 1960, Wainright et al. 1993) commonly found in aquatic and moist habitats as saprophytes on refractory materials, such as chitin, keratin, or cellulose, and as parasites of plants, algae, aquatic animals, and other fungi (Powell 1993, Sparrow 1960). Because of their simple thalli and the subcellular structure of their posteriorly uniflagellate zoospores produced within these thalli, chytridiomycetes are considered closely allied with members of Protista such as choanoflagellates (Barr 1992), an insight recently supported from molecular phylogenetic investigations (Wainright et al. 1993). In many proposed delimitations of eukaryotic kingdoms, chytridiomycetes are placed in the

* Correspondence and reprints: Department of Biology, James Madison University, Harrisonburg, VA 22807, U.S.A.

overlap zone of recognizable continuity between extant genera of the two Kingdoms, Protista and Fungi (Whittaker 1969). Because of structural similarities and possible phylogenetic links with Protista, it is important to include chytridiomycetes when we consider cell surface features of Protista for perspectives into possible homologous structures or insights into analogous morphologies converging with exposure of organisms to similar selective pressures.

Although a few chytridiomycetes are adapted for dispersal of propagules by insects or wind (Powell 1993) or possibly by passive water flow (Powell and Blackwell 1991), most depend on posteriorly uniflagellate zoospores as primary dispersal agents. After a period of rapid motility, zoospores encyst into round cells and generate thalli in one of several ways: (1) injecting its protoplast into a host cell and later fabricating a wall; (2) germinating with a germ tube which becomes the

Fig. 1. DIC. Chytridiomycete zoospore with single, posteriorly-directed flagellum (arrowhead) and prominent lipid globule (*L*). × 1,200

Fig. 2. DIC. Chytridiomycete zoospores exhibiting pleomorphic shapes. Encysted spores are visible as round cells (*RC*). × 1,200

Fig. 3. Longitudinal section of zoospore of *Chytriomyces hyalinus* (Chytridiales) with posterior flagellum (*F*) and nucleus (*Nu*). The zoospore body is surrounded with a cell coat (*CC*). Photograph courtesy of David Dorward and Martha Powell. × 20,000

Fig. 4. Detail of tangential section of cell coat demonstrating that the outer substructure consists of distinct rod-like components (arrows). Photograph courtesy of David Dorward and Martha Powell. × 49,000

Fig. 5. Detail of tangential section of cell coat illustrating that outer rod-like material is in rows (arrows). Photograph courtesy of David Dorward and Martha Powell. × 49,000

Fig. 6. Longitudinal section of zoospore of *Spizellomyces* sp. (Spizellomycetales) with posterior flagellum (*F*), single nucleus (*Nu*), and lipid globule(*L*). No zoospore cell coat is visible. × 16,000

Fig. 7. Silver methenamine reaction demonstrating presence of carbohydrates in cell coat (*CC*) of *Chytriomyces hyalinus*. Notice that the cell coat is discontinuous between the zoospore body (double arrows) and the flagellum (*F*). The flagellar sheath, corresponding to the region of the electron-dense core visible in Fig. 8. (single arrow), contains little reaction product. Reaction product, less densely deposited than on the body cell coat, is pesent on the rest of the flagellar sheath. Photograph courtesy of David Dorward and Martha Powell. × 80,000

Fig. 8. Detail of posterior end of zoospore as in Fig. 7. The body cell coat (*CC*) appears to terminate near the site where the flagellum emerges and folds back against the plasma membrane (double arrows). The single arrow points out the electron-dense core in the transition region between the kinetosome and flagellum (*F*). Photograph courtesy of David Dorward and Martha Powell. × 80,000

Fig. 9. Tangential section of flagellum (*F*) of zoospore of the algal parasite *Polyphagus euglenae* (Chytridiales) showing coat material along the surface of the flagellum. The flagellum extends from the kinetosome (*K*) which is in longitudinal section. × 80,000

Fig. 10. Cross-section of flagellum of *Polyphagus euglenae*. The presence of a cell coat is clearly visible. × 120,000

Fig. 11. Detail of cell coat (*CC*) on body of zoospore of *Polyphagus euglenae*. Notice the cored-vesicles (arrow) subtending the plasma membrane. × 100,000

Fig. 12. Cell coat on zoospore of *Rhizophydium spherotheca* (Chytridiales) stained in the osmium tetroxide-potassium ferrocyanide technique. × 30,000

Fig. 13. At the surface of *Monoblepharella* sp. (Monoblepharidales) zoospore body, the scattered fibers visible along the plasma membrane are probably remnants of a cell coat. × 20,000

Fig. 14. Details of the fenestrated cisterna (*FC*) which cups the lipid globule (*L*) in zoospore of *Chytriomyces hyalinus* (Chytridiales). Fine linkers (arrows) join the cisterna (*FC*) to the plasma membrane (*PM*). Notice the cell coat (*CC*). × 80,000

Fig. 15. In zoospore of *Monoblepharella* sp. (Monoblepharidales), the fenestrated cisterna (*FC*) is adjacent to the plasma membrane (*PM*). A microbody (*Mb*) is adjacent to the fenestrated cisterna and to mitochondria (*M*). × 75,000

Fig. 16. In zoospore of *Coelomomyces punctatus* (Blastocladiales), a simple cisterna (*C*) is adjacent to the plasma membrane (*PM*). Notice microbody-lipid globule complex organellar assemblage including the simple cisterna, microbody (*Mb*), lipid globules (*L*), and mitochondrion (*M*). × 65,000

Fig. 17. In the microbody-lipid globule complex of *Chytriomyces hyalinus* zoospore, notice the microtubular rootlet (far left arrow) grazing the microbody (*Mb*) and extending toward the fenestrated cisterna. A linker (middle arrow) extends between the fenestrated cisterna and microbody (*Mb*). The far right arrow points toward the center of the cell, away from the cell surface. × 68,000

Fig. 18. Potassium antimonate reaction for calcium in zoospore of *Chytriomyces hyalinus*. Reaction product is in the fenestrated disc portion of the fenestrated cisterna (*FC*) and also in the non-fenestrated portion (double arrows) which continues around the lipid globule (*L*). × 51,000

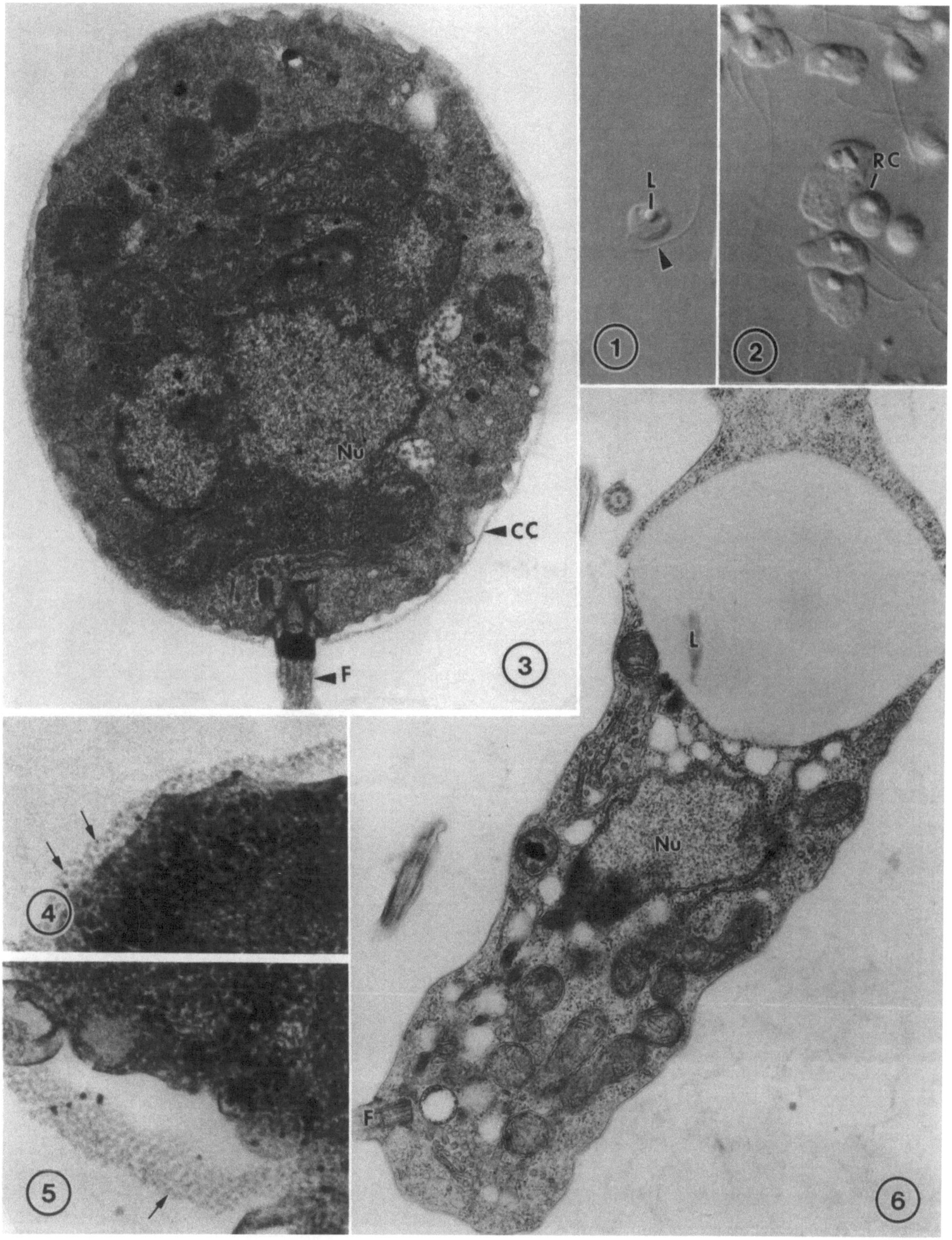

Figs. 1–6. Zoospores of chytridiomycetes

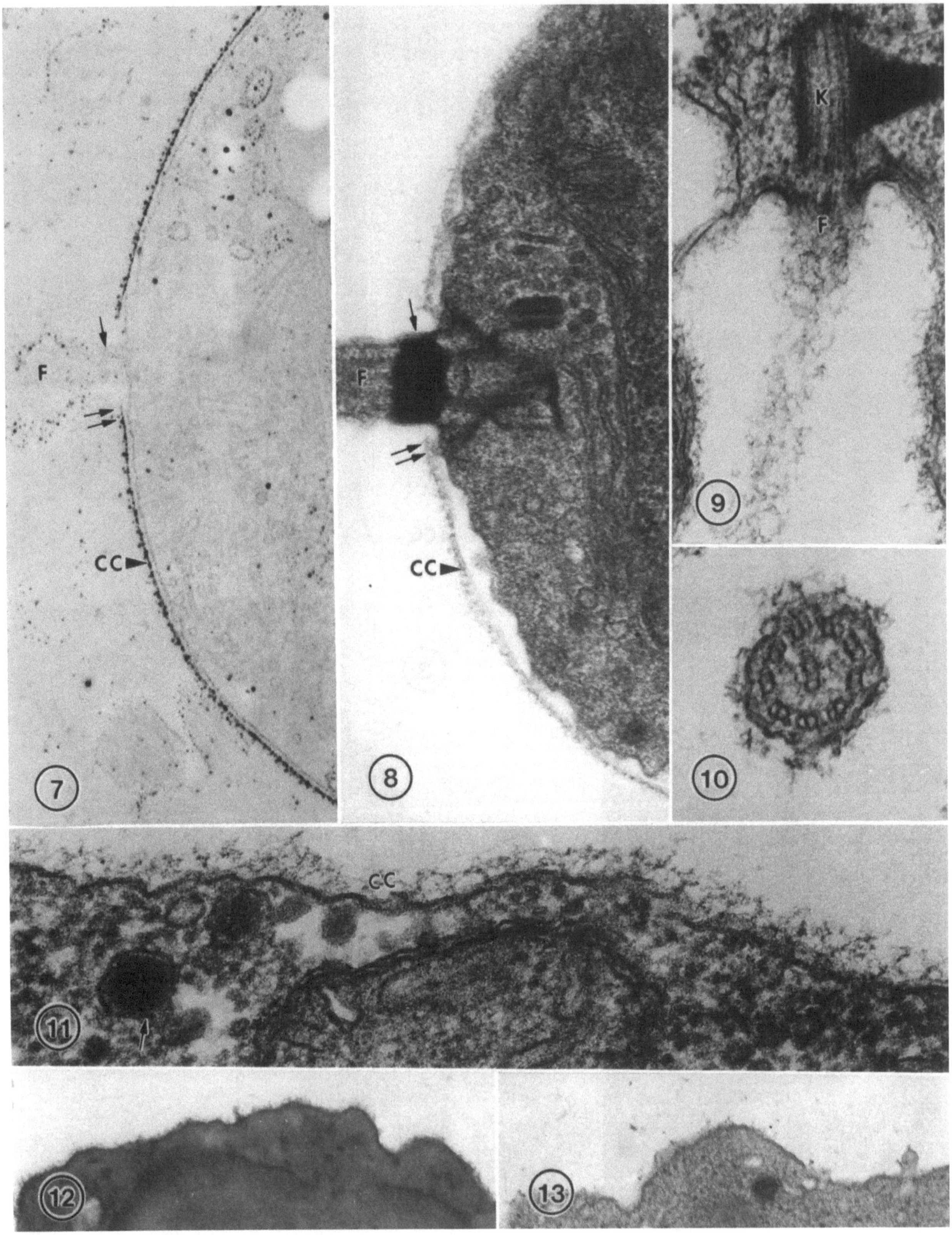

Figs. 7–13. Cell surface of zoospores of chytridiomycetes

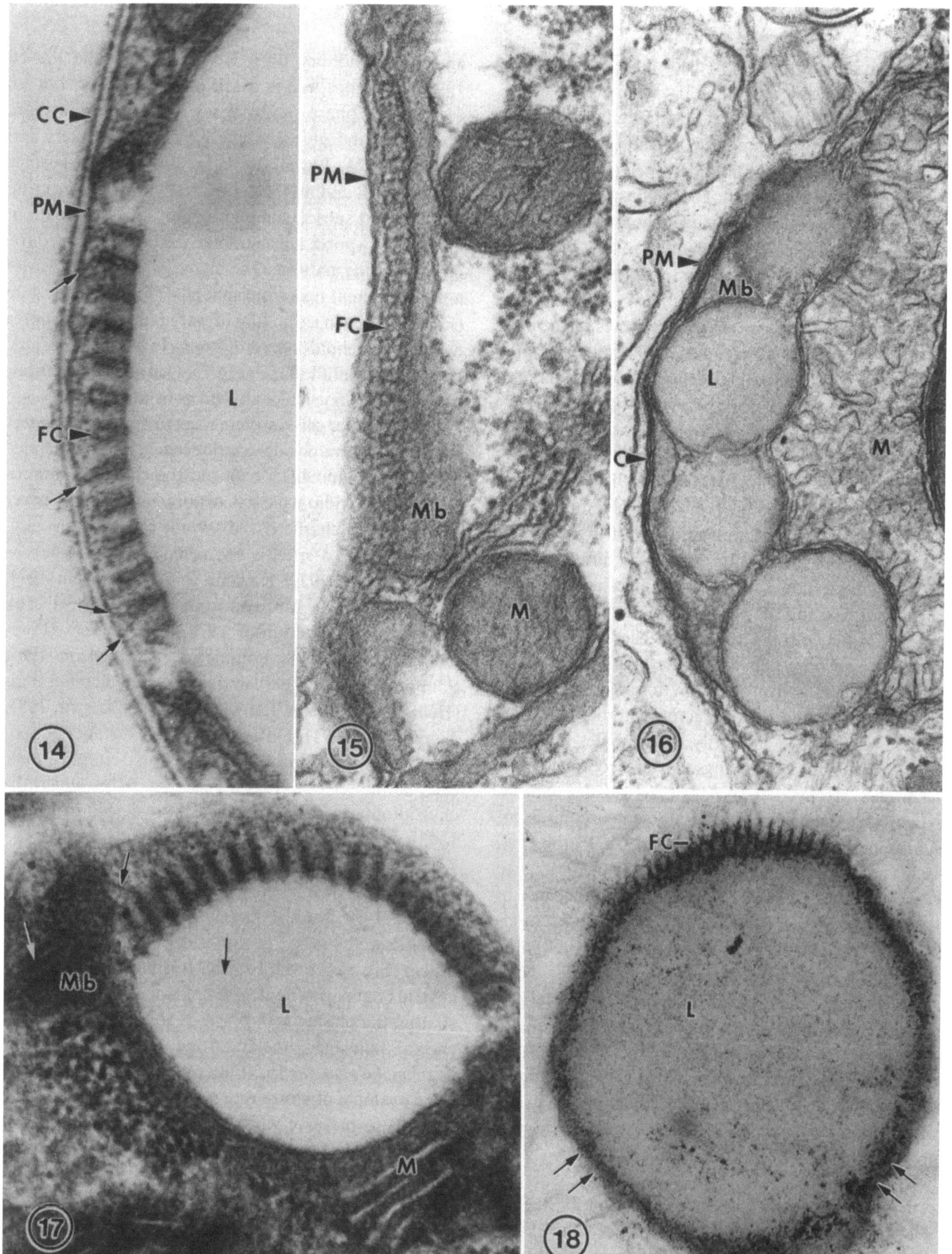

Figs. 14–18. Membrane cisternae of microbody-lipid globule complexes (an assemblage of mitochondria, microbodies, lipid globules, and membrane cisternae) spatially associated with the plasma membrane and cell surface of chytridiomycete zoospores

rhizoidal system as the encysted zoospore enlarges into a multinucleate sporangium; (3) germinating with a germ tube which becomes the sporangium with the rhizoids extending from the tip of the germ tube; or (4) germinating with a germ tube which elongates into a rhizomycelium or hypha. The thallus enlarges and eventually produces reproductive structures which will generate more zoospores. Production of new or stored extracellular material is involved with the transition of each developmental stage.

Ultrastructural studies of chytridiomycetes in the past two decades have concentrated on the zoospore and have revealed a remarkable variety of designs in subcellular architecture, both in terms of the flagellar apparatus (Barr 1981, 1990; Roychoudhury and Powell 1992) and organellar relationships, such as the microbody-lipid globule complex (Powell 1978, Powell and Roychoudhury 1992). However, our understanding of the structure and developmental role of extracellular features is more limited. In this article differences in cell surface structures in chytridiomycetes are illustrated, emphasizing the need to explore more fully the interaction between environmental cues, the cell surface, and zoospore and thallus development. Just as chytridiomycete zoospores show a range of cellular designs, so do extracellular structures on zoospores and thalli. I have drawn primarily on examples in which developmental understandings are most complete from the orders of chytridiomycetes (Chytridiales, Spizellomycetales, Blastocladiales, Monoblepharidales, Neocallimasticales and Harpochytriales). Heath et al. (1983) originally assigned the anerobic rumen fungi to a family in the order Spizellomycetales, but because of developmental differences (Powell 1993) and questions of their proper affinities (Bowman et al. 1992), the group has recently been placed in a new order, Neocallimasticales (Li et al. 1993). Because of variations and the incomplete nature of our knowledge of cell surface features among chytridiomycetes, it is necessary to be cautious in generalizing as to processes and mechanisms in the production and modifications of extracellular material from one organism to another, even among members of the same order.

Cell surface of zoospores

The small size of the spores, ranging from 3 to 10 µm, and their conspicuous lipid globules (Figs. 1 and 2) belie the intricate organization of organelles and membranes (Powell 1978) interacting in the precise control of zoospore motility and development (Barr 1990, Powell and Roychoudhury 1993, Roychoudhury and Powell 1993). The zoospore is a self-contained unit, not an assimilative phase, and carries within the zoospore body lipid and glycogen energy reserves (Myers and Cantino 1974, Powell 1978), preformed cell surface components, and mRNA (Lovett 1968) needed during motility and subsequent encystment and germination. Chytrid zoospores are characterized by a rapidly, darting swimming pattern. The shape of the spore varies among different taxa from spherical (Fig. 1) to elongate (Fig. 6). When in a thin film of water, chytrid zoospores exhibit amoeboid-like movement with pleomorphic changes in overall shape (Fig. 2). Cell surface features, therefore, do not make chytrid zoospores rigid.

Mechanisms for cell response to environmental triggers typically involve glycoprotein receptors at the cell surface (Sharon and Lis 1993), but the chemical composition of chytridiomycete zoospore surfaces has never been critically characterized. Motile cells of many chytrids respond to specific environmental cues such as light (Kazama 1972, Kazama and Schornstein 1977, Robertson 1972), nutrient sources (Muehlstein et al. 1988, Orpin and Bountiff 1978), pheromones (Pommerville 1977, 1978; Pommerville and Olson 1987; Pommerville et al. 1990), and host secreted compounds (Held 1974, Jansson and Thiman 1992, Kerwin 1983, Tunlid et al. 1991). Differences in cell surfaces of zoospores can have profound functional implications, such as the selective binding of viruses for which the chytridiomycetes serve as vectors (reviewed in Powell 1993).

A prominent cell coat (Figs. 3–5, and 7–11) is visible on zoospores of some taxa (Dorward and Powell 1983, Li et al. 1991, Manier and Loubes 1978, Munn et al. 1981, Powell 1981 a, Taylor and Fuller 1981) but not others (Fig. 6) (Powell 1981 b). It is most likely that all chytrids have varying degrees of extracellular material coating the plasma membrane of zoospores. The presence of mucilage at the anterior end of zoospores such as *Monoblepharella* sp. (Fuller and Reichle 1968) is a good example of a rare report of extracellular material difficult to preserve with conventional fixation methods. Fine fibrillar material at the zoospore plasma membrane surface (Fuller and Reichle 1968) frequently observed (Figs. 12 and 13) probably is extracellular material because cytochemical staining increases its visibility (Federici and Lucarotti 1986). When a cell coat is found on both the zoospore body and flagellum, it sometimes is morphologically the same (Figs. 9–11). On other zoospores it may vary in appearance and density and show discontinuity where the flagellum emerges from the zoospore body (Figs. 7 and 8) (Munn et al. 1981). The absence of a cell coat in this area of the flagellum corresponds to the location of the flagellar necklace identified in zoospores of one chytrid taxa (Kazama and Schornstein 1977).

As in two species of *Chytriomyces* (Figs. 4 and 5) (Dorward and Powell 1983) and several of the rumen fungi (Munn et al. 1981, Li et al. 1991), the surface layer on the cell body has intricate substructure with regular arrays. Cytochemical studies demonstrate that the coat material contains carbohydrates (Dorward and Powell 1983, Taylor and Fuller 1981). The zones of the cell coat may well differ in composition because blocking experiments showed that reaction product on the inner layer was due to reactive sulfhydryl groups rather than periodate oxidation of non 1,3 linked carbohydrates (Fig. 26) (Dorward and Powell 1983).

Lectin probes of cell coat sugar composition revealed that fucose was a commonly accessible sugar on the zoospore body and flagellum of rumen fungi (Guillot et al. 1990). Lectin binding pattern for different isolates even of the same genus, however, varied. For example the lectin binding pattern of some strains of *Neocallimastix* indicated D-mannose/D-glucose at the spore body surface or flagella, but this was not universal among all the isolates or of isolates of other genera (Guillot et al. 1990). These results suggest that cell coats of rumen fungal zoospores vary in composition and afford zoospores protection from acids and enzymes in the animal rumen. Differences in sugar composition may well adapt zoospores to differences in the rumen environment.

Glycoproteins at the cell surface of other chytrid zoospores bind lectins and have been theorized as potential sites involved in triggering encystment or in recognition of environmental cues (Dalley and Sonneborn 1982; Jen and Haug 1979 a, b). The most common lectin bound is concanavalin A, indicating available glycosyl and/or mannosyl groups.

Organellar associations with the zoospore cell surface

In zoospores of four orders (Chytridiales, Harpochytriales, Monoblepharidales, and Blastocladiales) of chytridiomycetes, there is typically a membrane cisterna spatially associated with the plasma membrane and the flagellar apparatus. This cisterna is a component of the microbody-lipid globule complex (MLC), an organellar assemblage (also including microbodies, lipid globules and mitochondria) implicated in the utilization of stored triglycerides as an endogenous energy source to power flagellar activity (Powell 1978, Powell and Roychoudhury 1992). In the Blastocladiales the cisterna is a simple cisterna which can be commonly envisioned as a sheet with irregularly spaced perforations cupping the other organelles of the MLC (Fig. 16) (Myers and Cantino 1974, Olson 1984). In the Chytridiales the cisterna covers the surface of the lipid globule directed toward the cell surface (Fig. 14). It can be either a simple cisterna or a fenestrated cisterna with a perforated disc where the pores are tightly spaced in a honey-comb arrangement (Powell and Roychoudhury 1992). Typically, the fenestrated portion of the cisterna is adjacent to the plasma membrane and joined to it with fine linkers (Figs. 14 and 17) (Dorward and Powell 1982). In the Harpochytriales and Monoblepharidales, the fenestrated cisterna lies along side a microbody rather than a lipid globule, but as in the Chytridiales is adjacent to the plasma membrane (Fig. 15) (Reichle 1972). Because of their specific location in zoospores and the fact that zoospores respond to environmental signals, the cisternae might participate in signal reception and transduction between the cell surface and the flagellar apparatus (Powell 1983). A role for storage of calcium in fenestrated cisternae of a member of the Chytridiales was demonstrated with the potassium antimonate precipitation technique (Fig. 18) (Powell 1983). Because calcium concentrations effect symmetry of flagellar beat and consequently direction in which the uniflagellate zoospore would swim, the fenestrated cisternae may regulate zoospore directional responses to environmental cues by regulating the concentration of calcium available in the region near the flagellar axonome (Powell 1983).

Additional support for a role of the cisterna in zoospore response to external stimuli comes from a freeze etch study of the cell surface of a positively phototactic chytrid zoospore (Kazama and Schornstein 1977). In the dark, membrane particles were randomly arranged in the plasma membrane opposite the fenestrated cisterna, but when zoospores were illuminated the randomly arranged particles clustered in the plasma membrane opposite the fenestrated cisterna. The role of the fenestrated cisterna in regulation of flagellar motion is also suggested in a chytrid zoospore with a compound type of MLC (Powell and Roychoudhury 1992). These zoospores contained multiple lipid globules, but the lipid globules in the posterior end of the spore and associated with the flagellar apparatus were the lipids surrounded by fenestrated cisternae. Associated with anteriorly located lipids were the microbodies and mitochondria, organelles which work in concert for energy production.

Organelles in cortical cytoplasm of zoospores and possible roles in encystment

When zoospores begin to encyst, they are typically sticky, attaching to substrates and to each other (Dalley and Sonneborn 1982, Olson 1984). This is not their property while free swimming zoospore. Thus during encystment, zoospore cell surface material is apparently displaced by new material with different properties. The fundamental question is how do these new extracellular materials form? Two sources are possible, either de novo synthesis or secretion from material prepackaged in vesicles.

There have been few studies under controlled developmental conditions analyzing events in zoospore encystment. In contrast to other fungal protists, such as oomycetes where we know which vesicles are involved in adhesion of the zoospore to a substrate and which form the cyst coat (Lehnen and Powell 1989), among chytridiomycetes such functions are largely surmised. One approach to elucidating which vesicles are involved in production of cell surface material is to determine which zoospore vesicles are no longer found in the encysted zoospore, inferring a role in the adhesion or encystment process. One problem with trying to generalize a common process is that morphologically classes of vesicles possibly involved in cyst adherence to substrates and coat production have not been enumerated for zoospores of chytridiomycetes.

Even within the same order, categories of vesicles located in cortical cytoplasm of zoospores vary. Some of the vesicles identified could also be involved with osmotic regulation. Zoospores of *Chytriomyces hyalinus* are excellent examples of many morphological types of vesicles being present in the cortical cytoplasm (Dorward and Powell 1983). At least four forms can be described. Cored-vesicles which McNitt (1974a) first described in another chytrid contain a compact central inclusion (Figs. 20 and 21). Coiled fibers (Dorward and Powell 1983) fill the matrices of fibrous-containing vesicles (Fig. 22). A third class of vesicles contains densely granular contents (Fig. 21). In addition there are smaller vesicles with electron-translucent matrices (Fig. 20) (Li et al. 1991). In zoospores of other genera, only cored vesicles have been reported (Figs. 23 and 24). Zoospores of some parasitic members of the Blastocladiales have vesicles with uniformly granular contents (Fig. 25) (Manier and Loubes 1978), but zoospores of other genera in the order lack these vesicles (Olson 1984). Cytochemical analysis demonstrates that some of these cortical vesicles contain carbohydrates (Fig. 26) (Dorward and Powell 1983, Federici and Lu-

Figs. 19–34. Vesicles in cortical cytoplasm of zoospores of chytridiomycetes and zoospore encystment

Fig. 19. Freeze-etch replica of cell surface of *Chytriomyces hyalinus* (Chytridiales) zoospore. The cell coat is present (*CC*) and vesicles (arrows) fuse with the plasma membrane. The sites of fusion (arrows) are clearly discerned by the absence of particles present in other areas of the plasma membrane. × 40,000

Fig. 20. Vesicles in cortical cytoplasm of *Chytriomyces hyalinus* zoospore. Notice the vesicles with electron-transparent contents (arrow) and the two cored vesicles. × 86,000

Fig. 21. Vesicles in cortical cytoplasm of *Chytriomyces hyalinus* zoospore. The arrow points out the vesicle with granular contents. A cored-vesicle is also present. × 86,000

Fig. 22. Vesicles in cortical cytoplasm of *Chytriomyces hyalinus*. The arrow points out a vesicle with fibrillar contents. × 86,000

Fig. 23. Cored vesicles (arrows) in zoospore of *Polyphagus euglenae* (Chytridiales). × 32,000

Fig. 24. Cored vesicles (arrows) in zoospore of *Spizellomyces* sp. (Spizellomycetales). × 32,000

Fig. 25. Adhesion vesicles (arrow) in zoospore of *Coelomomyces punctatus* (Blastocladiales). The large vesicle at the right contains electron-dense reaction product from the Gomori minus substrate reaction control for acid phosphatase activity. These results indicate that endogenous phosphorous is stored in this vesicle. × 30,000

Fig. 26. Silver methenamine reaction with iodoacetate used after periodate oxidation to block sulfhydryl-containing compounds. Reaction product (indicating location of non 1–3 linked carbohydrates) in zoospores of *Chytriomyces hyalinus* is deposited on the other zone of the cell coat. Reaction product is in fibrous-containing and cored vesicles (upper portion of picture) and electron-translucent vesicles (double arrows). There is very little reaction product in the granular vesicle (single arrow). × 37,000

Fig. 27. Portion of encysted zoospore with axoneme (*AN*) retracted into the cell. Vesicles present in the zoospore cannot be found in cytoplasm. × 50,000

Figs. 28–34. Flagellar retraction into zoospore of *Entophlyctis variabilis*. The zoospore bears a single flagellum (*F*). The axoneme becomes coiled at the tip of the flagellum as the flagellum progressively shortens (arrows; Figs. 29–31). The flagellar sheath fuses with the zoospore body, and the coiled axoneme is released into the cell (Figs. 32 and 33). The encysted zoospore then becomes a round cell (Fig. 34). × 1,500

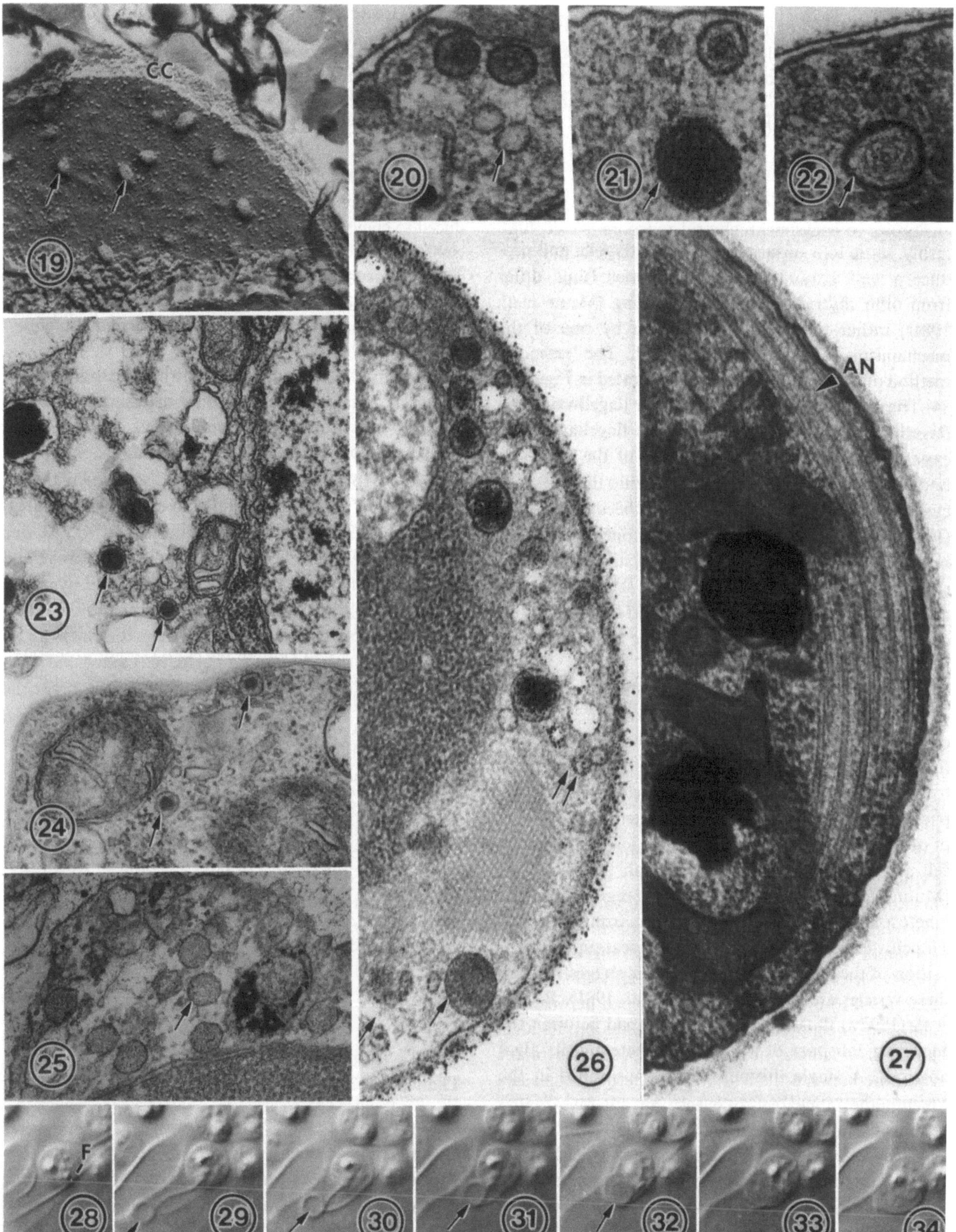
CC
19
20
21
22
23
24
25
26
AN
27
F
28
29
30
31
32
33
34

carotti 1986, Taylor and Fuller 1981). Attempts to enumerate morphological classes of vesicles in zoospores of chytrids are even more difficult because some vesicles contain storage products such as polyphosphate (Fig. 25) and protein (Hohn et al. 1984).

Zoospore encystment

Just prior to encystment, chytrid zoospores swim sluggishly, settle to a surface, retract their flagella and produce a wall around themselves. Rumen fungi differ from othe chytridiomycetes in sheading (Munn et al. 1981), rather than retracting flagella by one of the mechanisms Koch (1968) described. The vesicular method of axoneme retraction is illustrated in Figs. 28–34. The axoneme coils at the tip of the flagellum as the flagellum shortens (Figs. 28–30). The flagellar sheath expands into the plasma membrane of the zoospore body (Fig. 31), releasing the axoneme into the zoospore cytoplasm (Fig. 32), and the cell becomes round (Figs. 33 and 34). Thus profiles of axoneme microtubules are apparent in recently encysting zoospores (Fig. 27). Coordinated with flagellar retraction is the adhesion of zoospores to substrates and the appearance of a cyst coat displacing the zoospore cell coat which might be visible.

There are several clear examples of physical adhesion of encysted zoospores to substrates, and in each case certain morphological forms of cortical vesicles are not found later in the encysted zoospore. Deduction leads to the interpretation that these vesicles are the prepackaged sources of the adhesive material. Zoospores of one rumen fungus attach at the flagellar pole of the cell and join to the substrate with an adhesion pad (Munn et al. 1981). In the posterior region near the kinetosome of zoospores, there are electron-dense vesicles clustered near the kinetosome. Location and deposition of the adhesion pad are at least suggestive that these vesicles are secretory (Munn et al. 1981). Beakes et al. (1992 a) illustrated an adhesion pad between the encysting zoospore of a chytrid parasite and its algal host cell. A single, fibrous vesicle, positioned in the zoospore between the fenestrated cisterna and plasma membrane, appeared to release an adhesive matrix (Beakes et al. 1988). Developing thalli of monoblepharidalean fungi may have adhesive pads in the form of holdfasts (Reichle 1972). A broad plate of adhesive material glues encysting cells of a chytrid parasite of mosquitoes to host cuticles (Travland 1979 a, b). This extracellular material is released from numerous adhesion vesicles (Fig. 25) which stain for carbohydrates and which are totally absent from the encysted zoospore (Federici and Lucarotti 1986).

Similarly, cortical vesicles not found in the encysted zoospore could produce the new cyst coat. The absence of cortical vesicles in the recently encysted zoospore shown in Fig. 27 suggests that vesicles were involved with secretion of extracellular material. Freeze-etch preparations of the cell surfaces of zoospores induced to encyst demonstrate that vesicular fusion does occur during encystment (Fig. 19). Among other zoospores, a readily available population of encystment vesicles is not apparent, again demonstrating that there is not a single pattern for zoospore encystment in chytridiomycetes.

It is possible that most of the newly formed extracellular material arises de novo in some organisms. If this is true, then cortical vesicles would have a lesser role in transformation of zoospores into germlings. Some zoospores of chytridiomycetes contain vesicles with distinctive electron-opaque, cup-shaped inclusions called gamma particles (Myers and Cantino 1974). At one time they were believed to function in the storage and activation of chitin synthetase for cell wall synthesis during encystment (Barstow and Pommerville 1980). Other studies have failed to confirm the presence of chitin synthetase in gamma particles (Dalley and Sonneborn 1982, Hohn et al. 1984). Instead the organelle stores at least two major proteins (Hohn et al. 1984), and the bulk of zoospore chitin synthetase is in the plasma membrane (Dalley and Sonneborn 1982). Chitin synthetase regulation involves endoproduct inhibition (Selitrennikoff et al. 1980). There are not sufficient pools of the monomer sugar in zoospores to account for the quantities of chitin produced. As stored pools are used in the transformation of zoospores into germlings, preexisting enzymes are activated and synthetic pathways for monomer sugars become functional.

Thallus development

Germlings

Germ tubes of many chytridiomycetes become ensheathed with an extracellular matrix (Fig. 36). If saprophytic zoospores are allowed to encyst and germinate on the surfaces of acid cleaned walls (Fig. 35), the presence of this material can be readily detected (Fig. 36). Rather than an adhesive role, this material may well protect the germ tube as it enters the substrate.

Wall structure

As the encysted zoospore develops into a thallus, cell walls with microfibrillar substructure are often produced (Fig. 37), and this is typically the inner wall layer of the nonreproductive phase of the thallus. Like higher fungi, these walls contain microfibrillar chitin (Selitrennikoff et al. 1980). The architecture of the thallus wall is diverse (Figs. 37 and 39–41). Additional layers of non-microfibrillar material are frequently components of the cell wall (Fig. 39), and the ultrastructure of outer wall layers is distinctive, partially because it is the architecture of these layers that provides wall sculpturing (Fig. 40). Those layers with finely fibrillar structure are interpreted to consist of glycoproteins or mucopolysaccharides (Fig. 39).

It is common to find walls of chytridiomycetes which are composed of multiple layers (Figs. 40 and 51). This is clearly illustrated with resting sporangia of the plant parasite *Synchytrium endobioticum* (Powell 1993). In some chytridiomycetes, such as the mosquito parasite, *Coelomomyces* spp., the vegetative thallus lacks a cell wall but is robed with a granular sheath (Couch and Bland 1985, Powell 1976 b). Later it generates sculptured and internally channeled, but homogeneously electron-dense, cell walls around hyphal bodies which become resting sporangia (Figs. 41 and 53) (Powell 1976 b). Walls, particularly of resting sporangia, are thick (Figs. 43 and 53) (Couch and Bland 1985). Unlike wall biogenesis in plants and algae, nothing is known as to roles of cytoskeletal elements and plasma membrane enzyme complexes in wall biogenesis. It has been shown that endomembranes apparently are cytoplasmic templates controlling wall deposition and wall architecture (Powell 1976 b). Although we may not understand the selective advantages of the often bewildering array of thallus wall structure, comparative ultrastructure of wall features is valuable in the systematics of some groups of chytridiomycetes (Couch and Bland 1985).

Studies with lectins on rumen fungi show that different regions of thalli (either spore, sporangia or rhizoids) differ in surface sugar composition (Guillot et al. 1990) with an unexpected complexity and diversity of sugars at the surface of the thallus (Breton et al. 1991). Whether or not similar diversity in surface sugars exists for other chytridiomycetes has not been extensively explored.

One role of the cell wall is protection in interactions with other microbes. Chytridiomycetes often share habitats and substrates with bacteria which adhere to their surfaces (Fig. 38). On sporangial walls with cytoplasm present, there usually is no sign of decay (Fig. 38). Thus the cell walls encasing living protoplasm are resistant to bacterial activity. It appears that some chytridiomycetes have evolved a mechanism whereby the basal cell of a two-celled resting spore becomes devoid of cytoplasm, and bacteria degrade walls of this non-living compartment. The apical cell containing living cytoplasm is disarticulated from the basal cell and can be dispersed by capillary water in the soil (Powell and Blackwell 1991). When thalli are attacked by potential parasites (Held 1974, Powell 1982), their successful resistance is sometimes, but not universally, due to the extra deposition of extracellular material between the wall and plasma membrane as wall appositions (Fig. 45). Since prevention of invasion or biodegradation of the wall by other organisms requires living cytoplasmic activity, resistance is not merely the property of cell wall composition.

Thallus maturation

Development and maturation of the thallus include many changes in cell surface structures. Reproductive structures, either sporangia or resting sporangia, become delimited from nonreproductive units of the thallus with septa (Figs. 42–44). Septa are formed centripetally, and their ontogeny is associated with cored vesicles, resembling those found in zoospores (Powell 1974). This coincidence at least suggests a role for cored-vesicles in wall biogenesis.

Among members of the Chytridiales and Spizellomycetales, septa contain plasmodesmata (Fig. 44) (Powell 1974). Sometimes septa are quite thick and plasmodesmata are difficult to locate (Powell and Gillette 1987). With serial section analysis, however, it was possible to show that a thinner diaphragm region in such a septum contained plasmodesmata. Members of the Blastocladiales are reported to have inperforate septa at the bases of sporangia, but plasmodesmata can be located passing through septal and developing resting sporangial walls of some representatives (Fig. 43). Classical studies described cross-walls of some members as having a cartwheel appearance. Again, the structure of chytridiomycete extracellular material expresses a variety of morphologies.

In preparation for zoosporogenesis, there is the production of additional extracellular material and modification of preexisting material (Beakes et al. 1992 b, McNitt 1974 b, Porter and Smiley 1980, Powell 1976 a, Sewall et al. 1986, Taylor and Fuller 1981). In classification of chytridiomycetes, structure of the discharge apparatus has traditionally been considered of primary importance (Sparrow 1960). The Golgi apparatus appears actively involved in secretion of these materials in all orders of chytridiomycetes except for the Blastocladiales which lacks the typical Golgi apparatus with stacked cisternae. In the Blastocladiales, Golgi equivalent structures may function in analogous ways in these processes (Sewall et al. 1986). How synthesis, transport, and discharge of these extracellular struc-

---

Fig. 35. DIC. Germling of a chytridiomycete germinating on acid cleaned walls of the same fungus. The encysted zoospore remains at the surface as the germ tube (*GT*) penetrates the substrate. × 1,000

Fig. 36. TEM. Germling of a chytridiomycete germinating on acid cleaned wall of the same fungus. The germ tube (*GT*) penetrates the wall (*W*) and releases extracellular material (*EC*) along the germ tube. × 25,000

Fig. 37. Tangential section through sporangial wall of *Spizellomyces* sp. The arrows point out the microfibrillar substructure of the wall. The electron-dense areas are surface spines which can be seen in longitudinal section in Fig. 40

Fig. 38. In nature, bacteria (*B*) frequently form extensive films at the wall surface of sporangia of chytridiomycetes such as *Rhizophydium* sp. (three arrows). Cytoplasmic development proceeds with the cytoplasm cleaving into zoospores. × 27,000

Fig. 39. Developing sporangial wall of *Chytriomyces hyalinus*. The layer adjacent to the plasma membrane (*PM*) is dense, and the outer layer fibrillar. × 27,000

Fig. 40. In the sporangial wall of *Spizellomyces* sp., both the middle and the inner layer wall layers (adjacent to the plasma membrane, *PM*) have microfibrillar substructure. The arrow points out outer surface sculpturing into spines. × 30,000

Fig. 41. Resting sporangium wall of *Coelomomyces punctatus* is homogeneously electron-dense but traversed with openings. Extracellular material (arrow) coats the surface of the developing wall. × 30,000

Fig. 42. Thallus of *Entophlyctis variabilis* with multinucleate sporangium (*Sp*), a septum (*S*) delimiting the sporangium form the rhizoids, and a discharge tube with discharge plug (*DP*) formed. Cytoplasmic cleavage has not started. × 14,000

Fig. 43. Thick wall of *Allomyces javanicus* (Blastocladiales) resting sporangia (*RS*) forming within wall of hypha (*H*). A septum (*S*) forms between nascent resting spore and hyphal cytoplasm. Plasmodesmata (*PL*) traverse the septum (*S*) and the sculptured resting sporangium wall. × 20,000

Fig. 44. *Entophlyctis variabilis* (Spizellomycetales) rhizoid (*R*). A septum (*S*) with plasmodesmata (*PL*) delimits the rhizoids and sporangium. × 39,000

Fig. 45. *Allomyces javanicus* produces a wall apposition (*A*) along its hyphal (*H*) wall in response to perturbation by another parasitic chytridiomycete. × 18,000

Fig. 46. *Entophlyctis variabilis* sporangium in which, prior to cytoplasmic cleavage into zoospores and opening of the discharge pore, a discharge plug has formed (*DP*). Notice the row of vesicles (arrows) aligned along the plasma membrane adjacent to the discharge plug. Contents of vesicles and discharge plug are similar. × 15,000

Fig. 47. In the formation of an inoperculate discharge pore in *Entophlyctis variabilis* (Spizellomycetales), the apex of the discharge tube wall becomes homogeneous and expanded in appearance (double arrows). Notice the presence of the discharge plug (*DP*) and the similarity in texture between this material and contents (arrow) of vesicles that underlie the plasma membrane. × 15,500

Fig. 48. After the tip of discharge tube opens and cytoplasmic cleavage is complete, an endooperculum (*EO*) can appear between the discharge plug (*DP*) and interstitial material in sporangia of *Entophlyctis variabilis*. The arrows show that the endooperculum seals the cytoplasm and extends along the wall (*W*), tapering to an end (bottom arrow). Interstitial material (*IS*) fills the area between the cleaved zoospore plasma membrane and wall or endooperculum of the sporangium. × 39,000

Fig. 49. Formation of operculum (*O*) in *Chytriomyces hyalinus* (Chytridiales) sporangium. A discharge plug (*DP*) is present. Notice the erosion of the wall (*W*) at the site where the wall will break down (arrow) prior to formation of operculum and release of zoospores. × 35,000

Fig. 50. *Rhizophlyctus rosea* (Spizellomycetales) sporangium in which zoospores have been cleaved. The discharge tube (*DT*) is present. Material of the endooperculum (*EO*) continues all along the sporangial wall as a thin inner layer, like a sac. × 11,000

Fig. 51. Notice the layers of cell wall around the sporangium of *Rhizophlyctis rosea*. There is a discharge plug (*DP*) and endooperculum (*EO*) × 30,000

Fig. 52. Endooperculum (arrow) of *Rhizophlyctis rosea* sporangium which has been folded upward with zoospore release. × 29,000

Fig. 53. *Coelomomyces punctatus* (Blastocladiales) resting sporangium germination. The wall is opened with a dehiscence slit. The dome-shaped discharge plug (*D*), produced prior to complete cytoplasmic cleavage, tapers as it extends away from the dehiscence slit. After cleavage it will form a thin layer between the inner wall layer and interstitial material. The sculptured, outer wall layer (*W*) is electron-dense, and the inner layer is finely fibrillar. The discharge plug will expand into a discharge vesicle at release of zoospores. × 25,000

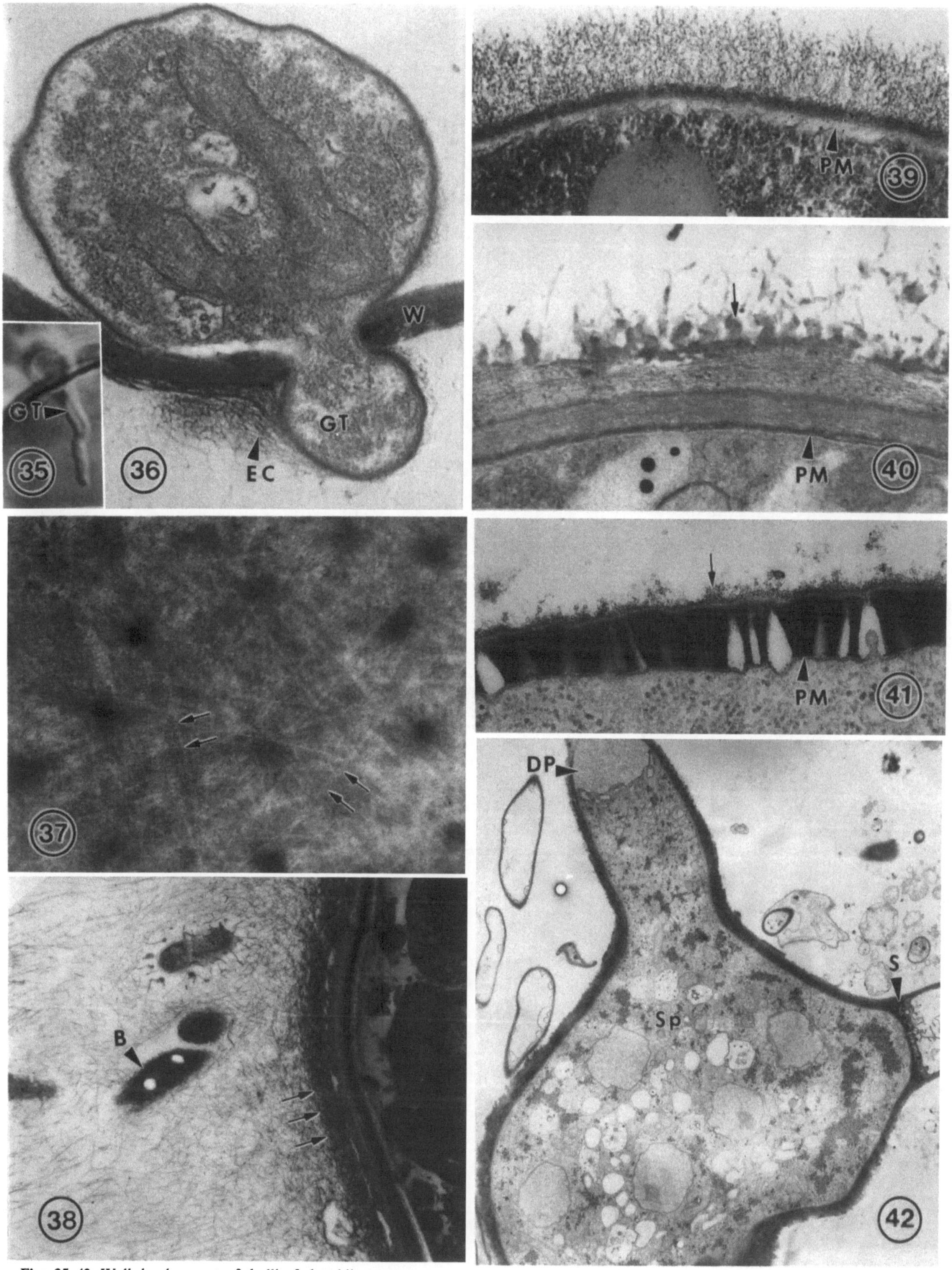

Figs. 35–42. Wall development of thalli of chytridiomycetes

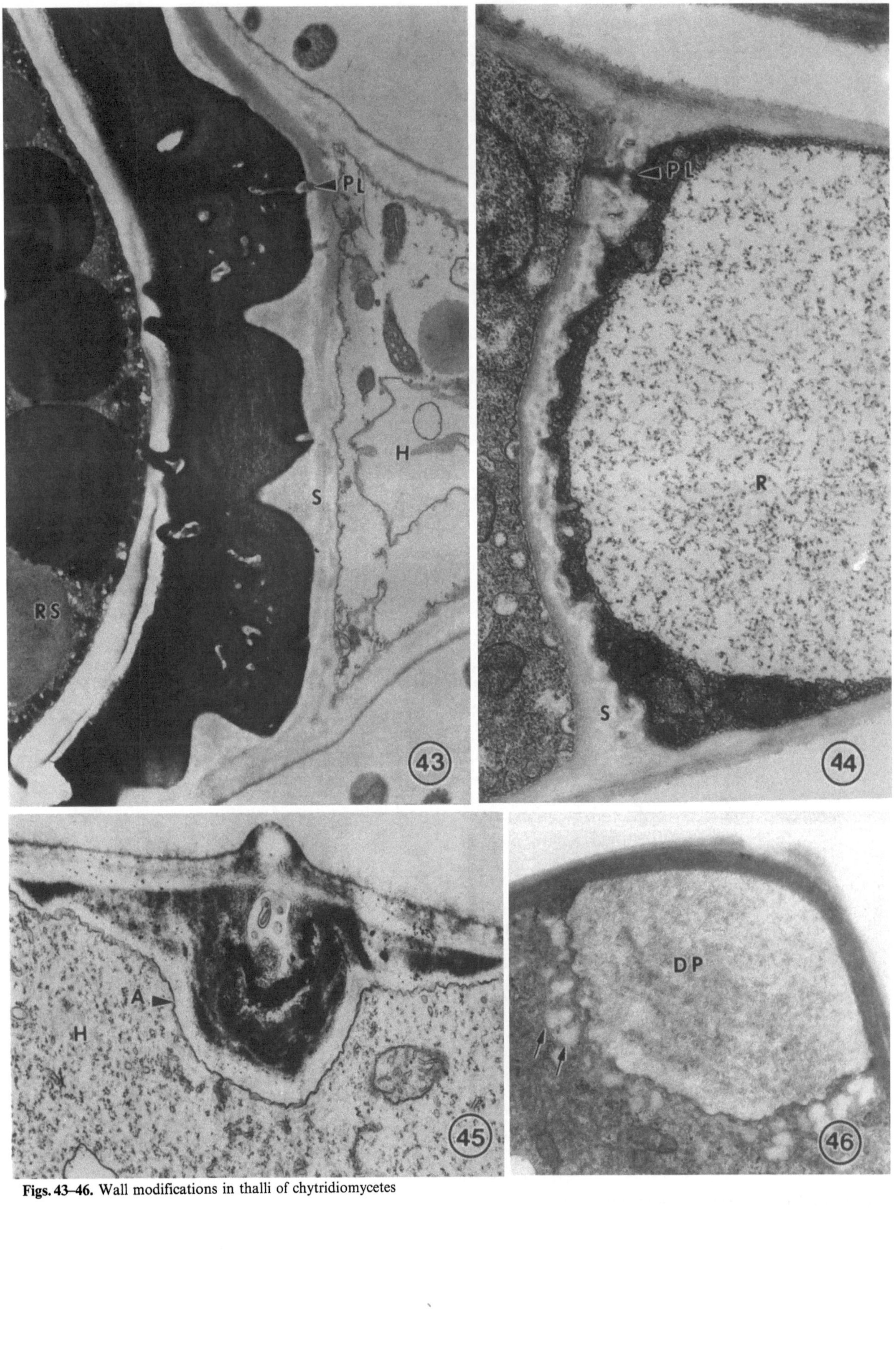

Figs. 43–46. Wall modifications in thalli of chytridiomycetes

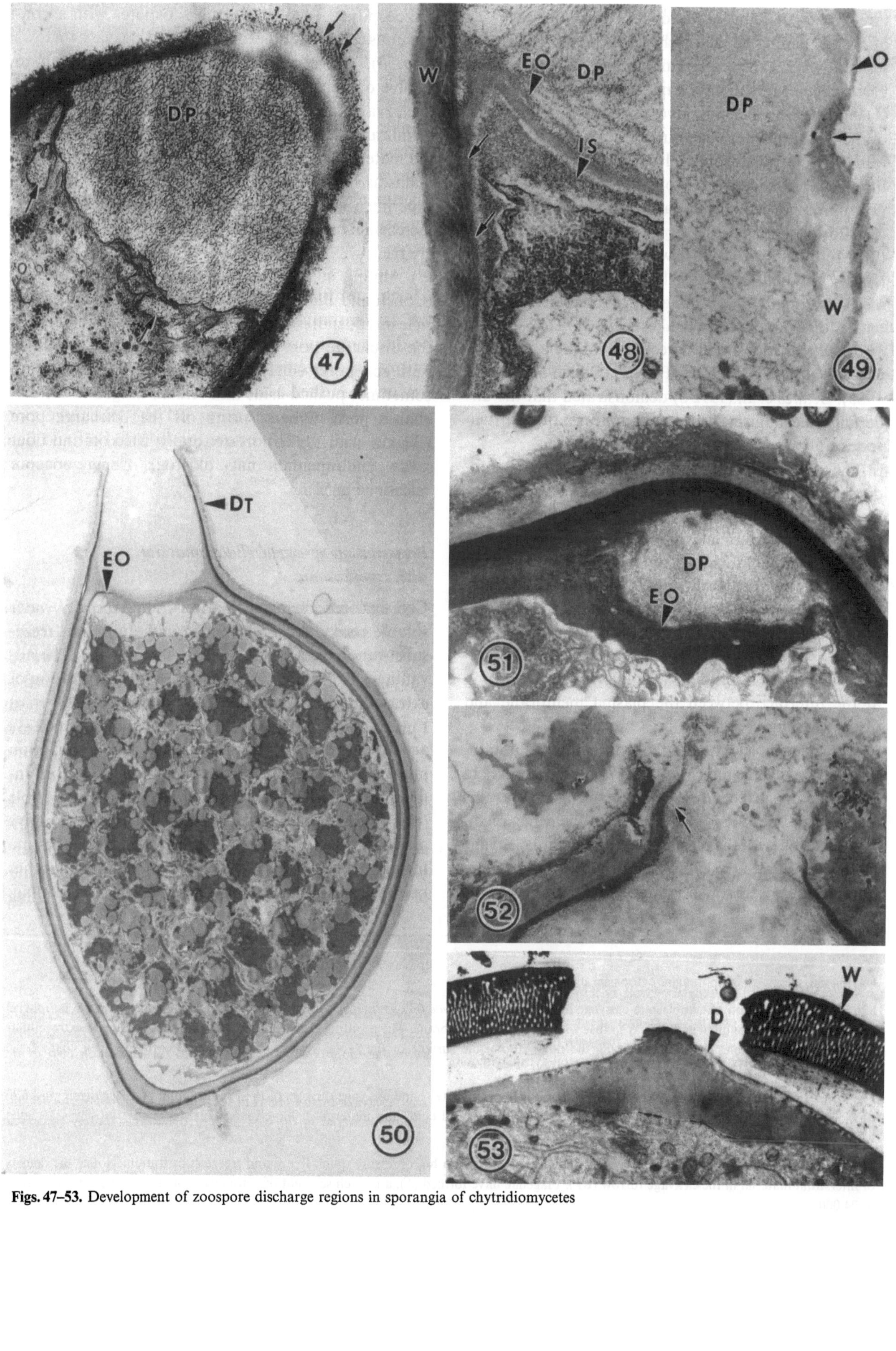

Figs. 47–53. Development of zoospore discharge regions in sporangia of chytridiomycetes

tures are regulated is not known. A possible sequence of events leading to the discharge of zoospores is described below. There are numerous variations on this basic scheme, and all stages may not exist in some taxa.
(1) Additional inner wall layers may appear. Material of the discharge plug accumulates below the site in the wall where an exit port will form (Fig. 42), allowing the release of zoospores. Golgi-derived vesicles transport and secrete this material to the site (Figs. 46 and 47) (Powell 1976 a).
(2) Modifications in the wall (localized digestion) (Figs. 47 and 49) eventually lead to the opening of the discharge area (Fig. 50) as an inorperculate or operculate discharge pore (Beakes et al. 1992 b, Powell 1976 a). With thick walled resting sporangia, a dehiscence split begins to erode (Fig. 53) (Couch and Bland 1985). Beakes et al. (1992 b) demonstrated that the underlying mechanisms for operculum formation in two species of the same genus were different.
(3) The discharge pore opens. At this stage an endooperculum may form between the discharge plug and plasma membrane. In some taxa a single layer of endoplasmic reticulum underlying the sporangial plasma membrane at the discharge site mirrors exactly the endooperculum, apparently templating this extracellular structure (Powell 1976 a). In another taxa the endooperculum is structurally different, and lines the sporangium like a closed sac (Fig. 50). McNitt (1974 b) described multiple layers of extracellular material forming at the interface between the plasma membrane and discharge plug.
(4) Cytoplasmic cleavage of zoospore initials containing single nuclei begins (McNitt 1974 b). Differences in the zoospore surface features can be recognized at this stage. Golgi-derived vesicles provide cleavage membrane, fusing with the plasma membrane until cleavage furrows push into the cytoplasm (Fig. 54). Membranes of the cleavage furrow become the plasma membrane of zoospores. Taxa producing zoospores with a conspicuous cell coat also produce coated cleavage furrows (Fig. 54). Those on which a cell coat is difficult to resolve on zoospores may contain abundant material in cleavage furrows, but there is a less distinct coat on membranes of the cleavage furrows (Fig. 55). Some of the material discharged into the cleavage furrows remains as interstitial material widely believed, but not experimentally proved, to have a hydroscopic nature, increasing turgor, and forcing zoospores out at discharge.
(5) Among some taxa, the discharge plug (Fig. 53) (Couch and Bland 1985) expands into a discharge vesicle as zoospores are released. In others, it remains in the discharge pore as a stopper and slowly expands until zoospores emerge (Powell 1976 a). Opercula are sometimes pushed aside and remain hinged to the discharge pore. Others spring off the discharge pore (Beakes et al. 1992 b) or are totally lifted off and float away. Endoopercula may fold back during zoospore release (Fig. 52).

Preservation of extracellular material with cryofixation

Conventional chemical fixations remove many water soluble compounds, whereas cryofixation and freeze-substitution extract less of these compounds. The advantages of cryofixation over conventional fixation of extracellular materials are illustrated in Fig. 56 versus Figs. 54 and 55. Interstitial material secreted in cleavage furrows densely fills the channels after cryofixation preparation (Fig. 56). With conventional fixation, interstitial material is less dense (Fig. 55) or is not visible (Fig. 54). Although this extraction allows one to resolve associations of extracellular material with the plasma membrane better (Fig. 54), representation of quantities of extracellular material on the membrane is misleading.

Figs. 54–56. Cleavage of sporangial cytoplasm into zoospores

Fig. 54. At the initiation of cytoplasmic cleavage in *Polyphagus euglenae* (Chytridiales), Golgi-derived vesicles (*GV*) cluster in the peripheral cytoplasm of the sporangium and appear to fuse with the plasma membrane. The resulting cleavage furrows (*CF*) are coated with extracellular material which resembles the material lining the non-cytoplasmic surface of the Golgi vesicles. This material forms the cell coat of the zoospore. The wall (*W*) is thin in this fungus. × 50,000

Fig. 55. Notice similarity of contents which fill Golgi-derived vesicles (*GV*) and cleavage furrows (*CF*) in sporangium of *Entophlyctis variabilis* (Spizellomycetales). Zoospores of this fungus lack a conspicuous cell coat, and material in the cleavage furrow is only loosely associated with membrane. × 73,000

Fig. 56. Cleaving sporangium of *Entophlyctis variabilis* preserved with high pressure quick freeze and freeze substitution. Notice the density of interstitial material in the cleavage furrow (*CF*). Flagellae have formed (*F*), and each segment of cytoplasm contains a single nucleus (*Nu*). × 24,000

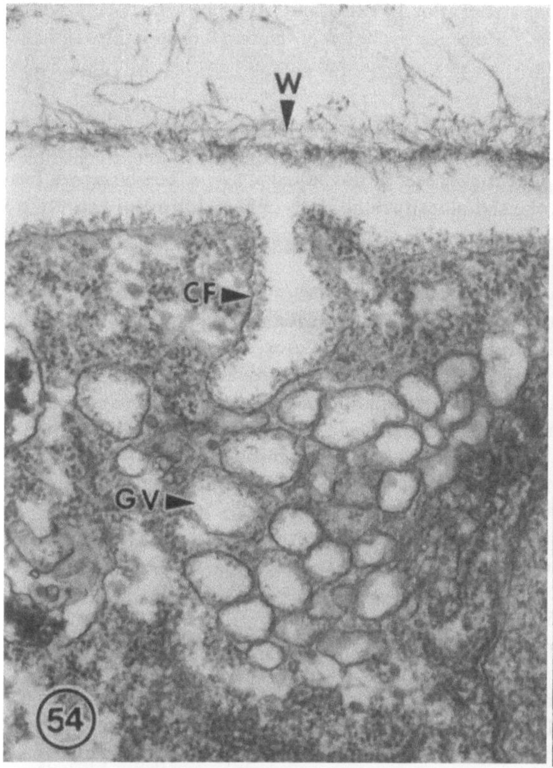
W
CF
GV
54

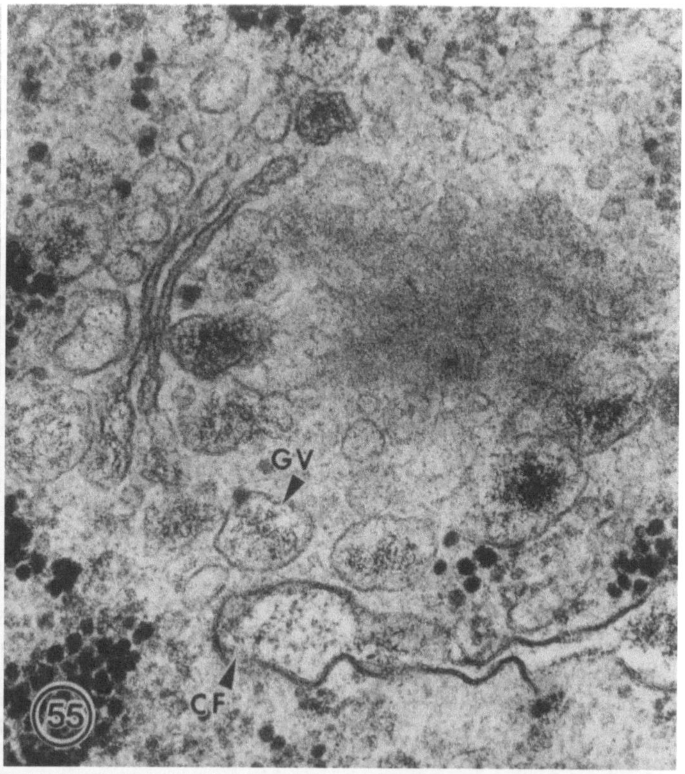
GV
CF
55

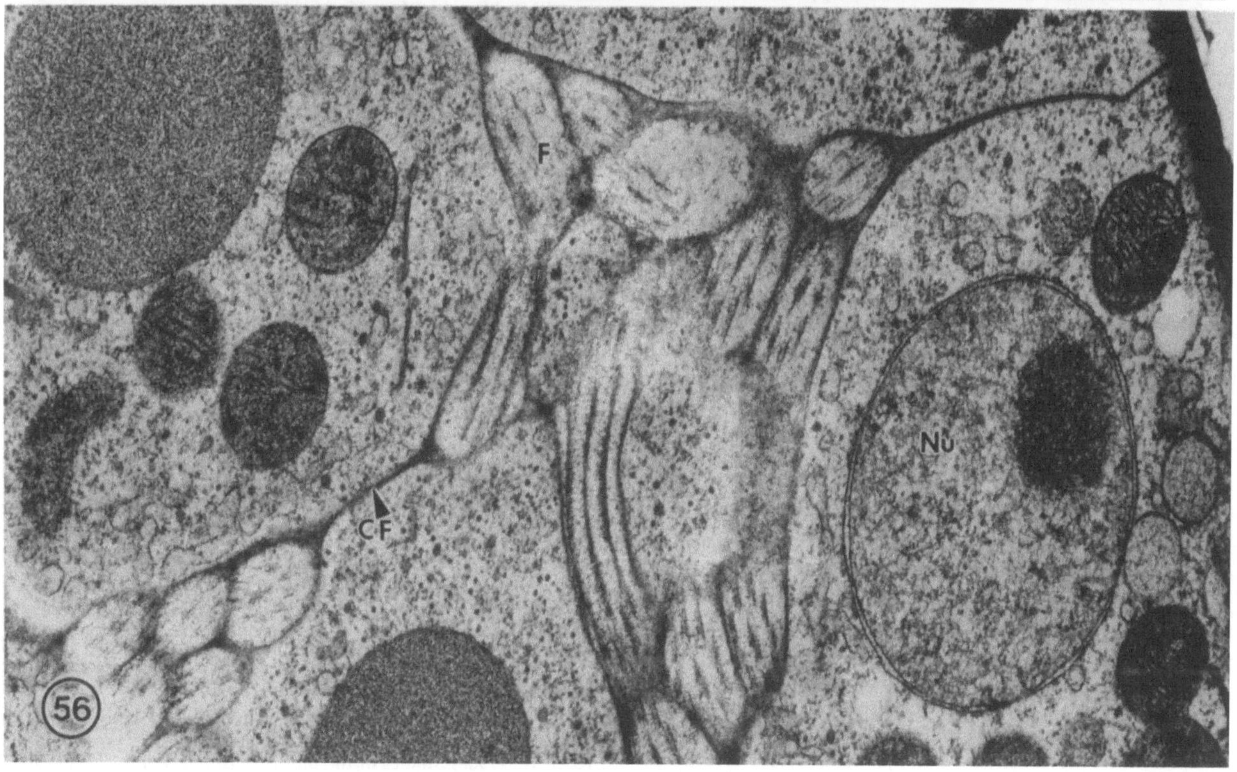
F
Nu
CF
56

Conclusion

Thus chytridiomycetes exhibit a range of patterns in the occurrence and subsequent modifications of extracellular materials, even for members within the same order. Such diversity in structure of chytridiomycete extracellular materials highlights that cell surface features are highly adaptive, rather than necessarily reflective of phylogenetic relationships among members within a class. It also cautions that if the same term is used for analogous structures, than homology should not be incorrectly assumed. Although there is a reasonable view of the types of extracellular material involved in chytridiomycete development, we have only limited understanding of their biogenesis and regulation. The most universally recognized role for these extracellular materials is protection, but their potential involvements in the interaction of the cell surface with the outside environment (Sharon and Lis 1993) and developmental signalling are areas awaiting more extensive investigation. Perhaps as Olson (1984) outlined, the generation and application of developmental mutants to questions of morphogenesis and function in chytridiomycetes are underutilized approaches which could provide valuable insights to the function of cell surface structures.

Acknowledgement

This study was supported in part by a Research Challenge Grant from the State of Ohio.

References

Baro DJS (1992) Evolution and kingdoms of organisms from the perspective of a mycologist. Mycologia 84: 1–11

– (1990) Phylum Chytridiomycota. In: Margulis L, Corliss JO, Melkonian M, Chapman DJ (eds) Handbook of Protoctista. Jones and Bartlett, Boston, pp 454–466

– (1981) The phylogenetic and taxonomic implications of flagellar rootlet morphology among zoosporic fungi. BioSystems 14: 359–370

Barstow WE, Pommerville J (1980) The ultrastructure of cell formation and gamma particles during encystment of *Allomyces macrogynus* zoospores. Arch Microbiol 128: 179–189

Beakes GW, Canter HM, Jaworski GHM (1988) Zoospore ultrastructure of *Zygorhizidium affluens* and *Z. planktonicum*, two chytrids parasitizing the diatom *Asterionella formosa*. Can J Bot 66: 1054–1067

– – – (1992 a) Comparative ultrastructural ontogeny of zoosporangia of *Zygorhizidium affluens* and *Z. planktonicum*, chytrid parasites of the diatom *Asterionella formosa*. Mycol Res 96: 1047–1059

– – – (1992 b) Ultrastructural study of operculation (discharge apparatus) and zoospore discharge in zoosporangia of *Zygorhizidium affluens* and *Z. planktonicum*, chytrid parasites of the diatom *Asterionella formosa*. Mycol Res 96: 1060–1067

Bowman BH, Taylor JW, Borwnlee AG, Lee J, Lu S-D, White TJ (1992) Molecular evolution of the fungi: relationships of basidiomycetes, ascomycetes, and chytridiomycetes. Mol Biol Evol 9: 285–296

Breton A, Dusser M, Gaillard-Martinie B, Guillot J, Millet L, Prensier G (1991) *Piromyces rhizinflata* nov.sp., a strictly anaerobic fungus from faeces of the Saharian ass: a morphological, metabolic and ultrastructural study. FEMS Microbiol Lett 82: 1–8

Couch JN, Bland CE (1985) The genus *Coelomomyces*. Academic Press, New York

Dalley NE, Sonneborn DR (1982) Evidence that *Blastocladiella emersonii* zoospore chitin synthetase is located at the plasma membrane. Biochim Biophys Acta 686: 65–76

Dorward DW, Powell MJ (1982) Cross-linking bridges associated with the microbody-lipid globule complex in *Chytriomyces aureus* and *Chytriomyces hyalinus*. Protoplasma 112: 181–188

– – (1983) Cytochemical detection of polysaccharides and the ultrastructure of the cell coat of zoospores of *Chytriomyces aureus* and *Chytriomyces hyalinus*. Mycologia 75: 209–220

Federici BA, Lucarotti CJ (1986) Structure and behavior of the meiospore of *Coelomomyces dodgei* during encystment on the copepod host, *Acanthocyclops vernalis*. J Invertebr Pathol 48: 259–268

Fuller MS, Reichle RE (1968) The fine structure of *Monoblepharella* sp. zoospores. Can J Bot 46: 279–283

Guillot J, Breton A, Damez M, Dusser M, Gaillard-Martinie B, Millet L (1990) Use of lectins for a comparative study of cell wall composition of different anaerobic rumen fungal strains. FEMS Microbiol Lett 67: 151–156

Heath IB, Bauchop T, Skipp RA (1983) Assignment of the rumen anaerobe *Neocallimastix frontalis* to the Spizellomycetales (chytridiomycetes) on the basis of its polyflagellate zoospore ultrastructure. Can J Bot 61: 295–307

Held AA (1974) Attraction and attachment of zoospores of the parasitic chytrid *Rozella allomycis* in response to host-dependent factors. Arch Microbiol 95: 97–114

Hohn TM, Lovett JS, Bracker CE (1984) Characterization of the major proteins in gamma particles, cytoplasmic organelles in *Blastocladiella emersonii* zoospores. J Bacteriol 158: 253–263

Jansson H-B, Thiman L (1992) A preliminary study of chemotaxis of zoospores of the nematode-parasitic fungus *Catenaria anguillulae*. Mycologia 84: 109–112

Jen CJ, Haug A (1979 a) Changes in surface properties of developing zoospores of *Blastocladiella emersonii*: binding of concanavalin A. J Gen Microbiol 115: 145–151

– – (1979 b) Concanavalin A-induced lysis of zoospores of *Blastocladiella emersonii*. Exp Cell Res 120: 425–428

Kazama FY (1972) Ultrastructure and phototaxis of the zoospores of *Phlyctochytrium* sp., an estuarine chytrid. J Gen Microbiol 71: 555–566

– Schornstein KL (1977) A freeze-etch study of the phototactic zoospores of *Phlyctochytrium* sp., a marine fungus. Protoplasma 91: 143–156

Kerwin JL (1983) Biological aspects of the interaction between *Coelomomyces psorophorae* zygotes and the larvae of *Culiseta inornata*: host-mediated factors. J Invertebr Pathol 41: 224–232

Koch WJ (1968) Studies of the motile cells of chytrids. V. Flagellar retraction in posteriorly uniflagellate fungi. Amer J Bot 55: 841–859

Lehnen LP Jr, Powell MJ (1989) The role of kinetosome-associated organelles in the attachment of encysting secondary zoospores of *Saprolegnia ferax* to substrates. Protoplasma 149: 163–174

Li J, Heath IB, Cheng K-J (1991) The development and zoospore ultrastructure of a polycentric chytridiomycete gut fungus, *Orpinomyces joyonii* comb.nov. Can J Bot 69: 580–589

– – Packer L (1993) The phylogenetic relationships of the anaerobic chytridiomycetous gut fungi Neocallimasticaceae and Chytridiomycota. II. Cladistic analysis of structural data and description of Neocallimasticales ord.nov. Can J Bot 71: 393–407

Lovett JS (1968) Reactivation of ribonucleic acid and protein synthesis during germination of *Blastocladiella* zoospores and the role of the ribosomal nuclear cap. J Bacteriol 96: 962–969

McNitt R (1974 a) Ultrastructure of *Phlyctochytrium irregulare* zoospores. Cytobiologie 9: 307–320

– (1974 b) Zoosporogenesis in *Phlyctochytrium irregulare*. Cytobiologie 9: 290–306

Manier J-F, Loubès C (1978) *Callimastix cyclopis* Weissenberg, 1912 (Phycomycète, Blastocladiale) parasite d'un *Microcyclops claus*, 1893 (Copépode, Cyclopoide) récolté au tcad: particularités ultrastructurales. Protistologica 14: 493–501

Muehlstein LK, Amon JP, Leffler DL (1988) Chemotaxis in the marine fungus *Rhizophydium littoreum*. Appl Environ Microbiol 54: 1668–1672

Munn EA, Orpin CG, Hall FJ (1981) Ultrastructural studies of the free zoospore of the rumen phycomycete *Neocallimastix frontalis*. J Gen Microbiol 125: 311–323

Myers RB, Cantino EC (1974) The gamma particle. In: Wolsky A (ed) Monographs in developmental biology. Karger, Basel, pp 1–117

Olson LW (1984) *Allomyces* – a different fungus. Op Bot 73: 1–96

Orpin CG, Bountiff L (1978) Zoospore chemotaxis in the rumen phycomycete *Neocallimastix frontalis*. J Gen Microbiol 104: 113–122

Pommerville J (1977) Chemotaxis of *Allomyces* gametes. Exp Cell Res 109: 43–51

– (1978) Analysis of gamete and zygote motility in *Allomyces*. Exp Cell Res 113: 161–172

– Olson LW (1987) Evidence for a male-produced pheromone in *Allomyces macrogynus*. Exp Mycol 11: 245–248

– Strickland JB, Harding KE (1990) Pheromone interactions and ionic communication in gametes of aquatic fungus *Allomyces macrogynus*. J Chem Ecol 16: 121–131

Porter D, Smiley R (1980) Development of the sporangium and discharge apparatus in a marine chytrid, *Phlyctochytrium* sp. Bot Mar 23: 99–115

Powell MJ (1974) Fine structure of plasmodesmata in a chytrid. Mycologia 66: 606–614

– (1976 a) Development of the discharge apparatus in the fungus *Entophlyctis*. Arch Microbiol 111: 59–71

– (1976 b) Ultrastructural changes in the cell surface of *Coelomomyces punctatus* infecting mosquito larvae. Can Bot 54: 1419–1437

– (1978) Phylogenetic implications of the microbody-lipid globule complex in zoosporic fungi. BioSystems 10: 167–180

– (1981 a) Ultrastructure of *Polyphagus euglenae* zoospores. Can J Bot 59: 2049–2061

– (1981 b) Zoospore structure of the mycoparasitic chytrid *Caulochytrium protostelioides* Olive. Amer J Bot 68: 1074–1089

– (1982) Ultrastructure of the host-parasite interface between *Allomyces javanicus* and its endoparasite *Catenaria allomycis*. Bot Gaz 143: 176–187

– (1983) Localization of antimonate-mediated precipitates of cations in zoospores of *Chytriomyces hyalinus*. Exp Mycol 7: 266–277

– (1993) Looking at mycology with a Janus face: a glimpse at chytridiomycetes active in the environment. Mycologia 85: 1–20

– Blackwell WH (1991) A proposed dispersal mechanism for *Septosperma rhizophydii*. Mycologia 83: 673–680

– Gillette L (1987) Septal structure of the chytrid *Rhizophlyctis harderi*. Mycologia 79: 635–639

– Roychoudhury S (1992) Ultrastructural organization of *Rhizophlyctis harderi* zoospores and redefinition of the type 1 microbody-lipid globule complex. Can J Bot 70: 750–761

Reichle RE (1972) Fine structure of *Oedogoniomyces* zoospores, with comparative observations on *Monoblepharella* zoospores. Can J Bot 50: 819–824

Robertson JA (1972) Phototaxis in a new *Allomyces*. Arch Microbiol 85: 259–266

Roychoudhury S, Powell MJ (1992) Precise flagellar configuration of the *Rhizophlyctis harderi* zoospore. Can J Bot 70: 762–771

Selitrennikoff CP, Dalley NE, Sonneborn DR (1980) Regulation of the hexosamine biosynthetic pathway in the water mold *Blastocladiella emersonii*: sensitivity to endproduct inhibition is dependent upon the life cycle phase. Proc Natl Acad Sci USA 77: 5998–6002

Sewall T, Olson L, Lange L, Pommerville J (1986) The effect of monensin on gametogenesis and zoosporogenesis in the aquatic fungus, *Allomyces macrogynus*. Protoplasma 133: 129–139

Sharon N, Lis H (1993) Carbohydrates in cell recognition. Scient Am 268: 82–89

Sparrow FK (1960) Aquatic phycomycetes, 2nd rev edn. University of Michigan Press, Ann Arbor

Taylor JW, Fuller MS (1981) The Golgi apparatus, zoosporogenesis and development of the zoospore discharge apparatus of *Chytridium confervae*. Exp Mycol 5: 35–59

Travland LB (1979 a) Initiation of infection of mosquito larvae (*Culiseta inornata*) by *Coelomomyces psorophorae*. J Invertebr Pathol 33: 95–105

– (1979 b) Structure of the motile cells of *Coelomomyces psorophorae* and function of the zygote in encystment on a host. Can J Bot 57: 1021–1035

Tunlid A, Nivens DE, Jansson H-B, White DC (1991) Infrared monitoring of the adhesion of *Catenaria anguillulae* zoospores to solid surfaces. Exp Mycol 15: 206–214

Wainright PO, Hinkle G, Sogin ML, Stickel SK (1993) Monophyletic origins of the Metazoa: an evolutionary link with fungi. Science 260: 340–342

Whittaker RH (1969) New concepts of kingdoms of organisms. Science 163: 150–160

Protoplasma (1994) 181: 142–163

PROTOPLASMA

Characterization of zoospore and cyst surface structure in saprophytic and fish pathogenic *Saprolegnia* species (oomycete fungal protists)

A. W. Burr and **G. W. Beakes***

Department of Biological and Nutritional Sciences, The University of Newcastle upon Tyne, Newcastle upon Tyne

Received August 5, 1993
Accepted March 30, 1994

Summary. The importance of the surface structure and chemistry in zoospores and cysts of oomycetes is briefly reviewed and the organelle systems associated with encystment described. The surface structure and chemistry of primary and secondary zoospores and cysts of *Saprolegnia diclina* (a representative saprophytic species) and *S. parasitica* (a representative salmonid fish pathogen) were explored using the lectins concanavilin A (Con A) and wheat germ agglutinin (WGA) and monoclonal antibodies (MAbs) raised against a mixed zoospore and cyst suspension of *S. parasitica*. The binding of lectins and antibodies to spores was determined using immunofluorescence microscopy with fluorescein isothiocyanate-labelled probes and with electron microscopy with gold-conjugated probes applied to spore suspensions post-fixation. In both species Con A, which is specific for glucose and mannose sugars, bound to both the surface of primary and secondary zoospores (the surface glycocalyx) and their cyst coats and readily induced zoospore encystment. The binding to the cysts appeared to be mainly associated with the matrix material released from the primary and secondary encystment vesicles and which appeared to diminish with time. No binding to germ tube walls was observed with this lectin. The MAb labelling showed a generally similar binding pattern to the primary and secondary cysts to that observed with Con A, although the binding to zoospores was more variable. Primary zoospores bound the antibodies but secondary zoospores appeared less reactive. It is suggested that the MAbs share a common epitope with one or more of the Con A-binding components. In both species WGA, which is specific for amongst other things the sugar N-acetyl glucosamine, bound to localised apical patches on the primary zoospores. This lectin also binds to the ventral groove region of secondary zoospores of *S. diclina*, which were induced to encyst by this lectin. In contrast secondary zoospores of *S. parasitica* were not induced to encyst by the addition of WGA and showed a patchy dorsal binding with this lectin. WGA also binds to both the inner wall of discharged primary cysts and the young germ tube walls of both species. These observations are discussed both in relation to other oomycete spores and to their possible functional and ecological significance.

Keywords: *Saprolegnia*; Lectins; Concanavalin A; Wheat germ agglutinin; Monoclonal antibodies; Ultrastructure; Pathogenesis.

Abbreviations: BSA bovine serum albumin; Con A Concanavalin A; DBA *Dolichos biflorus* agglutinin; ELISA enzyme-linked immunosorbent assay; EM electron microscope; EV encystment vesicles; FCS foetal calf serum; FITC Fluorescein isothiocyanate; FV peripheral fibrillar vesicles; G + F 0.2% glutaraldehyde and 2.0% formaldehyde primary fixative solution; 2G 2% glutaraldehyde primary fixative; LM light microscopy; MAbs monoclonal antibodies; LPV large peripheral vesicles; PBS phosphate buffered saline; PCV flattened peripheral cisternae; PEV primary encystment vesicle; PIPES piperazine-N,N1-bis(2-ethane sulfonic acid); PNA *Ricinus communis* agglutinin; RAM-FITC/Au_{10-20} Fluorescein isothiocyanate/gold (10 or 20 nm) labelled rabbit anti-mouse immunoglobulin; RCA *Ricinus communis* agglutinin; SEM scanning electron micrograph; SBA soybean agglutinin; SEV secondary encystment vesicles; TEM transmission electron micrograph; UEA I *Ulex europaeus* agglutinin; WGA wheat germ agglutinin.

Introduction

Fine-structural studies (see reviews by Beakes 1987, 1989) and molecular studies (Gunderson et al. 1987) have shown that oomycetes are more closely related to the chromophyte algae (chrysophytes, xanthophytes, and phaeophytes) and other heterokont protists than to higher fungi. Indeed it was the close morphological similarity between the coenocytic xanthophyte *Vaucheria* and the water mould *Saprolegnia*, which inspired Irene Manton to undertake her pioneering ultrastructural examination of their biflagellate zoospores (Manton et al. 1951). Water moulds such as *Saprolegnia* have a complex asexual cycle with two generations of both

* Correspondence and reprints: Department of Biological and Nutritional Sciences, Agriculture Building, The University of Newcastle upon Tyne, Newcastle upon Tyne NE1 7RU, U.K.

zoospores and cysts, the comparative structures of which have already been documented in detail previously (Holloway and Heath 1977, Beakes 1983). It is the stronger swimming reniform secondary zoospores of this group which are morphologically and probably phylogenetically equivalent to the zoospores of the important plant pathogenic peronosporalean oomycetes such as *Phytophthora* and *Pythium* (Beakes 1987).

Although the general fine-structure of a wide range of oomycete zoospores and cysts has been carefully described (e.g., Hemmes and Hohl 1971, Hoch and Mitchell 1972, Bimpong and Hickman 1975, Lunney and Bland 1976, Holloway and Heath 1977, Grove and Bracker 1978, Cerenius et al. 1984) the function of many of the described organelles associated with zoosporogenesis is either not known or only partially understood. One exception in *Saprolegnia* is the so called "bar-bodies" which form early in sporangium differentiation prior to septum formation (Heath and Greenwood 1970). These bodies eventually become distributed in the peripheral cytoplasm of primary zoospores and are released upon encystment to form the outer smooth, cyst coat layer in which tufts of pre-formed tubular spines are embedded (Heath and Greenwood 1970, Beakes 1983).

The transition from motile wall-less zoospore to nonmotile walled cyst (cystospore) is a critical event in the life cycle of many oomycetes, since it plays a central role in the initial interaction between the fungal propagule and its potential food substrate or host and has received special attention in many cytological studies (Bacic et al. 1985; Beakes 1983; Berbee and Kerwin 1993; Cho and Fuller 1989; Estrada-Garcia et al. 1990 a, b; Grove and Bracker 1978; Gubler and Hardham 1988; Hardham and Suzaki 1986; Heath and Greenwood 1970; Hoch and Mitchell 1972; Lehnen and Powell 1989; Lunney and Bland 1976; Sing and Bartnicki-Garcia 1975 a). Oomycete zoospore behaviour is modulated by the external environment often exhibiting a variety of responses such as negative geotaxis, chemotaxis and immobilisation (Cameron and Carlile 1977, 1980, Deacon 1988, Irving and Grant 1984, Jones et al. 1991). Zoospores can be induced to encyst by a wide range of environmental (e.g., agitation) and chemical triggers, such as metal ions, amino acids and carbohydrates (Bimpong and Hickman 1975; Byrt et al. 1982 a, b; Cameron and Carlile 1980, 1981; Kerwin et al. 1991; Paktitis et al. 1986; Smith et al. 1984). Specific sugars (L-fucose) in root mucilages were shown to bind to specific proteins on the zoospore surface of certain root pathogens (*Phytophthora* and *Pythium*), inducing encystment and mediating binding (Estrada-Garcia et al. 1990 b, Hinch and Clarke 1980, Longman and Callow 1987). However, it has recently been pointed out that other recognition mechanisms may exist between fungal spore and root surface (Mitchell and Deacon 1987). The mosquito pathogen, *Lagenidium giganteum*, has been shown to be specifically induced to encyst by derivatives of chitin and chitosan (Kerwin et al. 1991). It follows therefore that successful attachment and infection of a host might be regulated by the encystment process.

The first systematic attempt to identify and characterise the zoospore organelle(s) responsible for adhesion of *Phytophthora* zoospores to their substrates was carried out by Sing and Bartnicki-Garcia (1975 a, b). They demonstrated that Con A binding material released from what in *Phytophthora* are now generally known as the large peripheral vesicles (LPV), seemed to be responsible for the adhesion of *P. palmivora* zoospores to inert substrates such as plastic. A similar conclusion was also reached for *Pythium aphanidermatum* (Estrada-Garcia et al. 1990 a). Morphologically similar fibrillar vesicles (FV) occur in secondary zoospores of *Saprolegnia* (Hollway and Heath 1977, Beakes 1983, Lehnen and Powell 1989) and also appeared to be released by encysting secondary zoospores. It was therefore proposed by analogy with *Phytophthora* that in *Saprolegnia* this vesicle fraction probably also functioned as the source of adhesive and that such organelle populations should be referred to as "adhesive vesicles" (Beakes 1983, 1987). Bar bodies are also discharged from *Saprolegnia* during zoospore encystment and it was proposed that these should be re-named as "encystment vesicles" (EV) (Beakes 1987).

As far as could be told peronosporalean oomycetes, such as *Phytophthora*, lacked an amorphous outer coat and "encystment vesicle" equivalents (Beakes 1987). However, this interpretation has had to be radically re-evaluated in the light of the elegant studies of Hardham and her co-workers working with *P. cinnamomi* (Hardham et al. 1994). As well as examining the binding patterns of lectins to zoospores and cysts, panels of MAbs were produced to various zoospores and cyst organelle components (Hardham 1985, 1989; Hardham and Suzaki 1986, 1990; Hardham and Gubler 1990; Hardham et al. 1985, 1986, 1991 a, b) (see Table 3). These MAbs have proved invaluable as a means of discriminating between the various vesicular components that are associated with encystment (and those that are not) and also in facilitating their partial biochemical characterisation (e.g., Gubler and Hardham

1988, Hardham and Gubler 1990, Hardham et al. 1991 b). As a result of these studies the Con A binding LPVs were shown to contain high molecular weight storage glycoproteins which were not discharged upon encystment, although the LPVs were extremely prone to fixation-induced release (Gubler and Hardham 1988, 1989; Hardham and Gubler 1990). Initially the main vesicle fraction which was released from zoospores upon encystment was shown to be a "small peripheral vesicle" fraction (spv), which had electron-dense structured contents, which had formerly been considered to be peripheral microbodies (often called U-bodies in earlier accounts, see Powell and Bracker 1986). These vesicles contain a > 300 kDa glycoprotein which binds the lectin soybean agglutinin (SBA), indicating the presence of N-acetyl-D-galactose residues (Hardham et al. 1986). However, the raising of further MAbs subsequently revealed that the spv fraction in *P. cinnamomi* was in fact composed of two distinct populations whose distribution was predominantly polarised. The SBA-binding vesicles are mainly dorsally distributed and give rise to an amorphous outer cyst coat and are probably homologous to the EV fraction in saprolegnian spores (Table 4). However, the adhesive material was contained in a population of "small ventral vesicles" (vsv), which are largely clustered along the rims of the ventral flagellar grooves. The material released from these vesicles appears to be proteinaceous, periodate insensitive, and shows no lectin binding but is responsible for the adhesion of the zoospores to host root cells (Hardham and Gubler 1990). Almost co-incidentally with this discovery in *Phytophthora*, Lehnen and Powell (1989) demonstrated that *Saprolegnia* zoospores released an adhesive material from the clusters of large ventral K-bodies, which upon discharge release a fibrillar pad of material which binds wheat germ agglutinin (WGA), indicating the presence of N-acetyl-D-glucosamine residues. Cytochemical tests for carbohydrates also revealed these organelles were rich in sulfhydryl containing polysaccharides (Lehnen and Powell 1988, Sadowski and Powell 1990).

In addition, MAbs raised against oomycete spores show varied specificities ranging from isolate-, species-, or genus-specific as well as non-specific antibodies which bind to all oomycete spores (Estrada-Garcia et al. 1989; Hardham et al. 1985, 1991 a). They may therefore provide important tools for exploring taxonomic and phylogenetic relationships within oomycetes as well as the functional aspects being considered here.

Most of the studies on encystment have been on species of oomycetes which settle and attach to solid substrates or host surfaces such as roots (Estrada-Garcia 1990 b, Hardham and Gubler 1990). Encystment in zoospores of *S. parasitica*, isolated from lesions on live salmonid fish, appears to be triggered by significantly lower thresholds of inducing agents than spores of comparable saprophytic species such as *S. diclina* or *S. ferax* (Willoughby 1986). The one morphological feature which above all else characterises the spores of these fish lesion isolates are the bundles of long (< 12 µm) boathook spines on their secondary cysts (see Manton et al. 1951, Pickering et al. 1979, Beakes 1983). This species may depend upon the passive entanglement of encysted spores with the complex fish surface. This study therefore explores the surface chemistry of both primary and secondary zoospores and cysts of representative saprophytic (*S. diclina* E 3) and fish-lesion (*S. parasitica* TP 41) saprolegnias to see if any differences can be detected in their lectin binding patterns. In addition MAbs were raised against mixed zoospores and cyst suspensions of *S. parasitica* to try and refine the characterisation of the encystment apparatus in saprolegnian species. It is hoped that these studies will provide an insight into the likely mode of infection of the fish pathogen which is characterised by its ability to regularly colonise live, stressed fish. It has been previously shown using spores of the same isolates as studied here that the pathogen appears to be retained on the surface of live trout for longer than the saprophyte (Wood et al. 1988).

Materials and methods

General culture procedures and propagule production

The fungal cultures were maintained on modified *Achlya*-agar medium (Whiffen 1945). A representative isolate of *S. parasitica* Coker (TP 41, ATCC 42062) isolated from the brown trout *Salmo trutta* L. and *S. diclina* Humphrey (E 3, ATCC 36144) isolated from Lake Windermere (Willoughby 1978) were compared and contrasted throughout this investigation. Suspensions of zoospores and cysts were obtained by growing up cultures *Saprolegnia* in liquid *Achlya*-medium for 24–48 h. The mycelial mats were coarsely filtered and re-suspended in 150 ml of filtered Lake Windermere water in 250 ml flasks. After an interval of approximately 10–24 h (depending upon the isolate, ambient temperature, and the spore type required) a mixed suspension of primary and secondary zoospores and cysts was produced. Zoospores could be induced to encyst synchronously either by vortexing or by the addition of pectin or Ca^{2+} (Lehnen and Powell 1988).

Production of monoclonal antibodies (MAbs) against S. parasitica TP 41

The protocols developed for raising antibodies to *P. cinnamomi* (reviewed by Hardham et al. 1991 b) were used as the basis for monoclonal antibody production from *S. parasitica* TP41. A mixture of primary and secondary cysts and zoospores of isolate TP41 were fixed (15 min) in 1% glutaraldehyde in 50 mM piperazine-N,N1-bis(2-ethane sulfonic acid) (PIPES) buffer at pH 7.2 and suspended to give a concentration of around 2×10^6 spores in each 100 µl aliquot in PBS. Female BALB/c mice were immunised by sub-cutaneous injection into the back of the neck. Further booster injections of immunogen were given 14, 28, and 42 days after the initial immunisation with a final interperitoneal booster given after 66 days. Four days later the mice were killed and their spleens removed. The cell line P3-NS1-Ag4-1 (NS-1) used for hybridoma production was maintained at 37 °C at a cell concentration of between 5×10^4/ml and 5×10^5/ml. Antibody positive hybridoma culture medium was detected by either ELISA and/or by immunofluorescence assay. In the ELISA, an absorbance value of 0.4 (greater than corresponding control value) at 492 nm was chosen as the cut off point and cultures above this were considered to be antibody positive. During subsequent culture procedures, hybridomas were continually monitored by ELISA or immunofluorescence assay where suitable.

Fluorescent lectin and antibody labelling

Fluorescein isothiocyanate (FITC) conjugated lectins and their inhibitory saccharides were obtained from either Sigma Chemical Co. Ltd. (U.K.) or Vector Laboratories (U.K.). The labelled lectin conjugates were made up with PBS containing 0.5% bovine serum albumin (Sigma A 7638). Con A required the addition of 0.1 mM $CaCl_2$ and 0.01–0.05 mM $MnCl_2$. The MAbs were visualised using FITC conjugated rabbit anti-mouse immunoglobulin (Sigma). The lectins with their specific sugars tested in this study were concanavalin A/glucose, mannose (Con A), *Dolichos biflorus* agglutinin/α-N-acetylated galactosamine (DBA), *Ulex europaeus* agglutinin/L-fucose (UEA I), soybean agglutinin/α-N-acetylated galactosamine (SBA), peanut agglutinin/β-1,3-linked galactose and N-acetylated galactosamine and simple α and β galactose (PNA), *Ricinus communis* agglutinin/α and β galactose and unlinked N-acetylated galactosamine (RCA) and wheat germ agglutinin/α-N-acetylated galactosamine (WGA) all obtained from Sigma.

For these studies the mixed zoospore/cyst suspensions were first fixed with 0.2% glutaraldehyde and 2.0% formaldehyde (G + F) in 50 mM PIPES, pH 7.2 at room temperature for 30 min, which had been found to give the optimum specific fluorescence intensity and minimal autofluorescence in preliminary experiments (Burr 1991). After fixation, the cell suspensions were washed four times by centrifugation in 50 mM PIPES buffer. To inactivate residual aldehyde groups introduced by the fixative, the third wash (30 min) contained 75 mM glycine before a final wash in phosphate buffered saline (PBS). Fixed cell suspensions (about 2×10^5/ml) were either added (20 µl) to wells of 10-well tissue culture multitest slides (Flow Laboratories), air dried onto glass slides or were examined as freshly prepared cells.

Lectin solutions were mixed with equal volumes of fixed and concentrated zoospore suspensions or applied to dried spores to give a final lectin concentration of 20 µl/ml and incubated in the dark for one hour at room temperature. The following controls were used: (*i*) the relevant 0.1 M lectin-specific blocking saccharide which was added to the lectin solution and to the cell suspension 20 min prior to their labelled lectin incubation, (*ii*) addition of unlabelled native lectin to the cell suspension 20 min prior to incubation, or (*iii*) the omission of labelled lectin from the reaction sequence. The cells were then mounted in an "antifade" glycerol based mounting medium containing *p*-phenylenediamine dihydrochloride (Sigma) and sodium bicarbonate.

To determine the surface localisation of MAbs a suspension of zoospores and cysts were given an initial incubation in PBS containing 1% w/v BSA (45 min) followed by the addition of a 1 : 20 dilution of antiserum in PBS or neat tissue culture supernatent (TCS) for 45 min. Following primary antibody removal and rinsing, the slides were incubated with RAM-FITC (also containing 1% BSA) at dilutions of 1 : 20 or 1 : 100 in PBS before again rinsing with PBS and finally with nano-pure water. Slides were then mounted in "antifade" mountant. Each mouse sample was screened in relation to two controls: (*i*) no primary antibody, (*ii*) a non-immune serum.

Finally all slides were examined for their fluorescence using the appropriate filters for FITC fluorescence on a Leitz Diaplan photo microscope equipped with a 3-λ Ploemopak incident light fluorescence illuminator. Photomicrographs were taken using a Leitz Vario Orthomat camera system with either Kodak TMAX 800 uprated to 1600 ASA or with Kodak TMAX 3200.

Preparation of zoospores and cysts for electron microscopy

Suspensions of zoospores and cysts were added directly to an equal volume of fixative solution to give a final concentration of 2% glutaraldehyde (2G) in either 0.1 M cacodylate buffer or 50 mM PIPES buffer at pH 7.2 at ambient temperature (20–22 °C). After fixation (30 min) the cells were centrifuged at 170 **g** for 10 min and were washed three times by centrifugation in the relevant buffer. This pellet was then embedded by rapid centrifugation in 1.0% clear Oxoid No. 1 agar which had been liquefied and cooled to 45 °C. Once cooled the embedded spore suspension was cut into small 2–3 mm^3 blocks for further processing. Cells were then post-fixed by incubation for 1 h in buffered 1% OsO_4 (Agar Scientific). Each block was then dehydrated at room temperature for half an hour in each of a graded series of either ethanol or acetone solutions. After complete dehydration the blocks were embedded in Spurr's resin. Silver-grey sections were cut on a Reichert Ultracut ultramicrotome and were collected on formvar coated copper grids. Mounted sections were stained in an LKB automatic tissue stainer at 20 °C in aqueous uranyl acetate for 40 min and lead citrate for 6 min before examination in a Kratos Cora 300 or Philips 301 electron microscope, operated at 60 kV and 80 kV, respectively. Photomicrographs were taken on 70 mm Agfa film.

For examination of spores in the scanning electron microscope (SEM) some of the fixed cells were pipetted onto 9 mm diameter round glass coverslips (TAAB) where they were allowed to settle and adhere for 1 h. The coverslips with attached spores were then dehydrated in a graded acetone series followed by three washes in 100% acetone. Transfer of the acetone solution series was a gradual process whereby a constant volume of acetone solution was maintained in the dehydration vessel. The dehydrated samples were then loaded directly from the acetone into a Samdri critical-point dryer (EMScope). After critical-point drying the glass coverslip samples were mounted onto aluminium stubs with silver adhesive (TAAB) and each block was then gold shadowed in a Novotech Sputter Coater. Specimens were observed on a Cambridge Stereoscan 240 scanning electron microscope operated at between 5–12 kV and photomicrographs taken using Kodak TMAX100 75 mm film.

Pre-embedding lectin and antibody labelling

For determination of the surface binding of lectins at the ultrastructural level the cells were fixed with 0.2% glutaraldehyde and 2%

formaldehyde (G + F) in 50 mM PIPES buffer, pH 7.2, for 30 min. Zoospores of *S. parasitica* appeared to be more labile and prone to fixation damage than those of *S. diclina*. Con A conjugated to 20 nm (Con A-Au_{20}) gold particles and WGA-conjugated to either 10 nm (WGA-Au_{10}) or 20 nm gold particles (WGA-Au_{20}) were obtained from Sigma. The lectins and their gold conjugates were made up with PBS containing 0.5% bovine serum albumin (Sigma A 7638). Con A required the addition of 0.1 mM $CaCl_2$ and 0.01 mM $MnCl_2$. The gold labelled lectins were diluted to 1 : 20, 1 : 40 or 1 : 100 in PBS. When the cells had been washed and pelleted, the supernatant was discarded to leave 20 µl of pellet and its wash solution. 20 µl of the lectin conjugate, prepared at double concentration, was then added directly to the fixed re-suspended cell suspension and incubation took place in the dark at 20 °C for 1 h in 1 ml Eppendorf capsules. Cells were then washed three times in PBS and finally in distilled water (10 min each). The cells were further fixed for 1 h in 2% G in 50 mM PIPES and then processed as above.

The MAbs were visualised using gold particle labelled rabbit anti-mouse immunoglobulin (RAM-Au_{10} or -Au_{20}), together with BSA (Fraction V). Following washing steps the fixed spore suspensions were pelleted for immuno-electron microscopical applications. Suspensions were incubated with 1% w/v foetal calf serum (FCS) for 45 min to block non-specific protein binding sites. After three washes in PBS the cell suspensions were then incubated directly in undiluted tissue culture supernatant for a further 45 min at ambient temperature before washing a further three times in PBS. RAM-Au_{10} or -Au_{20} secondary antibody was then added to the suspension and a further incubation took place for 45 min in the dark. The material was then processed as described in the previous section.

Results

Surface-associated organelles in secondary zoospores and cysts

The primary and secondary zoospores and their corresponding cysts in the two *Saprolegnia* species under study were generally similar to those described previously (Holloway and Heath 1977, Beakes 1983). In the reniform secondary zoospores the flagella are inserted into a deep groove region on the ventral side of the spore (Figs. 1 and 2) whilst the dorsal surface is smooth and convex (Fig. 7). On the external surface of the zoospore is a "fuzzy coat" which is approximately 20–30 nm thick and covers the outer surface of the plasma membrane over the cell body and the flagella. This coat is extracted easily be acetone dehydration and so was only visualised after ethanol dehydration (see Fig. 21). Beneath the zoospore plasma membrane there are a series of flattened peripheral cisternae (PCV) which upon encystment rapidly break down into smaller vesicles (Fig. 4). The PCV underlie the whole of the dorsal surface and most of the ventral surface except in the region within the flagellar groove (Fig. 2) and adjacent to the water expulsion apparatus (not shown). Interspersed between the PCV are the other peripherally distributed vesicular components. Interestingly the PCV appears to be the only peripheral organelle which the primary zoospores lack (Holloway and Heath 1977, Beakes 1983). The main peripheral organelles associated with the encystment of secondary zoospores and the cyst surface up to the time of germling formation will now be described.

K-bodies

In primary zoospores the K_1-bodies are apically located and are smaller dumbbell-shaped structures (Holloway and Heath 1977, Beakes 1983). The K_2-bodies are

Fig. 1. SEM showing ventral face of a zoospore of *S. diclina* in which K_2-bodies have presumably discharged, releasing flocculant material (arrowhead) into the flagella groove. Note also the pits in the zoospore surface probably resulting from fixation induced discharge of the peripheral fibrillar vesicles. Bar: 3 µm

Fig. 2. Transverse section through the flagellar groove region of a secondary zoospore showing cross section of the smooth trailing flagellum (*f*) and profiles of two K_2 bodies (arrowed) showing their ventral location. × 30,000

Fig. 3. Discharging K_2-body (*K*) from a secondary zoospore of *S. parasitica*. The electron-dense vesicle lining material can be seen spreading from the vesicle margin (arrowheads). The contents of the K_2-body at this stage have already become transformed into a loose fibrillar and particulate reticulum. × 31,500

Fig. 4. Profile of the cell surface of an encysting secondary zoospore of *S. parasitica*. Note the conspicuous distension of the zoospore plasma membrane (*pm*) by the enlarged large peripheral fibrillar vesicle (star). The peripheral cisternum (*pc*) which underlies much of the zoospore plasma membrane has at this stage already begun to break up into smaller vesicles (arrowheads). × 38,400

Figs. 5 and 6. Profiles of peripheral fibrillar vesicles in zoospores of *S. parasitica* (Fig. 5) and *S. diclina* (Fig. 6) showing stages of the fusion (indicated by arrows) and apparent discharge of vesicle contents. × 18,700

Fig. 7. SEM view of part of the dorsal surface of a secondary zoospore of *S. diclina*. Fewer surface pits are present (small arrow) than on corresponding zoospores of *S. parasitica*. The plaque-like structure indicated by the thick arrow is possibly an underlying secondary encystment vesicle (SEV) immediately prior to discharge. Bar: 0.5 µm

Fig. 8. High power SEM view of the surface of a secondary zoospore of *S. parasitica* showing a pit formed by the rupture of a peripheral FV. The contents of this vesicle appear as a granular deposit around the rim of the pit. Bar: 50 nm

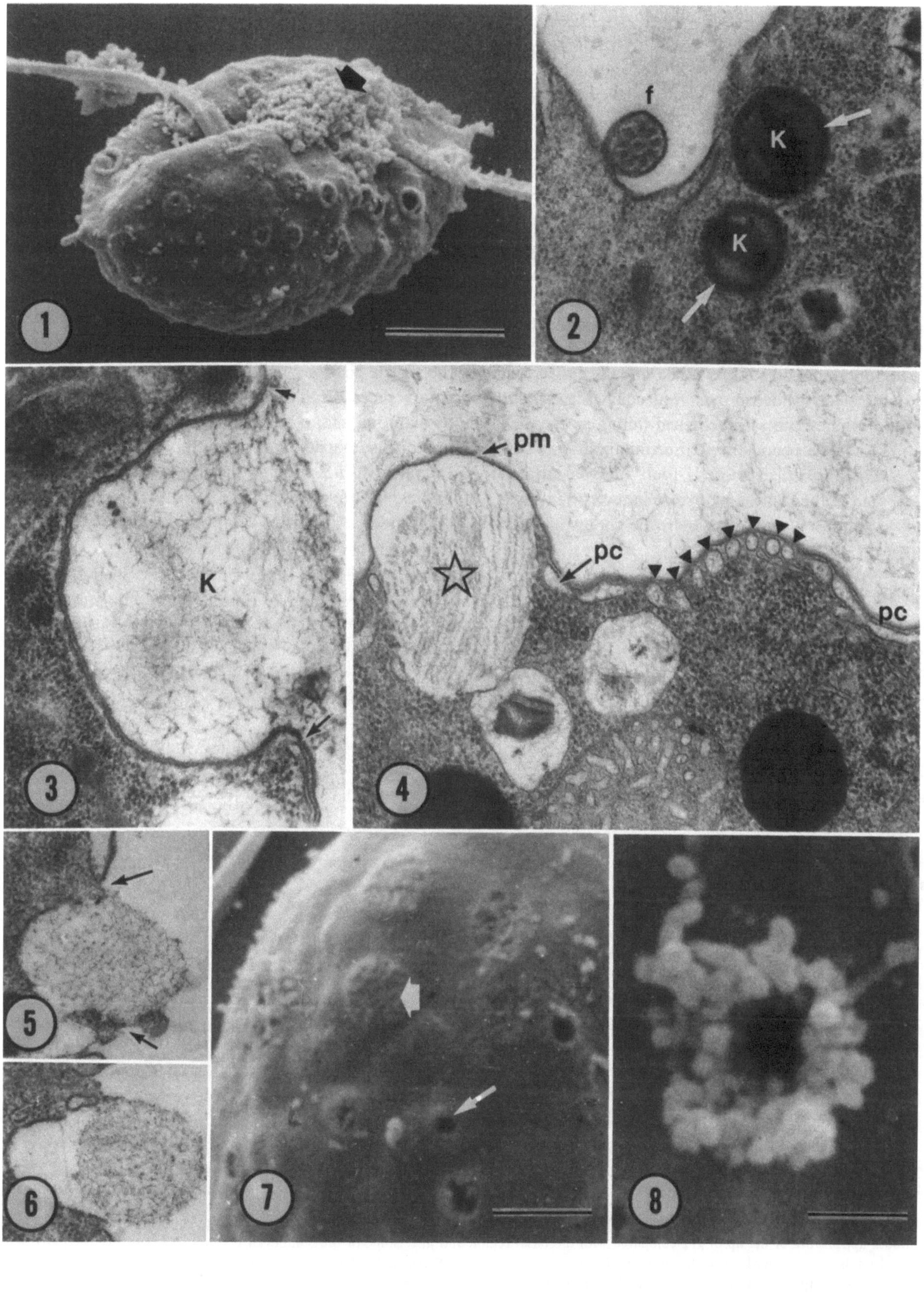
1
f
K
K
2
K
3
pm
pc
pc
4
5
6
7
8

formed in the primary cysts, following the pattern described in detail by Lehnen and Powell (1991). Their mature structure in *S. diclina* and *S. parasitica* does not differ significantly from that described in *S. ferax* by Lehnen and Powell (1988, 1989, 1991). K_2-bodies are round or elliptical in shape and are approximately 1 µm in diameter and are located just beneath the plasma membrane in the region associated with the nuclear beak and basal kinetosomes (Fig. 2). In the pectin-induced encystment of *S. diclina* the K_2-bodies appear to be the first organelle to be discharged. Fusion occurs with the plasma membrane and their contents rapidly disperse as described by Lehnen and Powell (1989) to form a diffuse fibrillar pad (Figs. 3 and 25). The peripheral lining to this vesicle spreads out to form the outer electron dense cyst wall coat (Fig. 3), suggesting it shares the same cortical lining as the SEV. In the SEM of a pectin-treated secondary zoospore of *S. diclina* (Fig. 1), a pad of K_2-body derived material can be seen in the ventral groove. Whatever the trigger, some K_2-bodies always remain in the spores after encystment (not illustrated).

Encystment vesicles

Primary encystment vesicles are narrower capsule shaped bodies, which have a similar cortical structure to their secondary counterparts but enclose a tuft of short straight tubular spines which are up to 0.5 µm in length in *S. diclina* and up to 1.5 µm long in *S. parasitica* (Beakes 1983).

In both species encystment of secondary zoospores is associated with the discharge of SEVs (Figs. 10–12) which gives rise to a thin electron-dense outer wall layer (Figs. 14 b and 22) in which the cyst coat spines are embedded. In addition a matrix derived from these vesicles appears to coat the entire cyst surface including the boathook spines (Fig. 14 b). In this species the SEVs range in size from 0.5–1.0 µm in width (Fig. 9) and up to 5.0 µm in length if sectioned longitudinally (Beakes 1983). These large structures first appear within a fenestrate membrane system which is apparently largely derived from ER, although contributions from the Golgi system may also be involved (Beakes 1983). Fully differentiated SEVs contain a structured peripheral cortex surrounding the cavity containing the bundle of regularly spaced spines, which are often coiled within the vesicle but may be interdigitated in the more elongate vesicles (Beakes 1983). The spines appear square to diamond shape in cross-section and are embedded in a diffuse fibrillar matrix material (Figs. 4 and 14 b). In contrast in *S. diclina*, the SEVs are capsule shaped bodies, 0.3 µm in width by 0.8–1.0 µm in length (not illustrated). A detailed comparative account of these structures is given by Beakes (1983).

Upon encystment areas of the SEV membrane and plasma membrane that are in contact, begin to fuse to form small vesiculate bodies (Fig. 10), and small disc-like spheres appear within the vesicle matrix (Fig. 10). The circular patch of rough membrane shown in the high power SEM view of the spore surface (Fig. 7) may represent the surface view of an early stage in this process. As these SEVs open out (compare Figs. 10 and 25 with Fig. 12), the electron dense layer which forms the outermost cortex of the SEV, appears to spread

---

Figs. 9–14. Organelles associated with zoospores and cysts of *Saprolegnia*

Fig. 9. Transverse profile of an SEV of *S. parasitica* illustrating the discrete peripheral cortex lining the vesicle, profiles of the bundle of boathook spines and the enclosed fibrillar central component (star). × 50,700

Fig. 10. Profile of a discharging SEV of *S. parasitica* illustrating the way in which the outer cortex appears to have broken into a series of small vesicular structures (arrowheads) as has the plasma membrane (the particles attached to this are Con A-Au_{20}). Disc-like inclusion bodies within this vesicle are arrowed. The outer membrane of the SEV has fused with the plasma membrane (thick arrow). Note how this fusing vesicle is flanked on either side by the peripheral cisternae, which form a collar around the neck of the vesicle (thick arrows). × 65,200

Fig. 11. SEM of part of the surface of an encysting secondary zoospore of *S. parasitica* showing the release of hoops of boathook spine bundles (arrows). Bar: 2 µm

Fig. 12. More or less fully opened SEV (edges arrowed) at the surface of an encysting zoospore of *S. parasitica*. Note the fibrillar vesicle (*fv*) is retained within the spore. × 31,100

Fig. 13. Encysted secondary zoospore of *S. parasitica* showing profile of a small germ tube (*g*). Note the orientation of lipid globules (*L*) on the side of the spore towards the germ tube. *N* Nucleus. × 8,200

Fig. 14. **a** SEM of an advanced germling of *S parasitica* showing germ tubes at differing stages of development (*gt 1* and *gt 2*). Note the smooth appearance of the tip of gt 2 compared with the basal region and gt 1. Some of the bundles of boathook spines have been arrowed. Bar: 4 µm. **b** TS of a bundle of boathook spines from a secondary cyst of *S. parasitica*. Note the regular spacing of the spines and the possible interconnecting fibrils between them (arrowed) and the matrix in which they are embedded. × 49,200

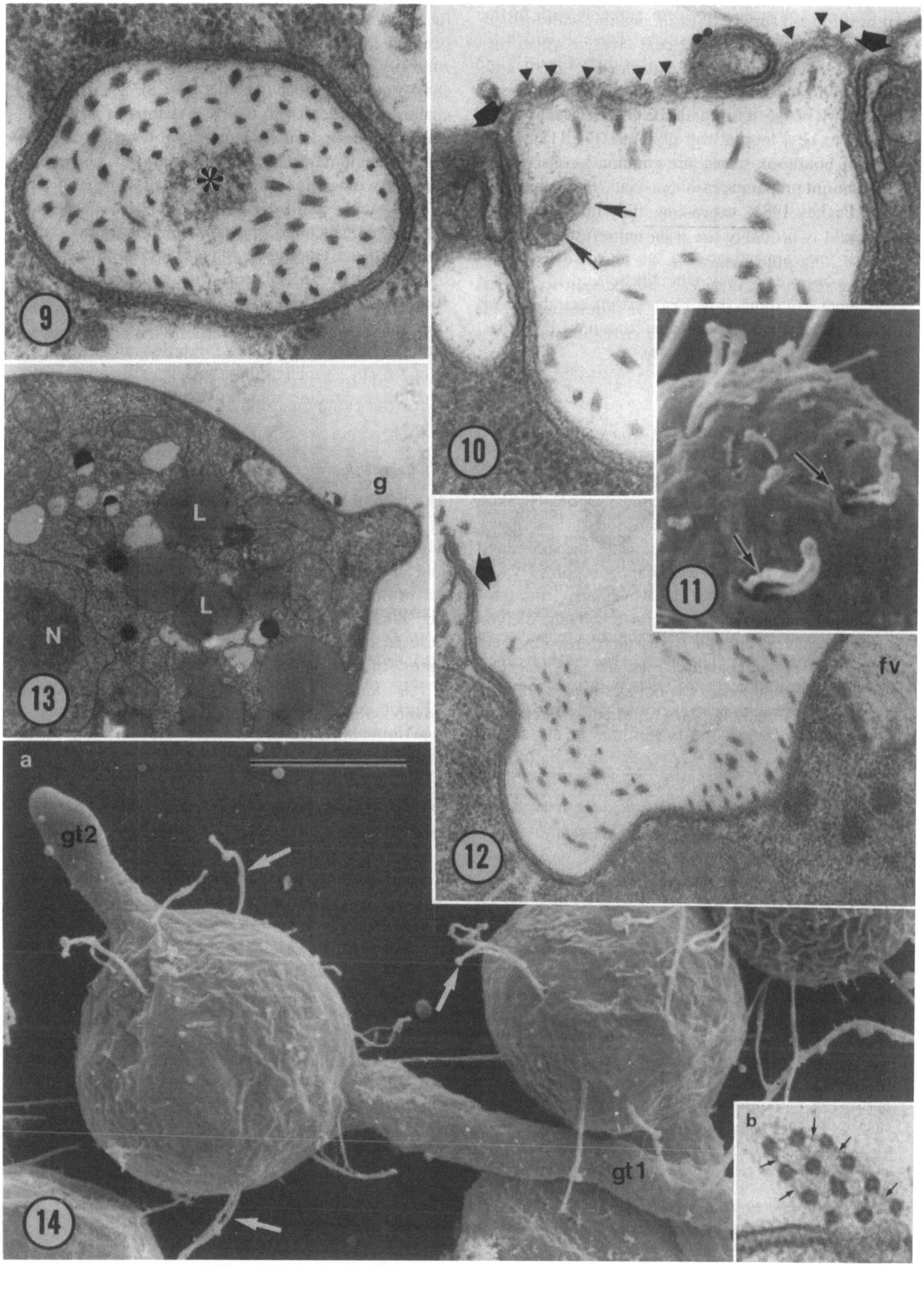
9
10
11
12
g
L
L
N
13
fv
a
gt2
gt1
b
14

out over the cell surface to form, possibly with further modifications, the outer cyst coat layer in which the boathook spine bundles are embedded (Figs. 12 and 40). Discharging SEV of *S. parasitica* can also be seen in the SEM, demonstrating the localised release of what appears to be a looped rod of spines (Fig. 11). These loops of boathook spines are commonly observed in whole mount preparations of cyst coats (Pickering et al. 1979, Beakes 1983) suggesting that this looped arrangement is probably the usual pattern of discharge of these long appendages. In the SEM, mature germinating secondary cysts can still be seen to possess rigid bundles of spines (Fig. 14 a). In thin section profile the individual spines appear to be cross-linked by cross bridges after discharge (Fig. 14 b).

Fibrillar vesicles

Primary zoospores also contain fibrillar type vesicles (FV) which appear smaller than their secondary zoospore equivalents and they do not appear as prone to osmotic distension as those in secondary spores (Beakes 1983). In the SEM the primary zoospore surface shows less (and smaller) pitting than shown by the secondary zoospores (Burr 1991). In secondary zoospores FV are spherical or ellipsoidal vesicles which vary from 0.5–1.0 µm in diameter and are generated by Golgi dictyosomes in primary cysts. They are characterised by their distinctive fibrillar contents (Figs. 4–6 and 12). The membrane of these vesicles is not as robust as that of the SEV and usually appears ruptured as a result of fixation (osmotic) damage. This results in these vesicles swelling and distending (Fig. 4) or even rupturing the zoospore plasma membrane (Fig. 5) and apparent discharge of the fibrillar contents (Fig. 6). This FV discharge was observed in G + F fixations as well as F fixation and was usually more pronounced in *S. parasitica* than in *S. diclina* (Figs. 1 and 6). The pits in the secondary zoospore surface visible in SEM (Figs. 1 and 7) are thought to arise as a result of the discharge of the FV. No evidence of spines can be seen arising from these pits and the globular material around the rims of these pits is never seen with discharging SEVs (Fig. 8). Peripheral FVs were detected in the cytoplasm during (Fig. 12) and after secondary encystment and retained until well after the formation of germ tubes (Fig. 13). In germlings these vesicles apparently fuse with the expanding somatic vacuole system (Beakes 1983).

Binding of fluorescence probes to zoospores and cysts

The pattern of binding of fluorescent probes (lectins or antibodies) to fungal spores need to be interpreted with care (Hardham 1985). The quality of membrane and organelle preservation during fixation protocols can profoundly influence the observed patterns of binding. Zoospores in particular are prone to osmotically induced rupture of their outer membrane to expose many of the underlying cytoplasmic components (Figs. 6 and 7). Fixation may also induce premature, possibly abnormal, encystment of zoospores.

Figs. 15–22. Patterns of binding of the lectin concanavalin A to zoospores and cysts of *Saprolegnia*

Fig. 15. Brightfield (**a**) and fluorescence micrographs (**b**) of primary zoospores of *S. diclina* showing uniform and intense Con A-FITC binding on the main body of the spore. Bar: 5 µm

Fig. 16. Fluorescence micrograph of a primary zoospore of *S. parasitica* showing distinct bright patches of staining at the cell surface (arrows). Magnification as in Fig. 15 a

Fig. 17 a–c. Micrographs of Con A-FITC binding to a mixture of secondary zoospores, cysts and germlings of *S. diclina* (**a**) and *S. parasitica* (**b** and **c**). The kidney shaped secondary zoospores all stain intensely (**a** and **c**). Some of the rounded cysts label intensely (arrowed in **b**) and are presumed to be newly encysted. However, the absence of binding to some cysts and germlings is also apparent (**b** and **c**). Bar: 20 µm

Fig. 18. Short exposure fluoresence micrograph demonstrating the binding of Con A-FITC to secondary zoospores of *S. diclina*. Binding of the lectin is patchy with a large patch associated with the flagellar groove (small arrows). The intense binding to one of the spores (large arrow) is probably an encysting spore prior to rounding up. Bar: 5 µm

Figs. 19 and **20.** Correlated light (**a**) and fluorescent (**b**) micrographs illustrating patterns of Con A-FITC binding to a primary zoospore (*PZ*), primary cyst (*PC*), encysting zoospore (*EZ*) and secondary zoospore (*SC*) of *S. parasitica*. No appreciable binding is observed to the primary cyst, whereas both the primary zoospore and secondary cysts and their bundles of boathook spines (white arrows) bind this lectin strongly. Bars: 10 µm

Fig. 21 a, b. Binding of Con A-Au_{20} to zoospore surfaces of *S. parasitica*. The sections are relatively thick and the gold labelling can be seen to be associated with an amorphous material coating the zoospore membrane. × 49,600; × 75,500, respectively

Fig. 22. Adjacent primary (left) and secondary (right) cyst surfaces of *S. parasitica* showing binding of Con A-Au_{20}. Note the uniform binding to the fibrillar coat (*fc*) of the primary cyst compared with the binding localized to spine bundles (*hb*) in the secondary cyst. × 23,400

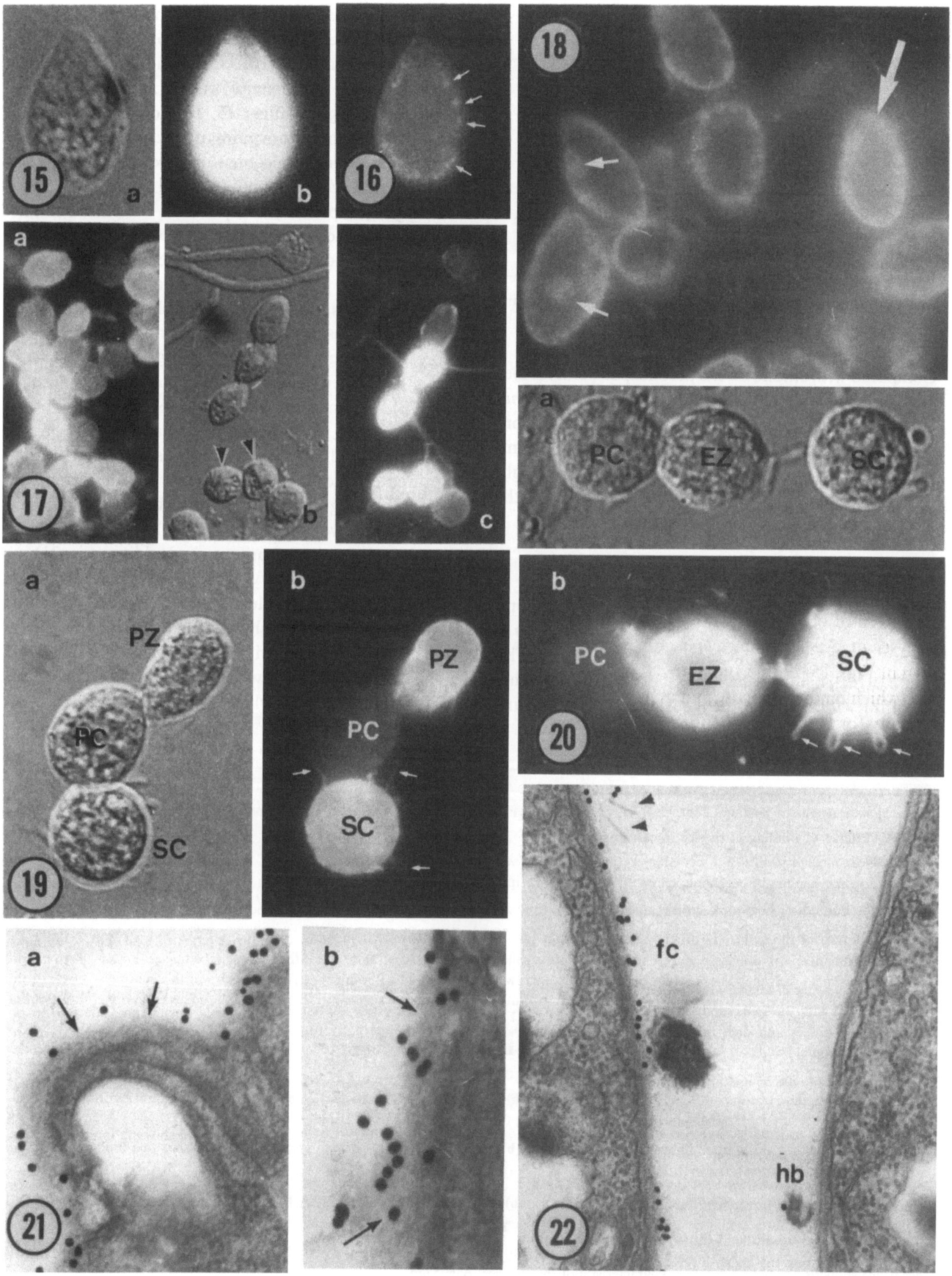
15
a
b
16
18
17
a
b
c
a
PC
EZ
SC
19
a
PZ
PC
SC
b
PZ
PC
SC
20
b
PC
EZ
SC
21
a
b
22
fc
hb

Table 1. Effect of lectins on per cent encystment of secondary zoospore suspensions

Lectins 50 µl/ml	*S. parasitica* control inhibitory saccharide	*S. parasitica* lectin	*S. diclina* control inhibitory saccharide	*S. diclina* lectin
Con A	25 ± 3	89 ± 4	23 ± 8	95 ± 2
WGA	10 ± 1	20 ± 5	21 ± 4	70 ± 5
SBA	12 ± 2	15 ± 3	25 ± 3	16 ± 3

Each reading is the mean of three replicate experiments (± standard error of mean)

Compared with studies on the peronosporales a further problem which arises when working with *Saprolegnia* is that suspensions contain a mixture of the four spore stages and it is often difficult or impossible to discriminate between primary and secondary cysts in the light microscope, particularly in *S. diclina* which lack the long bundles of secondary cyst spines.

Concanavalin A binding to zoospores and cysts

The pattern of binding shown by Con A to zoospores and cysts of *Saprolegnia diclina* and *S. parasitica* is shown in Figs. 15–27 and summarized in Table 3. This lectin (which binds to glucose and mannose sugars) has an affinity for many different surface components and consequently the analysis of the staining pattern is complex. It is a potent inducer of zoospore encystment in both species (Table 1). In both species Con A binds uniformly to the perimeter of primary zoospores except in the apical region (Figs. 15, 16, and 19). In surface focus of the primary zoospores of *S. parasitica* a patchy binding pattern of this lectin is observed which was not apparent in *S. diclina* (compare Fig. 16 with Fig. 15 b). The flagella of primary zoospores do not bind Con A. The entire surface of the secondary zoospore, including the flagella, binds Con A and the labelling intensity is greater than in any other cell type and consequently these zoospores are easily distinguished from the less intensely fluorescing cysts (Fig. 17) and primary zoospores. In *S. parasitica* short photographic exposures reveal the rather patchy nature of the Con A staining (Fig. 18). This may be due to fixation induced release or exposure of reactive internal organelles. However, little or no gold labelling was ever associated with the ruptured FV (not illustrated). Dense and uniform Con A-Au_{20} binding to the zoospore surface was observed when the cells were dehydrated in ethanol but not when dehydrated in acetone prior to embedding (Fig. 21). Although many primary cysts show intense Con A-binding capacity (as illustrated on secondary cysts; Figs. 19 and 20) others show virtually no binding at all (Figs. 19 b and 20 b). In thin-sectioned material Con A-binding is associated with the outermost layer of the cyst wall and particularly with the short primary cyst

Figs. 23–30. Patterns of binding of the lectin concanavalin A (Figs. 23–27) and wheat germ agglutinin (Figs. 28–30) to zoospores and cysts of *Saprolegnia*

Fig. 23. Near median profile of a secondary cyst of *S. parasitica* stained with Con A-Au_{20}. Note that the gold particles are almost entirely restricted to the bundles of boathook spines (arrowheads). L lipid, N nucleus; × 13,100

Fig. 24. TEM of part of the surface of an encysting secondary zoospore of *S. parasitica* showing a localized patch of Con A-Au_{20} binding material. × 26,600

Fig. 25. SEV undergoing discharge at secondary zoospore surface in *S. parasitica*. Note the rather sparse labelling with Con A-Au_{20} gold particles associated with the spines and vesiculated cortex (arrow) of the SEV. The fibrillar material at the top of this picture is derived from the discharged K_2-bodies and does not bind Con A. A ruptured FV (star) has apparently fused with an adjacent peripheral cisternum (arrowhead) × 21,000

Fig. 26. Discharging bundle of boathook spines demonstrating the association of Con A-Au_{20} particles with the spines and their associated matrix material. × 63,300

Fig. 27. Corresponding transmitted light (**a**) and fluorescence (**b**) micrographs of a secondary *S. parasitica* germling showing intense Con A-FITC binding to the cyst coat and spines, and lack of binding to germ tube wall. Bar: 10 µm

Fig. 28. Corresponding transmitted light (**a**) and fluorescence (**b**) micrographs of a *S. diclina* germling showing localized binding of WGA-FITC to small apical patch (arrowed). Bar: 5 µm

Fig. 29. Corresponding transmitted light (**a**) and fluorescence (**b**) micrographs illustrating WGA-FITC binding to secondary zoospores of *S. diclina* and demonstrating the highly localized binding to the region of the ventral groove. Bar: 5 µm

Fig. 30. Corresponding transmitted light (**a**) and fluorescence (**b**) micrographs illustrating WGA-FITC binding to secondary zoospores of *S. parasitica* showing the patchy labelling characteristic of this species. Bar: 5 µm

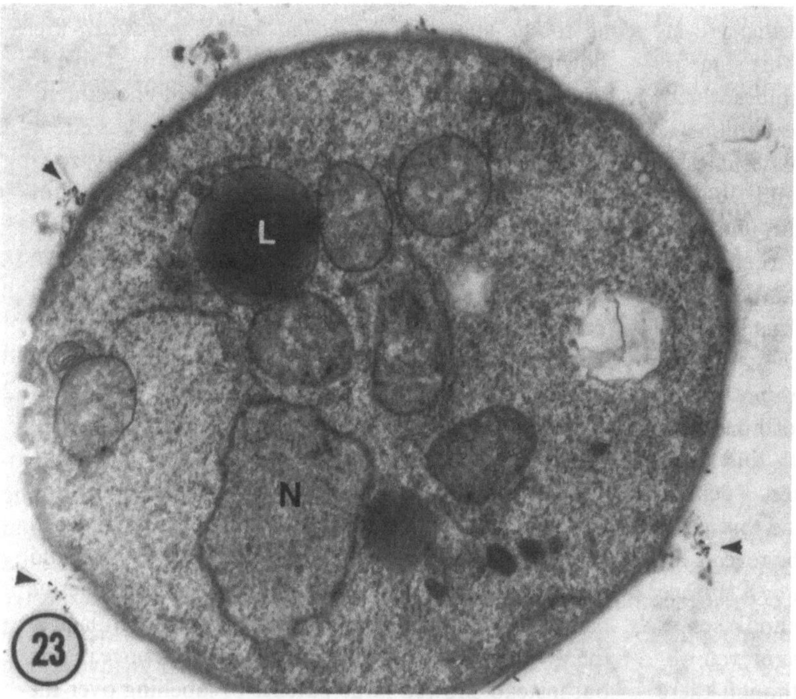
L
N
23

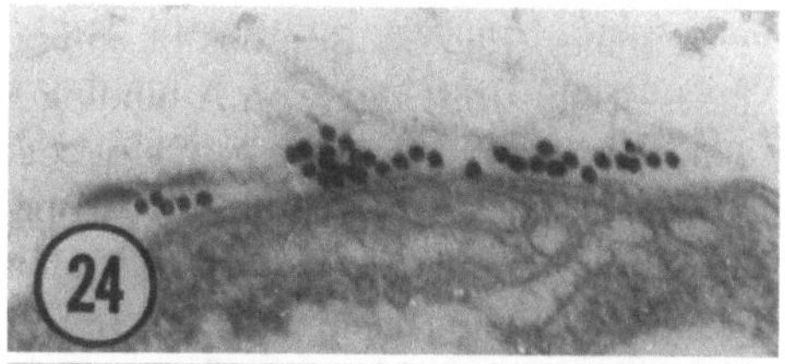
24

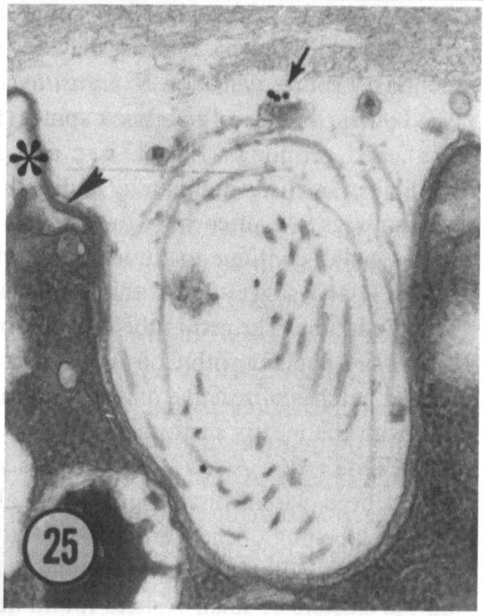
*
25

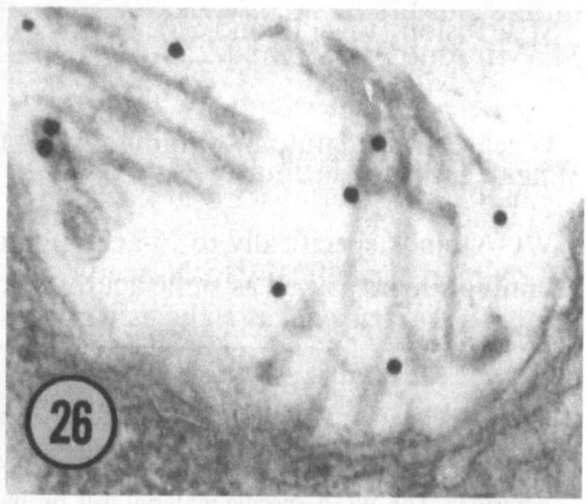
26

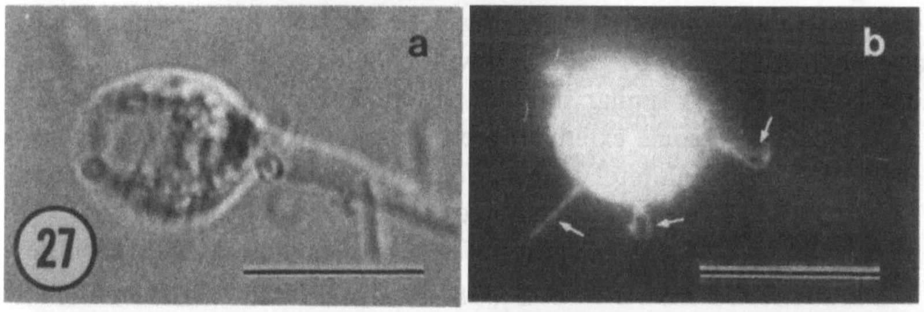
a
b
27

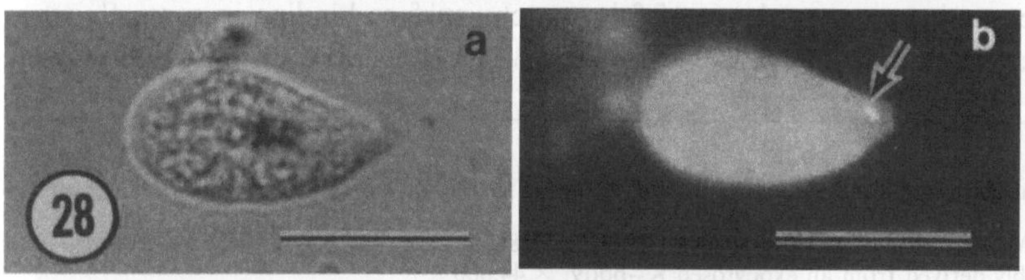
a
b
28

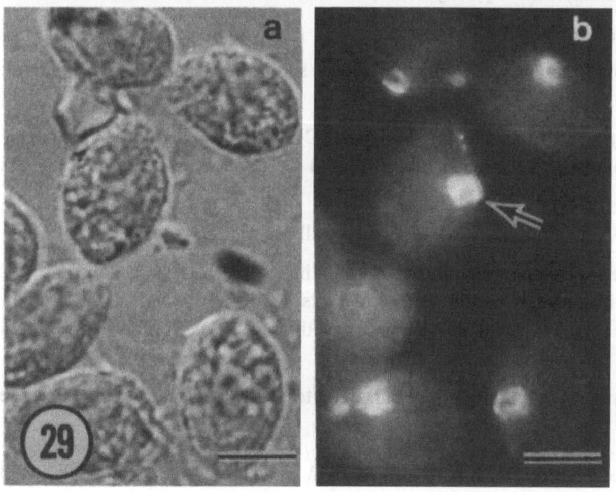
a
b
29

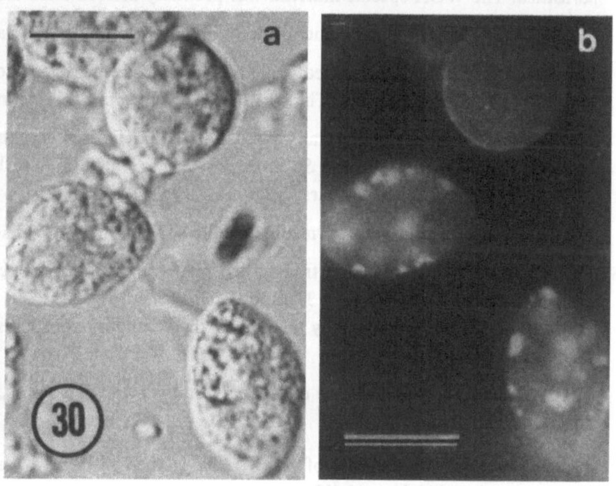
a
b
30

coat spines (Fig. 22). Sections of empty primary cyst cases demonstrate that Con A-binding at this stage is only associated with the tufts of cyst coat spines which are below the limits of light microscope resolution. Encysting secondary zoospores of both isolates release a Con A-binding material (Figs. 18–20) and fluorescence intensely, although often somewhat unevenly (Figs. 20 and 21). This binding of Con A to encysted secondary spores, which in *S. parasitica* can be recognised by the bundles of boathook spines (Figs. 19 b and 20 b), and germlings (Figs. 17 b, c and 27 b) is more variable in intensity and patchy in distribution. Indeed in germlings it is often only the bundles of boathook spines which continue to fluorescence with this lectin (Table 3) and suggests that this coat material deminishes with time. The germ tube does not bind this lectin (Fig. 27). Labelling with Con A-Au_{20} probe reveals localised patches of reactive material on the cyst surface and with the matrix surrounding the boathook spines (Figs. 22, 23, 26). A few gold particles were even observed deep within discharging SEV, although there was no labelling of the fibrillar material derived from the K_2-body (Fig. 25). Gold labelling is also observed on the clusters of small vesicles found at the points of SEV fusion (Figs. 10 and 25).

Wheat germ agglutinin-binding (WGA) to zoospores and cysts

WGA binds specifically to N-acetyl glucosamine containing sugars as well as sialic acid. Its pattern of binding to the spores of *S. diclina* and *S. parasitica* are shown in Figs. 28–36 and summarised in Table 3. Whereas this lectin induced high levels of encystment in zoospores of *S. diclina* it had significantly less effect on those of *S. parasitica* (Table 1). In primary zoospores of both species WGA binds to a small (< 0.5 µm diam.) localised apical region (Fig. 28), which lies close to the expected location of the K_1-bodies (Beakes 1983). In *S. diclina* WGA binding occurs in a distinct region (ca. 3 µm diam.) over the mid-ventral groove region of the zoospores (Fig. 29) as described by Lehnen and Powell (1989, 1993) in *S. ferax*. WGA-Au probes did not unequivocally link this lectin with discharging K-bodies since none of the latter were seen in spore profiles of this stained material. However, gold binding was associated with some rather dense fibrillar material in the flagellar groove of a spore (Fig. 31) and could represent the remnants of a K-body derived adhesive pad. The secondary zoospores of *S. parasitica* lacked the WGA-specific binding in the flagellar groove region but instead showed large patches of labelling over the dorsal surfaces (Fig. 30). This pattern of labelling could not be visualized with colloidal gold at the EM level. A similar staining pattern was also found to occur to some extent in pre-incubation controls with specific-inhibitory saccharide, although at much lower intensities (not illustrated).
Encysted spores of both species showed relatively little WGA-binding (Figs. 33–35) although there was evidence of faint patchy surface binding in some *S. parasitica* cysts (Fig. 30). Lehnen and Powell (1993) have

Figs. 31–40. Patterns of binding of the lectin wheat germ agglutinin (Figs. 31–36) and MAbs (Figs. 37–40) to zoospores and cysts

Fig. 31. Electron micrograph illustrating the WGA-Au_{20} with coarse fibrillar material in the flagellar groove of a secondary zoospores of *S. parasitica*. The WGA-specific material was probably derived from an exocytosed K_2-body. × 44,000

Fig. 32. WGA-Au_{10} binding to the inside of an empty primary cyst case. × 26,200

Fig. 33. Corresponding transmitted light and fluorescence micrograph of a secondary germling of *S. diclina* showing WGA binding to emerging germ tube (large white arrow). The cyst wall does not bind this lectin. × 20,000

Figs. 34 and 35. Corresponding transmitted light (**a**) and fluorescence (**b**) micrographs showing WGA-FITC binding to germ tubes of secondary cysts of *S. diclina* (Fig. 34) and *S. parasitica* (Fig. 36). The intense binding to the germ tube wall and its absence from the tube apex is illustrated (arrowheads in **b**). Bars: 5 µm

Fig. 36. TEM of a transverse germ tube profile showing heavy WGA-Au_{10} binding to the germ tube wall surface. × 45,300

Figs. 37 and 38. Corresponding transmitted light (**a**) and fluorescence (**b**) micrographs illustrating binding of MAb HB 6 to zoospore and cyst mixtures of *S. diclina* (Fig. 37) and *S. parasitica* (Fig. 38). This MAb binds to the surface of cysts (arrowheads) and to the primary zoospores (*PZ*) but only patchily to some of the secondary zoospores (Fig. 37). Note the absence of binding to the germ tube walls (arrows in Fig. 38). Bars: 5 µm

Fig. 39. The primary cyst surface of *S. diclina* stained with HB 5-Au_{10}, showing binding of gold particles to the fibrillar material on the spore surface. × 31,800

Fig. 40. Secondary cyst surface of *S. parasitica* stained with HB 5-Au_{20} demonstrating restricted binding to the secondary spine bundle and its associated material. × 46,500

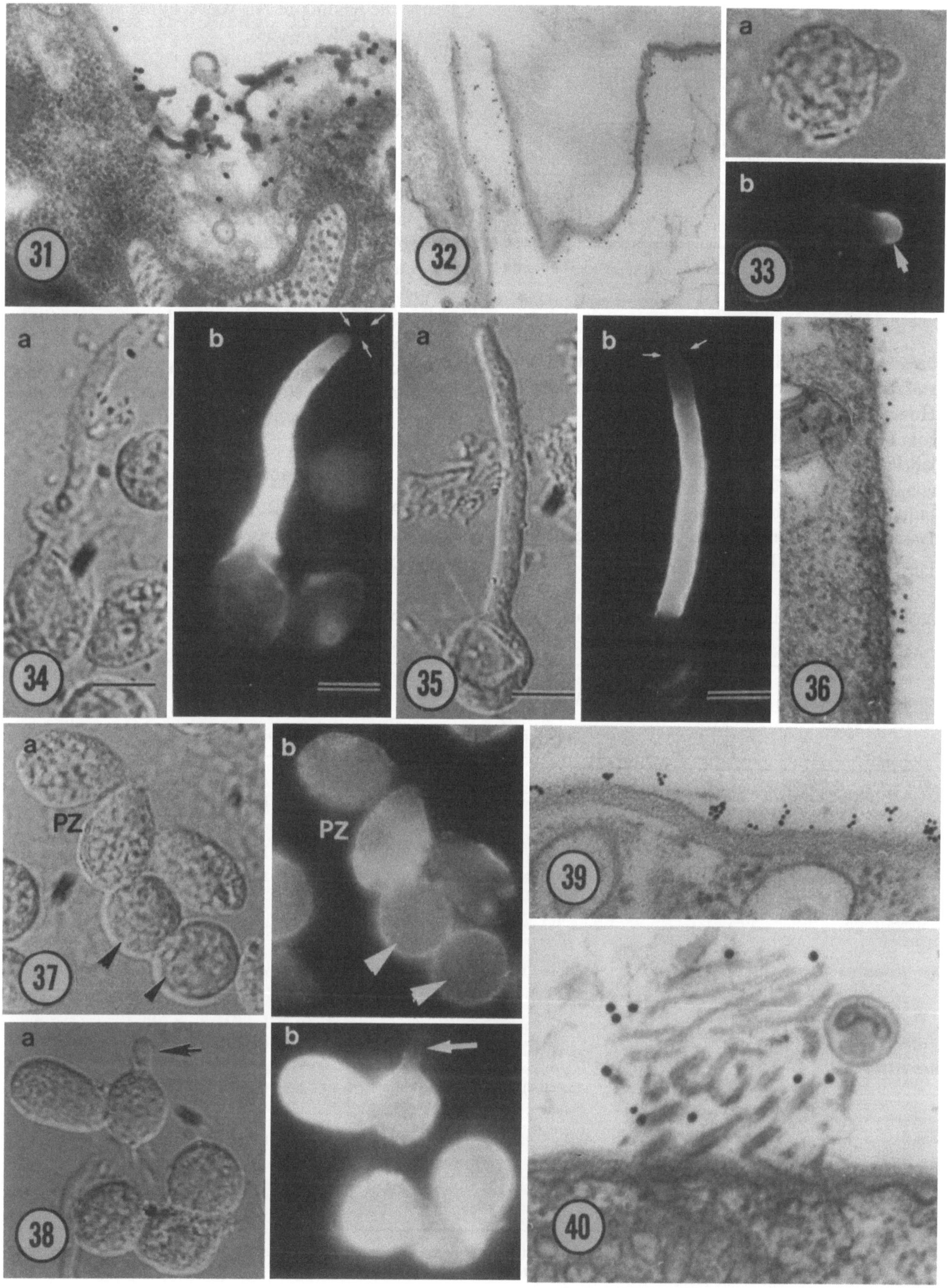
31
32
a
b
33
a
b
34
a
b
35
36
a
PZ
37
b
PZ
a
38
b
39
40

Table 2. Effect of monoclonal antibodies on per cent encystment of secondary zoospore suspensions

MAbs	*S. parasitica*		*S. diclina*	
	hybridoma medium	antibody undiluted	hybridoma medium	antibody undiluted
HB 5	55 ± 3	80 ± 5	60 ± 3	88 ± 6
HB 6	55 ± 3	60 ± 4	58 ± 3	60 ± 4

Each reading is the mean of three experiments (± standard error of mean)

demonstrated localised binding of this lectin to the surfaces of secondary cysts and less so to primary cysts. However, the inner surface of the empty primary cyst coat bound Au_{10}-conjugated WGA (Fig. 32), although labelling of empty cyst was not detected using FITC-conjugated WGA. The first 50–100 µm of germ tube initials, except at their extreme apices, fluoresced intensely, reflecting strong WGA binding along their surfaces (Figs. 33–35). Tangential sections show especially clearly the binding of Au-WGA to the surface of the germ tube walls (Fig. 36).

Other lectin binding to zoospores and cysts

The binding of SBA, DBA, PNA, and RCA to *Saprolegnia* spores (see Lehnen and Powell 1993) was also investigated. DBA bound very weakly to primary zoospores and possibly newly encysted primary cysts of both isolates. No binding was detected to other cysts or to any germling cell surface components. UEA I, specific for L-fucosyl residues, was only detected as binding to a ruptured zoospore and not to any other cell types. No specific binding was detected with any of the following lectins, SBA, PNA, and RCA. SBA did not induce encystment in zoospores of either *Saprolegnia* species (Table 1).

Binding of MAbs to zoospores and cysts

Five MAbs (HC 2, HB 2, HB 4, HB 5, and HB 6) raised against a mixed zoospore suspension of *S. parasitica* generally showed a similar overall pattern of immu-

Table 3. Summary of relative FITC-probe staining patterns conjugated with lectins and MAbs showing differences with spore types in *S. parasitica* and *S. diclina*

Spore type	Con A		WGA		MAbs (HB 5, HB 6)	
	S. parasitica	*S. diclina*	*S. parasitica*	*S. diclina*	*S. parasitica*	*S. diclina*
Prim. zoospore surface	+ + (+/− patchy)	+ +	+ (apical spot)	+ (apical spot)	+ + +/+ (+/− patchy)	+ + +/+ (some, with HB 5)
Flagella	−	−			−	
Prim. cyst coat	+ +/−	+ +/−	−/+ (localized patches)	−/+ (localized patches)	+ + +/+ (+/− patch)	+ + +/+
Spines					+ +	−
Sec. zoospore surface	+ + +	+ + +	+ (dorsal surface)	+ + + (flagellar groove)	+/− (patchy dorsal surface mainly with HB 5)	−
Flagella	+ +	+ +				
Sec. cyst coat	+ + +/+	+ + +/+	−/+ (localized faint patches)	−/+ (localized faint patches)	+ + +/+	+ + +/+
Spines	+ + +	−	−	−	+ +	−
Sec. germling coat	−/+	−/+	−	−	+ +/+ (often patchy)	+ +/+ (often patchy)
Spines	+	−	−	−	+	−
Germ tube wall	−	−	+ + +	+ + +	−	−

+ + +, + +, + Decreasing degrees of fluorescence intensity; − no detectable staining; +/− variable, with more spores stained than unstained; −/+ variable, with more spores unstained than stained

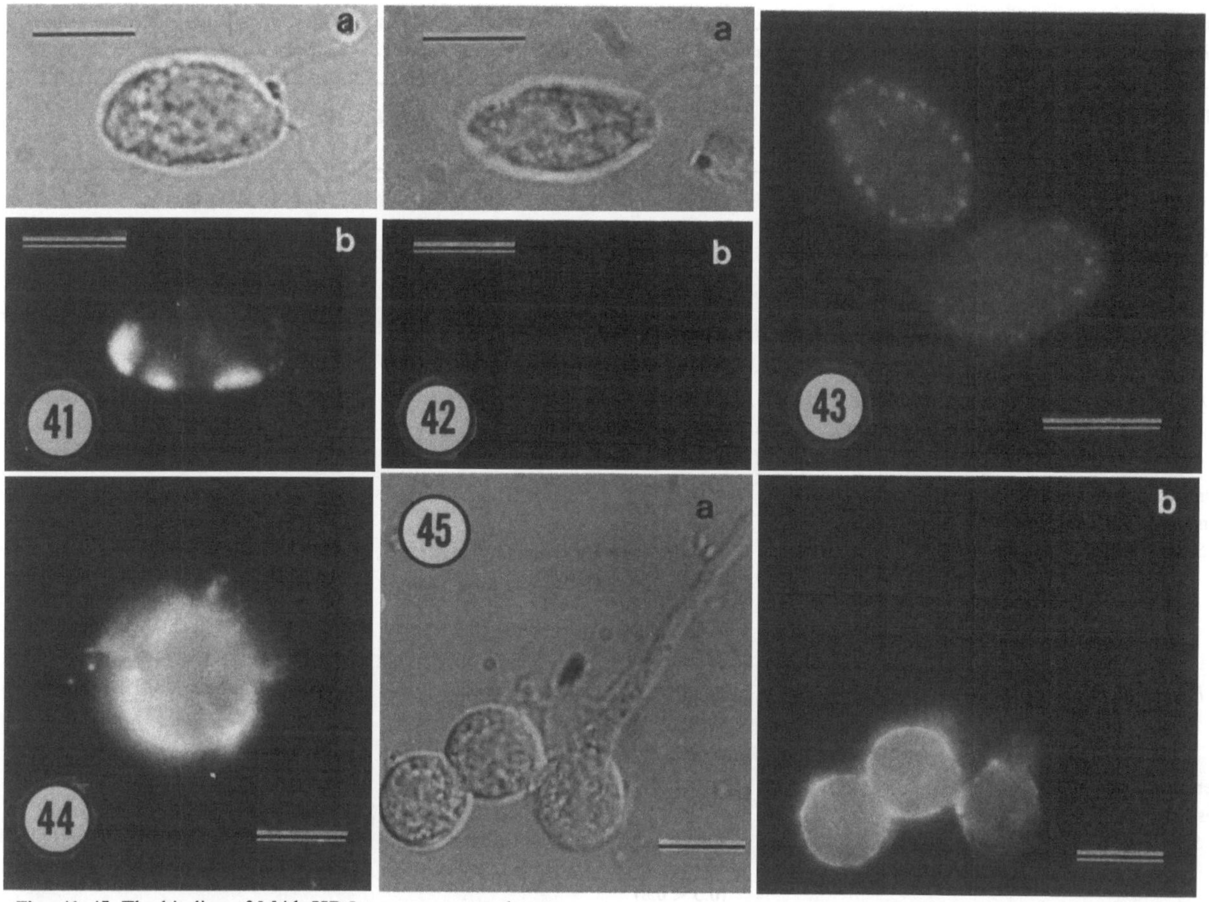

Figs. 41–45. The binding of MAb HB 5 to zoospores and cysts

Figs. 41 and **42.** Corresponding transmitted light (**a**) and fluorescence (**b**) photographs showing binding of the HB 5-FITC to primary (Fig. 41) and secondary (Fig. 42) zoospores of *S. diclina*. Bars: 5 µm

Fig. 43. Fluorescence micrograph showing binding of the HB 5-FITC to secondary zoospores of *S. parasitica*. Note small patches of binding to the periphery of these spores. Bar: 5 µm

Fig. 44. Fluorescence micrograph showing binding of the HB 5-FITC to secondary cyst of *S. parasitica*. Note intense binding and clear visualisation of the cyst coat spines (arrowheads). Bar: 5 µm

Fig. 45. Corresponding transmitted (**a**) light and fluorescence (**b**) photographs illustrating binding of HB 5-FITC to cysts and germling of *S. parasitica*. Note the absence of any binding to the germ tube and rather patchy binding to the germling cyst wall (arrow). Bar: 10 µm

nofluorescence binding to zoospores and cysts of both species (Figs. 37, 38, 41, 42, 44, and 45; Table 3). All induced encystment of zoospores, although HB 5 was marginally the most effective in this respect (Table 1). This may be due to a higher production of antibody from this particular hybridoma line. However, none of the MAbs was as effective at inducing zoospore encystment as Con A (Table 2). The MAbs bound to the cell body but not to the flagella of primary zoospores (Figs. 37 and 38), although in some *S. diclina* zoospores the labelling was confined to localized patches (Fig. 41). The MAbs labelled the entire primary cyst surfaces (Figs. 37 and 38) and at the ultrastructural level, the binding of Au-conjugated MAbs was shown to be localized to the outer cyst matrix and associated spines (Fig. 39). As with Con A labelling the MAb binding to cyst coats was rather variable and seemed to deminish in germlings (Fig. 45). Generally the MAbs did not bind to the secondary zoospores of either *S. diclina* or *S. parasitica* (Fig. 42) although a patchy binding to the secondary zoospores was observed (Fig. 43). Secondary cysts bind the MAbs over their entire surface (Figs. 38, 44, and 45) and in *S. parasitica* the bundles of boathook spines are also strongly reactive (Fig. 44). Germ tubes, however, showed no antibody binding (Figs. 38 and 45).

Table 4. Summary comparing the properties of the encystment apparatus in secondary type zoospores and cysts between Peronosporales (exemplified by *Phytophthora* spp., particularly *P. cinnamomi*) and Saprolegniales (exemplified by *S. parasitica/diclina/ferax*)

Believed function/role	Names	Usual abbrev.	Typical morphology and size	Lectin binding	Specific MAb	Biochemistry
General cyst coat generating vesicles						
Saprolegnia	encystment vesicles/ bar bodies/boathook vesicles	EV	capsule-shaped to rectangular or ovoid, tubular/hooked spines, peripheral cortex (0.5 × 1.0 < 1.5 × 5.0)	Con A (matrix component only?)	HC 2, HB 2, HB 4, HB 5, HB 6	unknown
Phytophthora	U-bodies microbodies small dorsal vesicles	Dsv	spherical/ovoid with dense granular contents (0.2 × 0.2 < 0.3 × 0.3)	SBA	Cpa-2	300 kDa glycoproteins rich in N-acetyl galactosamine
Ventral adhesive pad generating vesicles						
Saprolegnia	kinetosome associated-bodies	K_2-bodies	ovoid/conical; peripheral matrix with crystalline inclusions, plus cavity with tubules (0.75 × 1.0)	WGA	none available	glycoprotein containing N-acetyl glucosamine and rich in sulphyl-containing polysaccharides
Phytophthora	small ventral vesicles	Vsv	spherical with dense granular contents (0.15 × 0.2 < 0.03 × 0.3)	none	Vsv-I	> 200 kDa protein
Peripheral storage (?) vesicles						
Saprolegnia	large peripheral/ fibrillar vesicles	FV	spherical/ovoid fibrillar contents (0.5 < 0.8)	none?	none available	glycoprotein (partially digested by proteases)
Phytophthora	large peripheral vesicles	LPV	spherical to ovoid (0.3 < 0.6)	Con A	Cpw-1	high molecular weight glycoprotein

Discussion

Cell surface carbohydrates of fungal spores have been shown to play crucial roles both in their detection of and adhesion to plant host cells (Hinch and Clarke 1980, Longman and Callow 1987, Hamer et al. 1988). In addition surface saccharides have also been implicated in recognition phenomena such as the interaction between sperm and egg cells in *Fucus* which has been explored using both lectins (Bolwell et al. 1979) and MAbs (Jones et al. 1988). The development of MAb probes for a number of components of cysts and zoospores of *P. cinnamomi* by Hardham and her colleagues over the last decade (reviewed by Hardham et al. 1991 a) have helped clarify the role of many of the zoospore organelles. A more precise comparison of organelle properties (morphology, lectin and MAbs-labelling and their believed function) in *Saprolegnia* and *Phytophthora* species is now possible and has been summarised in Table 4. The complexity of binding patterns to zoospores and cysts makes it difficult to unequivocally identify the receptor(s) which might be responsible for governing attachment or specificity. We have demonstrated in this study that closely related species which have different ecological niches (general saprophyte and opportunist necrotroph compared with a more specialised fish pathogen) seem to show subtle differences in their WGA labelling pattern, even though their cell organelle complement is basically the same. The functional significance of the binding patterns reported in this paper will now be considered in more detail.

Con A binding sites

The occurrence of Con A binding saccharides (glucose and mannose) on zoospore surfaces has previously been described in the following oomycete species: *Lagenidium giganteum* (Kerwin et al. 1991, Berbee and Kerwin 1993), *Phytophthora* spp. (Bacic et al. 1985; Hardham

1985, 1989; Hardham and Suzaki 1990; Hardham et al. 1991 a; Sing and Bartnicki-Garcia 1975 a, b; Williams et al. 1984), *Pythium* (Estrada-Garcia et al. 1990), *Saprolegnia ferax* (Lehnen and Powell 1993) as well as the chytrid, *Blastocladiella* (Jen and Haug 1979). The main binding sites for this lectin in oomycete spores will now be discussed in turn.

Plasma membrane of primary and secondary zoospores

The Con A material is probably associated with the carbohydrate-rich glycocalyx which has been demonstrated in secondary type zoospores of *Aphanomyces euteiches* (Sadowski and Powell 1990), *Lagenidium giganteum* (Kerwin et al. 1991, Berbee and Kerwin 1993), *Phytophthora* spp. (Bacic et al. 1985; Powell and Bracker 1986; Sing and Bartnicki-Garcia 1975 a, b; Hardham et al. 1991 a) *Pythium aphanidermatum* (Estrada-Garcia et al. 1990), and *Saprolegnia ferax* (Lehnen and Powell 1988). Secondary zoospores of *Saprolegnia* lyse at concentrations of Con A above 100 µg/ml (Burr 1991) and similar findings were demonstrated by Sing and Bartnicki-Garcia (1975 a) in *P. palmivora*. In both *S. ferax* and *P. cinnamomi* Con A binds particularly strongly to the water expulsion vacuole (Lehnen and Powell 1993, Hardham and Suzaki 1990) which may be the main cause of this observed lysis.

Flagella of secondary zoospores

Con A binding receptors on the secondary zoospore flagella (although not on primary zoospore flagella) were also described in the closely related *S. ferax* (Lehnen and Powell 1993) and in secondary type zoospores of *Lagenidium* (Berbee and Kerwin 1993) and *Phytophthora* (Hardham 1985). They may correspond to proposed receptor complexes involved in tactic responses (Cameron and Carlisle 1981) and encystment triggering. The latter could explain why this lectin is so effective at inducing encystment in this species (Table 1).

EV matrix

The staining of the surface of encysted zoospores with Con A was rather variable both in intensity and distribution. Upon encystment the PEV and SEV are discharged and their matrix released to coat both the cyst surface and cyst tubules or boathook spines. It is thought that the Con A reactive material is associated with the matrix released from the EV. The small localized patches of Con A binding material observed in some *S. parasitica* secondary zoospores (Fig. 18) may also be the result of binding to disrupted SEV as a result of fixation-induced permeability of the plasma membrane (Hardham 1985). This material appears to be lost with time from both the surface of the primary and secondary cyst walls and some primary cysts of both species showed a complete lack of FITC-Con A binding. Secondary cysts of *S. diclina* and *S. parasitica* showed intense staining following pectin or vortex induced encystment. In addition to the outer cyst coat the long bundles of spines of *S. parasitica* also fluoresce with this lectin and in germlings of this species often only these bundles continue to bind this lectin. This is presumably as a result of the retention of matrix material entrapped between the bundles of cross-linked spines. This spine-bundle matrix material may act as an adhesive or binding agent and may increase the stickiness of these structures and so facilitate the attachment of these cysts to the fish or other surfaces. Secondary zoospores of *S. ferax* have quite small SEV and even the layer of outer electron dense material is often incomplete (Heath and Greenwood 1970). This may account for the apparent absence of Con A binding reported in secondary cysts *S. ferax* (Lehnen and Powell 1993). The cyst wall of *Lagenidium giganteum* also stain specifically with Con A, although it is not known from which internal vesicle fraction this material is derived (Berbee and Kerwin 1993). Although it is proposed in Table 4 that the SEV are functionally equivalent to the SBA-binding small dorsal vesicles of *P. cinnamomi* (Hardham et al. 1991 a) it is clear that these vesicles show significant differences, both biochemically and in their structural contribution to the cyst surface. Therefore these studies have revealed significant variations in cyst coat chemistry can occur both within the genus *Saprolegnia* itself as well as between other major genera of peronosporalean oomycetes.

Contents of the FV/LPV fraction

In *Phytophthora* (Sing and Bartnicki-Garcia 1975 b, Gubler and Hardham 1988) and *Pythium* (Estrada-Garcia et al. 1990) the contents of the LPV (proposed homologues to the FV, Table 4) have been shown to bind Con A. Upon encystment in *P. palmivora* a threefold increase in Con A-binding has been documented (Sing and Bartnicki-Garcia 1975 b) and Bacic et al. (1985) report an increase in *P. cinnamomi* from between two and sixteen-fold. Differences in the amounts of Con A labelling, achieved between Bacic et al. (1985) and Hardham and Suzaki (1990) for *P. cinnamomi*, have been explained by the latter authors as being dependent upon differences in the processing of cells and thus the removal of Con A-specific material from their surfaces. This Con A-binding is almost certainly to components of the LPV which are disrupted by formaldehyde fixation (Hardham and Suzaki 1990). There is now some doubt as to whether this Con A-binding material, which has been demonstrated to be released upon encystment from zoospores of both *P. palmivora* (Sing and Bartnicki-Garcia 1975 b) and *Py. aphanidermatum* (Estrada-Garcia et al. 1989), genuinely represents an adhesive material as originally proposed. Gubler and Hardham (1988, 1990) have demonstrated convincingly in *P. cinnamomi* that these vesicles are not normally discharged during encystment and proposed that their main function was as a repository for storage proteins. However, in the *Saprolegnia* species described here, pre-embedding labelling with Au-conjugated Con A showed little binding to the ruptured FV, even though comparable EV were stained. It therefore seems likely in *Saprolegnia* that most of the Con A material associated with cyst surface is derived from the PEV/SEV systems rather than the FV. Similarly in *L. giganteum* there was no direct evidence that the LPV fraction bound Con A, or indeed even contained much cytochemically detectable carbohydrate (Berbee and Kerwin 1993).

WGA-binding sites

FITC-WGA labelling and its inhibition with specific sugars, has demonstrated the presence of N-acetyl glucosamine residues on the surfaces of primary and secondary zoospores of both *S. diclina* and *S. parasitica* although the pattern of labelling of secondary zoospores was significantly different between the two species. This was reflected by the relative insensitivity of secondary zoospores of *S. parasitica* to be induced to encyst by this lectin (Table 1). In primary zoospores and the secondary zoospores of *S. diclina* the localised binding of WGA corresponded closely with the region of K-body location and fusion with the spore membranes. The small apical WGA-binding spot observed in primary zoospores of both species was in contrast with the more extensive binding of this lectin to primary zoospores of *S. ferax* (Lehnen and Powell 1993). However, the binding of this lectin to the ventral flagellar groove region in *S. diclina* secondary zoospores was similar to that reported in *S. ferax* (Lehnen and Powell 1989). In this species this material has been shown to be associated with material derived from K_2-bodies which forms an adhesive pad attaching these spores to their substrate (Lehnen and Powell 1989). In *P. cinnamomi* the ventral adhesive material which binds cysts to plant roots is a proteinaceous material secreted by the small ventral vesicle fraction (Hardham and Gubler 1990). The differences between this ventral fraction in *Saprolegnia* and *Phytophthora* (Table 4) indicates that the nature and origins of oomycete zoospore adhesive is variable. In *S. parasitica* the secondary zoospores demonstrated rather faint but distinct binding of FITC-WGA to large multiple patches on their dorsal surfaces which was also reported to some extent in *S. ferax* with another N-acetyl glucosamine-binding lectin *Griffonia simplicifololia* lectin II (GS II). The labelling in *S. parasitica* was not seen in hapten sugar controls suggesting this pattern of WGA-binding is specific, although it may be to internal spore components such as peripheral SEV.

In both *S. diclina* and *S. parasitica* Au-WGA also appeared to bind to the inner primary cyst wall after zoospore emergence. The binding of an N-acetyl glucosamine-binding lectin GS II to discharged primary cysts was also reported in *S. ferax* (Lehnen and Powell 1993). GS II binds to terminal N-acetyl glucosamine residues whereas WGA binds to both internal and terminal residues (see discussion by Lehnen and Powell 1993). It may be that there are subtle differences in the polymers containing this sugar between these closely related oomycete species. The other component which strongly binds WGA is the young germ tube wall. Oomycetes are generally characterised by the possession of cellulose rather than chitin as their main microfibrillar wall constituent although chitin does occur in the hyphal walls of *Apodachlya* (Lin and Aronson 1976) and in the close relative of *Saprolegnia, Achlya* (Dietrich and Campos 1978). It is also known from other oomycete species that significant differences occur in cell wall architecture between germ tubes and cysts (Hegnauer and Hohl 1973, Nyhlen and Unestam 1978). This WGA binding region corresponds closely to the sticky part of the hypha which binds to plastic surfaces such as petri dish bottoms (Willoughby 1986). This suggests that this WGA-specific material is also involved in the adhesion of the germ tubes to substrates. Whether this material is also derived from K_2-bodies or represents some other adhesive material is not known at present. *S. parasitica* produces extremely fast growing, rather narrow initial germ tubes, which become septate. These narrow germ tubes also bind WGA which may play a crucial role in securing the germling to the surface of the fish, even though it is a feature not unique to fish pathogens.

MAb binding patterns

As discussed above lectins are very useful as both cell surface and intracellular probes, however, they do have their limitations. Lectins will only bind to glycoconjugates containing accessible residues of the sugars and many are specific for sugars of certain orientation and position within the oligosaccharide (see Lehnen and Powell 1993). Specific antibodies, however, have the ability to bind to proteins and polysaccharides and their specificity can be used to probe for a single molecular species. At present the main practical obstacle to a more widespread use of immunocytological techniques in characterisation of the cell surfaces of pathogenic fungi is the lack of availability of antibodies which are specific for well-defined molecular components. All of the MAbs raised against *Saprolegnia* appear to react to various components of the immunodominant Con A-binding material which coats the surfaces of zoospores and cysts. The richness of mannose at the cell surfaces of many fungi has led to the production of antiserum primarily directed mostly against these mannose residues (Toroscantucci et al. 1990). Binding to the surface of secondary zoospores was not consistently observed suggesting the glycocalyx of primary and secondary zoospores is different from that of the cyst surface, even though both bind Con A. Likewise the antibodies were much less effective at inducing zoospore encystment than this lectin. Sometimes localised patchy binding of these MAbs to the secondary zoospore dorsal surfaces *S. parasitica* was observed. This is reminiscent to that shown by the MAb Cpa-2 in *Phytophthora cinnamomi* secondary-type zoospores (Hardham et al. 1986) and again may be due to the antibody binding to peripheral organelles as a result of the fixation induced permeability of the plasma membrane.

Oomycete protists clearly demonstrate how closely related species have evolved subtle modifications of their encystment apparatus and surface features. The fish pathogen clearly has adopted a "passive strategy", where its main infective spores are propably the secondary cysts with their long bundles of boathook spines. These spines are coated in a matrix material which may have some role in helping these spores to become localised on the surface of the fish. In contrast, the saprophytes rely more upon their motile zoospore stage. The discharge of a sticky adhesive pad arising from the K-bodies results in the "active attachment" of the zoospores to a suitable food substrate. These studies have revealed the complexity and diversity of surface components of the various spore types of *Saprolegnia* and other oomycete genera. Even within the genus *Saprolegnia* subtle differences in spore surface components were noted which may be related to the substrate preferences of these organisms.

Acknowledgements

AWB was a recipient of a Natural Environmental Research Council Studentship. We are grateful for the help and assistance given by Mr. Bob Hewit and Dr. Trevor Booth of the Biomedical EM Unit and to Dr. Geoff Tomms and co-workers for their help and advice with the hybridoma work. A generous grant from the Nuffield Foundation enabled us to undertake the hybridoma work.

References

Bacic A, Williams ML, Clarke AE (1985) Studies on the cell surface of zoospores and cysts of the fungus *Phytophthora cinnamomi*: nature of the surface saccharides as determined by quantitative lectin binding studies. J Histochem Cytochem 33: 384–388

Beakes GW (1983) A comparative account of cyst coat ontogeny in saprophytic and fish-lesion (pathologenic) isolates of the *Saprolegnia diclina-parasitica* complex. Can J Bot 61: 603–625

– (1987) Oomycete phylogeny: ultrastructural perspectives. In: Rayner ADM, Brasier CM, Moore D (eds) Evolutionary biology of the fungi. Cambridge University Press, Cambridge, pp 405–421 (British Mycological Society symposium volume)

– (1989) Oomycete fungi: their phylogeny and relationship to chromophyte algae. In: Green JC, Leadbetter BSC, Diver WL (eds) The chromophyte algae: problems and perspectives. Clarendon Press, Oxford, pp 323–340

Berbee ML, Kerwin JL (1993) Ultrastructural and light microscopic localization of carbohydrates and peroxidase/catalases in *Lagenidium giganteum* zoospores. Mycologia 85: 734–743

Bimpong CE, Hickman CJ (1975) Ultrastructural and cytochemical studies of zoospores, cysts, and germinating cysts of *Phytophthora palmirora*. Can J Bot 53: 1310–1327

Bolwell GP, Callow JA, Callow ME, Evans LV (1979) Fertilisation in brown algae. II. Evidence for lectin-sensitive complementary receptors involved in gamete recognition in *Fucus serratus*. J Cell Sci 36: 19–30

Burr AW (1991) Comparative diplanetic developmental processes of salmonid-pathogenic and saprophytic isolates of the *Saprolegnia diclina-parasitica* complex. PhD Thesis, Newcastle University, Newcastle upon Tyne, UK

Byrt P, Irving HR, Grant BR (1982a) The effect of cations on zoospores of the fungus *Phytophthora cinnamomi*. J Gen Microbiol 128: 1189–1198

– – – (1982b) The effect of organic compounds on the encystment, viability and germination of *Phytophthora cinnamomi*. J Gen Microbiol 98: 599–602

Cameron JN, Carlile MJ (1977) Negative geotaxis of zoospores of the fungus *Phytophthora*. J Gen Microbiol 120: 347–353

– – (1980) Negative chemotaxis of zoospores of the fungus *Phytophthora palmivora*. J Gen Microbiol 120: 347–353

– – (1981) Binding of isovaleraldehyde, an attractant, to zoospores of the fungus *Phytophthora palmivora* in relation to zoospore chemotaxis. J Cell Sci 49: 273–281

Cerenius L, Olson LW, Lange L, Söderhäll K (1984) The secondary zoospore of *Aphanomyces astaci* and *A. laevis* (Oomycetes, Saprolegniales). Nordic J Bot 4: 697–706

Cho CW Fuller MS (1989) Ultrastructural studies of encystment and germination in *Phytophthora palmivora*. Mycologia 8: 539–548

Deacon JW (1988) Behavioural responses of fungal zoospores. Microbiol Sci 5: 249–252

Dietrich SMC, Campos GMA (1978) Effect of polyoxin-D on *Achlya radiosa*. J Gen Microbiol 105: 161–164

Estrada-Garcia M-T, Callow JA, Green JR (1989) Monoclonal antibodies to cell surface components of zoospores and cysts of the fungus *Pythium aphanidermatum* reveal species specific antigens. Exp Mycol 13: 348–355

– Green JR, Booth JM, White JG, Callow JA (1990 a) Monoclonal antibodies to the adhesive cell coat secreted by *Pythium aphanidermatum* zoospores recognise 200 × 10^3 M_r glycoproteins stored within large peripheral vesicles. J Cell Sci 95: 199–906

– Ray TC, Green JR, Callow JA, Kennedy JF (1990 b) Encystment of *Pythium aphanidermatum* zoospores is induced by root mucilage polysaccharides, pectin and a monoclonal antibody to a surface antigen. J Exp Bot 41: 693–699

Grove SN, Bracker CE (1978) Protoplasmic changes during zoospores encystment and cyst germination in *Pythium aphanidermatum*. Exp Mycol 2: 51–98

Gubler F, Hardham AR (1988) Secretion of adhesive material during encystment of *Phytophthora cinnamomi* zoospores characterised by immunogold labelling with monoclonal antibodies to components of peripheral vesicles. J Cell Sci 90: 225–235

– – (1990) Protein storage in large peripheral vesicles in *Phytophthora* zoospores. Exp Mycol 14: 393–404

– – Duniec J (1989) Characterising adhesiveness of *Phytophthora cinnamomi* zoospores during encystment. Protoplasma 149: 24–30

Gunderson JH, Elwood H, Ingold A, Kindle K, Sogin ML (1987) Phylogenetic relationships between chlorophytes, chrysophytes, and oomycetes. Proc Natl Acad Sci USA 84: 5823–5827

Hamer JE, Howard RJ, Chumley FG, Valent B (1988) A mechanism for surface attachment in spores of a plant pathogenic fungus. Science 239: 288–290

Hardham AR (1985) Studies on the cell surface of zoospores and cysts of the fungus *Phytophthora cinnamomi*: the influence of fixation on patterns of lectin binding. J Histochem Cytochem 33: 110–118

– (1989) Lectin and antibody labelling of surface components of spores of *Phytophthora cinnamomi*. Aust J Plant Physiol 16: 19–32

– Gubler F (1990) Polarity of attachment of zoospores of a root pathogen and pre-alignment of the emerging tube. Cell Biol Int Rep 14: 947–956

– Suzaki E (1986) Encystment of zoospores of the fungus *Phytophthora cinnamomi* is induced by specific lectin and monoclonal antibody binding to the cell surface. Protoplasma 133: 165–173

– – (1990 Glycoconjugates on the surfaces of spores of the pathogenic fungus *Phytophthora cinnamomi* studied using fluorescence and electron microscopy and flow cytometry. Can J Microbiol 36: 183–192

– – Perkin JL (1985) The detection of monoclonal antibodies specific for the surface components on zoospores and cysts of *Phytophthora cinnamomi*. Exp Mycol 9: 264–268

– – – (1986) Monoclonal antibodies to isolate-, species-, and genus-specific components on the surface of zoospores and cysts of the fungus *Phytophthora cinnamomi*. Can J Bot 64: 311–321

– Gubler F, Duniec J (1991 a) Ultrastructural and immunological studies of zoospores of *Phytophthora*. In: Lucas JA, Shattock RC, Shaw DS, Cooke LR (eds) *Phytophthora* Cambridge University Press, Cambrige, pp 50–69

– – – Elliott J (1991 b) A review of methods for the production and use of monoclonal antibodies to study zoosporic plant pathogens. J Microsc 162: 305–318

– Cahill DM, Cope M, Gabor BK, Gubler F, Hyde GF (1994) Cell surface antigens of *Phytophthora* spores: biological and taxonomic characterization. Protoplasma 181: 213–232

Heath IB, Greenwood AD (1970) Wall formation in the Saprolegniales. II. Formation of cysts by zoospores of *Saprolegnia* and *Dictyuchus*. Arch Mikrobiol 75: 67–79

Hegnauer H, Hohl HR (1973) A structural comparison of cyst and germ tube walls in *Phytophthora palmivora*. Protoplasma 77: 151–163

Hemmes DE, Hohl HR (1971) Ultrastructural aspects of encystation and cyst-germination in *Phytophthora parasitica*. J Cell Sci 9: 175–191

Hinch JM, Clarke AE (1980) Adhesion of fungal zoospores to root surfaces is mediated by carbohydrate determinants of the root slime. Physiol Plant Pathol 16: 303–307

Hoch HC, Michell JE (1972) The ultrastructure of zoospores of *Aphanomyces euteiches* and of their encystment and subsequent germination. Protoplasma 75: 113–138

Holloway SA, Heath IB (1977) An ultrastructural analysis of the changes in organelle arrangement and structure between the various spore types of *Saprolegnia*. Can J Bot 55: 1328–1339

Irving HR, Grant BR (1984) The effects of pectin and plant root surface carbohydrates on encystment and development of *Phytophthora cinnamomi* zoospores. J Gen Microbiol 130: 1015–1018

Jen CJ, Haug A (1979) Concanavalin A-induced lysis of zoospores of *Blastocladiella emersonii*. Exp Cell Res 120: 425–428

Jones JL, Callow JA, Green JR (1988) Monoclonal antibodies to sperm surface antigens of the brown alga *Fucus serratus* exhibit region-, gamete-, species-, and genus-preferential binding. Planta 176: 298–306

Jones SW, Donaldson SP, Deacon JW (1991) Behaviour of zoospores and zoospore cysts in relation to root infection by *Pythium aphanidermatum*. New Phytol 117: 289–301

Kerwin JL, Grant DF, Berbee Ml (1991) Specific induction of encystment of *Lagenidium giganteum* zoospores by concanavalin A and derivatives of chitin and chitosan. Protoplasma 161: 43–51

Lehnen LP, Powell MJ (1988) Cytochemical localisation of carbohydrates in zoospores of *Saprolegnia ferax*. Mycologia 80: 423–432

– – (1989) The role of kinetosome-associated organelles in the attachment of encysting secondary zoospores of *Saprolegnia ferax* to substrates. Protoplasma 149: 163–174

– – (1991) Formation of K_2-bodies in primary cysts of *Saprolegnia ferax*. Mycologia 83: 163–179

– – (1993) Characterisation of cell surface carbohydrates on asexual spores of the water mold *Saprolegnia ferax*. Protoplasma 175: 161–172

Lin CC, Aronson JM (1976) Chitin and cellulose in the cell walls of the oomycete *Apodachlya* sp. Arch Mikrobiol 72: 111–114

Longman D, Callow JA (1987) Specific saccharide residues are involved in the recognition of plant root surfaces by zoospores of *Pythium aphanidermatum*. Physiol Mol Plant Pathol 30: 139–150

Lunney CZ, Bland CE (1976) Ultrastructural observations of mature and encysting zoospores of *Pythium proliferum* de Bary. Protoplasma 90: 119–137

Manton I, Clarke B, Greenwood AD (1951) Observations with the electron microscope on a species of *Saprolegnia*. J Exp Bot 2: 321–331

Mitchell RT, Deacon JW (1987) Differential adhesion of zoospore cysts of *Pythium* on roots of graminiceous and non-graminiceous plants. Trans Br Mycol Soc 88: 401–433

Nyhlen L, Unestam T (1978) Cyst and germ tube wall structure in *Aphanomyces astaci*, Oomycetes. Can J Microbiol 24: 1296–1299

Paktitis S, Grant B, Lawrie A (1986) Surface changes in *Phytophthora palmivora* zoospores following induced differentiation. Protoplasma 135: 119–129

Pickering AD, Willoughby LG, McGrory CB (1979) Fine structures of secondary zoospore cyst cases of *Saprolegnia* isolates from infected fish. Trans Br Mycol Soc 72: 427–436

Powell J, Bracker CE (1986) Distribution of diaminobenzidine reaction products in zoospores of *Phytophthora palmivora*. Mycologia 78: 892–900

Sadowski LA, Powell ME (1990) Cytochemical detection of polysaccharides in zoospores of *Aphanomyces euteiches*. Can J Bot 68: 1379–1388

Sing VO, Bartnicki-Garcia S (1975 a) Adhesion of *Phytophthora palmivora* zoospores: electron microscopy of cell attachment and cyst wall fibril formation. J Cell Sci 18: 123–132

– – (1975 b) Adhesion of *Phytophthora palmivora* zoospores: detection and ultrastructural visualisation of concanavalin A-receptor sites appearing during encystment. J Cell Sci 19: 11–20

Smith SN, Armstrong RA, Rimmer JJ (1984) Influence of environmental factors on zoospores of *Saprolegnia diclina*. Trans Br. Mycol Soc 82: 413–421

Toroscantucci A, Palma C, Boccamera M, Ausiella CM, Spagnioli CG, Cassone A (1990) Lymphoproliferative and cytotoxic responses in human peripheral blood mononuclear cells to mannoprotein constituents of *Candida albicans*. J Gen Microbiol 136: 2155–2163

Whiffen AJ (1945) Nutritional studies of representatives of five genera of Saprolegniaceae. J Eli Mitch Sci Soc 61: 114–123

Williams ML, Bacic A, Clarke AE (1984) A method for estimating lectin binding to fungal zoospores and cysts. Exp Mycol 8: 238–244

Willoughby LG (1978) *Saprolegnias* of salmonid fish in Windermere: a critical analysis. J Fish Dis 1: 51–67

– (1986) An ecological study of water as the medium for growth and reproduction of the *Saprolegnia* from salmonid fish. Trans Br Mycol Soc 87: 493–502

Wood SE, Willoughby LG, Beakes GW (1988) Experimental studies on uptake and interaction of spores of the *Saprolegnia diclina-parasitica* complex with external mucous of brown trout (*Salmo trutta*). Trans Br Mycol Soc 90: 63–67

Protoplasma (1994) 181: 164–190

PROTOPLASMA

Ultrastructure, biogenesis, and functions of extrusive organelles in selected non-ciliate protists

P. Kugrens[1,] *, **R. E. Lee**[2], and **J. O. Corliss**[3]

Departments of [1] Biology and of [2] Anatomy and Neurobiology, Colorado State University, Fort Collins, Colorado, and [3] Albuquerque, New Mexico

Received August 5, 1993
Accepted January 16, 1994

Summary. The ultrastructural features, biogenesis and functions of several selected protist extrusive organelles are discussed. Most of the review focuses on some common extrusive organelles that were not considered by Hausmann and several types which have been described since that review of 16 years ago. For convenience, extrusomes are categorized as projectile or mucocyst extrusomes. The projectile extrusomes are further subdivided into non-penetrating and cell penetrating extrusomes. This review is restricted to projectile extrusomes such as ejectisomes, the microsporidian invasion apparatus, and the gun cell of oomycetes. Mucocysts include the apicomplexan rhoptries, the K_2 bodies of oomycetes, and the spermatial vesicles and adhesive vesicles of red algae. The possible phylogenetic importance of some extrusive organelles is briefly considered.

Keywords: Extrusomes; Protists; Ejectisomes; Gun cells; Polar filaments; Rhoptries; K_2 bodies; Spermatial vesicles; Adhesion vesicles.

Introduction

The topic of extrusive organelles in protozoan and algal protists was reviewed extensively by Hausmann (1978). There has not been a great deal of research specifically on these protistan structures since the time of Hausmann's review, but there have been several discoveries of new extrusomes, descriptions of previously unrecognized extrusomes and expanded observations on existing types. The reason for the present paper is to provide additional information on extrusomes, perhaps stimulating renewed interest in the many fascinating and unresolved aspects of the subject.

Before ca. 1950, the term trichocyst, with appropriate modifiers, was commonly used for any kind of extrusome; and the explosive or spindle trichocyst of the ciliate *Paramecium* was the best known example in the literature. In fact, from a historical point of view, it was that organelle, relatively easy to see under light microscopy, that was the earliest described extrusome (Ellis 1769). It was redescribed and given the name "trichocyst" by Allman (1855) in the following century. Restricted reviews on progress made in determining the general structure and possible function of these ejectile organelles during early years of the present century were published by Tönniges (1914) and Krüger (1936), the latter worker well remembered for his introduction of darkfield microscopy into the study of trichocysts sensu lato.

Following the extensive use of electron microscopy during the mid-twentieth century in biological/protistological research (for review, see Pitelka 1963), interest expanded in the study of various cytoplasmic organelles of protists. Such advances with respect to trichocysts (extrusomes) are chronicled in the comprehensive papers by Dragesco (1984), Hausmann (1978), and Hovasse and Mignot (1975). In recent years, molecular and biochemical approaches are being applied in trichocyst research (Eperon et al. 1993, Garofalo and Satir 1984, Glas-Albrecht and Plattner 1990, Hausmann et al. 1988, Lima et al. 1989, Lumpert et al. 1990, Peterson 1991, Peterson et al. 1987, Preparata et al. 1989). Additional references are available in these papers, as well as in Adoutte (1988) and Perkins (1991).

In regard to cell surface components of protists it appears appropriate to review extrusomes based on the

* Correspondence and reprints: Department of Biology, Colorado State University, Fort Collins, CO 80523, U.S.A.

fact that, although these organelles are initially located and stored inside the cell, their periodic extrusion places them in an extracellular or surface position, thus contributing to surface components. Some types form adhesive or penetrating structures, while others may form a coating or "blanket" over the body surface of the protist species producing them.

Some of our alternative/synonymous terminology (e.g., the words cilioprotists, cryptoprotists) employed occasionally for certain well-known groups of protists (e.g., for the ciliates and for the cryptophytes/cryptomonads) follows the nomenclature proposed by Heywood and Rothschild (1987). Our general formal taxonomy follows that used by authors in the volumes edited by Parker (1982). References to papers on extrusomes of cilioprotists that are not specifically cited by us may be found in the combined bibliographies of the following works: Corliss (1979), Hausmann (1978), Lyn and Corliss (1991), and Raikov (1993).

Characteristics of extrusomes

One of the problems in naming extrusomes is whether or not they should be designated on the basis of structural similarity or function. In the past they have been named on the basis of structural similarities, since functions for most extrusomes were unknown or tenuous at best. This still remains true today; however, in this paper we are erecting general categories using functions. Based on the numerous variations exhibited by these discharging organelles, we propose the following characteristics, which refine or extend the definitions or characterizations given by Adoutte (1988), Corliss and Lom (1985), Hausmann (1978), and Vickerman et al. (1991):

Extrusomes are stored, dormant, membrane-bounded, and structured organelles in the cytoplasm that are formed from some membranous precursors such as the rough endoplasmic reticulum or the cell's Golgi complex. Generally, they occupy a specific position in the cell, often near the cell membrane. Some may dock or anchor at specific, predetermined sites on the membrane or cell wall. At maturity, they may have a highly complex internal structure; and after a dormant stage, are extruded to the outside, often very rapidly, in response to some mechanical, electrical or chemical stimulus. Typically, some structural changes are associated with discharge, ranging from simple elongation or expansion to actual morphological transformations; thus exhibiting an altered morphology or structure in the discharged state. Extrusomes are characteristic and constant features in the specific cells in which they occur. When they are discharged periodically and the cell remains intact and functional, they are replaced within the same cell shortly after discharge. In other instances, the discharges initiate some developmental event. These characteristics separate extrusomes from other organelles which do not exhibit dormancy and secrete extracellular materials on a continuing basis to form the cell's glycocalyx, scales, cyst walls, and cell walls.

Classification of extrusomes

We find it convenient for the purposes of this review to separate protistan extrusive organelles into two major categories. This separation is based on the type of discharge and function exhibited by the extrusive organelles and whether they penetrate other cells or not. The categories are the following.

Projectile extrusomes

These extrusomes are highly structured and discharge quickly and forcibly. Included in this category are those that do not penetrate cells and those that are cell penetrating extrusomes found in parasitic protists.

Non-penetrating extrusomes

This category consists of hair-like, ribbon-like or tubular extrusive organelles that do not penetrate cells of other organisms. This category would include the following: trichocysts of dinoflagellates or dinoprotists (Taylor 1987, 1990) and ciliates or cilioprotists; toxicysts of ciliates; ejectisomes of cryptoprotists, some chloroprotists (Morrall and Greenwood 1980) and colorless flagellates (Lee and Kugrens 1991, Vørs 1992); the acmocyst of the little known parasitic rhynchoid ciliates, the discobolocyst of ochromonad flagellates (Hibberd 1970); the mucocyst of *Peranema* (Hilenski and Walne 1984); the haptocyst of suctorian ciliates; the kinetocyst of some heterokont and heliozoan protists; the nematocyst (nematocyst-taeniocyst complex) of some dinoflagellates (Vickerman et al. 1991, Westfall et al. 1983); and the rhabdocysts (rhabdosomes) of some other dinoflagellates (Vesk and Lucas 1986) and some karyorelictid ciliates (Raikov 1993).

Cell penetrating extrusomes

Cell penetrating extrusive organelles are found in parasitic protists which attack or invade cells. The types included in this category are the attack apparatus of oomycetes and the polar filament (invasion apparatus) of microsporidians.

Mucocyst extrusomes

This second broad category consists of subsurface vesicles containing unstructured or crystalline material which is typically slowly extruded, often as amorphous mucilage. These are of widespread occurrence among protists. For this review, the following are added to the category of mucocysts: spermatial vesicles of red algae, adhesive vesicles of red algae, K_2 bodies of oomycetes and rhoptries of apicomplexan (sporozoan) protists.

Selection of extrusomes

Extrusomes for this paper were selected because (1) they are common and often are characteristic of an entire group, and they have not recently been reviewed elsewhere; (2) they have been described since Hausmann's paper; or (3) they were not recognized as extrusomes by Hausmann (1978). Rather than duplicating the discussion previously presented by Hausmann, it is our intent to provide information that has been published since then. A number of additional extrusomes could have been included but our selections had to be restricted; therefore, the treatments are not always as comprehensive or detailed as Hausmann's were.

Description of extrusive organelles

Non-penetrating extrusomes

Ejectisomes

These extrusive organelles were discussed by Hausmann (1978), but more information on the variation of these organelles is now available. The cryptoprotist ejectisomes (formerly called trichocysts) were the first to be described, but recently different types have been discovered in other organisms. Ejectisomes are characterized by their unique internal structure of one or two tightly spiraled, tapered ribbons which unroll on discharge (Figs. 1, 2, 5, and 6). They are found principally in certain chloroprotists (Morrall and Greenwood 1980), cryptoprotists (Kugrens and Lee 1991), and the colorless flagellates *Katablepharis* (Lee and Kugrens 1991) and *Leucocryptos* (Vørs 1992). Similar structures, called kappa particles, are found in the R-bodies of some cilioprotists, but there is evidence that these represent bacterial (prokaryotic) endosymbionts (Lee and Kugrens 1992).

Cryptoprotists contain two sizes of ejectisomes. Within each cell there are usually large ejectisomes near the gullet/furrow complex, whereas small ejectisomes occur elsewhere in the peripheral cytoplasm (Fig. 1). Both types of ejectisomes are comprised of two unequal sized components which are joined together and enclosed by a membrane. Each of these components consists of a tightly wound, tapered ribbon. The widest part of the tape is toward the outside of the ribbon. The smaller ribbon generally faces toward the outside of the larger ribbon. The tape has a crystalline substructure (Fig. 2) (Grim and Staehelin 1984, Morrall and Greenwood 1980). The ejectisomes discharge when the organism is irritated and form a long tube, with the short portion oriented at an angle to the long tube (Fig. 3).

Some *Pyramimonas* species (chloroprotists) and the colorless protists *Katablepharis* (Lee and Kugrens 1991) and *Leucocryptos* (Vørs 1992) also possess ejectisomes; however, they differ in structure from cryptoprotist ejectisomes (Figs. 4–7). In all three genera, only the large, tightly coiled ribbon makes up the ejectisome (Figs. 5 and 8). In *Leucocryptos marina* three morphological types of ejectisomes were described (Vørs 1992). Types I and II are differentiated on the basis of size,

Fig. 1. Transverse section of a *Chilomonas* cell showing two sizes of ejectisomes. Large ejectisomes (*LE*) are located near the furrow (*Fu*) and small ejectisomes (*SE*) are near the cell membrane

Fig. 2. Freeze fracture through large ejectisomes. Note the crystalline nature of the ribbons (arrows). One of the ejectisomes is shown in cross section (*), displaying the coiled ribbon

Fig. 3. Extruded ejectisome of a cryptoprotist. Note the small component at an angle (arrow) to the large tube

Fig. 4. Shadowed cell showing released ejectisomes of *Katablepharis*. Note the absence of a small component

Fig. 5. Longitudinal section of a *Katablepharis* ejectisome. A small component is absent

Fig. 6. Transverse section of a *Katablepharis* ejectisome showing an early stage in unwinding. Arrows point to the unrolling ribbon

Fig. 7. Tips of *Katablepharis* ejectisome showing the absence of a small component. The tip is in the process of rolling into a tube. Bars: 1 μm

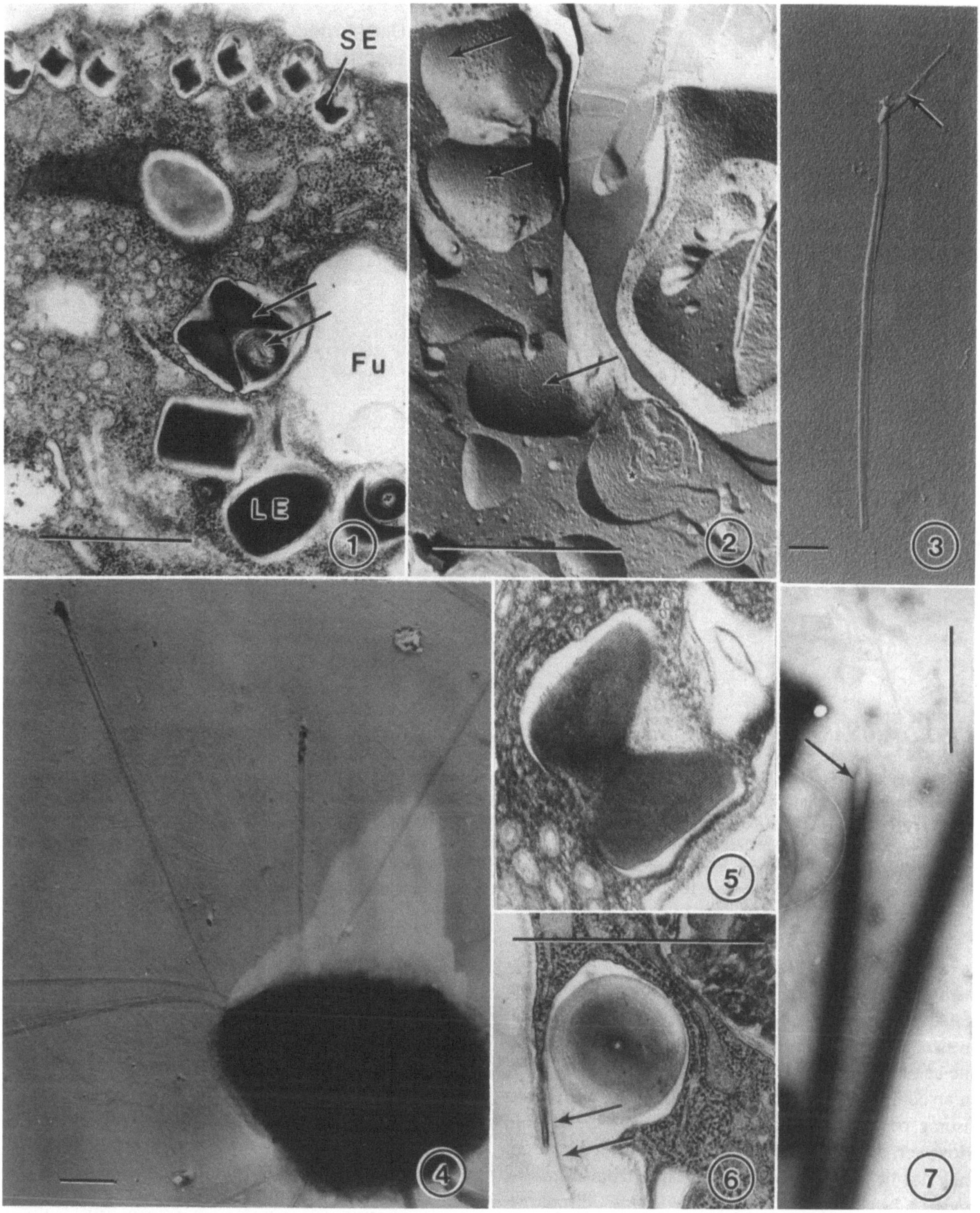
SE
Fu
LE
1
2
3
4
5
6
7

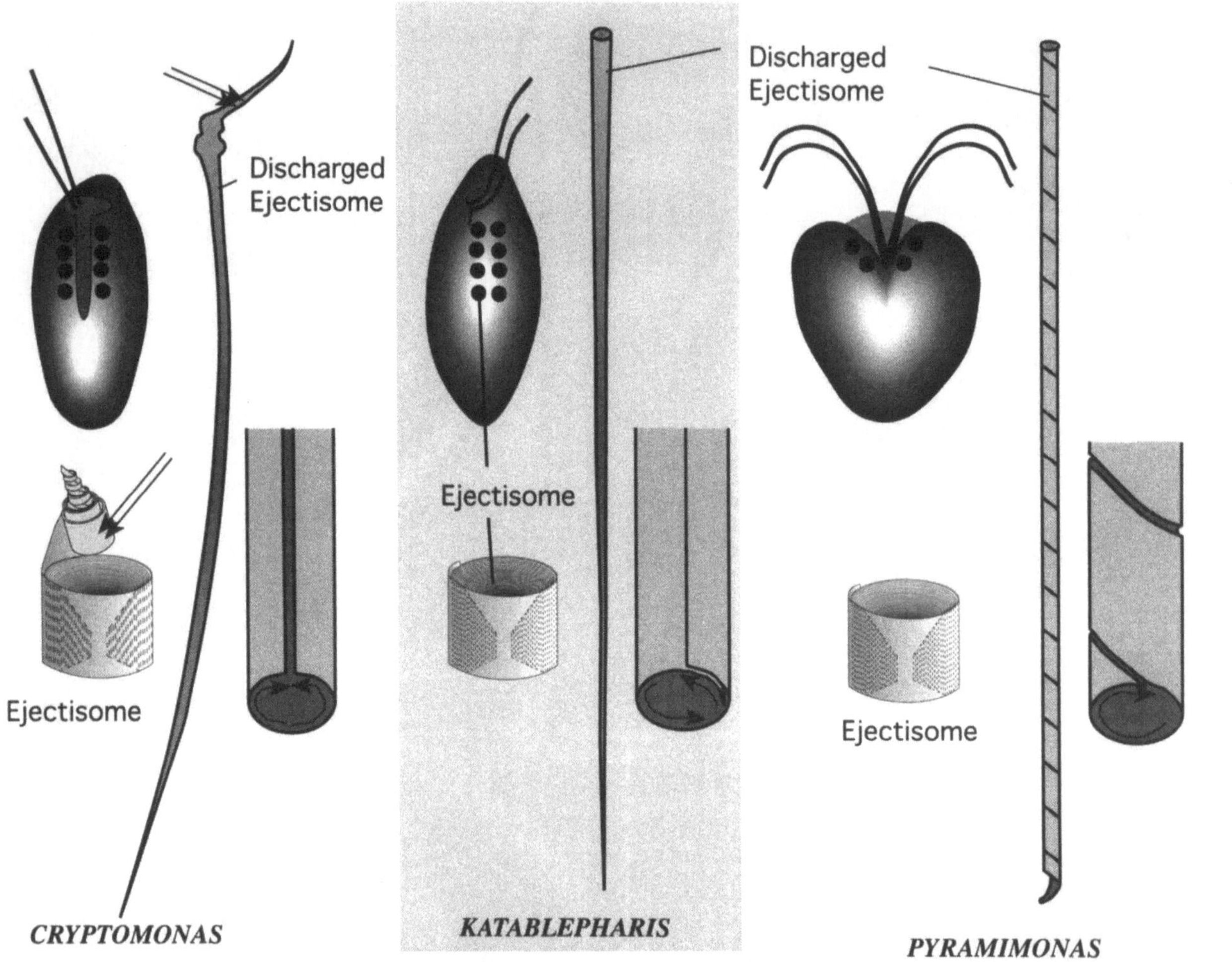

Fig. 8. Schematic diagrams of undischarged and discharged ejectisomes of *Cryptomonas, Katablepharis*, and *Pyramimonas*. Diagrams of each genus indicate the location of ejectisomes (black dots) in the cells. Undischarged ejectisomes in *Cryptomonas* consist of a large and small ribbon (double arrows) but in *Katablepharis* and *Pyramimonas* only the large ribbon is present. In *Cryptomonas* the discharged ejectisomes consist of a long and a short segment (double arrows), which is at an approximate 45° angle to the long segment. Only large components make up the discharged ejectisomes in *Katablepharis* and *Pyramimonas*. In *Cryptomonas* and *Katablepharis* the discharge ejectisomes roll up on themselves, whereas in *Pyramimonas* the ejectisomes form a tube by a spiral rolling of the ribbon

with the largest being Type I and the smallest designated as Type II. Type II ejectisomes are intermediate in size, and the widest part of the wound ribbon is in the center of the ejectisome. Due to these differences in structure, Vørs (1992) suggests that the term ejectisome be applied only to cryptoprotist extrusomes. However, as will be pointed out later, this designation may not be justified. When ejectisomes from *Katablepharis, Leucocryptos,* and *Pyramimonas* discharge (Figs. 4), the tape unwinds rapidly (Figs. 4 and 6) and all three types roll up to form a tube (Fig. 8), just as in cryptoprotists, but the small component is lacking (Figs. 4 and 7). However, in addition to morphological differences, the ejectisomes from the three groups of organisms display different ways of rolling into a tube. The structural differences and their different methods of forming tubes are depicted and briefly described in Fig. 8.

Biogenesis of ejectisomes. The ribbons are formed in Golgi-derived vesicles, where they originate as tightly wound ribbons or tapes. Initially, the tape consists of a few turns but the number increases as material is added. As the ejectisomes mature and enlarge, they are transported to the periphery of the cell, although no structural components accounting for this movement have been found in the cytosol. The ejectisome membranes contain intramembrane particles (IMPs) ar-

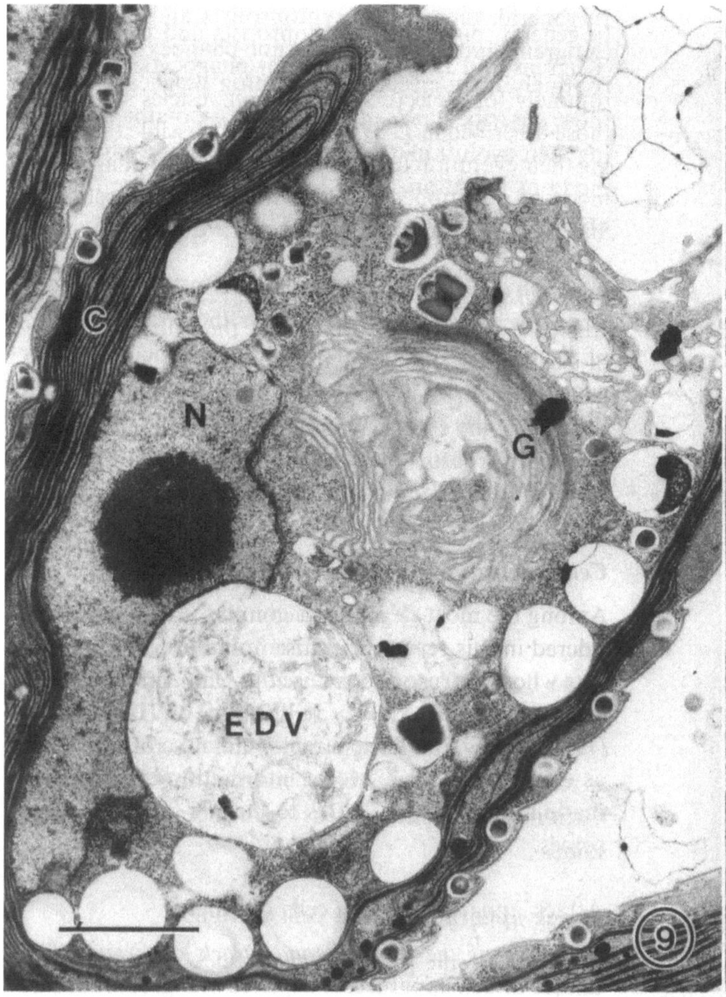

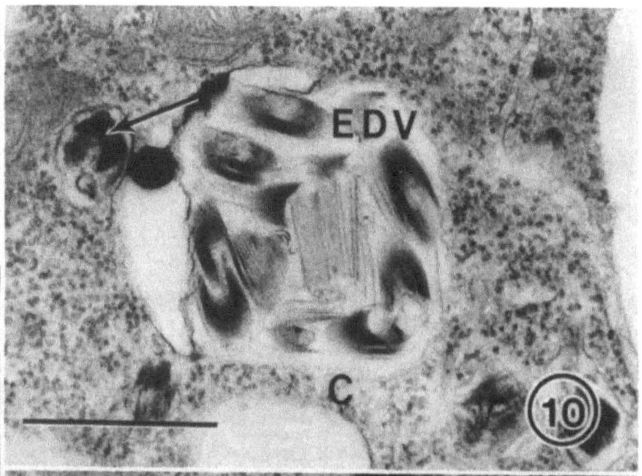

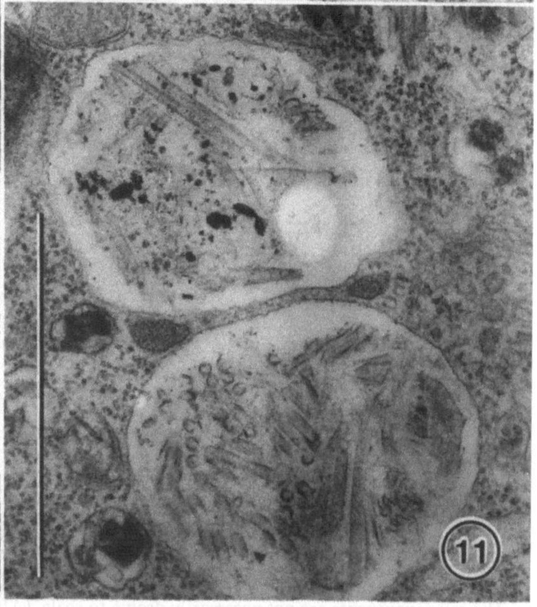

Fig. 9. *Chroomonas* cell with an ejectisome digestion vesicle (*EDV*) during the latter stages of digestion. *C* Chloroplast, *G* Golgi apparatus, *N* nucleus

Fig. 10. Early stage of ejectisome digestion vesicle forming. Several ejectisomes have fused and another is nearby (arrow)

Fig. 11. Later stages in ejectisome digestion vesicle formation. Ejectisomes have expanded and formed tubes prior to being digested. Bars: 1 µm

ranged in a rosette configuration (Grim and Staehelin 1984) that attach the ejectisome membrane to the cell membrane in a process called docking. In those species having a single periplast sheet, such as *Chilomonas*, there are numerous pores within the sheet, and the ejectisomes are also anchored to the sides of these pores (Grimm and Staehelin 1984).

Degradation of ejectisomes. Cryptoprotists and several colorless flagellates appear to have a special mechanism for recycling undischarged ejectisomes. Ultrastructural evidence indicates that some vesicles in cryptoprotists are specialized for ejectisome autolysis (Figs. 9–11) (Kugrens and Lee 1991). These vesicles originate from the fusion of several ejectisome chambers (Fig. 10) and continue to enlarge by the fusion of additional ejectisome chambers. Individual ejectisomes appear to disaggregate within the developing vesicles (Figs. 10 and 11). In older vesicles, components of expanded ejectisomes comprise most of the contents. In later stages most of the tubular, expanded components of ejectisomes are no longer recognizable and the contents appear fibrillar or granular (Fig. 9). The vesicle sizes are larger in older cultures and there may be several vesicles per cell. Golgi vesicles were observed to fuse with the existing vesicles. The vesicles apparently represent specific repositories for defective or surplus ejectisomes; thus they represent another unique component of cryptoprotist cells.

Vørs (1992) has described similar vesicles in the col-

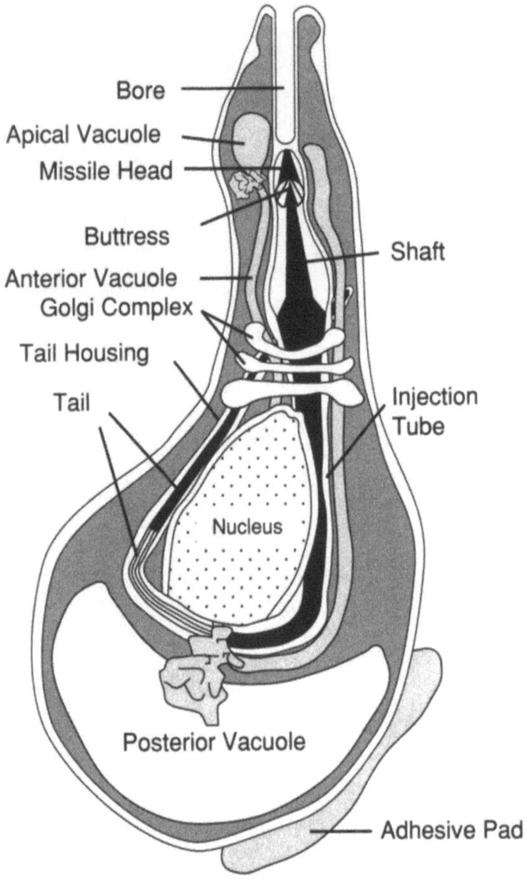

Fig. 12. Diagrammatic three-dimensional representation of the unfired gun cell, showing the relative positions of the major components of the attack apparatus (i.e., injection tube, missile, vacuolar system) as well as the relative positions of the nucleus and Golgi complex. The cell constituents are not necessarily drawn to scale. (After Robbin and Lee 1986 a)

orless flagellate *Leucocryptos* and she considered these to be food vacuoles. She proposes that the preponderance of expanded ejectisome material in these vacuoles is due to autophagocytosis of released ejectisomes which re-entered the cell during the ingestion of prey organisms. However, only one of her micrographs shows that material other than expanded ejectisomes is within the vacuole. The remainder show only ejectisomes present in the vacuoles and it is difficult to imagine that such a concentration of ejectisomes could re-enter the cell during phagocytosis. This vacuole in *Leucocryptos* thus may arise in the manner described in the preceding paragraph, but it may also function as a food vacuole in those organisms which are phagocytic. The ingested organisms would be transported to this vacuole for digestion. Consequently, the vacuole functions in digesting defective or superfluous ejectisomes, as well as ingested organisms.

In general, pigmented cryptoprotists and *Chilomonas* (Kugrens and Lee 1991) are not phagocytic, yet they routinely form ejectisome digestion vesicles similar to those described in *Leucocryptos*. It is possible that during their evolution to a photosynthetic, non-phagocytic mode of nutrition, these food vacuoles were retained and now function only in ejectisome digestion.

Function of ejectisomes. Ejectisomes routinely discharge when the water is disturbed near the cells. Discharge of ejectisomes is forceful and rapid, propelling the organism in a direction opposite from the discharge. Thus the ejectisome could function as an escape mechanism to avoid predators or it could be a defense mechanism to inflict damage to a potential predator.

Cell penetrating extrusomes

Among the most elegant and complex extrusomes considered in this review are those found in parasitic protists whose extrusomes are used to gain entry into host cells. Most were known at the time of Hausmann's (1978) review, but he apparently did not consider them as extrusomes. More recent information on their formation and function led us to include them as extrusomes.

Attack apparatus in gun cells of fungi

Structure of the gun cell and attack apparatus. The unique attack apparatus is depicted in Fig. 12. It occurs in the parasitic oomycete *Haptoglossa mirabilis*, which is endoparasitic in the rotifer *Adineta* (Robb and Barron 1982; Bobb and Lee 1986 a, b) but to date it has not been described for any other organism. At maturity, biflagellate zoospores are released from the thallus through special evacuation tubes. After a period of swimming, the zoospores encyst and within a short time each cyst germinates to form a gun cell. The mature gun cell remains attached to the empty cyst wall and is shaped like a miniature cannon which is aimed at an angle that facilitates host encounters. The gun cell's complexity is evident even with light microscopy (Barron 1980). A gun cell is approximately 17 µm long and pyriform in shape, the broad basal portion being anchored to the empty cyst by an adhesive pad. The anterior of the cell tapers then flares slightly to form an apical cone.

The attack apparatus is the most prominent cytological feature of the gun cell. It consists of a missile, the injection tube and a vacuolar system (Fig. 13). The injection tube and missile are extruded during dis-

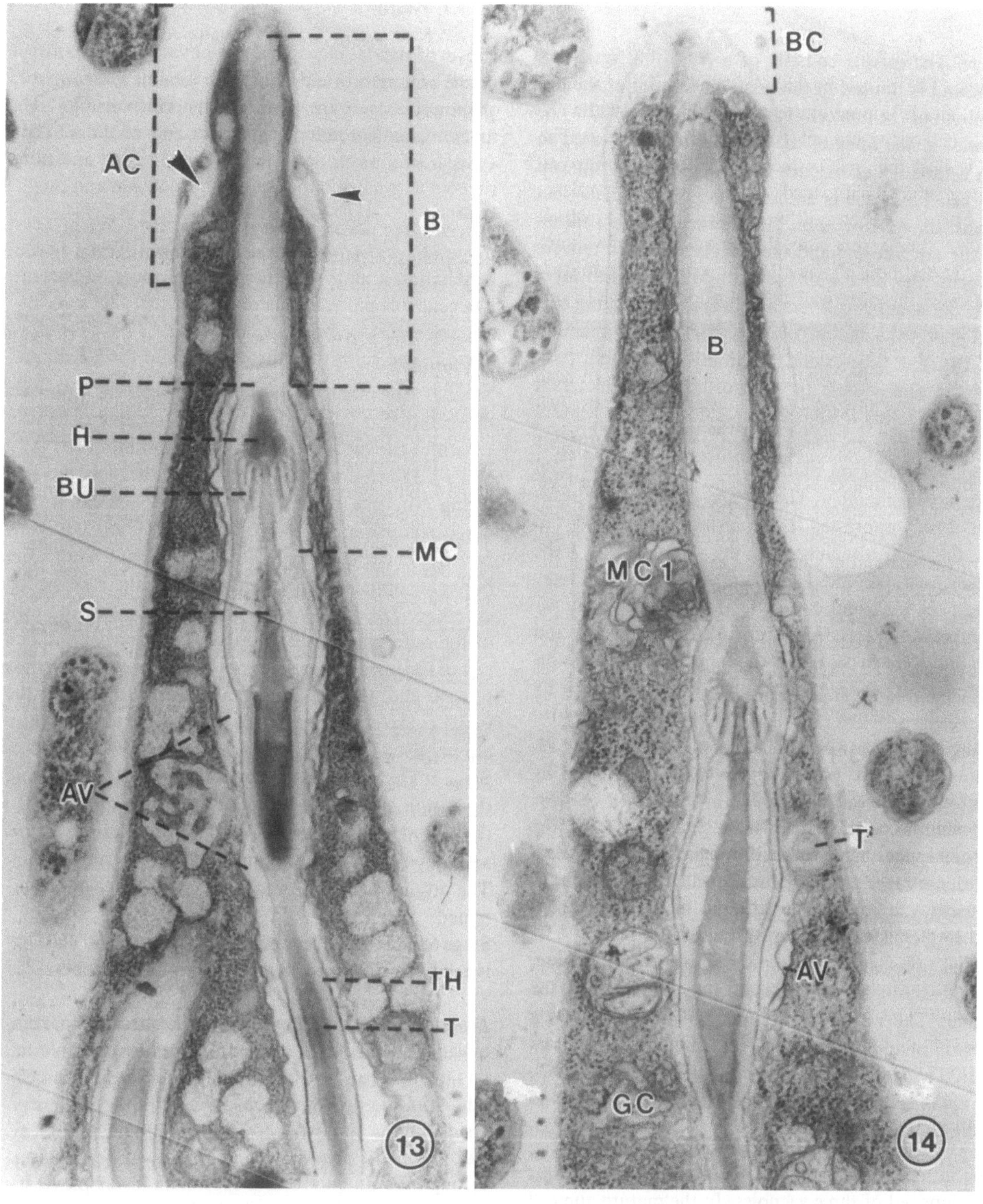

Fig.13. Longitudinal section through a portion of a gun cell through the injection tube and missile. Illustrated are part of the bore (*B*), the missile chamber (*MC*), part of the tail housing (*TH*) and tail (*T*), and the components of the missile itself: the head (*H*), the buttress (*BU*), and the shaft (*S*). The missile is sealed into the missile chamber by a plug (*P*) and the partially expanded anterior vacuole (*AV*) surrounds the injection tube from the plug to the coil. Note the basal, anterior, and neck portions of the apical cone (*AC*) and the thickened cell wall (arrowheads) encircling the neck region (From Robb and Lee 1986 a; original micrograph provided by Dr. Jane Robb)

Fig. 14. Longitudinal section through the anterior end of a gun cell in the latter stages of differentiation. The Golgi complex (*GC*) and the anterior membrane complex (*MC 1*) are terminal in position but the anterior vacuole (*AV*) is beginning to form. The terminal portion of the tail (*T*) is visible in a cross section of the tail housing. At this stage the muzzle is sealed by the bore cap (*BC*). (From Robb and Lee 1986 b; original micrograph provided by Dr. Jane Robb)

charge. The missile consists of a head, buttress, and shaft, and is housed in the missile chamber at the anterior end of the gun cell (Robb and Barron 1982). The cell wall at the apex of the gun cell is invaginated to form a tube which extends deeply into the cytoplasm (Fig. 14). This tube is differentiated into four distinct regions: the narrow bore, the expanded missile chamber, the tail housing and the tail structure. In mature gun cells, the bore wall has an additional wall layer inside the usual two layers and serves as plugging material that seals the missile into the missile chamber (Fig. 14). The tube extends in an almost straight path from the gun muzzle to the posterior vacuole. Just before contacting the posterior vacuole, it coils around the base of the nucleus and proceeds in an anterior direction, folding on itself, to end in a position level to the missile chamber. The wall forming the tail housing is closely appressed to the surface of the inner tail from the shaft attachment to the point at which the tube encircles the nucleus. At a position about halfway around the coil, and closest to the posterior vacuole, the tube wall parts from the surface of the tail and from this point to the end of the tube the two structures are separated by an electron lucid space traversed by fine electron dense fibrils. The housing wall also changes in appearance. The electron dense layer remains unchanged but the inner layer is traversed by fine electron dense fibrils similar in appearance to those traversing tube-tail space. At the end of the tube, the tube-tail space narrows and disappears. The outer electron dense layer of the housing wall becomes continuous with the base of the tail and the inner electron-lucid wall encloses the end of the tube.

The tail tip is solid and dark staining. When mature, the midsection forms a second inner tube within the housing. The wall of the midsection is dark staining (electron opaque) and attached to the housing by fine fibrils. The inner lumen is filled with fibrous material. The tail in the upper portion is solid where it joins the missile shaft. Since the tube and tail are continuous, they constitute the "injection tube".

The intracellular component of the attack apparatus is comprised of three vacuoles. In the mature gun cell the posterior vacuole occupies the basal quarter of the cell. The anterior vacuole forms a sleeve around the missile chamber and the upper portion of the tail housing. Two membrane complexes are associated with the anterior vacuole. The first consists of the tonoplast and plasma membrane and joins the anterior end of the gun cell just beneath the apical cone. The second complex which occurs at the extreme posterior end of the sleeve, protrudes into the posterior vacuole and consists of the cell membrane bounding the tail housing, and the tonoplasts of anterior and posterior vacuoles. The apical vacuole contacts the upper end of the anterior vacuole at a point opposite the missile head and plug.

Biogenesis of the attack apparatus. Stage 1: When a cyst germinates, a narrow germ tube is produced adjacent to a centriole pair, thus defining the anterior-posterior cell axis.

Stage 2: The germ tube then expands to form the gun cell initial.

Stage 3: The nucleus, centrioles and Golgi complex move into the elongating tube. A complete septum separates the cyst from the gun cell initial.

Stage 4: During migration into the germ tube, the maturing cisternae of the Golgi complex appear to undergo expansion. The wall of the anterior missile chamber, the enclosed missile head and buttress, and the bore plug appear to form within one of these expanded cisternae. The missile head and buttress, and the wall of the anterior missile chamber are formed first. The rest of the injection tube grows by addition of materials to each end. Current evidence indicates that both the Golgi complex and perinuclear endoplasmic reticulum are involved in this process.

Stage 5: The origin of the posterior vacuole results from the fusion and expansion of provacuolar organelles in the base of the gun cell initial, while the anterior vacuole arises by fusion and dilation of Golgi-derived vesicles. The origin of the anterior vacuole has not been determined.

Stage 6: This stage represents the mature gun cell that is ready to release its highly structured contents.

Function of the attack apparatus. The attack apparatus presumably functions to penetrate the host cells so that an infection can be established. Rapid changes in cellular osmotic pressure probably cause the sudden eversion of the hollow tube (Barron 1980). Rapid expansion of the posterior vacuole forces the cytoplasm through the tube to form a sporidium inside the host, after the needle-like projectile in the tube has breached the cuticle of the host (Robb and Barron 1982).

Polar filaments: the invasion or extrusion apparatus of microsporidians

Perhaps one of the most important and unusual extrusive organelles is the polar filament or the invasion apparatus in the Phylum Microspora. The presence of this organelle defines the phylum, and the variations in the invasion apparatus are used to delineate genera and species (Azvedo 1987; Bauer and Pankratz 1993; Beard et al. 1990; Becnel et al. 1986; Cali 1991; Cali and Owen 1988; Didier et al. 1991; Larsson 1986, 1992, 1993; Mitchell and Cali 1993; Moore and Brooks 1992). This structure was not considered to be an extrusome by Hausmann (1978) because the evidence available at that time was not convincing that these structures were enclosed by a membrane. Nevertheless, he still included a brief discussion of the invasion apparatus. A more recent review on microsporidians by Perkins (1991) included a discussion of this unique structure. The ultrastructural evidence indicates that the invasion apparatus, consisting of polar filaments and accessory components, is enclosed by a membrane (Larsson 1990, 1993; Calli and Owen 1988; Bauer and Pankratz 1993; Mitchell and Cali 1993) and is formed in association with the Golgi apparatus or endoplasmic reticulum. This type of extrusion apparatus is a diagnostic feature of the Phylum Microspora (Canning 1990, Canning and Lom 1986, Perkins 1991, Sprague 1982 a), whose members are commonly known as microsporidians since they form unicellular, small spores (usually < 4 µm in length) with no aperture in the wall. Microsporidia are intracellular, obligate parasites of all major animal groups (Canning 1990, Canning and Lom 1986, Perkins 1991) and several have recently been demonstrated as causative agents in several types of human diseases (Canning and Hollister 1992, Didier et al. 1991, Leitch et al.), including complications in AIDS patients (Bryan, Cali, Owen, and Spencer 1990, Cali et al. 1993, Cali and Owen 1990, Friedberg et al. 1990, Orenstein 1991, Shadduck 1989, Terada et al. 1987, Visvesvara et al. 1991). Microsporidia are among the most primitive eukaryotes (Vossbrinck et al. 1987) and they are characterized by their lack of mitochondria, flagella, and centrioles.

Structure of the polar filament. A diagrammatic representation of a generalized microspore is depicted in Fig. 15 and it serves as a basis for interpreting the electron micrographs in subsequent figures. Spores in most species are encased in two primary wall layers – the exospore and endospore (Fig. 16). The exospore is proteinaceous, electron-opaque and the endospore apparently is chitinous (Vávra 1976). The extrusive apparatus of microsporidians consists of three components – the polar filament, the anchoring disk, and the polaroplast (Figs. 15–18) – and is found exclusively in spores, although the formation may precede

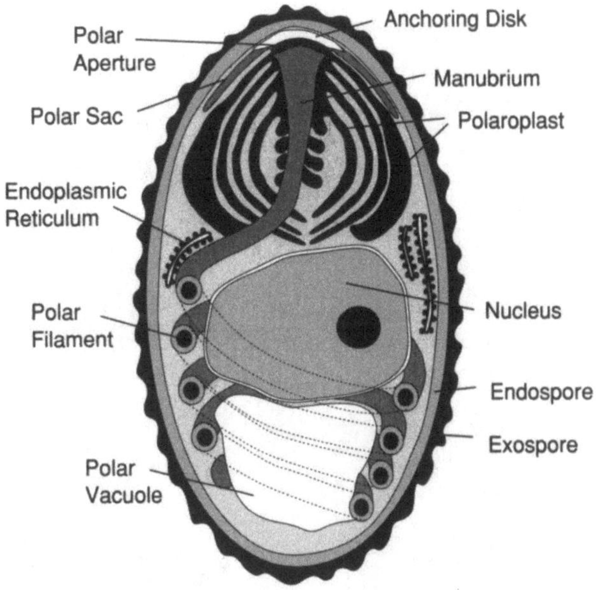

Fig. 15. Diagrammatic representation of a microsporidian spore showing the location of the different components of the polar filament

sporogenesis (Fig. 22) (Cali and Owen 1990). The anchoring disk is a dome shaped, membrane-enclosed structure which occupies the anterior end of the spore and is flanked by lateral extensions called the polar sac (Figs. 15–18). The anterior end of the polar filament emerges from the posterior portion of the anchoring disk and is oriented either straight or slightly oblique for one-half to two-thirds of the spore and then coils into loops (approximately 11 in most microsporidians). The straight portion of the polar filament is called the manubrium (Figs. 15 and 18). The membrane of the anchoring disk is continuous with the polar filament membrane throughout its entire length. The core of the polar filament is a solid cylinder of protein which usually consists of two or more concentric layers. The remaining component of the extrusion apparatus is the membranous polaroplast, which surrounds the manubrium. It consists of a stack of closely clustered, flattened vesicles (Fig. 16), expanded vesicles (Fig. 18) or in a honey-combed arrangement (Figs. 17 and 22) (Bauer and Pankratz 1990, Beard et al. 1990, Larsson 1993). The polaroplast membranes encircle the manubrium, except in the posterior end where it is less organized. A polar vacuole occupies the end opposite the anchoring disk in the spore (Figs. 15 and 21). It has not been conclusively demonstrated whether the membranes of the polaroplast and polar filament are continuous but, according to Weidner (1972), the membranes are continuous; and Azvedo (1987) suggests that

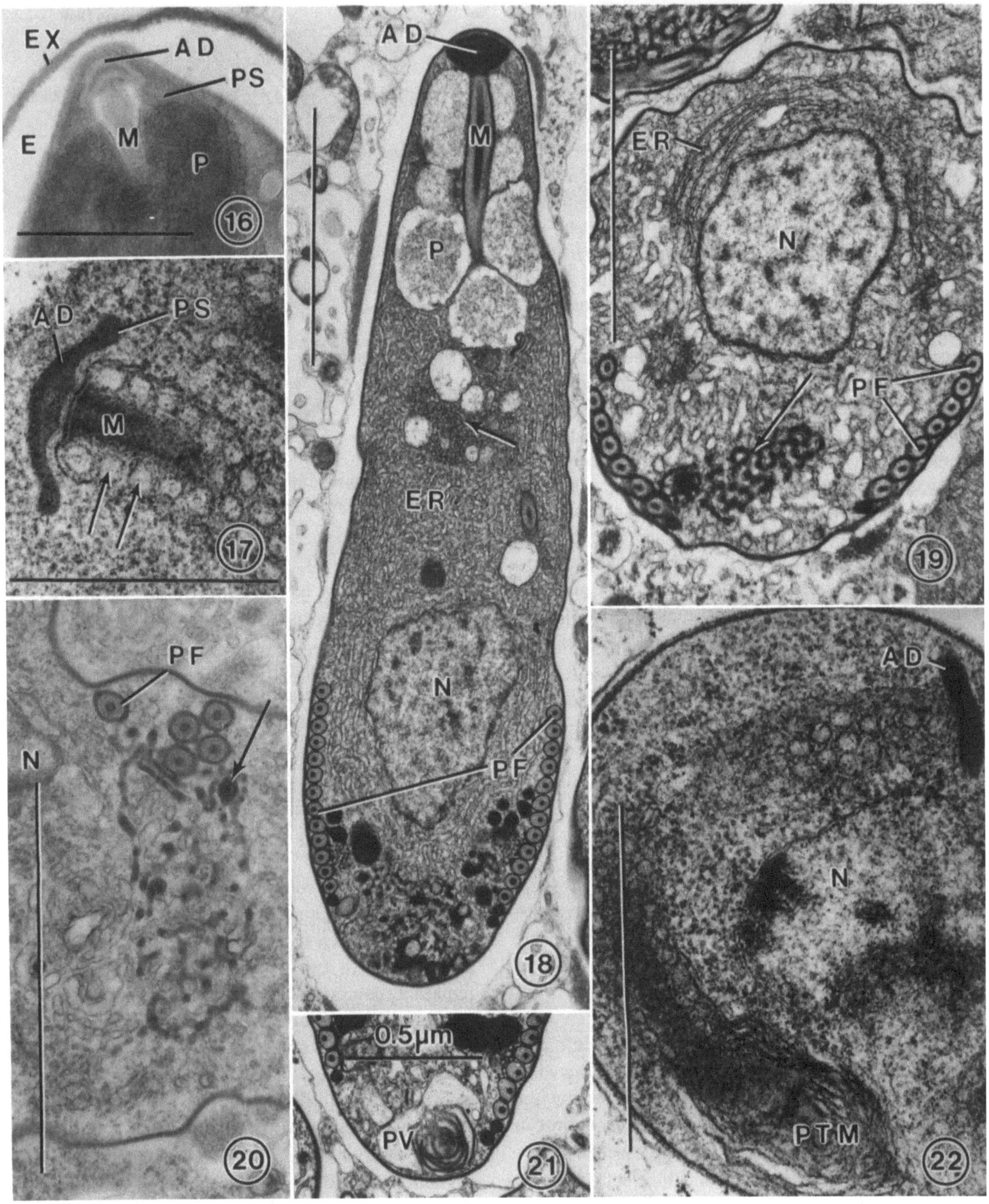
EX
AD
PS
E
M
P
16
AD
PS
M
17
AD
M
P
ER
N
PF
18
ER
N
PF
19
PF
N
20
0.5μm
PV
21
AD
N
PTM
22

they become continuous when the spores mature. Thus, all of the preceding components can be considered to be one organelle, which constitutes a complex extrusion or invasion apparatus, but this complex of components generally is referred to as the polar filament.

Biogenesis of the extrusion apparatus. The sequence of development for the extrusive apparatus has not been as well documented as it has for the other extrusomes discussed in this paper. Furthermore there are apparently several ways in which this complex can develop. The extrusive apparatus is usually formed within the spore after cleavage of the plasmodial cell (Cali et al. 1993, Larsson 1993). The polar filament and the polar sac appear to be generated by the Golgi apparatus (Cali et al. 1993, Larsson 1993, Mitchell and Cali 1993), and this appears to be the most common way of forming the invasion apparatus. When polar filament formation begins, membranous vesicles aggregate and coalesce to form a reticulum of fused vesicles, which are interpreted to be a primitive Golgi apparatus, although it appears that this complex represents fused endoplasmic reticulum cisternae. Endoplasmic reticulum is extensive in developing spores (Figs. 18 and 19) at this time and there is no evidence that the classical Golgi apparatus of other eukaryotes is present. After vesicle fusion, the cisternae form darker staining material whose staining properties and appearance are similar to the developing polar filaments (Fig. 20). During polar filament development, other membranous vesicles in the posterior end of the spore expand to form the polar vacuole (Fig. 21). Visual evidence also exists that a mass of membranous material separates and produces the polar filament in some genera (Fig. 22).

However, Cali and Owen (1990) describe a different biogenetic pathway for *Enterocytozoon*, a microsporidian unique to AIDS patients (Fig 23). Unlike other microsporidians, the formation of the polar filament, manubrium, anchoring disk and polaroplast is precocious and occurs in the uncleaved plasmodial stages. Small electron opaque disks form in conjunction with rough endoplasmic reticulum in the sporogonial plasmodium. Presumably these disks are enclosed within a membrane. Whether the endoplasmic reticulum is involved in their formation was not determined. The flattened disks then aggregate into stacks, followed by a fusion of individual disks to form arcs, and with continued fusion they form coiled structures which represent the incipient polar filament or polar tubule. Anchoring disks and polaroplasts form at the anterior end of the manubrium, but the details of this formation were not discussed. Cytokinesis occurs during this sequence of developmental events. Each polar filament apparently surrounds a single nucleus in the plasmodium (Cali and Owen 1990).

Function of the extrusion apparatus. Polar filaments, which are solid cores of protein (Widner 1982), convert to tubes upon extrusion from the cell. In effect, the extrusion of a polar tube represents the infectious stage

---

Fig. 16. Anterior portion of a microspore from *Nosema scripta* showing a portion of the invasion apparatus. *AD* Adhesion disk, *E* endospore, *Ex* exospore, *M* manubrium, *P* polaroplast, *PS* polar sac. (From Bauer and Pankratz 1993; original electron micrograph provided by Dr. Leach S. Bauer)

Fig. 17. Spore from the microsporidian *Nolleria* having a "honey-comb" polaroplast membrane system (arrows) associated with the polar filament (*PF*). *AD* Adhesion disk, *M* manubrium, *PS* polar sac. (From Beard et al. 1990; original micrograph provided by Dr. James J. Becnel)

Fig. 18. Microspore from *Culicospora magna* showing the formation of the polaroplast (*P*), manubrium (*M*), and adhesion disk (*AD*). Polaroplast formation appears to involve a membrane complex that is located at the anterior end (arrow). Note the extensive endoplasmic reticulum (*ER*) in the cell. Polar filaments (*PF*) have been formed in the posterior of the cell. (Electron micrograph provided by Dr. James J. Becnel)

Fig. 19. *Culicospora* cell in the early stages of polar filament formation. A reticulum of membranes (arrow) near the forming polar filaments contain dark staining material which probably fuse to form the polar filaments (*PF*). *ER* Endoplasmic reticulum, *N* nucleus. (Electron micrograph provided by Dr. James J. Becnel)

Fig. 20. Higher magnification of early polar filament formation in *Culicospora*. Membranes from endoplasmic reticulum or Golgi vesicles appear to be fusing and forming polar filaments. Note the cross section of one of these vesicles which contains an inner dark core of material similar to the polar filaments (*PF*). *N* Nucleus. (Electron micrograph provided by Dr. James J. Becnel)

Fig. 21. Polar vacuole formation in *Culicospora*. Note the enlarging polar vacuole (*PV*) within membrane precursors. (Electron micrograph provided by Dr. James J. Becnel)

Fig. 22. Putative formation of a polar filament from a polar tube membrane system (*PTM*) In *Nolleria*. *AD* Adhesion disk, *N* nucleus. (From Beard et al. 1990; original electron micrograph provided by Dr. James J. Becnel). Bars in Figs. 16–20 and 22: 1 μm

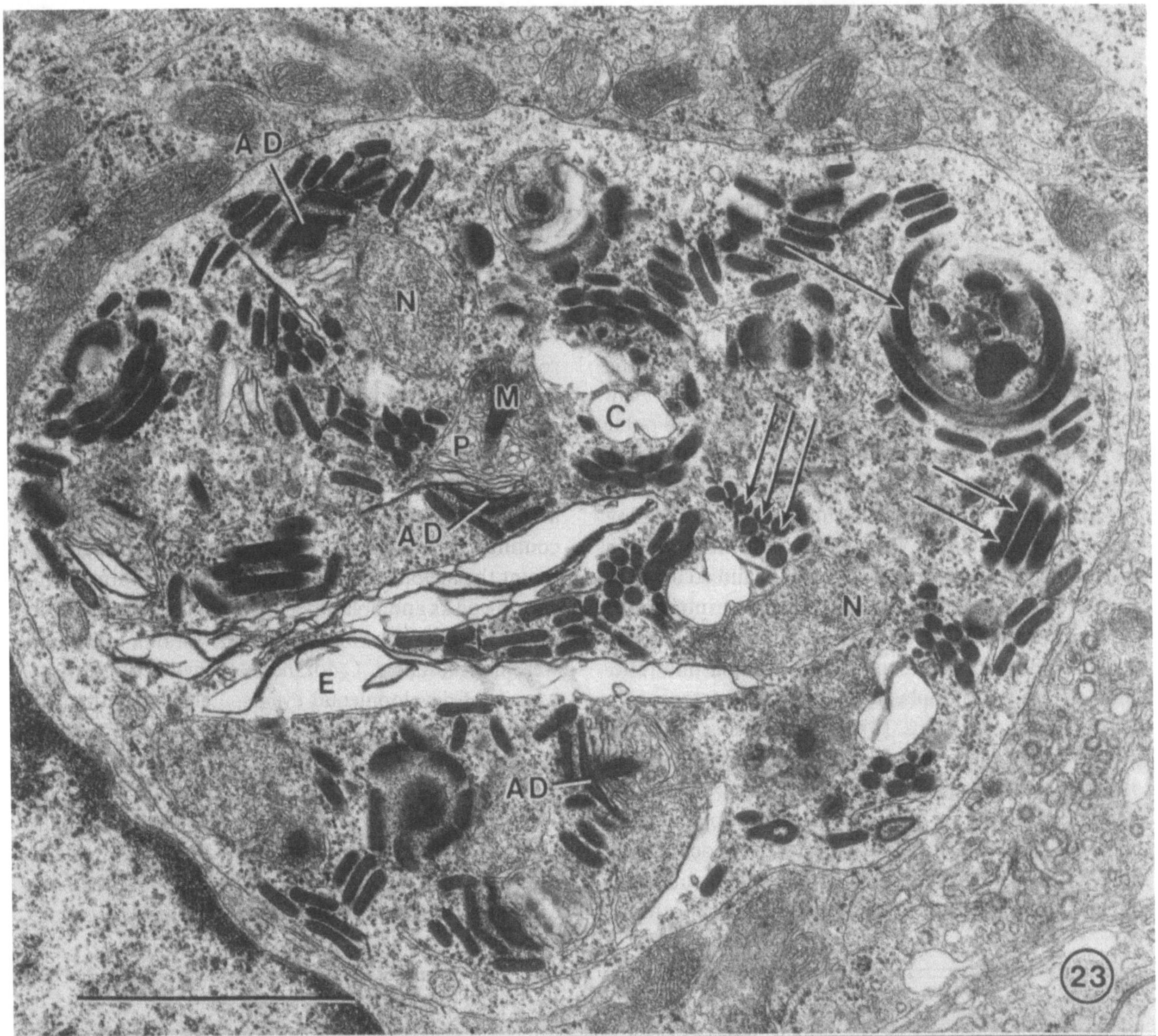

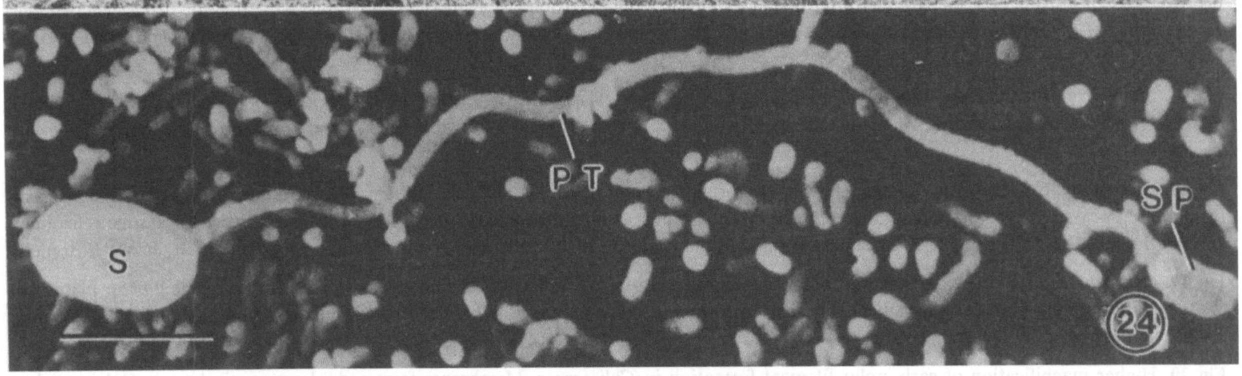

Fig. 23. Late sporogonial plasmodium with advanced stages of polar filament formation. Fused dark-staining disks are seen in coiled (single arrow), stacked (double arrows) and cross sectional (triple arrows) profiles in the cytoplasm. Anchoring disks (*A*), manubria (*M*) and associated polaroplast (*P*) membranes appear at this stage. Electron translucent regions (cleavage furrows?) are seen in elongated (*E*) and cross section views (*C*). *N* Nucleus. (From Cali and Owen 1990; electron micrograph provided by Dr. Robert L. Owen)

Fig. 24. Scanning electron micrograph of a discharged spore (*S*) with a long polar tube (*PT*) and sporoplasm (*SP*) being released at the end of the tube. (From Visvesvara et al. 1991). Bars: 1 μm

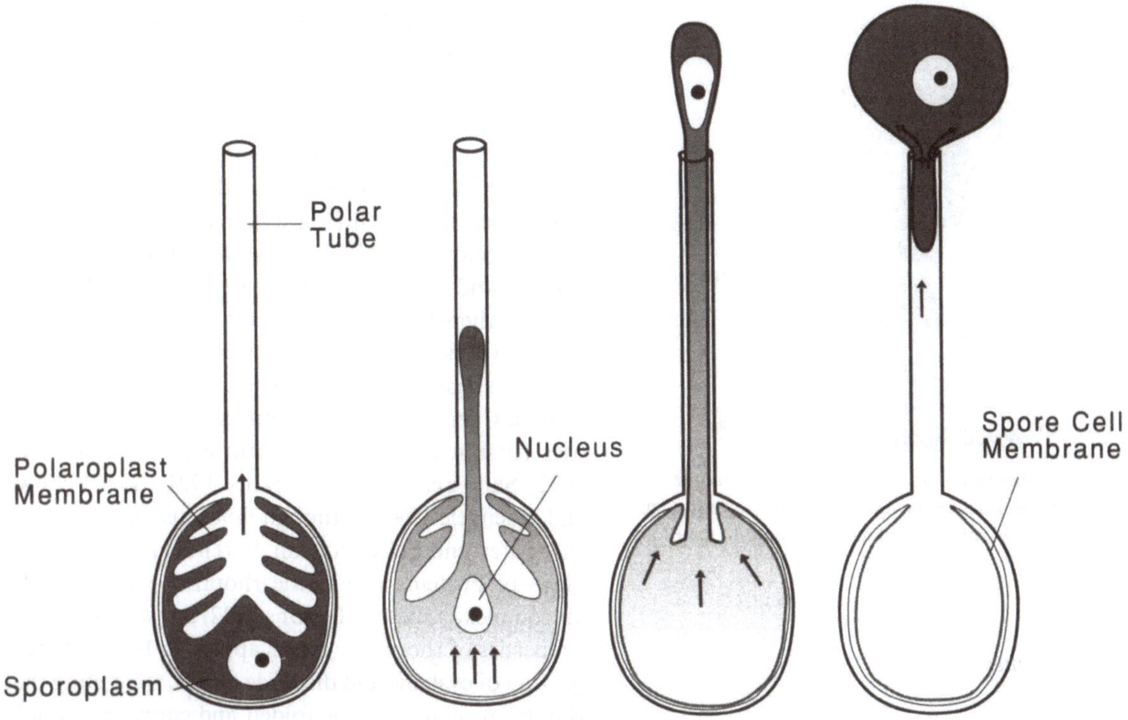

Fig. 25. Diagram showing the involvement of the polaroplast membrane in forming the new cell membrane in the host after injection of the cytoplasm into the host cell. The polaroplast membrane surrounds the sporoplasm which is migrating through the polar tube, eventually forming the new cell membrane of the sporoplasm after it is injected into a host cell. The spore membrane remains with the microspore wall. (After Weidner et al. 1984)

of spore germination (Fig. 24). It is this mechanism by which the cytoplasm and nucleus of a microsporidian are injected into the host cell. The filament everts with an acceleration of up to 500 $\mu m/s^2$ (Frixione et al. 1992, Leitch et al. 1993), which is sufficient to penetrate a host cell. The force from this eversion ruptures the host cell membrane and then injects the cytoplasmic contents (sporoplasm) of the parasite.

The cytoplasm of the spore (sporoplasm) enters the host after intact spores are ingested by the host, triggering a rapid and explosive emergence of the polar filaments from the anterior end of the spore and forming a tube which penetrates the host cell. After the polar filament has been everted to become a tube, the microsporidian's cytoplasm and nucleus migrate through the lumen of the tube within 5–30 s. The cell membrane of the spore remains with the spore wall and a new cell membrane is formed from the polaroplast membrane which surrounds the cytoplasm that migrated through the polar tube (Fig. 25).

Mechanism of extrusion. The process of polar filament extrusion and eversion has been studied by several investigators (Frixione et al. 1992, Leitch et al. 1993, Visvesvara et al. 1991, Weidner 1982, Weidner and Byrd 1982, Weidner et al. 1984, Pleshinger and Weidner 1985, Scarbrough-Bull and Weidner 1985), and a role for the polaroplast membrane in this injection process has been proposed (Weidner 1984). Filament extrusion and eversion are a rapid process which apparently begins by a pressure build-up in the polaroplast matrix, causing the swelling of the polaroplast. This swelling appears to be stimulated by an increase in pH, the presence of polyanions such as polyglutamate or mucin and a displacement of Ca^{++} from the spore wall/plasma membrane to the polaroplast, presumably into its matrix (Weidner 1982, Pleshinger and Weidner 1985). The anchoring disk at the anterior end of the spore also swells and explodes, rupturing the anterior, thin part of the spore wall. Evidence indicates that polar tubes assemble by a flow of material which emerges at the tip (Weidner 1982). The materials within the membrane of the polar filament flow out through a short sheath created by the anchoring disk material. The principal component of the polar filament and tube is protein (Weidner 1982).

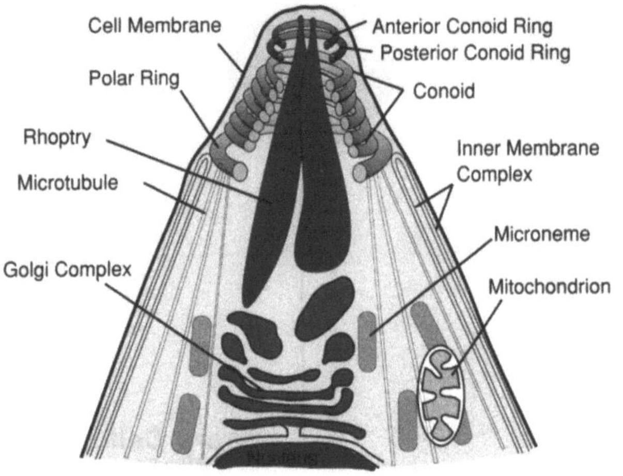

Fig. 26. Schematized diagram of an apicomplexan showing the formation of the rhoptries from Golgi vesicles and their location and the structure of the apical complex and its relationship to the rhoptries

Mucocyst extrusomes

Rhoptries

Although these structures were known prior to Hausmann's (1978) review article, he did not recognize them as extrusomes. Since more information regarding their function has been published (Bannister and Mitchell 1989; Jensen and Edgar 1976; Kimata and Tanabe 1987; Leriche and Dubremetz 1991; Levine 1982, 1985, 1988; Navdeep et al. 1993; Neill et al. 1989; Nichols et al. 1983; Nichols and Chiappino 1987; Perkins 1991; Roger et al. 1988; Safer and Schwartzman 1991; Stewart et al. 1986), it is obvious that they meet the extrusome criteria proposed in the introduction and could best be designated as mucocysts which function in host cell invasion.

Structure of rhoptries. The rhoptry is a club-shaped (rho in Greek means club), membrane-bounded, electron opaque (dark-staining) organelle found in apicomplexan parasites and, in fact, is one of the unifying structures of the Phylum Apicomplexa (Fig. 26). To some extent rhoptries resemble mucocysts and toxicysts of other protists such as the ciliates *Tetrahymena* and *Paramecium* (Vickerman et al. 1991). Rhoptries are located in the anterior of infectious cells and they are implicated in having a central role in host cell invasion (Nichols et al. 1983). They have been found in the motile invasive stages of all coccidian parasites. They form a part of a complex apical configuration (apical complex) from which the apicomplexans derive their name. The apical complex consists of rhoptries and a conoid (Figs. 27–29), which is a truncated, hollow cone with openings at both ends. The anterior end of the conoid consists of two preconoidal rings. Posterior to these rings are helically wound tubular units which are attached to a posterior polar ring. 20 to 70 microtubules are attached to the polar ring and they serve as a cytoskeleton for the cell (Fig. 28). The number of rhoptries per cell is variable. For instance, in zoites of *Plasmodium* and *Sarcocystis* there are only two rhoptries per cell whereas in *Toxoplasma* there are at least eight per cell (Perkins 1992). Rhoptries are absent in non-invasive stages of these parasites.

Complex proteins are the major constituents of rhoptries (Etzion et al. 1991, Ossorio et al. 1992, Perkins 1992, Sam-Yellowe and Perkins 1991, Sam-Yellowe and Ndengele 1993), although lipids also may be present (Safer and Schwartzman 1991). The number and kinds of proteins within the rhoptries vary with the apicomplexan species (Perkins 1992) but some common properties of rhoptry proteins appear to be a high proportion of cysteine and disulfide bonds, suggesting that rhoptry proteins may be folded and compact (Cooper et al. 1988, Sam-Yellowe et al. 1988, Perkins 1992). One of the proteins isolated causes lysis of host cells and enhances penetration by the invading parasite (Perkins 1992).

Biogenesis of rhoptries. Rhoptries are formed before the schizont cleaves into merozoites (Perkins 1992) but the mechanism by which they are more or less equally partitioned to each merozoite remains to be determined. Rhoptries presumbably are formed by vesicles derived from the Golgi apparatus which enlarge as they become filled primarily with proteins, although some lipids are also present (Perkins 1992). The proteins are believed to be post-translationally inserted into the endoplasmic reticulum (Perkins 1992) and then sorted to the Golgi apparatus.

Function of rhoptries. Rhoptries discharge through the conoid when the parasite makes contact with a host cell surface (Figs. 27 and 29). It has been proposed that rhoptries facilitate host cell penetration by injecting membrane materials (Fig. 27) into the host membrane of an erythrocyte. The 140/130/110 kDa protein complex has been implicated in this response (Perkins 1992; Sam-Yellowe 1992; Sam-Yellowe and Perkins 1981, 1990; Sam-Yellowe et al. 1988; Sam-Yellowe and Ndengele 1993). Ultrastructural studies have shown the presence of lamellar whorls that resemble membranes (Fig. 29) being secreted from the rhoptry during host cell invasion (Sam-Yellowe et al. 1988, Bannister and

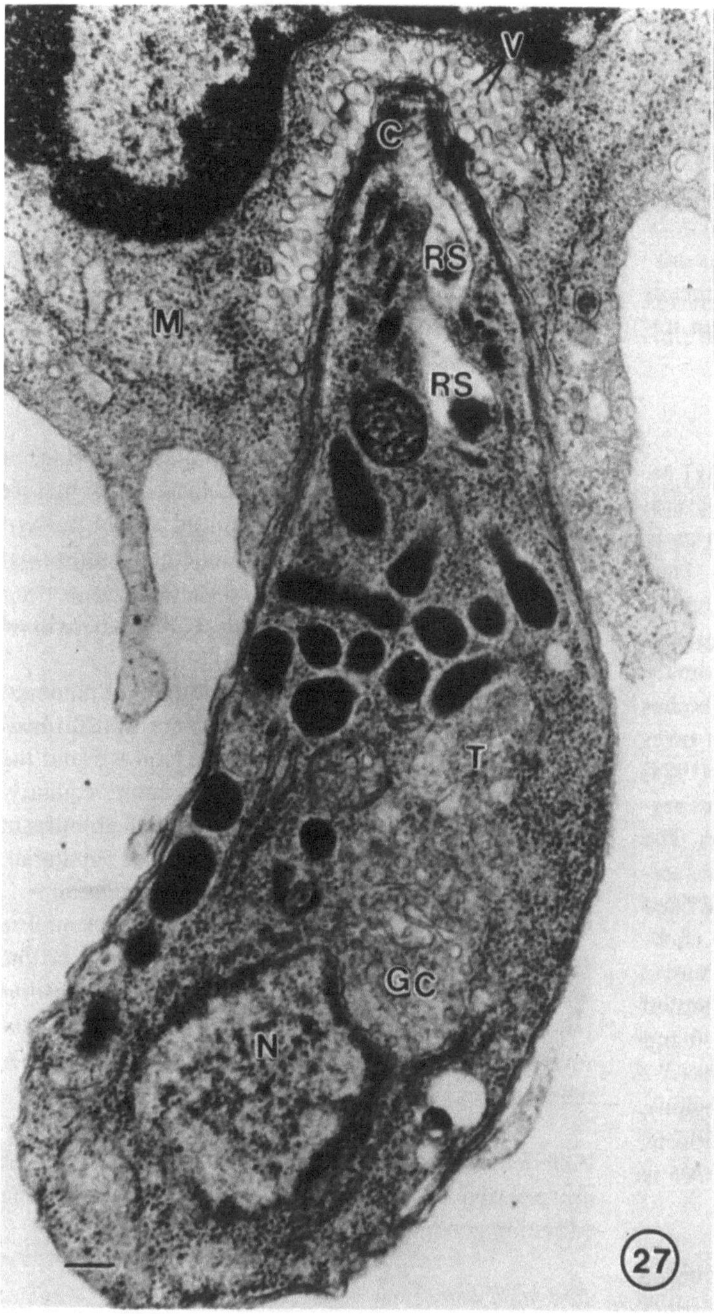

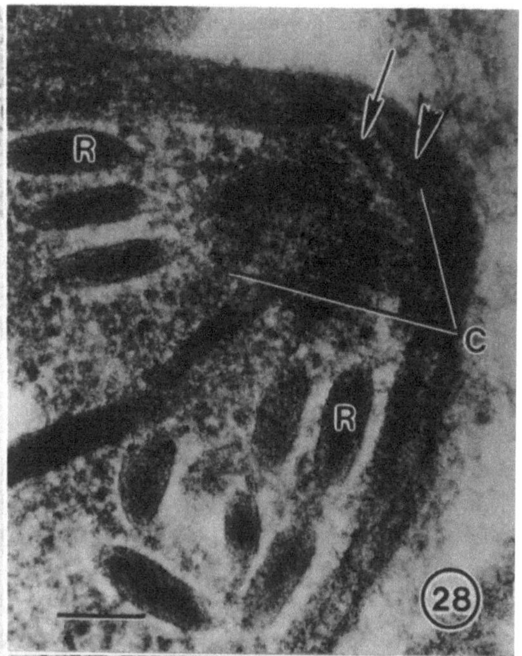

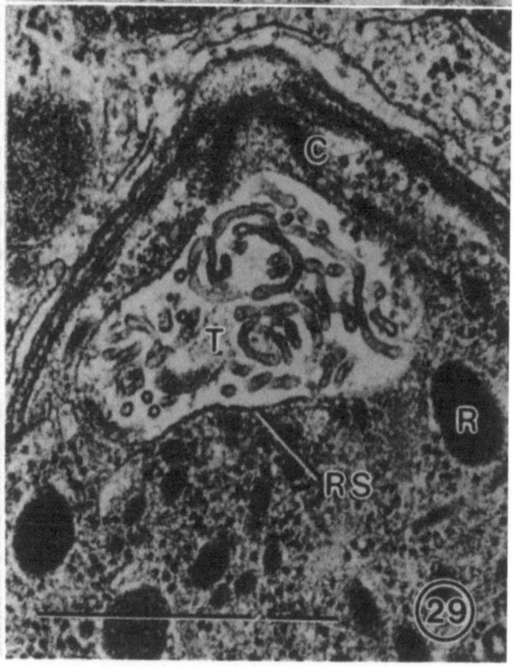

Fig. 27. A toxoplasma (*T*) invading a macrophage (*M*). Vesicles (*V*) are seen outside the toxoplasma. Rhoptries have transformed into short sac-like structures (*RS*) and some dense material remains inside. The conoid (*C*) is extended during invasion. *N* Nucleus, *GC* Golgi complex. (From Nichols et al. 1983)

Fig. 28. Tangential section through the conoid (*C*) showing spiral elements. Anterior (arrowhead) and posterior (arrow) conoidal rings are evident. Note the orientation of the rhoptries (*R*). (From Neill et al. 1989; electron micrograph provided by Dr. Pamela Neill)

Fig. 29. Sac-like rhoptry (*RS*) containing masses of convoluted tubules (*T*) resembling those commonly found outside the cell membrane of the invading cell. These may be the contents of the rhoptries that will contribute to the host cell membrane. *R* Undischarged rhoptry. (From Nichols et al 1983). Bars: 1 μm

Mitchell 1989, Nichols et al. 1983, Nichols and Chiappino 1987). The protein complex may form the membrane-like lamellae. The organization of rhoptry proteins into a lamellar structure could facilitate their integration into the host membrane. The insertion of the proteins may induce the formation of the parasitophorous vacuole membrane by expanding the erythrocyte membrane inwardly. As a consequence this mixture of host and parasite membrane may be a means by which the parasite avoids the initiation of an immunogenic response by the host.

K_2 bodies

Structure of K_2 bodies. Kinetosome (basal body) associated organelles are found in zoospores of several oomycetous protists, and have been designated as K bodies by Lehnen and Powell (1988, 1989, 1991). They have further been separated into K_1 and K_2 bodies based on whether they occur in the primary or secondary zoospores. Upon emergence from the primary cyst, secondary zoospores contain several K_2 bodies per cell, indicating that the K body formation takes place in the primary cyst. Lehnen and Powell (1988) demonstrated that K_2 bodies in *Saprolegnia* were regulated secretory organelles, similar to extrusomes. The K_2 body is particularly distinctive in *Saprolegnia* secondary zoospores. The matrix consists of a tubule filled hemispherical cavity containing carbohydrates (Lehnen and Powell 1988, 1989) and a granular matrix composed of proteins. The matrix is further separated into four distinctive and complex regions (Figs. 30 and 31). There is a region of coarse, clustered globules (A 2 region), a shell just beneath the membrane (A 3 region), a region containing microtubules (A 4), and a fibrous region (A 5). The precursor to regions A 4 and A 5 is A 1.

Biogenesis of K_2 bodies. The developmental sequence of K_2 bodies (Fig. 32) can be separated into five distinct stages (Lehnen and Powell 1991):

(1) Small precursor vesicles originate from smooth-surfaced fenestrated cisternae which are near the center of the cyst. The vesicles have coated outer surfaces while attached to the cisternae but lose the coat when they are released. The vesicles contain fine granular-fibrillar matrices, an electron-opaque globule and fragmented plates. Presumably the cisternae arise from endoplasmic reticulum or Golgi cisternae.

(2) The precursor vesicles are transported to the peripheral cytoplasm where they fuse with each other, forming larger vesicles called nascent K_2 bodies. The precursor vesicles consist of three regions of the matrix (A 1, A 2, A 3) and, after fusion, the three regions become more extensive and concentrated.

(3) Coated patches form on the cytoplasmic surface of nascent K_2 bodies after they have achieved their mature size, when the shell becomes organized. As the nascent K_2 bodies mature, the matrix undergoes additional modification and becomes more structured. A portion of these changes causes the nascent K_2 body to assume a spherical shape.

(4) The matrix configuration continues to undergo changes. Region A 1 apparently becomes transformed into two other regions, specifically region A 5 and the tubules of A 4. The tubules become more regularly arranged and thicker. The coarse, clustered globules in A 2 spread out between the shell (A 3) and A 5 material. The cell thickens due to addition of A 2 material.

(5) The nascent K_2 body matures and attains its mature size and spherical shape. It is transported back to the region where the flagellar rootlets develop. Microtubules probably participate in the relocation of the mature K_2 bodies. The involvement of actin in this movement also cannot be excluded.

Function of K_2 bodies. The contents of the K_2 bodies are released to form pads of material for attaching encysting zoospores.

Red algal mucocysts

Red algae are known for their production of large quantities of mucilage, particularly by their reproduc-

Fig. 30. Three K_2 bodies (*KB*) are near the center of the cell. Peripheral vesicles (*PV*) also have formed at this time. *N* Nucleus, *K* kinetosome (= basal body). (From Lehnen and Powell 1991; original electron micrograph provided by Dr. Martha Powell; reprinted by permission of The New York Botanical Garden)

Fig. 31. Mature K_2 bodies adjacent to the kinetosome (*K*). All regions except A 1 are distinguishable. *K* Kinetosome, *M* mitochondrion. (From Lehnen and Powell 1991; original electron micrograph provided by Dr. Martha Powell; reprinted by permission of The New York Botanical Garden): Bars: 1 μm

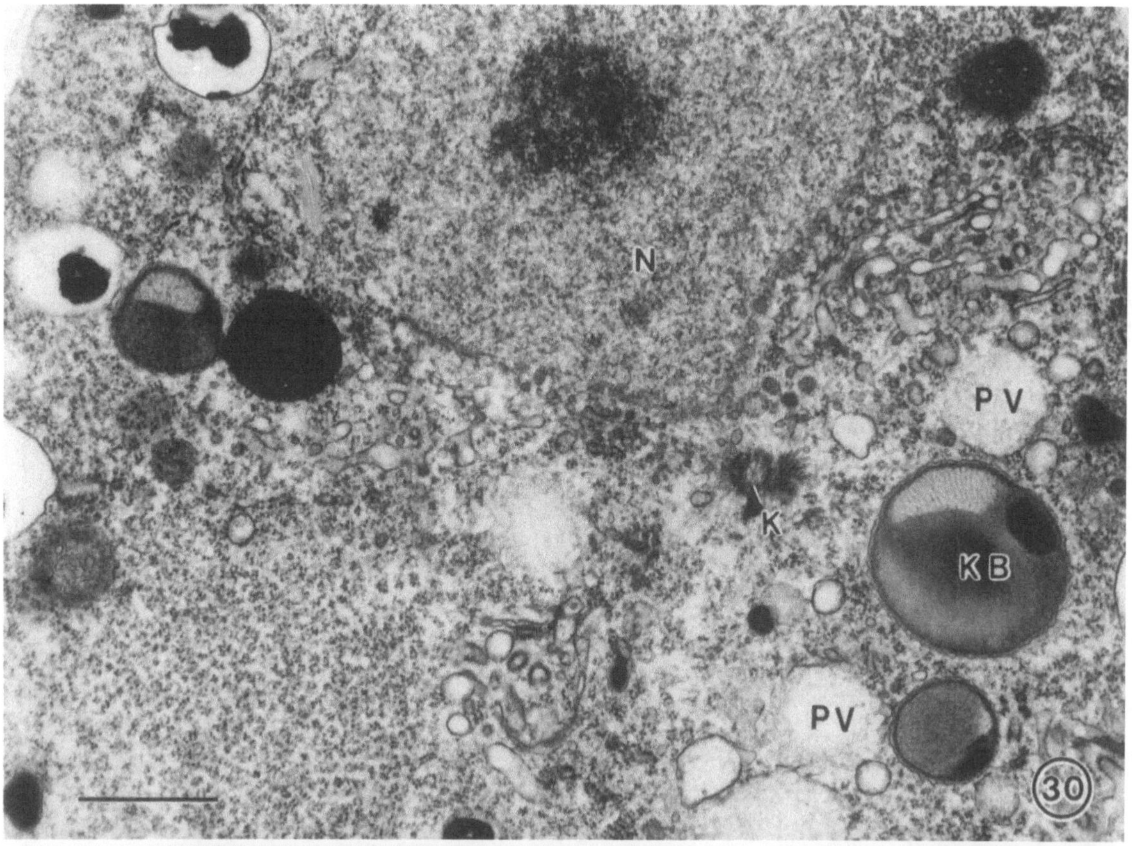
N
PV
K
KB
PV
30

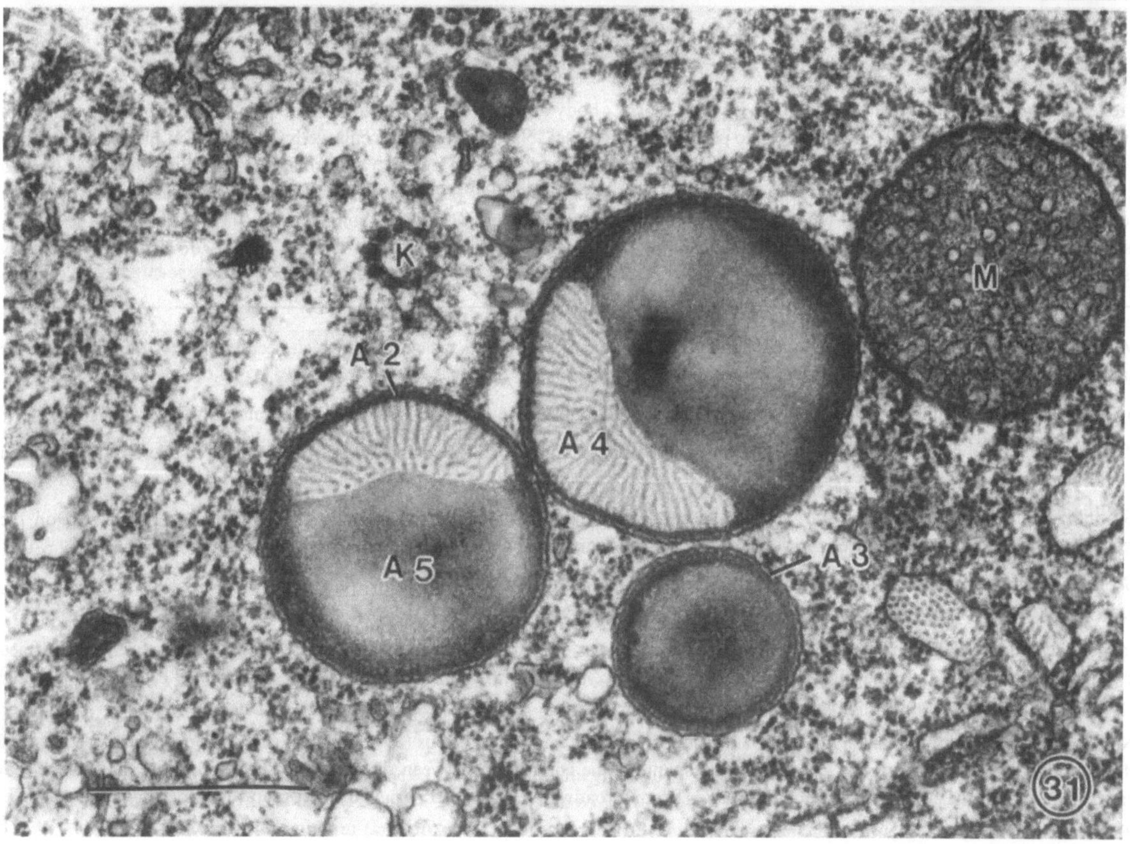
K
M
A 2
A 4
A 3
A 5
31

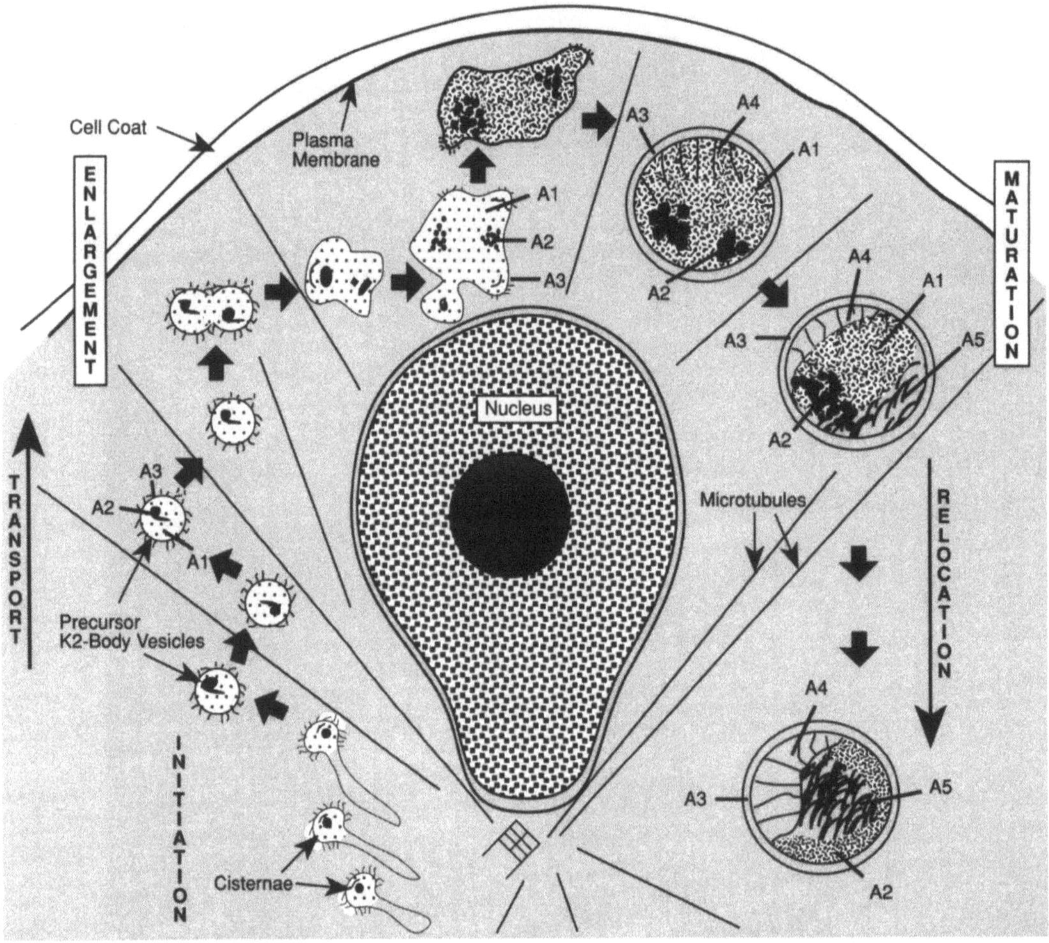

Fig. 32. Model of K_2 body formation in a primary cyst. Precursor K_2 body vesicles (containing zones A 1, A 2, and A 3) are initiated from smooth-surfaced, fenestrated cisternae and move to the peripheral cytoplasm. Nascent K_2 bodies enlarge by the fusion of precursor K_2 body vesicles. As nascent K_2 bodies mature, the matrix is additionally modified by morphological transformation of existing material. Coated patches are scattered on the cytoplasmic surface of nascent K_2 bodies when the shell (A 3) becomes organized. Later the nascent organelle loses the coated patches and two new zones (A 4, A 5) appear to arise from the material of the A 1 zone. During the maturation of the matrix, K_2 bodies are relocated to the center of the cell near the developing kinetosomes. This is the position where K_2 bodies are located in secondary zoospores. (After Lehnen and Powell 1991)

Fig. 33. Mature red algal spermatium (*Janczewskia*) showing the prominent spermatial vesicles (*SV*) which occupy most of the cell. Note the fibrous nature of the matrices. *N* Nucleus

Fig. 34. Spermatial vesicle forming by a fusion of endoplasmic reticulum cisternae (*ER*). *N* Nucleus, *SV* spermatial vesicle

Fig. 35. Phase contrast micrograph of *Spyridia* spermatium with six appendages. (From Broadwater et al. 1991; original micrograph provided by Dr. Joseph L. Scott)

Fig. 36. Scanning electron micrograph of two *Spyridia* spermatia with different lengths of appendages. Note the terminal knob on the appendages (arrows). (From Broadwater et al. 1991; original electron micrograph provided by Dr. Joseph L. Scott)

Fig. 37. *Spyridia* spermatium forming the early spermatial vesicle (*SV*) by fusion of endoplasmic reticulum. (From Broadwater et al. 1991; original micrograph provided by Dr. Joseph L. Scott)

Fig. 38. Tangential section through a mature spermatium of *Spyridia* just prior to spermatial vesicle release. The contents have been transformed into four spherical inclusions, but one lobe still contains concentric, fibrillar striations (arrow). (From Broadwater et al. 1991; original micrograph provided by Dr. Joseph L. Scott)

Fig. 39. Newly released spermatium, which has not rounded up, containing an apical nucleus (*N*) starch (*S*) and two appendages with terminal knobs (*A*). A bridge-like structure (arrow) can be detected spanning the connection between the stalk and the knob. Note the fibrous coat of the terminal knob and the fibrillar nature of the stalk. (From Broadwater et al. 1991; original micrograph provided by Dr. Joseph L. Scott). Bars: 1 μm

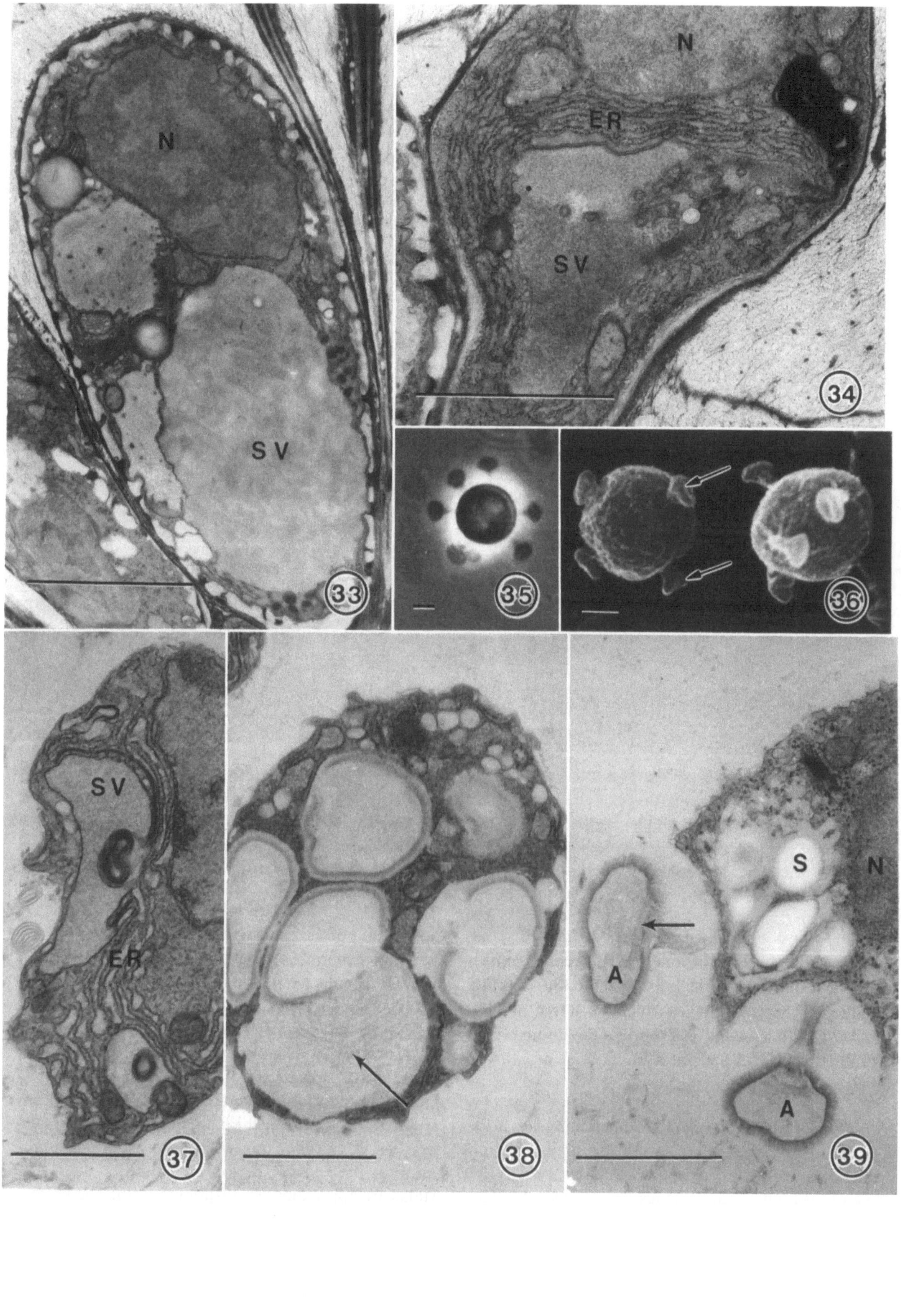
N
SV
33
N
ER
SV
34
35
36
SV
ER
37
38
S
N
A
A
39

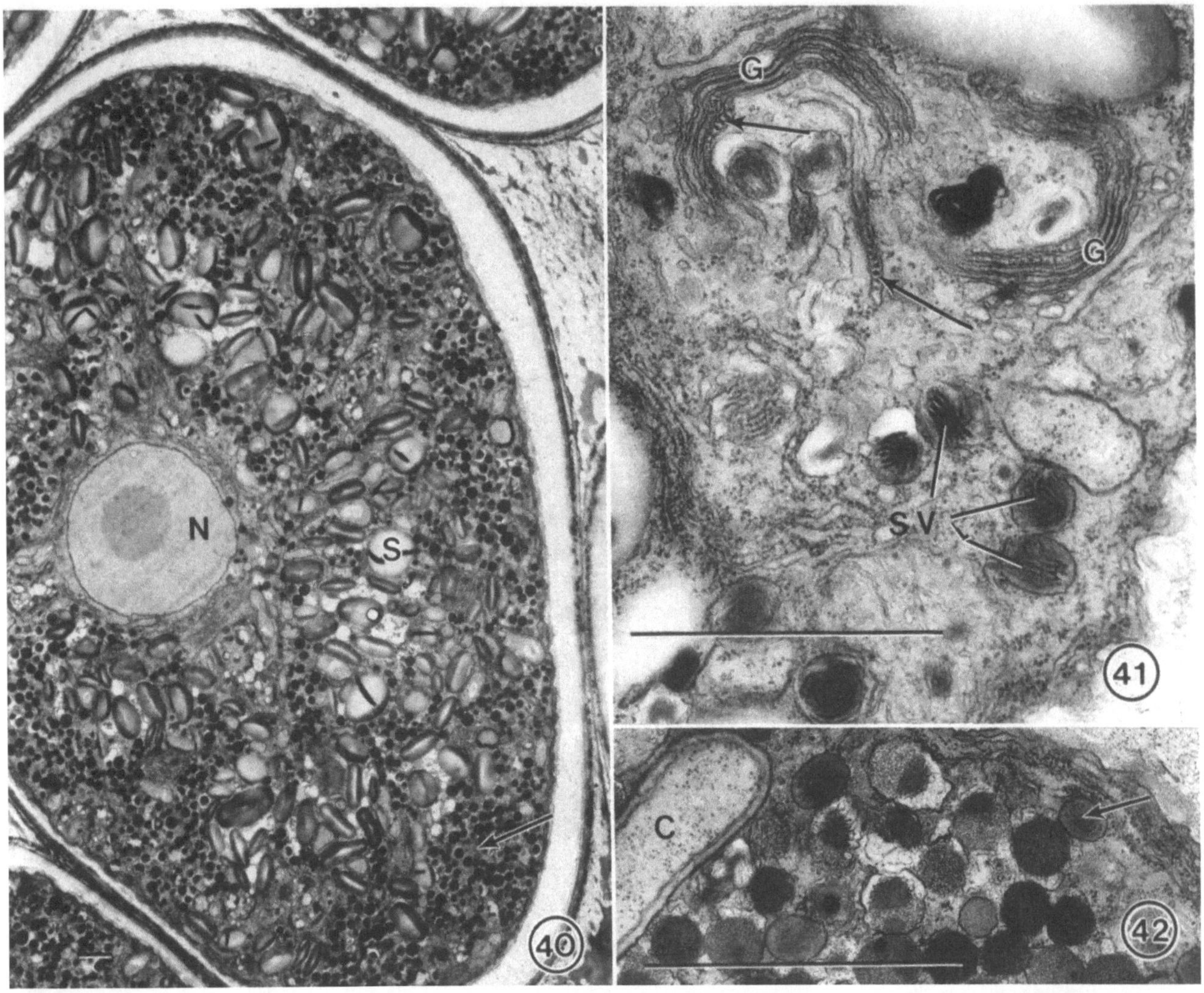

Fig. 40. Mature carpospore of the red algal parasite *Plocamiocolax* filled with large amounts of starch (*S*) and dark staining adhesive vesicles (arrow). *N* Nucleus

Fig. 41. Stage four Golgi complexes (*G*) showing the production of striated adhesive vesicles (*SV*). Striae are visible within the cisternae of the maturing faces (arrows). Note the dark-staining of the striated adhesive vesicles

Fig. 42. Section of a carpospore (*C*) showing mature adhesion vesicles. They are granular/fibrillar. Arrow indicates the striations still evident in one of the adhesive vesicles. Bars: 1 μm

tive cells. Until this time no one has designated some of these as extrusive organelles, but there are several mucilage producing organelles which are stored in the cytoplasm and released at appropriate developmental or germination stages and meet the criteria of extrusomes. Two types are discussed.

Red algal spermatial vesicles and appendages

Structure of spermatial vesicles. Spermatia are male gametes of red algae characterized by large posterior vesicles that contain mucilage which is extruded at the time of spermatial release (Figs. 33, 37, and 38). The appearance of the mature vesicle contents varies among different genera (Fetter and Neushul 1981, Kugrens 1980, Magruder 1984) and in most genera the released contents form elongate strands of mucilage while in others the contents may become highly structured. The most structured extruded mucilage was described in *Spyridia* (Broadwater et al. 1991). In *Spyridia* the extruded contents form structured spermatial appendages (Figs. 35, 36, and 39), which are not limited by a membrane. The extruded appendages consist of a terminal knob at the end of an elongated fibrillar stalk. The stalks are approximately 1.0 μm in length. The knob is also fibrillar but the exterior of the knob consists of a conspicuous fibrous coat (Fig. 39). The entire released spermatium has the appearance of a "sputnik" satellite (Broadwater et al. 1991). However, in most red algal spermatia the contents generally are fibrillar (Fig. 33), and they may be compressed to appear convoluted or they may appear as distinct hemispherical inclusions.

Biogenesis of spermatial vesicles. Initially, young spermatia are spherical and the nucleus is centrally situated in the spermatogenous cell. When endoplasmic reticulum accumulates at the basal end, it causes an elongation of the cell. Spermatial vesicles form from a fusion of endoplasmic reticulum (Figs. 34 and 37) and they continue to enlarge by a continued fusion of endoplasmic reticulum cisternae. The contents of the spermatial vesicle undergo several morphological changes (Figs. 37 and 38). At first, the spermatial vesicles appear granular and uniform in content (Figs. 34 and 37). The granular appearance of the contents is transformed into light and dark bands and the contents become fibrillar. The fibrillar vesicles represent mature vesicles which will be extruded upon spermatial release.

In *Spyridea* the vesicular contents become concentric (Fig. 35). The latter change is associated with the development of large spherical regions within the spermatial vesicle which have electron-translucent centers and moderately dark staining fibrous coats. Mature spermatia contain 6–8 of these spherical inclusions. Immediately after spermatial release, the spherical inclusions are extruded (Figs. 35, 36, and 39), thus forming the external spermatial appendages which have prominent knobs at their tips.

Function of spermatial vesicles appendages. Multiple functions have been postulated for spermatial vesicles. It is currently theorized that spermatial vesicles may produce spermatial coats, they may play a role in spermatial expulsion by the hydration of contents (Kugrens 1980, Magruder 1984, Mine and Tatewaki 1994) or they may serve to attach spermatia to female plants (Fetter and Neushul 1981, Magruder 1984) or they may function in attachment of spermatia to the trichogyne of the female gamete or carpogonium (Kugrens 1980, Magruder 1984, Mine and Tatewaki 1994). Another possible function was suggested by West and Calumpong (1989). They proposed that the appendages could change the specific gravity, shape and surface area of *Spyridia* spermatia, thus affecting and enhancing their movement in the water column.

Red algal adhesion vesicles

Structure of adhesion vesicles. In many of the higher red algae that have been studied, reproductive spores such as carpospores and tetraspores contain dark-staining vesicles (Figs. 40–42) termed adhesion vesicles (Pueschel 1979, Kugrens and Delivopoulos 1986). They are small and spherical, and they contain compacted material. These vesicles are stored in clusters throughout the spore cytoplasm until spore germination (Fig. 40).

Biogenesis of adhesion vesicles. During the latter stages of spore formation, a characteristic type of Golgi apparatus is formed which produces vesicles with a structured matrix (Fig. 41). Depending on the genus, either cored vesicles or striated vesicles are produced. These vesicles measure approximately 0.2–0.3 μm in diameter. Following their detachment from the forming face of the Golgi apparatus, their striated/fibrillar matrix becomes granular (Fig. 42). The granular material compacts, reducing the size of the vesicles, and in the process they become darker staining (electron opaque).

Functions of adhesion vesicles. The vesicles are stored within the cytoplasm until spore germination. It has been proposed that the vesicular contents are released at the time of spore germination to produce adhesive material for spore attachment (Pueschel 1979), or they may represent a compacted wall material to allow for rapid wall deposition during the initial phases of germination (Kugrens and Delivopoulos 1986). This is particularly advantageous in parasitic types where rapid growth during penetration is required.

Evolutionary considerations using extrusomes

Any phylogenetic considerations are handicapped by the fact that we do not know whether similar extrusomes are homologous, and we rely strictly on structural aspects to determine possible evolutionary relationships. The major argument favoring the use of similar types of extrusomes in phylogeny is that it is difficult to imagine that such similar structures could have arisen independently in diverse groups. Consequently, certain extrusomes have been used to imply affinities among groups of protists. For instance, the similarity in trichocyst and mucocyst structure (Hausmann 1978; Dodge and Gruet 1987; Taylor 1987, 1990) could be used to imply a relationship between the ciliates and dinoflagellates, even suggesting that dinoflagellates may be ancestral to the ciliates (Corliss 1972, 1988; Lee and Kugrens 1992).

Ejectisomes occur in all cryptoprotists (Kugrens and Lee 1991), some chloroprotists (Morrall and Greenwood 1981) and in the colorless flagellates *Katablepharis* (Lee and Kugrens 1991) and *Leucocryptos* (Vørs 1992). There are some slight structural and post-release developmental differences (cf. Fig. 8); nevertheless, the similar structure and development of these organelles may indicate a relationship among these three groups. Manton (1968) suggested that a completely independent origin for the cryptoprotists and *Pyramimonas* seemed improbable but she also cautioned about the use of similarities in phyletic speculation. Taylor (1976, 1978) stressed that ejectisome similarity was evidence of affinity. Conversely, Morrall and Greenwood (1980) doubted the relationship between the cryptoprotists and *Pyramimonas* based on what they perceived as significant structural differences in the ejectisomes. As a result they suggested that these two similar appearing structures may have evolved separately. However, molecular data (Andersen 1992, Douglas et al. 1991) indicate that the nuclear encoded rRNA in cryptoprotists apparently is related to the green algal lineage. Thus ejectisome similarity appears to be a valid indicator of a relationship between the two groups. When molecular data become available, it will be interesting to see whether *Katablepharis* and *Leucocryptos* also have some phylogenetic affinities with the cryptoprotists and/or chloroprotists, perhaps representing the ancestral types for either of these groups.

Robb and Lee (1986) suggest that the attack mechanism of *Haptoglossa mirabilis* apparently is similar to the microsporidian polar tube and the Stachel-Rohr-Schlauch attack apparatus in the plant parasite *Plasmodiophora* (Myxomycota). The microsporidian spore also contains a vacuolar system which is closely analogous in structure, origin and presumed functions to the attack mechanism (Vávra 1978). The similarities in the extrusion apparatuses probably represent an analogous situation since these organisms are, at best, distantly related.

Similarities with the Plasmodiophorales, likewise, are based on the attack apparatus which consists of an electron dense, bullet-like structure (Stachel), which is formed within a tube (Rohr) that also possesses a short, narrow tail (Schlauch) inside the zoospore cyst. According to Robb and Lee (1986), these obvious similarities in structure and function of the attack apparatus, in apparently diverse parasites, might reflect an evolutionary relationship.

Microsporidian systematics is based primarily on the variations in polar filaments and their associated structures. Some of these variations were noted in this review. Perhaps these differences in the structure and development of polar filaments could be useful in determining phylogenetic relationships among the microsporidians, pending the accumulation of a comprehensive molecular data base.

Conclusions

With the exception of ejectisomes, this review has introduced several types of extrusomes that have not been reviewed as extrusomes elsewhere and as a result, the list of organelles that can be considered as extrusomes and their structural diversity has been expanded. One of the most significant departures from Hausmann's review is the inclusion of the complex extrusive organelles of parasitic protists and the mucocysts of red algae and oomycetous protists. The accumulated information on different invasion apparatuses should be informative and useful to individuals who are not acquainted with protistan parasitology. Microsporidians were emphasized in this paper because they occupy a pivotal position in phylogeny, and they are becoming increasingly important in human affairs. Unlike the extrusomes in non-parasitic protists, the functions of most of the extrusive organelles described in this review are well established and in some cases detailed biochemical information is available to supplement visual evidence.

In this review we were unable to include the wide range of extrusomes currently known. For instance, we excluded conocysts of the ciliate *Loxophyllum*, crystallocysts of the ciliate *Conchophthirus*, pigmentocysts of certain heterotrich ciliates, the unique extrusomes of karyorelictid ciliates (see review by Raikov 1993), sagenogens of the labyrinthuloid protists, scale-containing vesicles of many protists (especially algal flagellates but also one cilioprotist species and several sarcodinoid amoebae), solenocysts – in addition to the well-known haptocysts – in the tentacles of some suctorian ciliates (Benwitz 1984), the Stachel (penetrating organelle) of certain plasmodiophorans, polar capsules of the Myxozoa (Lom 1990; Lom et al. 1982; Sprague 1982b; Supamattaya et al. 1991, 1993; Perkins 1991; Yamamoto and Sanders 1979), and stalk-producers in several ciliates and some pigmentless flagellates. General references to the literature on these organelles may be found in the references provided or in the combined bibliographies of appropriate chapters in the following three compendia: Harrison and Corliss (1991), Lee et al. (1985) and Margulis et al. (1990). For additional information on extrusomes that were discussed by Hausmann, we refer the reader to publications by Adoutte (1988), Lynn and Corliss (1991), Perkins (1991), and Vickerman et al. (1991). There is a considerable opportunity for additional research on extrusive organelles and, hopefully, this paper has provided some information which will provide impetus for additional research on these unusual and exciting organelles in protists. Many of the extrusomes listed in this paragraph, but not presented in this paper, nor discussed by Hausmann, perhaps could form the basis of a future review.

Acknowledgements

We would like to express our gratitude to the following individuals who granted their permission to reprint their electron micrographs or who supplied original micrographs or negatives for use in this chapter: Dr. James C. Becnel, USDA/ARS, Medical and Veterinary Research Laboratory, Gainesville, Fl.; Dr. Ann Cali, Rutgers University; Dr. Barbara A. Nichols, F.I. Proctor Foundation for Research in Ophthalmology, University of California, San Francisco; Dr. Pamela Neil, Baylor College of Medicine; Dr. Robert L. Owen, M.D., University of California, San Francisco; Dr. Martha J. Powell, Miami University of Ohio; Dr. Jane Robb, Guelph University; and Dr. Joseph C. Scott, College of William and Mary. Permission to use various micrographs in this publication by the following journals and publishers is gratefully acknowledged: The Journal of Protozoology/Eukaryotic Microbiology (Figs. 16, 17, 22–24, and 28); Mycologia/The New York Botanical Garden (Figs. 30 and 31); Canadian Journal of Botany/Research Council of Canada (Fig. 14); Academic Press (Figs. 27 and 29); and Phycologia/Blackwell Scientific Publications (Figs. 35–39). We also would like to thank Mr. Brec Clay and Mr. Chuck Meyer who helped in numerous ways during the latter stages of manuscript preparation.

References

Adoutte A (1988) Exocytosis: biogenesis, transport and secretion of trichocysts. In: Görtz H-D (ed) *Paramecium*. Springer, Berlin Heidelberg New York Tokyo, pp 325–362

Allen RD (1988) Cytology. In: Görtz H-D (ed) *Paramecium*. Springer, Berlin Heidelberg New York Tokyo, pp 4–40

Allman GJ (1855) On the occurrence among the infusoria of peculiar organs resembling thread cells. Q J Microsc Sci 3: 177–179

Andersen RA (1992) Diversity of eukaryotic algae. Biodivers Conserv 1: 267–292

Azvedo C (1987) Fine structure of the microsporidian *Abelspora portulacalensis* gen. n. sp. n. (Microsporida) parasite of the hepatopancreas of *Carcinus maenas* (Crustacea, Decapoda). J Invertebr Pathol 49: 83–92

Bannister LH, Mitchell GH (1989) The fine structure of secretion by *Plasmodium knowlesi* merozoites during red cell invasion. J Protozool 36: 362–367

Barron GL (1980) A new *Haptoglossa* attacking rotifers by rapid injection of an infective sporidium. Mycologia 72: 1186–1194

Bauer LS, Pankratz HS (1993) *Nosema scripta* n.sp. (Microsporidia: Nosematidae), a microsporidian parasite of the cottonwood leaf beetle, *Chrysomela scripta* (Coleoptera: Chrysomelidae). J Protozool 40: 135–141

Beard BB, Butler JF, Becnel JJ (1990) *Nolleria pulicis* n.gen., n.sp. (Microsporidia: Chytridiopsidae), a microsporidian parasite of the cat flea, *Ctenocephalides felis* (Siphonaptera: Pulicidae). J Protozool 37: 90–99

Becnel JJ, Hazard EI, Fukuda T (1986) Fine structure and development of *Pilosporella chapmani* (Microspora: Telohaniidae) in the mosquito, *Aedes triseriatus* (Say). J Protozool 33: 60–66

Benwitz G (1984) Die Entladung der Haptocysten von *Ephelota gemmipara* (Suctoria, Ciliata). Z Zellforsch 39 c: 812–817

Broadwater ST, Scott JL, West JA (1991) Spermatial appendages of *Spyridia filamentosa* (Ceramiaceae, Rhodophyta). Phycologia 30: 189–195

Brugerolle G (1985) Des trichocytes chez bodonidés, un caractère phylogénétique supplémantaire entre Kinetoplastida et Euglenida. Protistologica 21: 339–348

Bryan RT, Cali A, Owen RL, Spencer KC (1990) Microsporidia. Opportunistic pathogens in patients with AIDS. Prog Clin Parasitol 2: 1–26

Cali A (1991) General microsporidian features and recent findings on AIDS isolates. J Protozool 38: 625–630

– Owen RL (1988) Microsporidiosis. In: Balows A, Hauser WJ, Lennette EH (eds) The laboratory diagnosis of infectious diseases: principles and practice, vol 1. Springer, Berlin Heidelberg New York Tokyo, pp 929–950

– – (1990) Intracellular development of *Enterocytozoon*, a unique microsporidian found in the intestine of AIDS patients. J Protozool 37: 145–155

– Kotler DP, Orenstein JM (1993) *Septata intestinalis* n.g, n.sp., and intestinal microsporidian associated with chronic diarrhea and dissemination in AIDS patients. J Protozool 40: 101–112

– Orenstein JM, Kotler DP, Owen RL (1991) A comparison of two microsporidian parasites in enterocytes of AIDS patients with chronic diarrhea. J Protozool 38: 96S–98S

Canning EU (1990) Phylum Microspora. In: Margulis L, Corliss JO, Melkonian M, Chapman DJ (eds) Handbook of Protoctista. Jones and Bartlett, Boston, pp 53–72

– Hollister WS (1992) Human infections with microsporidia. Rev Med Microbiol 3: 35–42

– Lom J (1986) The microsporidia of vertebrates. Academic Press, New York

Cooper JA, Ingram LT, Fardoulys CA, Stenzel D, Schofield L, Saul LJ (1988) The 140/105 kilodalton protein complex in the rhoptries of *Plasmodium falciparum* consists of discrete polypeptides. Mol Biochem Parasitol 29: 251–260

Corliss JO (1972) The ciliate protozoa and other organisms: some unresolved questions of major phylogenetic significance. Am Zool 12: 739–753

– (1979) The ciliated protozoa: characterization, classification and guide to the literature, 2nd edn. Pergamon, London

– (1988) The quest for the ancestor of the Ciliophora: a brief review of the continuing problem. BioSystems 7: 338–349

– Lom J (1985) An annotated glossary of protozoological terms. In: Lee JJ, Hutner SH Bovee EC (eds) An illustrated guide to the Protozoa. Society of Protozoologists, Lawrence, KS, pp 576–602

Didier PJ, Didier ES, Orenstein JM, Shadduck JA (1991) Fine structure of a new human microsporidia, *Encephalitozoon hellem*, in culture. J Protozool 38: 502–507

Dodge JD, Gruet C (1987) Dinoflagellate ultrastructure and complex organelles. In: Taylor FJR (ed) The biology of dinoflagellates. Blackwell, Oxford, pp 93–142

Douglas SE, Murphy CA, Spencer DF, Gray MW (1991) Cryptomonad algae are evolutionary chimaeras of two phylogenetically distinct unicellular eukaryotes. Nature 350: 148–151

Dragesco J (1984) Organites sous-cuticulaires éjectables: éjectisomes (trichocystes). In: Grassé P-P (ed) Traité de Zoologie, vol 2, fasc 1. Masson, Paris, pp 181–199

Ellis J (1969) Observations on a particular manner of increase in the animalcula of vegetable infusions. Philos Trans Soc Lond [Biol] 59: 138–152

Eperon S, Vigues B, Peck RK (1993) Immunological characterization of trichocyst proteins in the ciliate *Pseudomicrothorax dubius*. J Eukar Microbiol 40: 81–91

Etzion Z, Murray MC, Perkins ME (1991) Isolation and characterization of rhoptries of *Plasmodium falciparum*. Mol Biochem Parasitol 47: 51–62

Fetter R, Neushul M (1981) Studies on the development and released spermatia in the red alga *Tiffaniella snyderae* (Rhodophyta). J Phycol 17: 141–159

Friedberg DN, Stenson SM, Orenstein JM, Charles N (1990) Microsporidial keratoconjunctivities in the acquired immunodeficiency syndrome. Arch Ophthalmol 108: 504–508

Frixione E, Ruiz L, Santillan M, de Vargas LV, Tajero JM (1992) Dynamics of polar filament discharge and sporoplasm expulsion by microsporidian spores. Cell Motil Cytoskeleton 22: 38–50

Garofalo RS, Satir BH (1984) *Paramecium* secretory granule content: quantitative studies on in vitro expansion and its regulation by calcium and pH. J Cell Biol 99: 2193–2199

Glas-Albrecht R, Plattner H (1990) High yield isolation procedure for intact secretory organelles (trichocysts) from different *Paramecium tetraurelia* strains. Eur J Cell Biol 53: 164–172

Grim JN, Staehelin LA (1984) The ejectisomes of the flagellate *Chilomonas paramecium*: visualization by freeze-fracture and isolation techniques. J Protozool 3: 259–267

Harrison FW, Corliss JO (eds) (1991) Protozoa. Wiley-Liss, New York [Harrison FW (ed) Microscopic anatomy of invertebrates, vol 1]

Hausmann K (1978) Extrusive organelles in protists. Int Rev Cytol 52: 197–276

– Fok AK, Allen RD (1988) Immunocytochemical analysis of trichocyst structure and development in *Paramecium*. J Ultrastruct Mol Struct Res 99: 213–225

Heywood P, Rothschild LJ (1987) Reconciliation of evolution and nomenclature among the higher taxa of protists. Biol J Linn Soc 30: 91–98

Hibberd DJ (1970) Observations on the cytology and ultrastructure of *Ochromonas tuberculatus* sp.nov. (Chrysophyceae), with special reference to the discobolocysts. Br Phycol J 5: 119–143

Hilenski LL, Walne PL (1983) Ultrastructure of mucocysts in *Peranema trichophorum* (Euglenophyceae). J Protozool 30: 491–496

Hovasse R, Mignot JP (1975) Trichocystes et organites analogues chez les protistes. Ann Biol 14: 397–422

Jensen JB, Edgar SA (1976) Possible secretory function of the rhoptries of *Eimeria magna* during penetration of cultured cells. J Parasitol 62: 988–992

Kimata I, Tanabe K (1978) Secretion by *Toxoplasma gondii* of an antigen that appears to become associated with parasitophorous vacuole membrane upon invasion of the host cell. J Cell Sci 88: 231–239

Krüger F (1936) Die Trichocysten der Ciliaten im Dunkelfeldabbild. Zoologica 34: 1–82

Kugrens P (1980) Electron microscopic observations on the differentiation and release of spermatia in the marine red alga *Polysiphonia hendryi* (Ceramiales, Rhodomelaceae). Amer J Bot 67: 519–528

– Delivopoulos SG (1986) Ultrastructure of the carposporophyte and carposporogenesis in the parasitic red alga *Plocamiocolax pulvinata* Setch (Gigartinales, Plocamiaceae). J Phycol 22: 8–21

– Lee RE (1991) Organization of cryptoprotists. In: Patterson DJ, Larsen J (eds) The biology of free-living heterotrophic flagellates. Clarendon Press, Oxford, pp 219–233 (Systematics Association special volume 45)

Larsson JIR (1986) Ultrastructure, function and classification of microsporidia. Prog Protistol 1: 325–390

– (1992) The light and electron microscopic cytology of *Iancekia adpophila* n.sp. (Microspora, Tuzetiidae), a microsporidian parasite of *Ptychoptera paludosa* (Diptera, Ptychopteridae). J Protozool 39: 432–440

– (1993) Description of *Chytridiopsis trichopterae* n.sp. (Microspora, Chytridiopsidae), a microsporidan parasite of the caddis fly *Polycentrophus flavomaculatus* (Trichoptera, Polycentropodidae), with comments on relationships between families Chytridiopsidae and Metchnikovellidae. J Protozool 40: 37–48

Lee JJ, Hutner SH, Bovee EC (eds) (1985) An illustrated guide to the Protozoa. Society of Protozoologists, Lawrence, KS

Lee RE, Kugrens P (1991) *Katablepharis ovalis*, a colorless flagellate with interesting cytological characteristics. J Phycol 27: 505–513

– – (1992) Relationship between the flagellates and the ciliates. Microbiol Rev 56: 529–542

Lehnen LP, Powell MJ (1988) Cytochemical localization of carbohydrates in zoospores of *Saprolegnia ferax*. Mycologia 80: 423–432

– – (1989) The role of kinetosome associated organelles in the attachment of secondary zoospores of *Saprolegnia ferax* to substrates. Protoplasma 149: 163–174

– – (1991) Formation of K_2 bodies in primary cysts of *Saprolegnia ferax*. Mycologia 8: 163–179

Leitch GJ, Qing H, Wallace S, Visvesvara GS (1993) Inhibition of the spore polar filament extrusion of the microsporidium, *Encephalocytozoon hellem*, isolated from an AIDS patient. J Eukar Microbiol 40: 711–717

Leriche MA, Dubremetz JF (1991) Characterization of the protein contents of rhoptries by subcellular fractionation and monoclonal antibodies. Mol Biochem Parasitol 45: 249–260

Levine ND (1982) Apicomplexa. In: Parker SP (ed) Synopsis and classification of living organisms. McGraw-Hill, New York, pp 570–587

– (1985) Phylum II. Apicomplexa Levine, 1970. In: Lee JJ, Hutner SH, Bovee EC (eds) An illustrated guide to the Protozoa. Society of Protozoologists, Lawrence, KS, pp 322–374

– (1988) The protozoan Phylum Apicomplexa, vol I. CRC Press, Boca Raton

Lima O, Gulik-Krzywicki T, Sperling L (1989) *Paramecium* trichocysts isolated with their membranes are stable in the presence of millimolar Ca^{2+}. J Cell Sci 93: 557–564

Lom J (1990) Phylum Myxozoa. In: Margulis L, Corliss JO, Melkonian M, Chapman DJ (eds), Handbook of Protoctista. Jones and Bartlett, Boston, pp 36–52

– Dyková I, Lhotáková S (1982) Fine structure of *Sphaerospora renicola* Dyková & Lom, 1982. A myxosporean from carp kidney and comments on the origin of pansporoblasts. Protistologica 19: 489–502

Lumpert CJ, Kersken H, Plattner H (1990) Cell surface complexes (corticles) isolated from *Paramecium tetraurelia* cells as a model system for analyzing exocytosis in vitro in conjunction with microinjection studies. Biochem J 269: 639–645

Lynn DH, Corliss JO (1991) Phylum Ciliophora. In: Harrison FW, Corliss JO (eds) Protozoa. Wiley-Liss, New York, pp 333–467 [Harrison FW (ed) Microscopic anatomy of invertebrates, vol 1]

Magruder WH (1984) Specialized appendages on spermatia from the red alga *Aglaothamnion neglectum* (Ceramiales, Ceramiaceae) specifically bind with trichogynes. J Phycol 20: 436–440

Manton I (1968) Tubular trichocysts in a species of *Pyramimonas parkeae*. Oesterr Bot Z 116: 378–397

Margulis L, Corliss JO, Melkonian M, Chapman DJ (eds) (1990) Handbook of Protoctista. Jones and Bartlett, Boston

Mine I, Tatewaki M (1994) Attachment and fusion of gametes during fertilization of *Palmaria* sp. (Rhodophyta). J Phycol 30: 55–66

Mitchell MJ, Cali A (1993) Ultrastructural study of the development of *Vairimorpha necatrix* (Kramer 1965) (Protozoa, Microsporida) in larvae of the corn earworm, *Heliothis zea* (Boddie) (Lepidoptera, Noctuidae) with emphasis on sporogony. J Eukar Microbiol 40: 701–710

Moestrup Ø (1984) Further studies on *Nephroselmis* and its allies (Prasinophyceae). II. *Mamiella* gen.nov., Mamiellaceae fam.nov., Mamiellales ord.nov. Nord J Bot 4: 109–121

Moore CB, Brooks WM (1992) An ultrastructural study of *Vairimorpha necatrix* (Microspora, Microsporida) with particular reference to episporontal inclusions during octosporogony. J Protozool 39: 392–398

Morrall S, Greenwood AD (1980) A comparison of the periodic structures of the trichocysts of the Cryptophyceae and Prasinophyceae. BioSystems 12: 71–84

Navdeep SJ, Rozario C, Ridley RG, Perkins ME (1993) Biogenesis of rhoptry organelles in *Plasmodium falciparum*. Mol Biochem Parasitol 57: 269–280

Neill PJG, Smith JH, Box ED (1989) Pathogenesis of *Sarcocystis falculata* (Apicomplexa : Sarcocystidae) in the budgerigar (*Melopsittacus undulatus*). IV. Ultrastructure of developing, mature and degenerating sarcocysts. J Protozool 36: 430–437

Nichols BA, Chiappino ML (1987) Cytoskeleton of *Toxoplasma gondii*. J Protozool 34: 217–226

– – O'Conner GR (1983) Secretion from the rhoptries of *Toxoplasma gondii* during host-cell invasion. J Ultrastruct Res 83: 85–98

Orenstein JM (1991) Microsporidiosis in the acquired immunodeficiency syndrome. J Parasitol 77: 843–864

Ossorio PN, Schwartzman JD, Boothroyd JC (1992) A *Toxoplasma gondii* protein associated with host cell penetration has unusual discharged symmetry. Mol Biochem Parasitol 50: 1–16

Parker SP (ed) (1982) Synopsis and classification of living organisms, vols. 1, 2. McGraw-Hill, New York

Perkins FO (1991) "Sporozoa": Apicomplexa, Microsporidia, Paramyxea, Myxosporidia and Actinosporidia. In: Harrison FW, Corliss JO (eds) Protozoan. Wiley-Liss, New York, pp 261–331 [Harrison FW (ed) Microscopic anatomy of invertebrates, vol 1]

Perkins ME (1992) Rhoptry organelles of apicomplexan parasites. Parasitol Today 8: 28–32

Peterson JB (1991) Small GTP-binding proteins associated with secretory vesicles of *Paramecium*. J Protozool 38: 495–501

– Heuser JE, Nelson DL (1987) Dissociation and reassociation of trichocyst proteins: biochemical and ultrastructural studies. J Cell Sci 87: 3–25

Pueschel CM (1979) Ultrastructure of tetrasporogenesis in *Palmaria palmata* (Rhodophyta). J Phycol 15: 409–424

Pitelka DR (1963) Electron microscopic structure of Protozoa. Pergamon, London

Pleshinger J, Weidner E (1985) The microsporidian spore invasion tube. IV. Discharge activation begins with pH triggered Ca^{2+} influx. J Cell Biol 100: 1834–1838

Powell MJ, Lehnen LP, Bortnick RN (1985) Microbody-like organelles as taxonomic markers among oomycetes. BioSystems 18: 321–334

Preparata RM, Meyer EB, Preparata FB, Simon EM, Vossbrink CR, Nanney DL (1989) Ciliate evolution: the ribosomal phylogenies of tetrahymenine ciliates. J Mol Evol 28: 427–441

Raikov IB (1993) The orthonematocyst, a new type of extrusome found in *Remanella rugosa* and *Remanella brunnea* (Ciliophora: Karyorelictida). Eur J Protistol 29: 81–87

Robb J, Barron GL (1982) Nature's ballistic missile. Science 218: 1221–1222

– Lee B (1986 a) Developmental sequence of the attack apparatus of *Haptoglossa mirabilis*. Protoplasma 135: 102–111

– – (1986 b) Ultrastructure of mature and fired gun cells of *Haptoglossa mirabilis*. Can J Bot 64: 1935–1947

Roger N, Dubremetz JF, Delplace P, Fortier B, Tronchin B, Vernes A (1988) Characterization of 225-kilodalton rhoptry protein of *Plasmodium falciparum*. Mol Biochem Parasitol 27: 135–142

Safer LD, Schwartzman JD (1991) A soluble phospholipase of *Toxoplasma gondii* associated with host cell penetration. J Protozool 38: 454–460

Sam-Yellowe TY (1992) Molecular factors responsible for host cell recognition and invasion in *Plasmodium falciparum*. J Protozool 39: 181–189

– Ndengele MM (1993) Monoclonal antibody epitope mapping of *Plasmodium falciparum* rhoptry proteins. Exp Parasitol 76: 46–58

– Perkins ME (1981) Interaction of the 140/130/110 kDa rhoptry protein complex of *Plasmodium falciparum* with the erythrocyte membrane and liposomes. Exp Parasitol 73: 161–171

– – (1990) Binding of *Plasmodium falciparum* rhoptry proteins to mouse erythrocytes and their possible role in invasion. Mol Biochem Parasitol 39: 91–100

– – (1991) Interaction of the 140/130/110 kDa rhoptry protein complex of *Plasmodium falciparum* with erythrocyte membrane and liposomes. Exp Parasitol 73: 161–171

– Shio H, Perkins ME (1988) Secretion of *Plasmodium falciparum* rhoptry protein into the plasma membrane of host erythrocytes. J Cell Biol 106: 1507–1513

Scarborough-Bull A, Weidner E (1985) Some properties of discharged *Glugea hertwigi* (Microsporidia) sporoplasms. J Protozool 32: 284–289

Shadduck JA (1989) Human microsporidiosis in AIDS. Rev Infect Dis 11: 203–207

– Meccoli RA, Davis R, Font RL (1989) First isolation of a microsporidian from a human patient. Infect Dis 162: 773–776

Sprague V (1982 a) Microspora. In: Parker SP (ed) Synopsis and classification of living organisms. McGraw-Hill, New York, pp 589–594

Sprague V (1982 b) Myxozoa. In: Parker SP (ed) Synopsis and classification of living organisms. McGraw-Hill, New York, pp 595–597

Stewart MJ, Schulman S, Vanderberg JP (1986) Rhoptry secretion of membranous whorls by *Plasmodium falciparum* merozoites. Am J Tropical Med 35: 37–44

Supamattaya K, Fischer-Scherl T, Hoffman RW, Boonyaratpalin S (1991) *Sphaerospora epinepheli* n.sp. (Myxosporea: Sphaerosporidae) observed in grouper (*Epinephelus malabaricus*). J Protozool 38: 448–454

– – – – (1993) Light and electron microscopic observations on presporogonic and sporogonic stages of *Sphaerospora epinepheli* (Myxosporea) in grouper (*Epinephelus malabaricus*). J Protozool 40: 71–80

Taylor FJR (1976) Flagellate phylogeny: a study in conflicts. J Protozool 23: 28–30

– (1978) Problems in the development of an explicit hypothetical phylogeny of the lower eukaryotes. BioSystems 10: 67–69

– (1987) The biology of dinoflagellates. Blackwell, Oxford

– (1990) Dinoflagellata. In: Margulis L, Corliss JO, Melkonian M, Chapman DJ (eds) Handbook of Protoctista. Jones and Bartlett, Boston, pp 419–437

Terada S, Reddy R, Jeffers LJ, Cali A, Schiff ER (1987) Microsporidian hepatitis in a patient with the acquired immunodeficiency syndrome. Ann Intern Med 107: 61–62

Tönniges G (1914) Die Trichocysten von *Frontonia leucas* und ihr chromidialer Ursprung. Arch Protistenk 32: 298–378

Vavra (1976) Structure of the microsporidia. In: Bulla LA Jr, Cheng TC (eds) Comparative pathobiology, vol 1, biology of microsporidia. Plenum, New York, pp 1–85

Vesk M, Lucas IAN (1986) The rhabdosome: a new type of organelle in the dinoflagellate *Dinophysis*. Protoplasma 134: 62–64

Vickerman K, Brugerolle G, Mignot J-P (1991) Mastigophora. In: Harrison FW, Corliss JO (eds) Protozoa. Wiley-Liss, New York, pp 13–159 [Harrison FW (ed) Microscopic anatomy of invertebrates, vol 1]

Visvesvra GS, Leitch GJ, Moura H, Wallace S, Weber R, Bryan RT (1991) Culture, electron microscopy and immunoblot studies on a microsporidian parasite isolated from the urine of a patient with AIDS. J Protozool 38: 105S–113S

Vørs N (1992) Ultrastructure and autecology of the marine, heterotrophic flagellate *Leucocryptos marina* (Braarud) Butcher 1967 (Katablepharidaceae/Kathablepharidaceae), with a discussion of the genera *Leucocryptos* and *Katablepharis*/*Kathablepharis*. Eur J Protistol 28: 369–389

Vossbrinck CR, Maddox JV, Friedman S, Debrunner-Vossbrinck BA, Woese CR (1987) Ribosomal RNA sequence suggests microsporidia are extremely ancient eukaryotes. Nature 326: 411–414

Weidner E (1972) Ultrastructural study of microsporidian invasion into cells. Z Parasitenk 40: 227–242

– (1976) The microsporidian spore invasion tube. The ultrastructure, isolation and characterization of the protein comprising the tube. J Cell Biol 71: 23–34

– (1982) The microsporidian invasion tube. III. Tube extrusion and assembly. J Cell Biol 93: 976–979

– (1984) Microsporidian spore discharge and the transfer of polaroplast organelle membrane into plasma membrane. J Protozool 31: 208–213

– Byrd W (1982) The microsporidian spore invasion tube. II. Role of calcium in the activation of invasion tube discharge. J Cell Biol 93: 970–975

– Byre A, Scarborough J, Pleshinger J, Sibley D (1984) Microsporidian spore discharge and the transfer of the polaroplast organelle membrane into plasma membrane. J Protozool 31: 195–198

West JA, Calumpong HP (1989) On the reproductive biology of *Spyridia filamentosa* (Wulfen) Harvey (Rhodophyta) in culture. Bot Mar 32: 379–387

Westfall JA, Bradbury PC, Townsend JW (1983) Ultrastructure of the dinoflagellate *Polykrikos*. I. Development of the nematocyst-taeniocyst complex and morphology of the site for extrusion. J Cell Sci 63: 245–261

Yamamoto T, Sanders JE (1979) Light and electron microscopic observations of sporogenesis in the myxosporidian *Ceratomyxa shasta* (Noble, 1950). J Fish Dis 2: 411–428

Protoplasma (1994) 181: 191–201

PROTOPLASMA

Biology of the *Leishmania* surface: with particular reference to the surface proteinase, gp 63

D. G. Russell*

Department of Molecular Microbiology, Washington University Medical School, St. Louis, Missouri

Received August 5, 1993
Accepted March 21, 1994

Summary. The protistan parasite *Leishmania* is a dimorphic cell that survives as a motile promastigote in the insect digestive tract, and a non-motile, amastigote from within the phagolysosomal compartment of the vertebrate host's phagocytes. The surface of *Leishmania* must interface with a range of differing environments and facilitate uptake of nutrients, whilst protecting the parasite from various host defence mechanisms. This review discusses the organization of the leishmanial cell, and the biology of its major surface constituents, the lipophosphoglycan and the surface proteinase, gp 63.

Keywords: Flagellar pocket; Surface proteinase gp 63; *Leishmania*; Lipophosphoglycan.

Introduction

The genus *Leishmania* contains the etiologic agents for a diverse range of diseases ranging from a simple cutaneous lesion; through the more severely disfiguring mucocutaneous diseases; to visceral leishmaniasis, or Kala-Azar (Molyneux and Ashford 1983, Alexander and Russell 1992). The diseases inflict 12 million people in 86 countries of the world, and have proven difficult to control due to the lack of any prophylactic agent, the toxicity of effective anti-leishmanials, and the innate ability of the parasite to infect many species of mammals. Treatment of these diseases appears effective only in presence of an intact immune system, and it is likely that the drugs employed, such as pentavalent antimonial compounds, function because they help provide an environment conducive to the expansion of a cellular immune response capable of activating the host-macrophages leading to death of the parasite.

* Correspondence and reprints: Department of Molecular Microbiology, Washington University Medical School, 660 South Euclid Avenue, St. Louis, MO 63110, U.S.A.

Cellular organization and the *Leishmania* life cycle

All *Leishmania* species alternate between a phlebotimid vector and a vertebrate host, where they reside inside the parasitophorous vacuole of the host's phagocytes (Molyneux and Ashford 1983, Alexander and Russell 1992, Russell et al. 1992) (Figs. 1 and 2). The alternation between hosts induces a differentiation from the uniflagellate, motile, promastigote form in the insect midgut to the smaller, non-motile, amastigote form, with a vestigial flagellum, that is found inside the host's macrophages. The sandfly vector is a pool feeder and releases infective promastigotes during feeding. The cycle is complete when a non-infected sandfly ingests infected macrophages from a vertebrate host.

The shape of the leishmanial cell, as with all trypanosomatids, is dictated by a corset of pellicular microtubules that originate at the flagellar pocket (Russell et al. 1984, Sherwin et al. 1989, Hemphill et al. 1991). These microtubules are highly crosslinked by different families of accessory proteins (Fig. 3) (Schneider et al. 1988 a, b; Woods et al. 1989; Hemphill et al. 1992), although these cells do not show the morphological plasticity found in other, superficially similar, flagellates such as *Euglena*. The flagellar pocket of *Leishmania* species has combined with the cytostome (Balber 1990, Webster and Russell 1993). This region lacks an underlying microtubular lattice and is the only area of the cell in which endocytosis and exocytosis has been shown to occur.

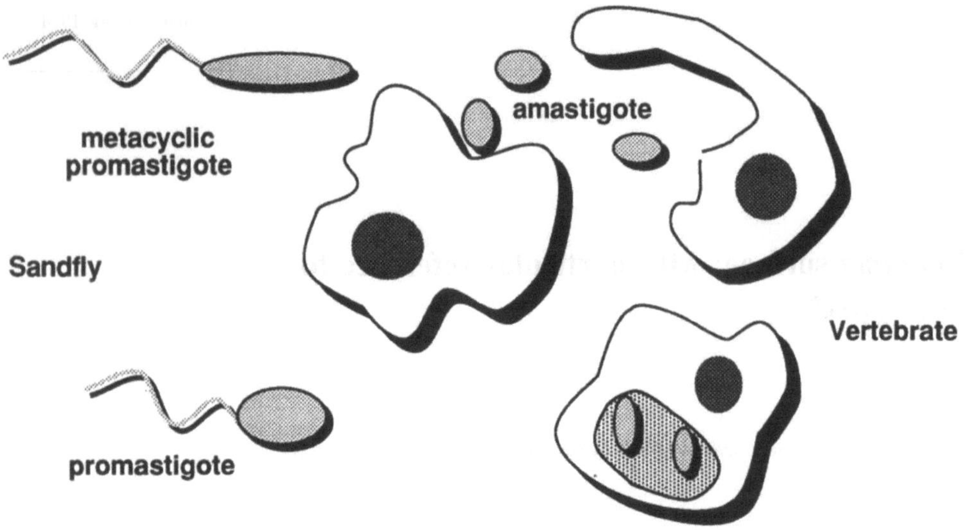

Fig. 1. Diagram of the life cycle of *Leishmania* species. The parasite exists as an actively-dividing, motile, flagellated promastigote stage in the insect digestive tract. It differentiates into an elongate, infective, metacyclic promastigote during its migration to the sandfly proboscis. The parasite is then introduced into the vertebrate host whilst the sandfly takes a bloodmeal. Once in the vertebrate skin tissue, the parasite adheres to, and enters the host's macrophages. Inside the macrophage the promastigote absorbs its flagellum and transforms into the smaller, non-motile, amastigote form. The amstigotes are, in theory, transmitted to fresh host cells on lysis of the host macrophage once the parasite burden becomes too great.

The leishmanial cell divides by binary fission from anterior to posterior. The initial indication is a duplication of the basal bodies of the flagellum, followed by an elongation of the kinetoplast (Robinson and Gull 1991), and the initiation of nuclear division (Fig. 3). Nuclear division is closed, and there is no obvious haploid stage during the parasite's life cycle. Analyses of populations of *Leishmania* suggest that the parasite is capable of genetic exchange, although this occurs infrequently, and many of the populations studied have a clonal structure (Tibayrenc et al. 1990).

The *Leishmania* surface

The surface of leishmanial parasites is required to protect the cell from two hydrolytic environments; the digestive tract of the sandfly, and the phagolysosome of the macrophage, and to facilitate transfer between these two locations. In addition to protecting the parasite against hydrolytic attack, its surface must also be resistant to complement and mediate attachment to both the sandfly midgut and the host macrophage.

The surface of the promastigote stage of all *Leishmania* species examined to date is dominated by two highly abundant molecules; the glycolipid lipophosphoglycan (LPG) (Turco and Descoteaux 1992), and the surface proteinase gp 63 (Colomer-Gould et al. 1985, Etges et al. 1985, Bordier 1987). Both molecules are retained in the plasmalemma by a phosphatidylinositol-glycan, and occur at 2–5 $\times$ 10^6 LPG and 5 $\times$ 10^5 gp 63 copies per cell. The surface of the amastigote stage has been somewhat neglected in its characterization, however, it is relatively protein-poor, and possesses structurally-distinct forms of phosphorylated lipophosphoglycans (Glaser et al. 1991, McConville et al. 1991, Turco et al. 1991). Obviously the surface of the two stages have other proteins such as transporter molecules and pumps, as well as other surface-exposed enzymes such as 3′ and 5′ nuclotidase activity and a membrane-bound acid phosphatase, recently reviewed by a Schneider et al. (1992), however, this article will restrict its attention to the major surface components mentioned above.

Lipophosphoglycan components of the Leishmania surface

The structure of lipophosphoglycan varies between *Leishmania* species, however, in all isolates characterized it exhibits a tripartite structure comprising repeated, phosphorylated saccharide units linked via a mannose-rich carbohydrate core to a lysoalkylphosphatidylinositol lipid anchor (Turco and Descoteaux 1992). Lipophosphoglycan expresses many characteristics important to the success of the leishmanial par-

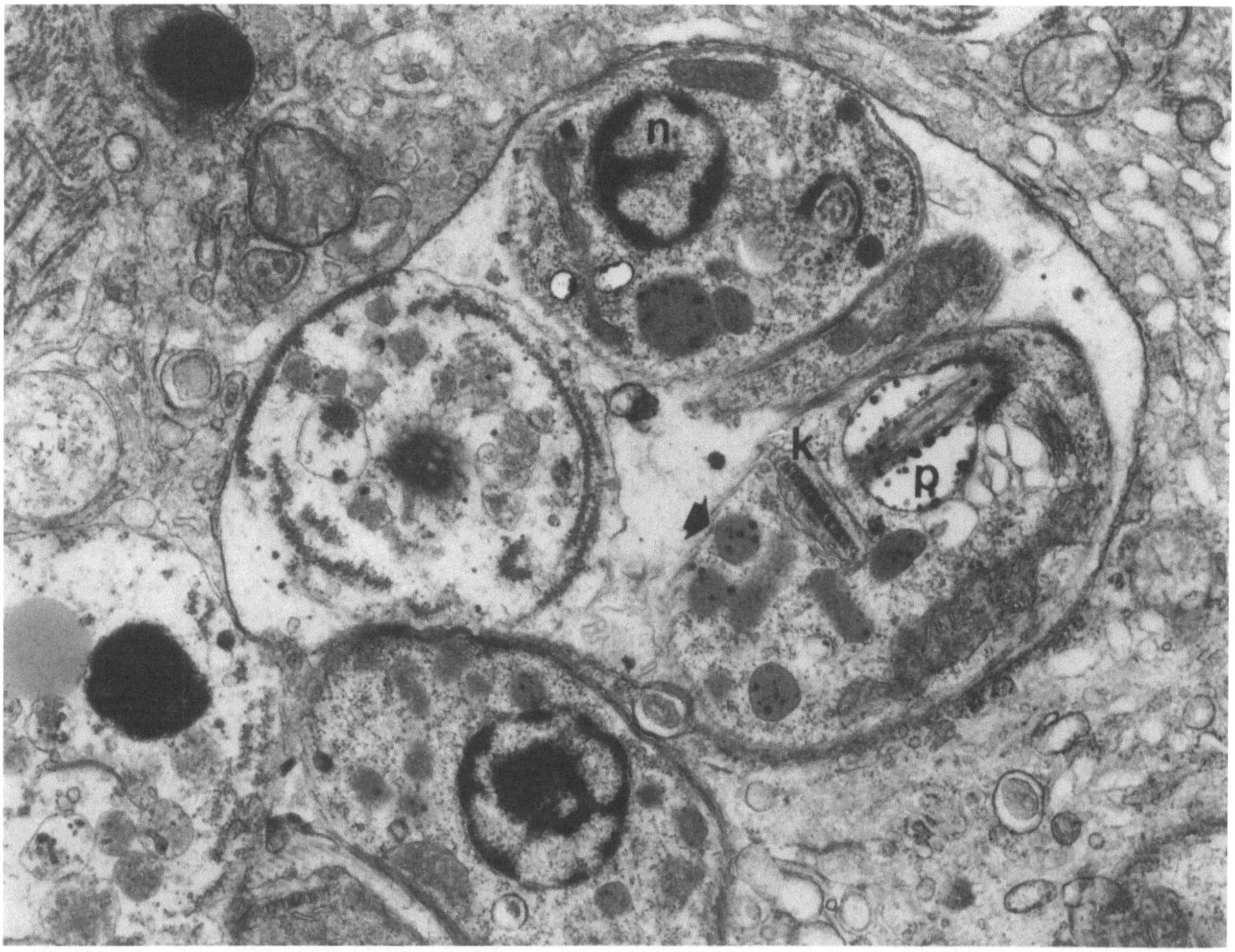

Fig. 2. Electron micrograph from the lesion of a CBA mouse infected with *Leishmania mexicana* showing the amastigote stage inside its parasitophorous vacuole of the host macrophage. The nucleus (*n*), kinetoplast (*k*), and flagellar pocket (*p*) structures are clearly visible. The megasomes, which contain amastigote gp 63, are arrowed

asite. The structure of this lipophosphoglycan has been shown to undergo structural alterations during the parasite's life cycle which markedly alter the parasite's behavior (Fig. 4). *L. major* promastigotes modulate the structure of their LPG by varying the number of saccharide repeats, and their terminal sugars, increasing the length of LPG during transformation from non-infective, dividing promastigotes to infective, non-dividing metacyclic promastigotes (Sacks et al. 1990). This alteration is responsible for the release of the promastigotes from the midgut, and their ability to re-adhere to the mouthparts of the sandfly host (Pimenta et al. 1992). The LPG on the metacyclic promastigotes also confers greater complement resistance on the parasite by preventing access of the membrane attack complex to the plasmalemma whilst, allowing complement deposition which aids in host cell entry (Da Silva et al. 1989, Puentes et al. 1990). LPG is capable of binding directly to the $\beta 2$ family of phagocyte integrins (Talamas-Rohana et al. 1990, Van Strip et al. 1993), although the affinity of this interaction is enhanced by complement.

LPG is released in appreciable quantities during entry of the promastigote into the macrophage, and persists inside the macrophage during the early stages of infection (Fig. 4). LPG has been shown to actively modulate macrophage behavior by interfering with macrophage activation events and scavenging reactive oxygen radicals harmful to the parasite. It likely plays a crucial role in the establishment of infection inside the macrophage (Handman et al. 1986; McNeely et al. 1989, 1990; Descoteaux et al. 1991, 1992). However, once transformed into the amastigote stage the parasite does not synthesize the same form of LPG. The surface

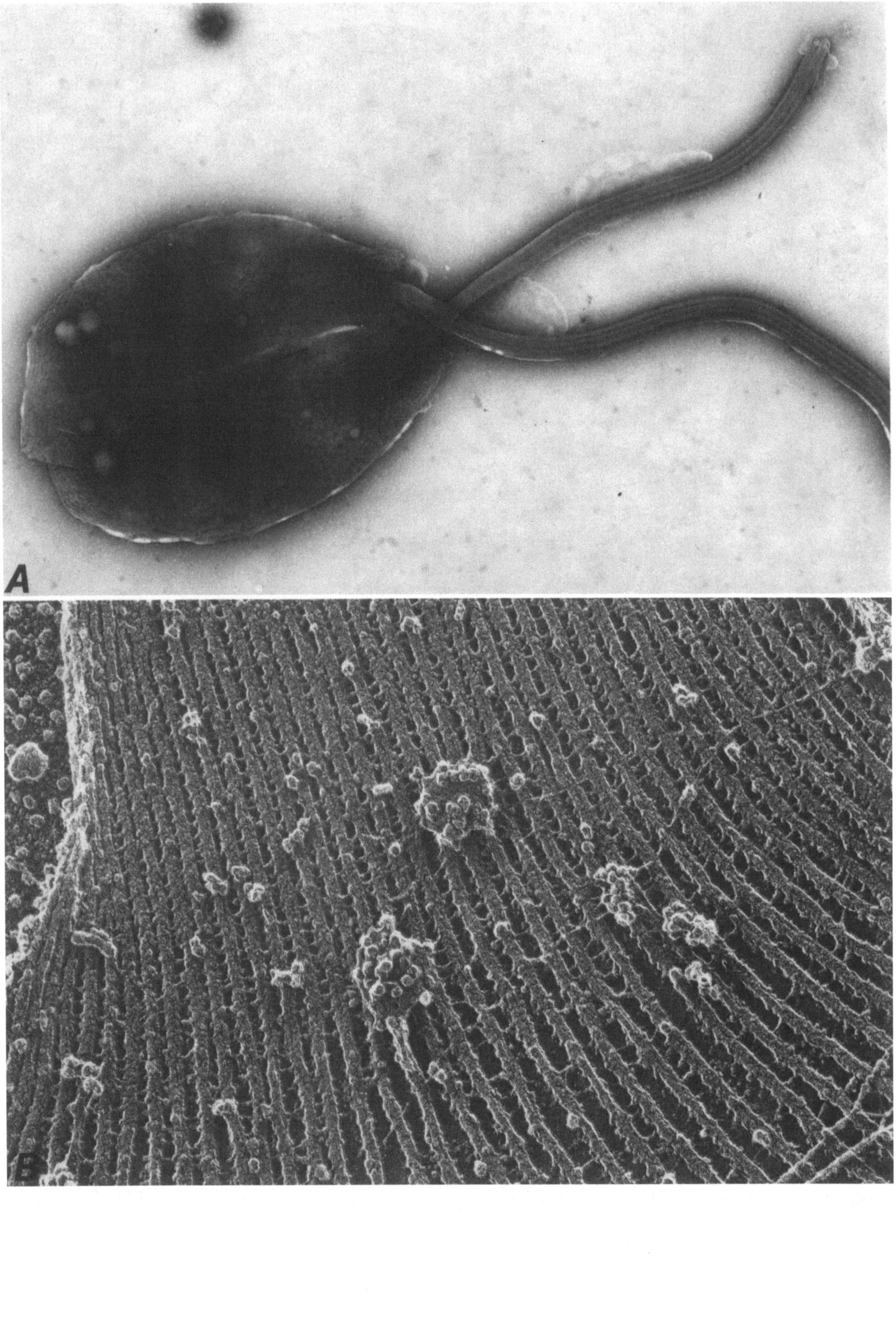
A
B

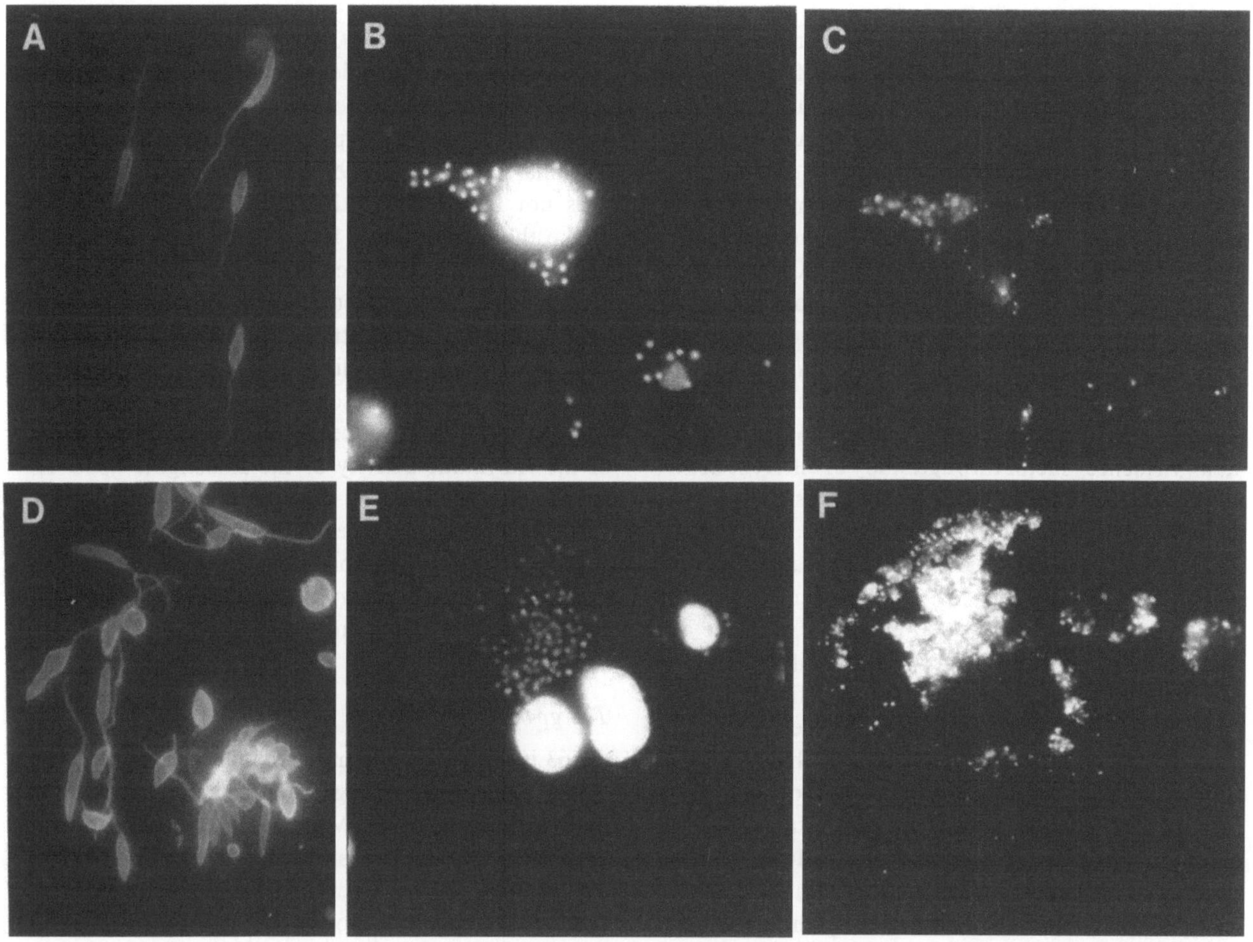

Fig. 4. Immunofluorescence of *Leishmania mexicana* promastigotes (**A** and **D**) and amastigotes (**B, C, E,** and **F**) labelled with anti-gp 63 and anti-LPG monoclonal antibodies. **A** and **C** Probed with L 3.8 anti-gp 63 monoclonal antibody and **B** Hoechst dye to label the nuclei. The antibody decorated the surface of the promastigotes and produced a punctate pattern, now known to correspond to the megasomes, in the amastigote stage. **D** and **F** Probed with WIC 108.3 anti-LPG monoclonal antibody and **E** Hoechst dye to label the nuclei. Again the promastigote surface is strongly labelled. Although this form of LPG is not made in the amastigote stage, the lipophosphoglycan persists in the early infection. This macrophage was processed 24 h after infection with promastigotes. In addition to LPG, this antibody recognizes a carbohydrate epitope present on a small number of amastigote proteins (Ilg et al. 1991)

of the amastigote has not been examined in the same depth although this stage is known to display a truncated lipophosphoglycan or "GIPL" (glycosylinositolphospholipid) (McConville and Blackwell 1991), which shares many structural similarities with LPG. Studies into the biology of this family of molecules are just beginning.

Biology of the Leishmania surface proteinase, pg 63

In the promastigote stage, virtually all the gp 63 in the cell is exposed at the cell surface (Fig. 4). The protein is post-translationally modified by the addition of a phosphatidylinositol-glycan at an asparagine residue exposed by removal of the carboxyl 17 amino acids, following synthesis of the polypeptide (Schneider et al. 1990). Gp 63 is a metallo endopeptidase complexed with 1 mol of Zn^{2+} per mol of gp 63 (Bouvier et al. 1989, Jongeneel et al. 1989). The Zn-binding motif in the gp 63 sequence shows close structural homology to both prokaryote and higher eukaryote Zn-metalloproteinases.

Analysis of the pH optimum for gp 63's activity has

Fig. 3. Electron micrographs from *Leishmania mexicana* promastigotes illustrating the overall organization of the subtending, pellicular cytoskeleton. **A** Negative stained preparation processed during cell division. This is in the late stages of division, the flagellum has duplicated and the cell body is in the process of cleaving from anterior to posterior. **B** Freeze-etch micrograph of a promastigote following treatment by sonication. This micrograph is looking at the cytoplasmic face to the subpellicular microtubules which appear heavily crosslinked with themselves and the plasmalemma

produced contradictory data, however, most laboratories find that the promastigote form of gp 63 has a pH optimum between pH 7–7.5 (Chaudhuri et al. 1988, Bouvier et al. 1990, Ip et al. 1990, Tzinia et al. 1991). Although gp 63 has been shown capable of cleaving a range of different substrates, from synthetic peptides (Bouvier et al. 1990, Ip et al. 1990) to the third component of complement (Chaudhuri and Chang 1988), its relative activity on substrates of biological significance remains unconvincing. In addition, the structure and abundance of LPG, see the previous section, will compromise the access of gp 63 to macromolecular substrates. Recent studies have attempted to define gp 63 as a "virulence" factor through study of different *Leishmania* isolates expressing differing levels of the proteinase (Chang et al. 1990, Murray et al. 1990), or the induction of raised gp 63 levels through transformation of *Leishmania* (Liu et al. 1992). However, because of the attenuation of *Leishmania* in culture, and the absence of a gp 63-null background it is difficult to put these studies into perspective, and a defined role for gp 63 in the promastigote remains elusive.

Despite a report to the contrary (Schneider et al. 1992 b), gp 63 is expressed in the amastigote of at least some *Leishmania* species (Medina-Acosta et al. 1989, 1993 a; Frommel et al. 1990; Bahr et al. 1993) (Fig. 4). However, the protein is the product of a different set of gp 63 genes and exhibits markedly different post-translational processing and intracellular location (Medina-Acosta et al. 1989, 1993 a; Bahr et al. 1993). The most extensive analysis of gp 63 in the amastigote stage has been conducted on *L. mexicana*. In this species, gp 63 is expressed in the amastigote at 10–30% the level found in the promastigote (Medina-Acosta et al. 1989, Bahr et al. 1993). The gp 63 genes expressed in the amastigote encode an extended carboxyl-terminus, 41 amino acids longer than that of the gp 63 genes predominantly expressed in the promastigote (Medina-Acosta et al. 1993 a). The amastigote gp 63 does not fractionate with the cell membranes, and, unlike the promastigote form of the protein, lacks a phosphatidylinositol membrane anchor (Medina-Acosta et al. 1989, Ilg et al. 1993). Immunoelectron-microscopy has demonstrated that the protein in the amastigote stage is localized to the megasome (Bahr et al. 1993), which are lysosomal bodies specific to this complex of *Leishmania* species. Recent characterization of the enzymatic activity of the amastigote gp 63 revealed that it had an acid pH optimum (Ilg et al. 1993), which is logical for its new location, and may explain some of the contradictory data from earlier studies because the amastigote gp 63 genes are also expressed to a lower level in the promastigote form of the parasite.

The gp 63 gene family

Gp 63 is encoded by a multigene family, whose members are tandemly arranged in the species analysed to date (Button et al. 1989, Murray et al. 1990, Hanekamp et al. 1991, Medina-Acosta et al. 1993 b). In most species studied, the gp 63 gene are present at single locus on a similarly-sized chromosome (Button et al. 1989, Medina-Acosta et al. 1993 b), however, isolates from the *L. braziliensis* complex in particular exhibit greater heterogeneity with respect to the numbers and location of their gp 63 genes (Hanekamp and Langer 1991, Steinkraus et al. 1991). The gp 63 gene family of *L. mexicana* contains 3 structurally distinct classes of

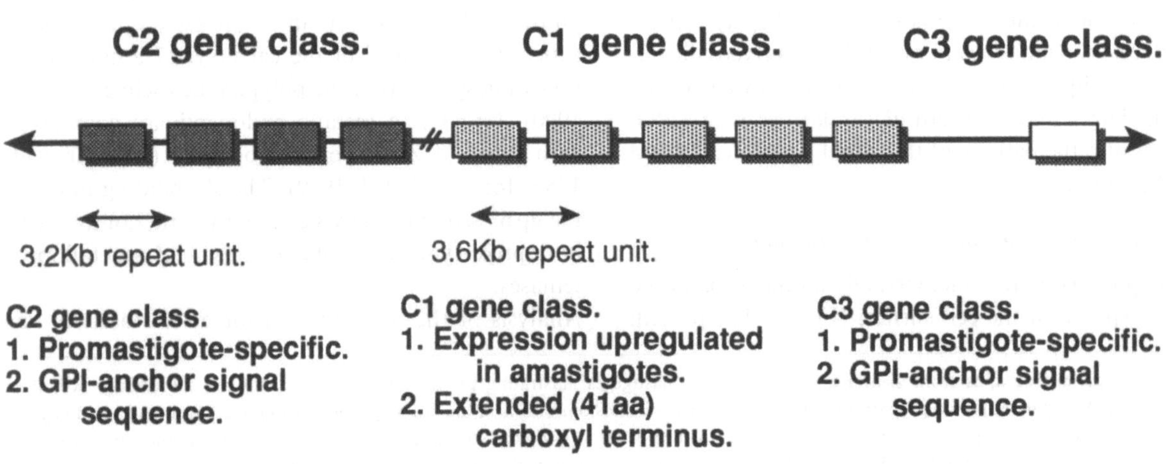

Fig. 5. Diagram of the gp 63 gene locus of *Leishmania mexicana*. Physical map of the 80 kb Hin d III fragment

genes referred to as C 1, C 2, C 3, containing 5, 4, and 1 gene copies, respectively (Medina-Acosta et al. 1993 b), summarized in Fig. 5. The C 1 gene class encodes the extended carboxyl-terminus and is the only gene class transcribed in the amastigote. A gp 63 gene sequence comparable to the C 1 gene class of *L. mexicana* has also been described in *L. donovani* (Ramamoorthy et al. 1992). The C 2/C 3 gene classes encode the PI-signal peptide characterized previously in *L. major* (Schneider et al. 1990), and are expressed only in the promastigote form of the parasite. The mechanism behind the differential expression of the different gp 63 gene classes is nuclear. The C 1 gene class is situated in between the C 2 and C 3 gene classes. As transcription in trypanosomatids tends to proceed in multi-genic units, it is likely that the level of expression is controlled through processing and degradation of multi-gene transcripts, rather than differential transcription.

We recently sequenced a gp 63 homologue from the monoxenous trypanosomatid *Crithidia fasciculata* (Inverso et al. 1993). The sequence contains many of the structural features of the *Leishmania* gp 63, most notably the Zn-binding site. The demonstration of a homologue for leishmanial gp 63 in *Crithidia* suggests that the ancestral function of the enzyme lies within the insect vector and that the existence of the metalloproteinase predated the invasion of the vertebrate host. However, this argument does not preclude gp 63 from fulfilling a function within the vertebrate, and it may turn out to be a useful preadaptation to survival in the mammal.

Extensive analysis of the sequence divergence within, and between *Crithidia* and 5 other *Leishmania* spp. revealed that the 5′ and 3′ ends of the gp 63 genes formed different linkage patterns between species, indicating that the genes had evolved in a mosaic fashion, most likely through gene conversion or unequal cross-over within the different classes of gp 63 gene (Medina-Acosta et al. 1993 a). It seems likely that the common ancestor of *Crithidia* and *Leishmania* had already evolved a multigene gp 63 locus prior to the divergence of the species.

Leishmania surface antigens and the immune response

Despite the fact that some forms of leishmaniasis, such as Kala-Azar, are normally marked by a high antibody titre against surface components of *Leishmania*, these antibodies play little role in determining the outcome of the infection, see Karp et al. (1991) for example. It is the cellular immune response that appears all important in limiting leishmanial infections. Several laboratories report induction of protective immunity in mice by transfer of T-cells of the T helper 1 phenotype, which are capable of inducing macrophage activation by secretion of cytokinesis such as interferon-γ (Howard 1986, Locksley and Louis 1991, Locksley and Scott 1992).

T cell responses have been reported against both the glycolipid LPG (Moll et al. 1989, Moll and Rollinghof 1991) and gp 63 (Russo et al. 1991, 1993), and some studies report successful protection of mice by vaccination with these components (Handman and Mitchell 1985, Russell et al. 1988). However, it is becoming more accepted that the ability of LPG to induce T cell proliferation (Moll et al. 1989) is dependent on an associated peptide, rather than the lipophosphoglycan itself (Mendonca et al. 1991, Russo et al. 1992). In contrast, studies with gp 63 continue to look promising and the recent vaccination with the attenuated mycobacteria BCG expressing gp 63 induced strong protective immunity in some mouse strains, in response to challenge with cutaneous leishmaniasis strains (Connell et al. 1993).

The flagellar pocket: a specialized region of the leishmanial surface

Given that most trypanosomatids are exposed to potentially hostile environments in their vertebrate and invertebrate hosts, it is perhaps predictable that internalization of macromolecular nutrients is restricted to a specialized region of the surface. The flagellar pocket of *Leishmania*, and other trypanosomatids, lacks the subtending microtubular corset, and is instead subtended by a complex array of membranous compartments that comprise both the endocytic and exocytic network for the cell (Fig. 6) (Balber 1990, Webster and Russell 1993).

Experiments of different trypanosomatids have demonstrated that the flagellar pocket region is the site of internalization of a range of fluid phase markers, and ligands. Both transferrin, low-density lipoprotein and other endocytic markers could be followed into the flagellar pocket, and into the endocytic network (Coppens et al. 1987, 1988; Webster 1989; Webster et al. 1989; Schell et al. 1991 a, b; Grab et al. 1992). In addition, biotin-labelled plasmalemma components can also be found internalized by the cell. As this appears to be the sole site for addition of components to the plasmalemma, and internalization and recycling of surface receptors and constituents, the rate of turnover within the pocket region is calculated to be extremely

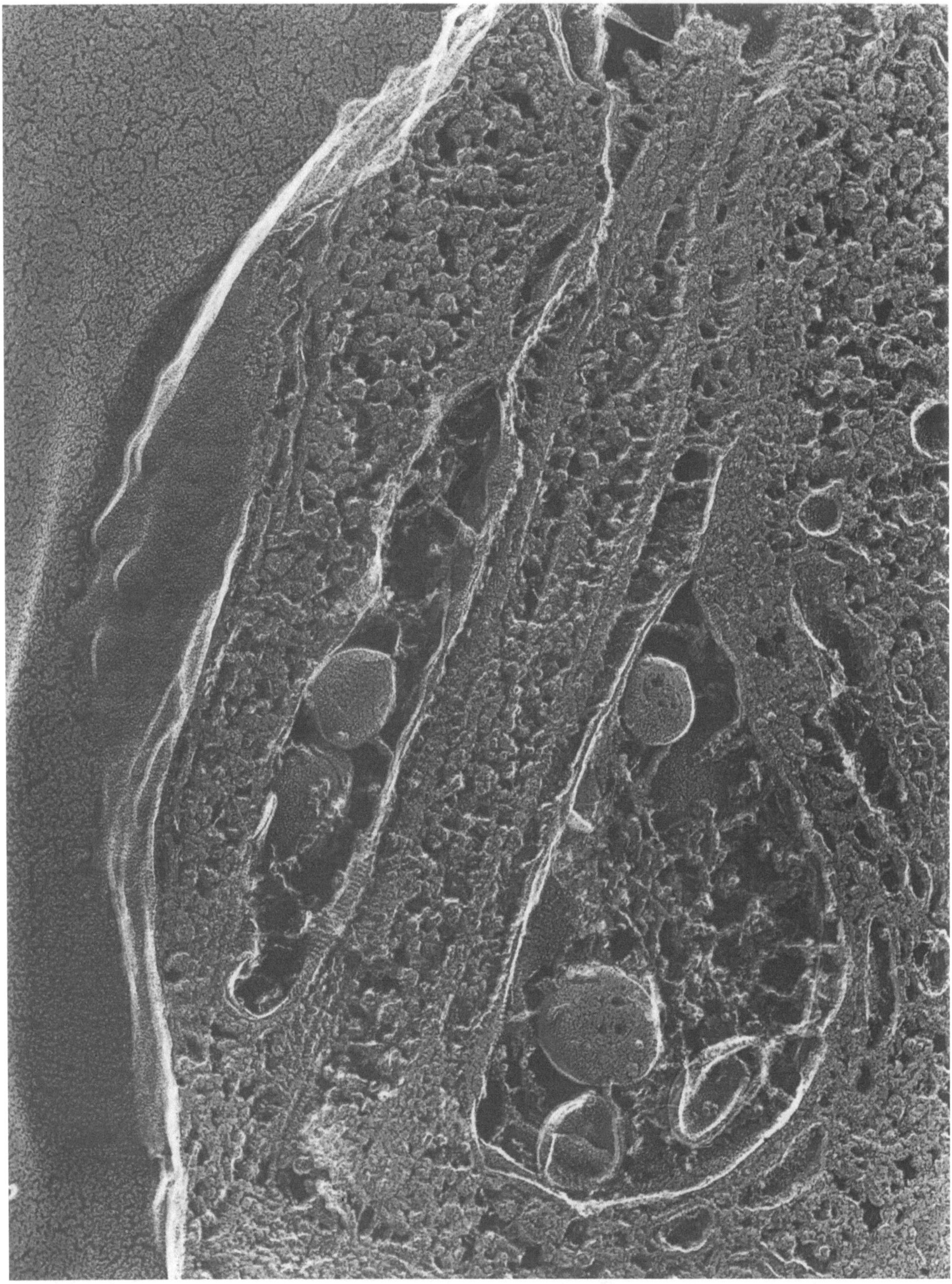

Fig. 6. Freeze-etch micrograph for the flagellar pocket region of *Leishmania mexicana*. The pocket region is subtended by a network of endocytic and exocytic vesicles responsible for protein trafficking into and out of the pocket. The lumen of the pocket is filled with filamentous material that contains the secreted acid phosphatase from this species of *Leishmania* (Stierhof et al. 1991)

high (Webster and Russell 1993). The factors controlling access to the pocket, and the signal requirements for trafficking into and out of the pocket region have not been defined but are the subject of an increasing number of studies.

Concluding remarks

We now have a reasonable level of understanding about the biochemistry of the surface-exposed molecules of *Leishmania*, however, our comprehension of how these molecules function inside the insect and vertebrate host is only just developing. LPG mediates binding to the insect midgut and mouthparts, and plays a role in complement resistance and host cell entry. The surface proteinase, gp 63, also mediates attachment to the macrophage, however, the biological role of its enzymatic activity has yet to be established. The mechanism by which these, and other parasite-derived molecules, influence macrophage behaviour and down-regulate its microbicidal responses is also unknown. In fact, microbial pathogenesis as a field, is only now feeling comfortable enough with eukaryote cell biology to start addressing the intriguing questions of host cell-modulation by intracellular parasites. Surface molecules constitute the first contact between the pathogen and its host environment, and, as such will be found to play key roles in the determination of host behaviour.

Acknowledgements

DGR is supported by US Public Health Service grants AI 26889 and AI 34207 from the National Institutes of Health. The freeze-etch electron micrographs in Figs. 2 and 5 were processed by Robyn Roth in John Heuser's laboratory, Department of Cell Biology and Physiology, Washington University. The author is extremely grateful to Dr. Heuser for his continued support and interest in this research.

References

Alexander J, Russell DG (1992). The interaction of *Leishmania* species with macrophages. Adv Parasitol 31: 175–254

Bahr V, Stierhof Y-D, Ilg T, Demar M, Overath P (1993) Expression of lipophosphoglycan and glycoprotein 63 in promastigotes and amastigotes of *Leishmania mexicana*. Mol Biochem Parasitol 58: 107–122

Balber AE (1990) The pellicle and the membrane of the flagellum, flagellar adhesion zone, and flagellar pocket: functionally discrete surface domains of the bloodstream form of African trypanosomes. Crit Rev Immunol 10: 177–201

Bordier C (1987) The promastigote surface protease of *Leishmania* Parasitol Today 3: 151–153

Bouvier J, Bordier C, Vogel H, Reichelt R, Etges R (1989) Characterization of the promastigote surface protease of *Leishmania* as a membrane-bound zinc endopeptidase. Mol Biochem Parasitol 37: 235–245

– Schneider P, Etges R, Bordier C (1990) Peptide substrate specificity of the membrane-bound metalloprotease of *Leishmania*. Biochemistry 29: 10113–10119

Button LL, Russell DG, Klein HL, Medina-Acosta E, Karess RE, McMaster WR (1989) Genes encoding the major surface glycoprotein in *Leishmania* are tandemly linked at a single chromosomal locus and are constitutively transcribed. Mol Biochem Parasitol 32: 271–283

Chang K-P, Chaudhuri G, Fong D (1990) Molecular determinants of *Leishmania* virulence. Annu Rev Microbiol 44: 449–529

Chaudhuri G, Chang KP (1988) Acid protease activity of a major surface membrane glycoprotein (gp 63) from *Leishmania mexicana* promastigotes. Mol Biochem Parasitol 27: 43–52

Colomer-Gould V, Galvao Quintao L, Keithly J, Nogueira N (1985) A common surface antigen on amastigotes and promastigotes of *Leishmania* species. J Exp Med 162: 902–916

Connell N, Medina-Acosta E, McMaster WR, Bloom BR, Russell DG (1993) Effective immunization against cutaneous leishmaniasis with recombinant BCG expressing the *Leishmania* surface proteinase (gp 63). Proc Natl Acad Sci USA 90: 11473–11477

Coppens I, Opperdoes FR, Courtoy PJ, Baudhuin P (1987) Receptor-mediated endocytosis in the bloodstream form of *Trypanosoma brucei*. J Protozool 465–473

– Baudhuin P, Opperdoes FR, Courtoy PJ (1988) Receptors for the host low density lipoproteins on the hemoflagellate *Trypanosoma brucei*: purification and involvement in the growth of the parasite. Proc Natl Acad Sci USA 85: 6753–6757

Da Silva RP, Hall BF, Joiner KA, Sacks DL (1989) CR1, the C3b receptor, mediates binding of infective *Leishmania major* metacyclic promastigotes to human macrophages. J Immunol 143: 617–622

Descoteaux A, Turco SJ, Sacks DL, Matlashewski G (1991) *Leishmania donovani* lipophosphoglycan selectively inhibits signal transduction in macrophages. J Immunol 146: 2747–2753

– Matlashewski G, Turco SJ (1992) Inhibition of macrophage protein kinase C-mediated protein phosphorylation by *Leishmania donovani* lipophosphoglycan. J Immunol 149: 3008–3015

Etges RJ, Bouvier J, Hoffman R, Bordier C (1985) Evidence that the major surface proteins of three *Leishmania* species are structurally related. Mol Biochem Parasitol 14: 141–151

Frommel TO, Button LL, Fujikura Y, McMaster WR (1990) The major surface glycoprotein (GP 63) is present in both life stages of *Leishmania*. Mol Biochem Parasitol 38: 25–32

Glaser TA, Moody SF, Handman E, Bacic A, Spithill TW (1991) An antigenically distinct lipophosphoglycan on amastigotes of *Leishmania major*. Mol Biochem Parasitol 45: 337–344

Grab DJ, Wells CW, Shaw MK, Russo DCW, Webster P (1993) Endocytosed transferrin in African trypanosomes is delivered to lysosomes and may not be recycled. Eur J Cell Biol 59: 398–404

Handman E, Mitchell GF (1985) Immunization with *Leishmania* receptor for macrophages protects mice against cutaneous leishmaniasis. Proc Natl Acad Sci USA 82: 5910–5914

– Schnur LF, Spithill TW, Mitchell GF (1986) Passive transfer of *Leishmania* lipopolysaccharide confers parasite survival in macrophages. J Immunol 137: 3608–3613

Hanekamp T, Langer PJ (1991) Molecular karyotype and chromosomal localization of genes encoding two major surface glycoproteins, pg 63 and gp 46/M 2, hsp 70, and beta-tubulin in cloned strains of several *Leishmania* species. Mol Biochem Parasitol 48: 27–37

Hemphill A, Seebeck T, Lawson D (1991) The *Trypanosoma brucei* cytoskeleton: ultrastructure and localization of microtubule-associated and spectrin-like proteins using quick-freeze, deep-etch, immunogold electron microscopy. J Struct Biol 107: 211–220

Hemphill A, Affolter M, Seebeck T (1992) A novel microtubule-binding motif identified in a high molecular weight microtubule-associated protein from *Trypanosoma brucei.* J Cell Biol 117: 95–103

Howard JG (1986) Immunological regulation and control of experimental leishmaniasis. Int Rev Exp Pathol 28: 79–116

Ilg T, Menz B, Winter G, Russell DG, Etges R, Schell D, Overath P (1991) Monoclonal antibodies of *Leishmania mexicana* promastigote antigens. I. Secreted acid phosphatase and other proteins share epitopes with lipophosphoglycan. J Cell Sci 99: 175–180

– Harbecke D, Overath P (1993) The lysosomal gp 63-related protein in *Leishmania mexicana* amastigotes is a soluble metalloproteinase with an acidic pH optimum. FEBS Lett 327: 103–107

Inverso JA, Medina-Acosta E, O'Connor J, Russell DG, Cross GAM (1993) *Crithidia fasciculata* contains a transcribed leishmanial surface proteinase (gp 63) gene homologue. Mol Biochem Parasitol 57: 47–54

Ip HS, Orn A, Russell DG, Cross GAM (1990) *Leishmania mexicana mexicana* gp 63 is a site-specific neutral endopeptidase. Mol Biochem Parasitol 40: 163–172

Jongeneel CV, Bouvier J, Bairoch A (1989) A unique signature identifies a family of zinc-dependent metallopeptidases. FEBS Lett 242: 211–214

Karp CL, Turco SJ, Sacks DL (1991) Lipophosphoglycan masks recognition of the *Leishmania donovani* promastigote surface by human kala-azar serum. J Immunol 147: 680–684

Liu X, Chang KP (1992) Extrachromosomal genetic complementation of surface metalloproteinase (gp 63)-deficient *Leishmania* increases their binding to macrophages. Proc Natl Acad Sci USA 89: 4991–4995

Locksley RM, Louis JA (1992) Immunology of leishmaniasis. Curr Opin Immunol 4: 413–418

– Scott P (1991) Helper T-cell subsets in mouse leishmaniasis: induction, expansion and effector function. Immunol Today 12: 61–64

McConville MJ, Blackwell JM (1991) Developmental changes in the glycosylated phosphatidylinositols of *Leishmania donovani.* Characterization of the promastigote and amastigote glycolipids. J Biol Chem 266: 15170–15179

McNeely TB, Turco SJ (1990) Requirement of lipophosphoglycan for intracellular survival of *Leishmania donovani* within human monocytes. J Immunol 144: 2745–2750

– Rosen G, Londner MV, Turco SJ (1989) Inhibitory effects on protein kinase C activity by lipophosphoglycan fragments and glycosylphosphatidylinositol antigens of the protozoan parasite *Leishmania.* Biochem J 259: 601–604

Medina-Acosta E, Karess RE, Schwarz H, Russell DG (1989) The promastigote surface protease (gp 63) of *Leishmania* is expressed but differentially processed and localized in the amastigote stage. Mol Biochem Parasitol 37: 263–273

– – Russell DG (1993 a) Structurally distinct genes for the surface protease of *Leishmania mexicana* are developmentally regulated. Mol Biochem Parasitol. 57: 31–45

– Beverley SM, Russell DG (1993 b) Evolution and expression of the *Leishmania* surface proteinase (gp 63) gene locus. Infect Agents Dis 2: 25–34

Mendonca SC, Russell DG, Coutinho SG (1991) Analysis of the human T cell responsiveness to purified antigens of *Leishmania*: lipophosphoglycan (LPG) and glycoprotein 63 (gp 63). Clin Exp Immunol 83: 472–478

Moll H, Mitchell GF, McConville MJ, Handman E (1989) Evidence of T-cell recognition in mice of a purified lipophosphoglycan from *Leishmania major.* Infect Immun 57: 3349–3356

– Rollinghoff M (1991) T-cell reactivity to purified lipophosphoglycan from *Leishmania major*: a model for analysis of the cellular immune response to microbial carbohydrates. Behring Inst Mitt 88: 161–169

Molyneux DH, Ashford RW (1983) The biology of *Trypanosoma* and *Leishmania*, parasites of man and domestic animals. Taylor and Francis, London

Murray PJ, Handman E, Glaser TA, Spithill TW (1990) *Leishmania major*: expression and gene structure of the glycoprotein 63 molecule in virulent and avirulent clones and strains. Exp Parasitol 71: 294–304

Pimenta PF, Turco SJ, McConville MJ, Lawyer PG, Perkins PV, Sacks DL (1992) Stage-specific adhesion of *Leishmania* promastigotes to the sandfly midgut. Science 256: 1812–1815

Puentes SM, da Silva RP, Sacks DL, Hammer CH, Joiner KA (1990) Serum resistance of metacyclic stage *Leishmania major* promastigotes is due to release of C5b-9. J Immunol 145: 4311–4316

Ramamoorthy R, Donelson JE, Paetz KE, Maybodi M, Roberts SC, Wilson ME (1992) Three distinct RNAs for the surface protease gp 63 are differentially expressed during development of *Leishmania donovani chagasi* promastigotes to an infectious form. J Biol Chem 267: 1888–1895

Robinson DR, Gull K (1991) Basal body movements as a mechanism for mitochondrial genome segregation in the trypanosome cell cycle. Nature 352: 731–733

Russell DG, Alexander J (1988) Effective immunization against cutaneous leishmaniasis with defined membrane antigens reconstituted into liposomes. J Immunol 140: 1274–1279

– Miller D, Gull K (1984) Tubulin heterogeneity in the trypanosome *Crithidia fasciculata.* Mol Cell Biol 4: 779–790

– Xu S, Chakraborty P (1992) Intracellular trafficking and the parasitophorous vacuole of *Leishmania mexicana*-infected macrophages. J Cell Sci 103: 1193–1210

Russo DM, Burns JJ, Carvalho EM, Armitage RJ, Grabstein KH, Button LL, McMaster WR, Reed SG (1991) Human T cell responses to gp 63, a surface antigen of *Leishmania.* J Immunol 147: 3575–3580

– Turco SJ Burns JJ, Reed SG (1992) Stimulation of human T lymphocytes by *Leishmania* lipophosphoglycan-associated proteins. J Immunol. 148: 202–207

– Jardim A, Carvalho EM, Sleath PR, Armitage RJ, Olafson RW, Reed SG (1993) Mapping human T cell epitopes in *Leishmania* gp 63. Identification of cross-reactive and species-specific epitopes. J Immunol 150: 932–939

Sacks DL, Brodin TN, Turco SJ (1990) Developmental modification of the lipophosphoglycan from *Leishmania major* promastigotes during metacyclogenesis. Mol Biochem Parasitol 42: 225–233

Schell D, Borowy NK, Overath P (1991 a) Transferrin is a growth factor for the bloodstream form of *Trypanosoma brucei.* Parasitol Res 77: 558–560

– Evers R, Preis D, Ziegelbauer K, Kiefer H, Lottspeich F, Cornelissen AW, Overath P (1991 b) A transferrin-binding protein of *Trypanosoma brucei* is encoded by one of the genes in the variant surface glycoprotein gene expression site. EMBO J 10: 1061–1066

Schneider A, Eichenberger W, Seebeck T (1988 a) A microtubule-binding protein of *Trypanosoma brucei* which contains covalently bound fatty acid. J Biol Chem 263: 6472–6475

– Hemphill A, Wyler T, Seebeck T (1988 b). Large microtubule-associated protein of *T. brucei* has tandemly repeated, near-identical sequences. Science 241: 459–462

– Ferguson MA, McConville MJ, Mehlert A, Homans SW, Bordier C (1990) Structure of the glycosyl-phosphatidylinositol membrane anchor of the *Leishmania major* promastigote surface protease. J Biol Chem 265: 16955–16964

– Bordier C, Etges R (1992 a) Membrane proteins and enzymes of *Leishmania*. Subcell Biochem 18: 39–72

– Rosat JP, Bouvier J, Louis JA, Bordier C (1992 b) *Leishmania major*: differential regulation of the surface metalloprotease in amastigote and promastigote stages. Exp Parasitol 75: 196–206

Sherwin T, Gull K (1989) The cell division cycle of *Trypanosoma brucei* brucei: timing of event markers and cytoskeletal modulations. Philos Trans R Soc [Biol] 323: 573–588

Steinkraus HB, Panitz S, Langer PJ (1991) Isolation and electrophoretic karyotype analysis of cellular clones of several *Leishmania* species. Ann Trop Med Parasitol 85: 467–469

Stierhof Y-D, Schwarz H, Menz B, Russell DG, Quinten M, Overath P (1991) Monoclonal antibodies to *Leishmania mexicana* promastigote antigens. II. Cellular localization of antigens in promastigotes and infected macrophage. J Cell Sci 99: 181–186

Talamas-Rohana P, Wright SD, Lennartz MR, Russell DG (1990) Lipophosphoglycan from *Leishmania mexicana* promastigotes binds to members of the CR 3, p 150,95 and LFA-1 family of leukocyte integrins. J Immunol 144: 4817–4824

Tibayrenc M, Kjellberg F, Ayala FJ (1990) A clonal theory of parasitic protozoa: the population structures of *Entamoeba, Giardia, Leishmania, Naegleria, Plasmodium, Trichomonas*, and *Trypanosoma* and their medical and taxonomical consequences. Proc Natl Acad Sci USA 87: 2414–2418

Turco SJ, Descoteaux A (1992) The lipophosphoglycan of *Leishmania* parasites. Annu Rev Microbiol 46: 65–94

– Sacks DL (1991) Expression of a stage-specific lipophosphoglycan in *Leishmania major* amastigotes. Mol Biochem Parasitol 45: 91–99

Tzinia AK, Soteriadou KP (1991) Substrate-dependent pH optima of gp 63 purified from seven strains of *Leishmania*. Mol Biochem Parasitol 47: 83–89

Van Strip JAG, Russell DG, Tuomanen E, Brown EJ, Wright SD (1993). Ligand specificity of purified complement receptor type three (CD11b/CD18, MAC 1); indirect involvement of an Arg-Gly-Asp (RGD) sequence. J Immunol 151: 3324–3336

Webster P (1989) Endocytosis by African trypanosomes. I. Three-dimensional structure of the endocytic organelles in *Trypanosoma brucei* and *T. congolense*. Eur J Cell Biol 49: 295–302

– Fish WR (1989) Endocytosis by African trypanosomes. II. Occurrence in different life-cycle stages and intracellular sorting. Eur J Cell Biol 49: 303–310

– Russell DG (1993) The flagellar pocket of trypanosomatids. Parasitol Today 9: 201–206

Woods A, Baines AJ, Gull K (1989) Evidence for a M_r 88,000 glycoprotein with a transmembrane association to a unique flagellum attachment region in *Trypanosoma brucei*. J Cell Sci 93: 501–508

Protoplasma (1994) 181: 202–212

PROTOPLASMA

Adhesion of fungal spores and germlings to host plant surfaces

E. J. Braun* and **R. J. Howard**

Department of Plant Pathology, Iowa State University, Ames, Iowa and Central Research and Development, the DuPont Company, Wilmington, Delaware

Received August 5, 1993
Accepted January 16, 1994

Summary. Firm adhesion of fungal plant pathogens to their hosts is critical at several stages in the host-parasite interaction. Spores of many fungal species are capable of rapid, non-specific attachment to various surfaces. This early adhesion, which often occurs well before germ tube emergence, prevents spores from being blown or washed from the host surface before infection can take place. Adhesion is critical for proper sensing of topographic signals involved in thigmotropic responses and for differentiation and function of appressoria. Four fungal pathogens which exhibit a variety of adhesion mechanisms have been selected for discussion.

Keywords: Fungal adhesion; Appressoria; Plant disease; *Cochliobolus heterostrophus; Magnaporthe grisea; Nectria haematococca; Uromyces* spp.

Abbreviations: EMC extracellular matrix; FSTEM freeze-substitution transmission electron microscopy; Con A concanavalin A; CryoSEM cryo scanning electron microscopy; MTM macroconidial tip mucilage; STM spore tip mucilage.

Introduction

The initiation of infection by a typical plant pathogenic fungus involves a complex sequence of events (Aist 1981). Fungal spores must first be dispersed to a suitable host plant by wind, water, or insect vectors. Spores then remain quiescent until conditions become favorable for germination. Of the many factors known to trigger spore germination, moisture and nutrient availability are among those most commonly believed to affect plant pathogenic fungi. At germination, a filamentous germ tube grows along the plant surface until, in response to physical and/or chemical signals, a swelling called an appressorium develops (Wynn and Staples 1981). The appressorium attaches tightly to the plant surface and produces a narrow penetration peg which enters the plant via physical pressure and/or enzymatic action (Emmett and Parbery 1975). The fungus then produces infection hyphae and begins colonizing its host.

Adhesion of fungal pathogens to the plant surface is thought to be essential for normal prepenetration development and successful infection (Nicholson and Epstein 1991). Adhesion enables spores, which usually germinate in water films, to develop without being washed from the leaf surface. Proper differentiation and function of appressoria is also dependent on firm adhesion. Many fungal pathogens are known to secrete extracellular matrix materials (ECM) which are believed to act as adhesives. Excellent recent reviews are available which discuss adhesion in a wide range of pathogenic fungi and provide historical context (Nicholson and Epstein 1991, Hardham 1992). For present considerations, we have chosen to focus on four host-parasite systems in which adhesion has been critically examined. We hope these examples will provide a glimpse of the variety of adhesion mechanisms which plant pathogenic fungi have evolved, and demonstrate the importance of adhesion in pathogenesis.

Cochliobolus heterostrophus

C. heterostrophus (anamorph *Bipolaris maydis*) is an ascomycete which causes the southern corn leaf blight disease. Inoculum consists of conidia (asexual spores) which are produced from necrotic host tissues and are dispersed to healthy maize plants by wind and splashing

* Correspondence and reprints: Department of Plant Pathology, Iowa State University, Ames, IA 50011-1020, U.S.A.

rain. Conidia germinate rapidly in water and produce germ tubes which are surrounded by ECM. Appressoria differentiate at the junctions between epidermal cells (Wynn and Staples 1981) and send penetration pegs through the underlying plant cell walls. The pegs give rise to infection hyphae which rapidly proliferate in epidermal and mesophyll tissues. Penetration can occur as early as 3 h after inoculation (Braun unpubl.). Non-specific attachment of *C. heterostrophus* conidia and germlings to a wide variety of surfaces has recently been demonstrated (Braun and Howard 1994). Conidia in water began adhering to a glass surface just before germ tube emergence, about 20 min after hydration. Over 90% of the germinating conidia were firmly attached within a 1 h incubation period. The fungus attached in a similar fashion to maize leaves and a wide variety of artificial surfaces including Teflon, cellophane, Mylar, polystyrene, and polyethylene. There was no obvious relationship between surface hydrophobicity and degree of adhesion. Attachment was prevented by the metabolic inhibitors sodium azide and cycloheximide indicating that respiration and protein synthesis were required for adhesion.

The temporal relationship between matrix secretion and attachment was also examined, establishing a role for ECM in adhesion (Braun and Howard 1994). Just before and during emergence of the germ tubes, ECM was secreted from the spore tips (Figs. 1 and 3). Conidia became tethered to the surface by pads of this adhesive. As the germ tubes developed, they remained covered by a thin layer of this material which could be stained with a variety of histological reagents (Evans et al. 1982) (Fig. 2). A much broader outer layer of ECM was also produced around germ tubes and appressoria. The outer layer was only visible with the light microscope when germlings were negatively stained using India ink (Evans et al. 1982) (Fig. 2). This layer could also be visualized with cryoSEM (Fig. 5). When germlings were removed from leaves, the leaf surface waxes were removed from areas which were in contact with germ tubes and appressoria (Fig. 4), but not from regions which contacted conidia. While the inner layer of ECM is clearly responsible for early adhesion of conidia, it is likely that both layers contribute to the attachment of germ tubes and appressoria. The composition of the ECM associated with *C. heterotrophus* germlings has not been determined although, based on staining characteristics, it is likely that at least some component of the inner sheath layer is proteinaceous (Braun and Howard 1994, Evans and Stempen 1986).

Magnaporthe grisea

M. grisea (anomorph *Pyricularia grisea*), an ascomycete, causes rice blast, a devastating disease of rice found worldwide. It has recently been demonstrated that adhesion plays at least two critical roles in the preinfection development of this pathogen. *M. grisea* has evolved a remarkable mechanism for rapid attachment of conidia to leaves, and its appressoria have been shown to bind very tightly to the surfaces on which they differentiate.

Hamer et al. (1988) demonstrated that *M. grisea* conidia were capable of attachment to a variety of surfaces immediately upon hydration. Within 20 min after inoculation, before germ tube emergence, over 90% of the conidia were firmly attached to a Teflon surface. Conidia were found to adhere more tightly to Teflon than to a more hydrophilic glass surface. Spore attachment was prevented by treatment with the lectin Concanavalin A (Con A). When conidia were hydrated in the presence of fluorescein isothiocyanate-conjugated Con A, a globule of material at the spore apex became strongly labeled. This material, which Hamer et al. called spore tip mucilage (STM), was shown to be the adhesive which bound conidia to leaves and other surfaces. In dormant conidia, STM was stored in a periplasmic compartment at the conidial apex (Fig. 6). When the conidia were hydrated, the spore wall ruptured and STM was immediately released (Figs. 7 and 11). Upon contact, the droplet of STM tethered the conidium to the surface (Fig. 12). The process of STM release and spore attachment was not dependent on fungal metabolism and occurred in the presence of sodium azide and cycloheximide. The composition of STM has not been determined but the binding of Con A indicates the presence of α-mannoside or α-glucoside (Figs. 8–10).

M. grisea penetrates host tissues from spherical, darkly-pigmented appressoria which develop at the ends of relatively short germ tubes (Fig. 13). A layer of melanin lines the inside surface of the appressorial walls except for a portion of the bottom which is in contact with the host (Bourett and Howard 1990). It has been shown that the melanized appressorial walls are differentially permeable to water and solutes. Efflux of solutes is blocked by the melanin layer allowing the buildup of osmotically generated hydrostatic pressures of considerable magnitude (Howard and Ferrari 1989, Howard et al. 1991). Appressorial turgor pressures in excess of 8.0 MPa have been measured (Howard et al. 1991). This pressure provides the necessary force to push the pen-

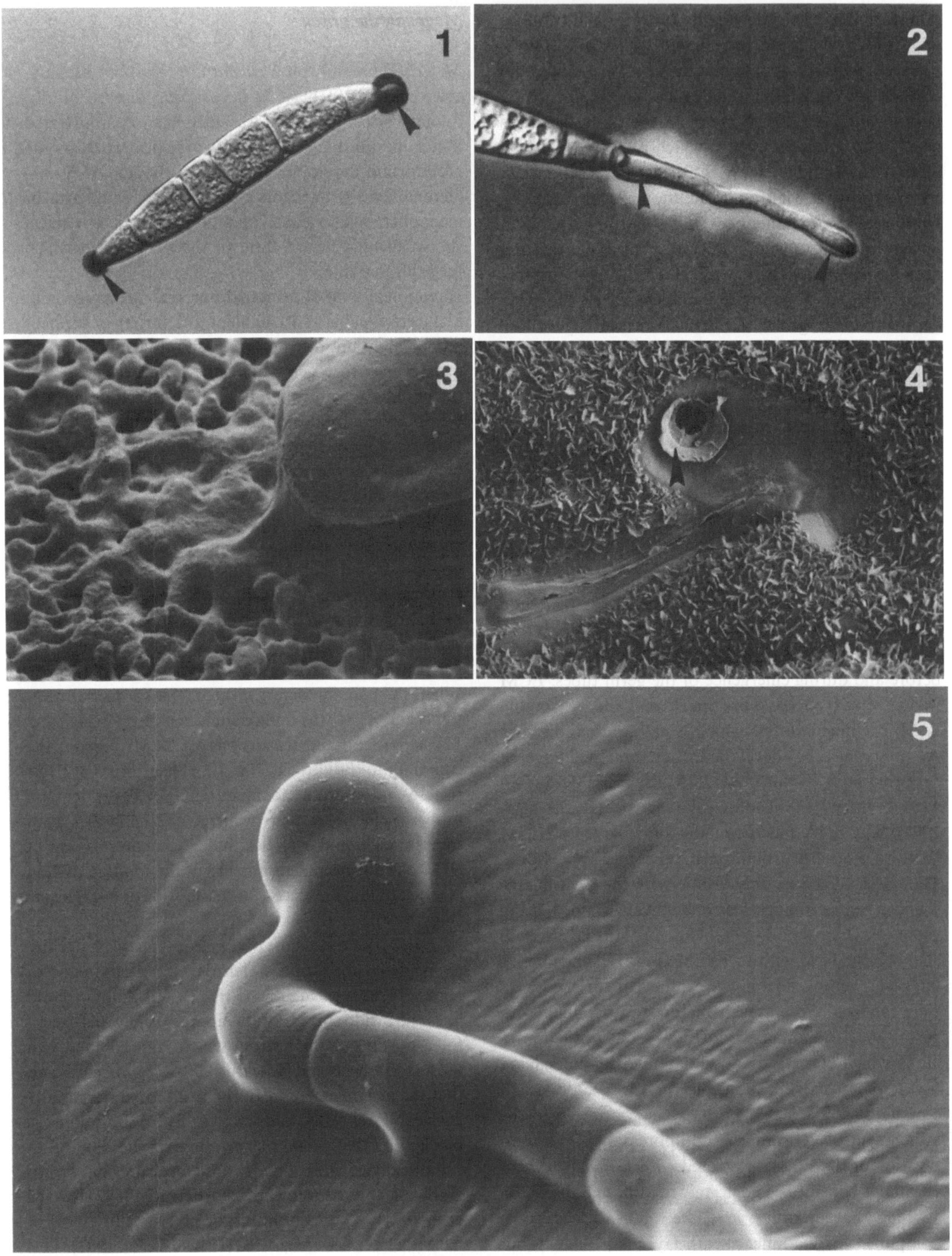
1
2
3
4
5

etration peg through the host cuticle and cell wall or into hard artificial surfaces (Fig. 14). In order for this pressure-driven penetration mechanism to function, the appressoria must be very tightly sealed to the host surface. Extracellular matrix and a specialized ring of wall material at the base of the appressorium are thought to function in sealing the appressorium to the substratum (Fig. 14). Appressoria which formed on plastic coverslips remained attached through extensive sonication and exposure to extremely high shear forces (Howard and Ferrari 1989).

Uromyces appendiculatus/U. viciae-fabae

Of the many important rust pathogens, those in the genus *Uromyces* have been investigated most rigorously with respect to adhesion. *U. appendiculatus* and *U. viciae-fabae* are two closely related species causing diseases of *Phaseolus vulgaris* and *Vicia faba*, respectively. Attachment of urediospores to host and artificial surfaces has recently been examined in *U. viciae-fabae*. Deising et al. (1992) found that when urediospores were placed on host leaves, misted with water, and maintained in a moist environment, a thin film of matrix material could be seen on the spore surface. When these spores were allowed to dry, the material appeared to collect between the spore and the leaf surface (Fig. 15). If the spore was then removed with adhesive tape, the material, referred to as an adhesion pad, remained on the leaf surface (Fig. 16). Adhesion pads were also formed by autoclaved spores, however, when the dead spores were removed from the leaf surface, the pads remained attached to the spores. When living urediospores were exposed to the serine-esterase inhibitor diisopropyl fluorophosphate, the pads also failed to remain attached to the leaf surface. Two esterases and a cutinase were isolated from the surface of the urediospores. When autoclaved spores were incubated in the presence of these enzymes, the pads were again able to adhere to the leaf surface. On the basis of this and other evidence, the authors felt that the esterases and cutinase were actively involved in spore adhesion. Although pad formation appeared to be a passive process, firm adhesion of the pads to the leaf surface was dependent on the enzyme activity. The progressive increase in adhesion which they observed over a 2 h incubation period was attributed to the activity of the esterases and cutinase associated with the spore surface. Beckett and co-workers (Beckett et al. 1990, Clement et al. 1993 a) have also studied attachment of *U. viciae-fabae* urediospores and have suggested that hydrophobic interactions and ECM together are responsible for adhesion. They hypothesized that dry urediospores first attached to leaves and synthetic surfaces via hydrophobic interactions. Once the spores were hydrated and germlings began to develop, ECM released from the spore surfaces and that secreted by the developing germlings also contributed to adhesion (Fig. 18). They also noted an increase in germling adhesion during a 2 h incubation period and indicated that this increase was closely correlated with increased ECM production. Beckett's group questioned the importance of cutinase and esterase in urediospore attachment (Clement et al. 1993 a). They pointed out that Deising et al. (1992) performed their experiments at 30 °C, an atypically high temperature. The *Uromyces* strains used by Beckett's group failed to germinate at these high temperatures. They observed that adhesion progressively increased during a 2 h incubation period when spores were placed on several synthethic surfaces which would not be effected by cutinase and esterase enzymes (Clement et al. 1993 a). They felt that this increase in non-specific adhe-

Fig. 1. *Cochliobolus heterostrophus* conidium on glass surface 30 min after inoculation, stained with nigrosin. Adhesive materials (arrows) are present at each end of the conidium where germ tubes are emerging. Differential interference contrast, × 575. [Figures 1–5 are from Braun and Howard (1994), with permission of "Experimental Mycology"]

Fig. 2. Germling 2 h after inoculation onto a glass surface, stained with nigrosin and India ink. Germ tube is surrounded by two-layered sheath of extracellular matrix material. Inner sheath layer is stained by nigrosin (arrows), outer layer excludes India ink and appears as a halo surrounding the germ tube. Differential interference contrast, × 750

Fig. 3. Secreted matrix material at conidial apex prior to germ tube emergence, 40 min after inoculation onto a Gelman filter. CryoSEM, × 5900

Fig. 4. Imprint of appressorium and germ tube on surface of maize leaf following removal of germling 4 h after inoculation. Cone of fungal wall material (arrow) surrounds hole at site of penetration. Crystals of leaf surface wax have been removed along with the germling. CryoSEM, × 2500

Fig. 5. Germ tube and appressorium covered with extracellular matrix 4 h after inoculation onto cellophane. CryoSEM, × 3450

sion could best be explained by the increase in ECM production which occurs during this period. They concluded that, although cutinase and esterase enzymes may play a role in adhesion to the plant cuticle under some conditions, this mechanism is certainly not the only one which can explain the phenomenon of urediospore adhesion.

There has been considerable speculation that cutinases and esterases released from spores can cause erosion of the cuticle and waxes on plant host surfaces (Nicholson and Epstein 1991). On occasion, such observations have been made with rust spores (Fig. 17). Although these findings are certainly provocative, more work needs to be done in order to determine their significance.

One of the most fascinating aspects of rust biology concerns the prepenetration development of urediospore germlings. Germ tubes exhibit directional growth patterns which increase the likelihood that they will contact stomata of the host leaf. Once a germ tube contacts the stomate, an appressorium is formed and the penetration peg enters the leaf through the stomatal pore. Both directional growth and appressorium induction are mediated by physical contact of the germ tube with specific topographical features of the host (Wynn snd Staples 1981). The topographical signals recognized by the bean rust fungus, *U. appendiculatus*, have been precisely defined through the use of microfabricated synthetic surfaces (Hoch et al. 1987).

It is now clear that firm attachment is essential for proper recognition of the topographic signals for oriented growth and appressorium differentiation. *U. appendiculatus* germlings grown in the presence of the non-specific protease, pronase E, showed reduced levels of adhesion, directional growth and appressorium differentiation compared to controls treated with heat-denatured enzyme (Epstein et al. 1987). Pronase E treatments also removed ECM from the germ tube-substratum interface. Germlings which were firmly attached to parafilm or leaf surfaces could be removed by treatment with very high concentrations of pronase E (500 µg/ml). One or more adhesive proteins seem to be of critical importance for adhesion and differentiation of *U. appendiculatus* germlings.

Terhune and Hoch (1993) have recently examined the effect of surface hydrophobicity on *U. appendiculatus* germlings. By coating glass or quartz substrates with organosilanes they were able to create surfaces with a range of hydrophobicities. Microfabricated surfaces capable of inducing appressorium formation were also coated. Urediospore and germling attachment, area of germling-substratum contact, and appressorium formation were all greater on the more hydrophobic surfaces. The hydrophobicity of the bean leaf surface was similar to the test surfaces which proved favorable for adhesion and germling development.

Adhesion continues to be important as *Uromyces* penetrates and colonizes host tissues. *U. appendiculatus* appressoria are not pigmented and only modest increases in turgor pressure have been noted during their development (Terhune et al. 1993). The appressoria do, however, bind tightly to the substratum and are capable of generating enough force to deform cuticular ridges on guard cells as well as ridges on microfabricated polystyrene and polycarbonate surfaces (Terhune et al. 1993). Once the fungus enters the substomatal chamber, its hyphae develop intercellularly, generally in contact with host cell walls (Fig. 19). A layer of ECM, probably acting as an adhesive, is present between the fungal hyphae and the host cells (Beckett and Woods 1987). A cytochemically different matrix material is found between adjacent hyphae. Some fungal cells differentiate into haustorial mother cells which function very much like appressoria, binding tightly to the host wall and producing penetration pegs which develop into haustoria within the host cells. The haustorial

---

Fig. 6. Dry *Magnaporthe grisea* conidium prepared by freeze substitution. Periplasmic compartment at the tip of the conidium contains spore tip mucilage (STM), the material responsible for surface attachment of *M. grisea* spores. Arrow denotes plasmalemma bordering the periplasmic compartment. Freeze-substitution transmission electron microscopy (FSTEM), × 34,400. **Inset** Low magnification of the same conidium. × 400. [Figures 6 and 7 are from Hamer et al. (1988), with permission of "Science"]

Fig. 7. Hydrated conidium in which spore wall (arrow) has ruptured due to expansion of STM. FSTEM, × 36,000

Fig. 8. Globule of STM at tip of hydrated conidium. STM has been labeled with gold-conjugated Concanavalin A (Con A). FSTEM, × 21,500. [Figures 8–10 are from Bourett et al. (1993), with permission of "Experimental Mycology"]

Fig. 9. Adjacent section to Fig. 8, preincubated with hapten α-D-mannopyranoside before treatment with Con A-gold. Labeling has been reduced dramatically, FSTEM, × 21,500

Fig. 10. Tip of conidium substituted in osmium tetroxide/ethanol rather than osmium tetroxide/acetone resulting in more electron dense STM. Labeled with Con A-gold. FSTEM, × 21,500

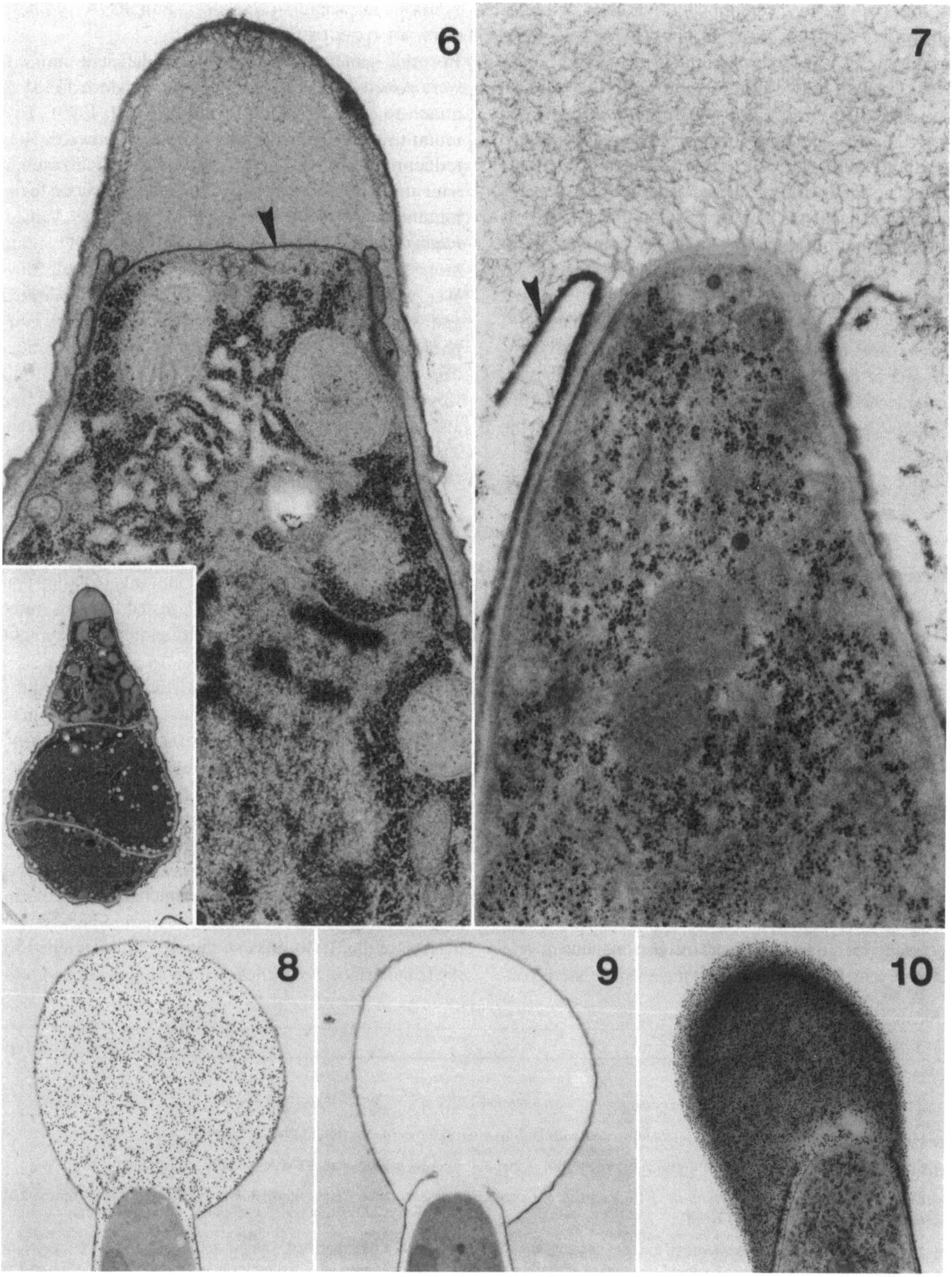
6
7
8
9
10

mother cells are also surrounded by a layer of ECM. Efforts are now underway to characterize the ECM produced by *Uromyces* germlings (Moloshok et al. 1993, Clement et al. 1993 b). Culture filtrates were assayed for total proteins, carbohydrates, ions, pH, and many enzyme activities, and proteins were separated by SDS-PAGE. The information gained from these studies so far is of a preliminary nature, but detailed biochemical analyses have potential to provide important information about the adhesive components of ECM. Both of these research groups stress that ECM may serve other important functions in addition to a likely role in adhesion. It has been suggested that ECM may protect fungi against toxic host metabolites, slow the rate of dessication, provide a favorable milieu for enzyme delivery and function, etc. (Nicholson and Epstein 1991). Moloshok et al. (1993) provide convincing evidence that the ECM of *Uromyces* functions in maintaining a favorable microenvironment for germling growth.

Nectria haematococca

N. haematoccoca (anamorph *Fusarium solani* f.sp. *cucurbitae*) is an ascomycete which attacks cucurbits. Jones and Epstein (1989, 1990) have characterized attachment with this pathogen and have carried out genetic studies to assess the importance of adhesion as a virulence factor. Conidia were found to attach to tissues of host and non-host plants and both hydrophilic and hydrophobic artificial surfaces. Conidial adhesion reached maximum levels on polystyrene in about 20 min although germ tube emergence did not occur until 3–7 h after inoculation. The attachment process was inhibited by sodium azide and cycloheximide indicating that respiration and protein synthesis were necessary for adhesion. Tunicamycin, monensin, and actinomycin D had no effect, suggesting that glycosylation, intracellular transport, and RNA synthesis were not required for attachment.

For the genetic studies, adhesion-deficient mutants were selected by enriching for spores which failed to attach to polystyrene (Jones and Epstein 1990). Two mutant strains selected for further study showed a 50% reduction in adhesion to both polystyrene and zucchini fruit surfaces. The authors acknowledge that since the mutants were not backcrossed to wild-type *N. haematococca*, the adhesion-deficient strains might contain more than a single mutation. They determined, however, that the mutants were not generally debilitated and were similar to the wild-type strain in spore morphology, germination, sporulation, biomass production, prototrophy, and fertility. When inoculated into wounded zucchini fruits, the mutants caused the same amount of tissue maceration as the wild-type pathogen. The adhesion-deficient mutants were, however, greatly reduced in virulence when inoculated onto non-wounded zucchini fruits. Ultrastructural analysis revealed that as wild-type conidia became adhesion competent, a new outer wall layer appeared which was lacking in the adhesion-reduced mutants (Ceasar-Ton That and Epstein 1991). They suggested that this outer wall layer, which covered the entire spore surface, might be responsible for adhesion.

There is no question of the power which genetic analysis can bring to studies of fungal adhesion. However, there are inconsistencies which make these studies with *N. haematococca* very difficult to interpret. Most important, the mutants were adhesion-reduced, meaning that a significant proportion of the conidia were still capable of attachment. The data presented indicates that 30–50% of spores produced by the mutant strains did attach to polystyrene and zucchini fruit surfaces. In virulence assays, inoculum droplets contained 20,000 macroconidia. If 30–50% of these conidia attached to the fruit surface, a significant amount of disease should

---

Fig. 11. Hydrated *M. grisea* conidium on conidiophore with a globule of STM at the tip. CryoSEM, × 3,400

Fig. 12. Conidium glued to surface of a plastic coverslip (poly vinyl chloride) by a droplet of STM. CryoSEM, × 3,800

Fig. 13. Mature spherical appressorium on a Mylar surface. Conidium and germ tube have collapsed. CryoSEM, × 3,950. (Unpublished micrograph, RJ Howard and ND Read)

Fig. 14. Interface between an appressorium and the polyethylene surface on which it differentiated. Germling and most of the appressorium have been removed by sonication. Exposed remnant of smooth melanized wall material (*mw*) is bound tenaciously to the substratum and is surrounded by matrix material which may act as an adhesive. Hole through polyethylene surface was made by the penetration peg. CryoSEM, × 26,400

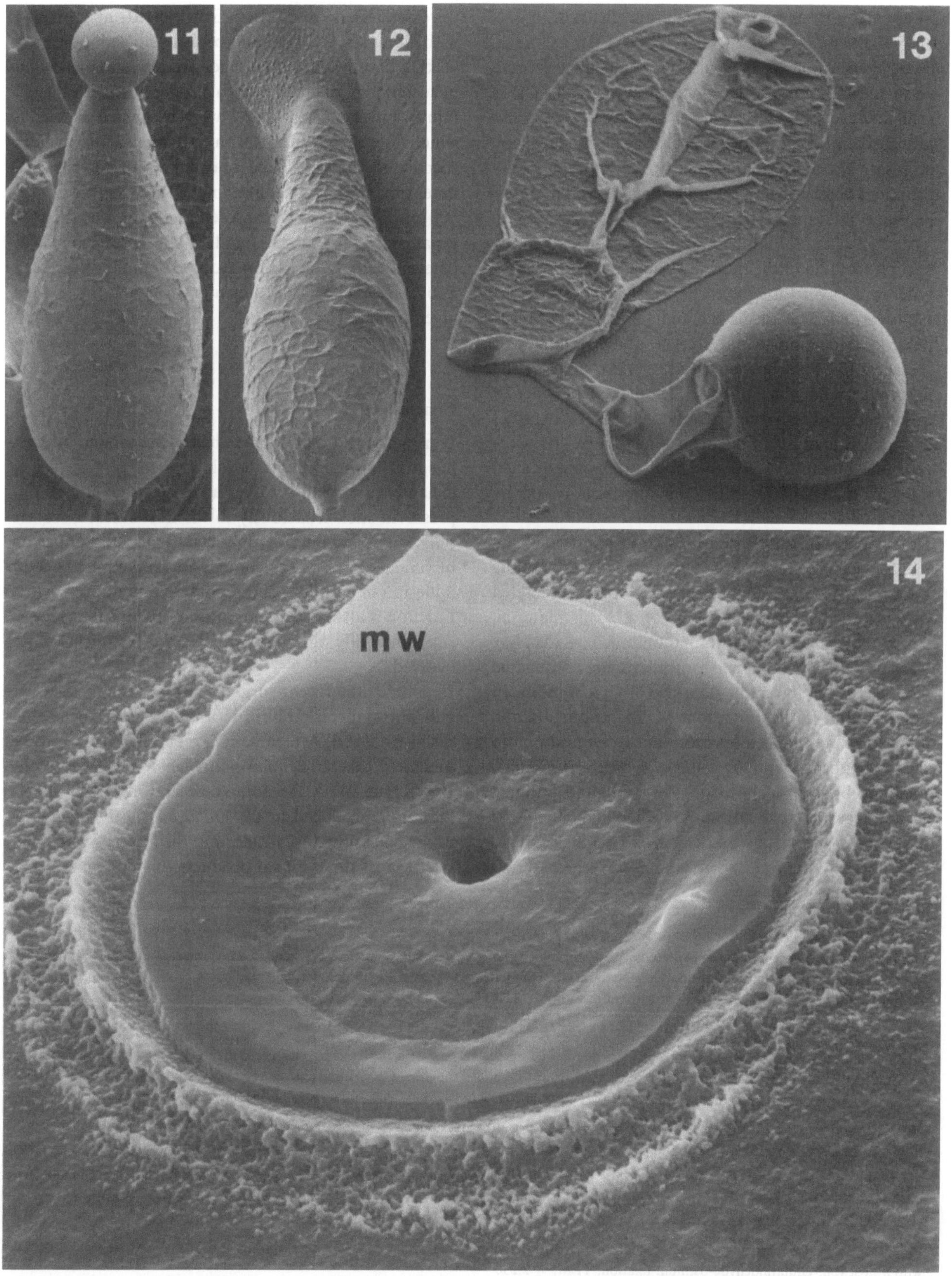
11
12
13
14
mw

be apparent 12 days after inoculation. Instead, in one assay the mutants produced roughly 10% as much tissue maceration as the wild-type, and in a second assay the mutants produced no disease at all. These results suggest that either (1) the mutants were compromised in other important pathogenicity traits not necessarily related to adhesion, or (2) the type of attachment observed with the adhesion-reduced mutants had little relevance with respect to pathogenesis.

The ultrastructural information is also difficult to interpret. Conidia of the mutant strains lacked a normal outer wall layer thought to be involved in adhesion. However, 30–50% of these conidia should still be capable of attaching to polystyrene and zucchini fruit surfaces. It would be instructive to view some of the mutant conidia which did stick to surfaces in order to assess the importance of the outer wall layer. A careful cytological analysis of cucurbit infection by the wild-type and mutant *N. haematococca* strains might shed light on many of the important questions raised by these studies.

Recently, Kwon and Epstein (1993) have provided evidence that adhesion of *N. haematococca* macroconidia is correlated with production of a glycoprotein secreted at the spore tips prior to germination. This material, which they called macroconidial tip mucilage (MTM), was labeled by fluorescein isothiocyanate-conjugated Con A. Incubation in the presence of Con A inhibited adhesion but not germination. MTM production and adhesion were induced by zucchini extract and potato dextrose broth, but not by water, V-8 broth, or minimal medium. Spores incubated in V-8 broth and minimal medium germinated normally, which indicated that MTM was specifically associated with adhesion and and not germination. When spore surface materials were separated by SDS-PAGE, blotted, and probed with Con A, a 90 kDa band became labeled. Since the 90 kDa band was obvious when spores were incubated in zucchini extract and potato dextrose broth, but only faintly present when spores were incubated in media which did not induce adhesion competence, the authors speculated that the band may represent MTM. Adherent macroconidia were removed by treatment with 500 µm/ml pronase E or proteinase K, as was the material present in 90 kDa band. It will be of interest to determine if the adhesion-reduced mutants discussed above are capable of producing MTM and the 90 kDa glycoprotein.

Conclusions

During the last decade there has been renewed interest in fungal adhesion and its role in pathogenesis. While it has been known for over a century that fungal appressoria stick tenaciously to the host surface, we are only now beginning to appreciate that rapid adhesion of fungal spores is critical as well. The importance of adhesion in germling differentiation and appressorium function is also now apparent. It is obvious that several fundamentally different adhesion mechanisms have evolved.

In spite of the recent surge of interest in adhesion, we still know remarkably little about the actual mechanisms involved. Attachment is generally believed to be mediated by extracellular matrix materials secreted by the fungus. With the notable exception of the matrix materials produced in *Colletotrichum graminicola* acervuli, we know very little about the function and composition of these fungal products (Nicholson and Ep-

---

Fig. 15. *Uromyces viciae-fabae* urediospore on the surface of a *Vicia faba* leaf. Spore was misted, incubated in a moist atmosphere and allowed to dry. Adhesion pad composed of matrix material has accumulated where spore is in contact with the leaf. CryoSEM, × 4,650. [Figures 15 and 16 are from Deising et al. (1992), with permission of "The Plant Cell"]

Fig. 16. Adhesion pad which remained on the leaf surface after urediospore, as in Fig. 15, was removed with adhesive tape. CryoSEM, × 3,075

Fig. 17. Naturally collapsed urediospores of *Puccinia hordei* on the surface of a barley leaf. This cell collapse is a reversible phenomenon which allows spores to reduce their water content. Erosion of cuticle and leaf wax around the spores may have been caused by enzymes released from the ungerminated spores. CryoSEM, × 850. [From Read (1991), with permission of Springer-Verlag]

Fig. 18. *Uromyces vignae* germling on *Vigna sinensis* leaf. Germ tube surrounded by extracellular matrix material (arrows) is approaching a stomatal complex. CryoSEM, × 950. (Unpublished micrograph, ND Read)

Fig. 19. Freeze-fractured broad bean leaf revealing *Uromyces viciae-fabae* hyphae in substomatal chamber. Matrix material can be seen where hypha has been fractured away from host palisade mesophyll cell (arrows). CryoSEM, × 550. [Figures 19 and 20 from Beckett and Woods (1987), with permission of the "Canadian Journal of Botany"]

Fig. 20. *U. viciae-fabae* passing through stomatal pore and into substomatal chamber. Haustorial mother cell (arrow) can be seen in contact with a host cell. Glutaraldehyde-osmium fixation, TEM, × 2,250

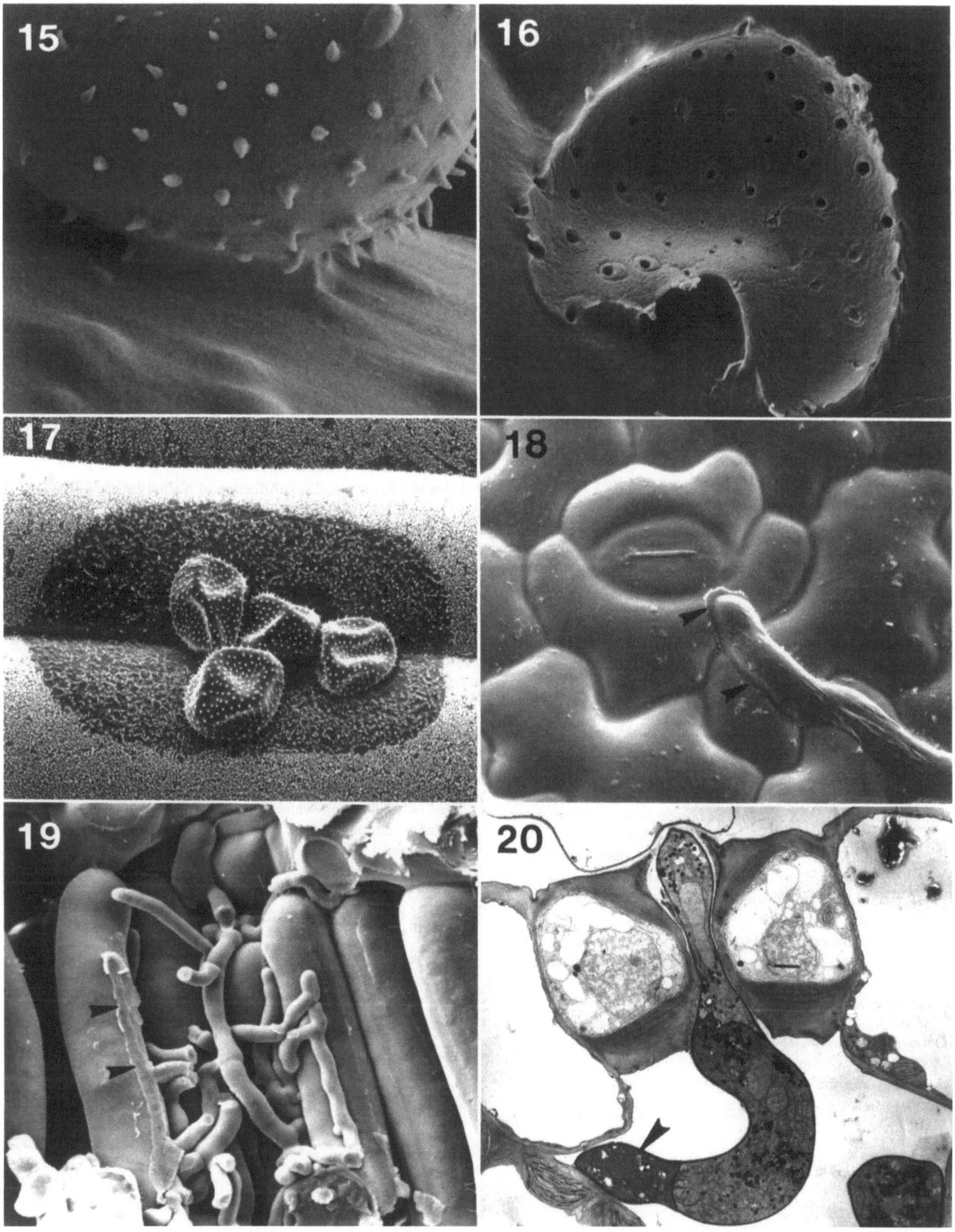
15
16
17
18
19
20

stein 1991). Intensive efforts need to be directed towards the isolation and biochemical characterization of the adhesive materials produced by fungal pathogens. Finally, except for the investigations of *N. haematococca* by Epstein and co-workers discussed above, few have taken a genetic approach to studying adhesion of fungal plant pathogens. Genetic analysis has greatly increased our understanding of many complex developmental processes and it needs to be fully exploited in studies of fungal adhesion. Investigations which combine the strengths of genetics, molecular biology, biochemistry and cell biology are likely to yield important insights into the mechanisms and significance of adhesion in host-parasite interaction.

References

Aist JR (1981) Development of parasitic conidial fungi in plants. In: Cole GT, Kendrick B (eds) Biology of conidial fungi, vol 2. Academic Press, New York, pp 75–110

Beckett A, Woods AM (1987) The pattern of colony development and the formation of the uredinium of *Uromyces viciae-fabae* on *Vicia faba*. Can J Bot 65: 1998–2006

– Tatnell JA, Taylor N (1990) Adhesion and pre-invasion behaviour of urediniospores of *Uromyces viciae-fabae* during germination on host and synthetic surfaces. Mycol Res 94: 865–875

Bourett TM, Howard RJ (1990) In vitro development of penetration structures in the rice blast fungus *Magnaporthe grisea*. Can J Bot 68: 329–342

– Picollelli MA, Howard RJ (1993) Postembedment labeling of intracellular concanavalin A-binding sites in freeze-substituted fungal cells. Exp Mycol 17: 223–235

Braun EJ, Howard RJ (1994) Adhesion of *Cochliobolus heterostrophus* conidia and germlings to leaves and artificial surfaces. Exp Mycol (in press)

Ceasar-TonThat TC, Epstein L (1991) Adhesion-reduced mutants and the wild-type *Nectria haematococca*: an ultrastructural comparison of the macroconidial walls. Exp Mycol 15: 193–205

Clement JA, Martin SG, Porter R, Butt TM, Beckett A (1993 a) Germination and the role of extracellular matrix in adhesion of urediniospores of *Uromyces viciae-fabae* to synthetic surfaces. Mycol Res 97: 585–593

– Butt TM, Beckett A (1993 b) Characterization of the extracellular matrix produced in vitro by urediniospores and sporelings of *Uromyces viciae-fabae*. Mycol Res 97: 594–602

Deising H, Nicholson RL, Haug M, Howard RJ, Mendgen K (1992) Adhesion pad formation and the involvement of cutinase and esterases in the attachment of uredospores to the host cuticle. Plant Cell 4: 1101–1111

Emmett RW, Parbery DG (1975) Appressoria. Annu Rev Phytopathol 13: 147–167

Epstein L, Laccetti LB, Staples RC, Hoch HC (1987) Cell-substratum adhesive protein involved in surface contact responses of the bean rust fungus. Physiol Mol Plant Pathol 30: 373–388

Evans RC, Stempen H (1986) Localization of protein in the hyphal sheath of *Bipolaris maydis* race T. Phytopathology 76: 792–794

– – Frasca P (1982) Evidence for a two-layered sheath on germ tubes of three species of *Bipolaris*. Phytopathology 72: 804–807

Hamer JE, Howard RJ, Chumley FG, Valent B (1988) A mechanism for surface attachment in spores of a plant pathogenic fungus. Science 239: 288–290

Hardham AR (1992) Cell biology of pathogenesis. Annu Rev Plant Physiol Plant Mol Biol 43: 491–526

Hoch HC, Staples RC, Whitehead B, Comeau J, Wolf ED (1987) Signaling for growth orientation and cell differentiation by surface topography in *Uromyces*. Science 235: 1659–1662

Howard RJ, Ferrari MA (1989) Role of melanin in appressorium function. Exp Mycol 13: 403–418

– – Roach DH, Money NP (1991) Penetration of hard substrates by a fungus employing enormous turgor pressures. Proc Natl Acad Sci USA 88: 11281–11284

Jones MJ, Epstein L (1989) Adhesion of *Nectria haematococca* macroconidia. Physiol Mol Plant Pathol 35: 453–461

– – (1990) Adhesion of macroconidia to the plant surface and virulence of *Nectria haematococca*. Appl Environ Microbiol 56: 3772–3778

Kwon YH, Epstein L (1993) A 90 kDa glycoprotein associated with adhesion of *Nectria haematococca* macroconidia to substrata. Mol Plant Microbe Interact 6: 481–487

Moloshok TD, Leinhos GME, Staples RC, Hoch HC (1993) The autogenic extracellular environment of *Uromyces appendiculatus* urediospore germlings. Mycologia 85: 392–400

Nicholson RL, Epstein L (1991) Adhesion of fungi to the plant surface: prerequisite for pathogenesis. In: Cole GT, Hoch HC (eds) The fungal spore and disease initiation in plants and animals. Plenum, New York pp 3–23

Read ND (1991) Low-temperature scanning electron microscopy of fungi and fungus-plant interactions. In: Mendgen K, Lesemann D-E (eds) Electron microscopy of plant pathogens. Springer, Berlin Heidelberg New York Tokyo, pp 17–29

Terhune BT, Hoch HC (1993) Substrate hydrophobicity and adhesion of *Uromyces* urediospores and germlings. Exp Mycol 17: 241–252

– Bojko RJ, Hoch HC (1993) Deformation of stomatal guard cell lips and microfabricated artificial topographies during appressorium formation by *Uromyces*. Exp Mycol 17: 70–78

Wynn WK, Staples RC (1981) Tropisms of fungi in host recognition. In: Staples RC, Toenniessen GH (eds) Plant disease control: resistance and susceptibility. Wiley, New York, pp 45–69

Protoplasma (1994) 181: 213–232

PROTOPLASMA

Cell surface antigens of *Phytophthora* spores: biological and taxonomic characterization

A. R. Hardham*, D. M. Cahill, M. Cope, B. K. Gabor, F. Gubler⁺, and G. J. Hyde⁺⁺**

Plant Cell Biology Group, Research School of Biological Sciences, Australian National University, Canberra, ACT

Received October 25, 1993
Accepted January 2, 1994

Summary. The oomycetes are a class of protists that produce biflagellate asexual zoospores. Members of the oomycetes have close phylogenetic affinities with the chromophyte algae and are widely divergent from the higher fungi. This review focuses on two genera, *Phytophthora* and *Pythium*, which belong to the family Pythiaceae, and the order Peronosporales. These two genera contain many species that cause serious diseases in plants. Molecules on the surface of zoospores and cysts of these organisms are likely to play crucial roles in the infection of host plants. Knowledge of the properties of the surface of these cells should thus help increase our understanding of the infection process. Recent studies of *Phytophthora cinnamomi* and *Pythium aphanidermatum* have used lectins to analyse surface carbohydrates and have generated monoclonal antibodies (MAbs) directed towards a variety of zoospore and cysts surface components. Labelling studies with these probes have detected molecular differences between the surface of the cell body and of the flagella of the zoospores. They have been used to follow changes in surface components during encystment, including the secretion of an adhesive that bonds the spores to the host surface. Binding of lectin and antibody probes to the surface of living zoospores can induce encystment, giving evidence of cell receptors involved in this process. Freeze-substitution and immunolabelling studies have greatly augmented our understanding of the synthesis and assembly of the zoospore surface during zoosporogenesis. Synthesis of a variety of zoospore components begins when sporulation is induced. Cleavage of the multinucleate sporangium is achieved through the progressive extension of partitioning membranes, and a number of surface antigens are assembled onto the zoospore surface during cleavage. Comparisons of antibody binding to many isolates and species of *Phytophthora* and *Pythium* have revealed that surface components on zoospores and cysts exhibit a range of taxonomic specificities. Surface antigens or epitopes may occur on only a few isolates of a species; they may be species-specific, genus-specific or occur on the spores of both genera. Spore surface antigens thus promise to be of significant value for studies of the taxonomy and phylogeny of these protists, as well as for disease diagnosis.

Keywords: Oomycetes; *Pythium; Phytophthora;* Monoclonal antibodies; Surface antigens; Immunocytochemistry.

Abbreviations: MAbs monoclonal antibodies; ConA Concanavalin A; SBA soybean agglutinin; WGA wheat germ agglutinin; gps glycoproteins.

Introduction

Organisms in the class oomycetes (Oomycota) (Dick 1989) are recognised by their fungal-like growth habit, aseptate mycelium and their production of biflagellate, heterokont zoospores. These characters have led the oomycetes to be commonly included in the fungal kingdom (e.g. Arms and Camp 1979, Ray et al. 1983). There is, however, much evidence that indicates that the phylogenetic affinities of this group of protists lie with the algae rather than with fungi. A diverse range of structural, biochemical, and molecular characteristics found in the oomycetes differ from their counterparts in the higher fungi and in the uniflagellate chytridiomycetes, but are similar to their counterparts in the chromophyte algae. These features are summarised in Table 1, and include the structure of the flagellar apparatus and spindle pole; the morphology of mitochondrial cristae; the occurrence of cellulose rather than chitin as the

* Correspondence and reprints: Plant Cell Biology Group, Research School of Biological Sciences, Australian National University, Canberra, ACT 2601, Australia.

** Present address: PetoSeed Co., Woodland, California, U.S.A.

⁺ Present address: Cooperative Research Centre for Plant Science, Research School of Biological Sciences, Australian National University, Canberra, ACT, Australia.

⁺⁺ Present address: Department of Biology, York University, Ontario, Canada.

Table 1. Comparison of structural, biochemical and molecular characters linking Oomycetes to the chromophyte algae and distinguishing them from Chytridiomycetes and higher fungi

Characters	Higher fungi[a]	Chytridiomycetes	Oomycetes	Chromophyte algae	References
Flagellation	none	one posterior whiplash flagellum	heterokont with anterior tinsel and posterior whiplash flagella	heterokont with anterior tinsel and posterior whiplash flagella	Barr 1983
Mastigonemes	none	none	tubular, tripartite	tubular, tripartite	Cavalier-Smith, 1986
Flagellar rootlets	none	microtubules radiate laterally or posteriorly from basal bodies	most microtubules radiate distally from basal body	most microtubules radiate distally from basal body	Barr 1981, 1983
Connection of basal bodies to plasma membrane	none	non-microtubular props	microtubules	microtubules	Barr 1981, 1983
Mitochondrial cristae	plate-like	plate-like	tubular	tubular	Cavalier-Smith 1981
Mitotic spindle poles	amorphous or multi-layered plaques	centrioles	centrioles	centrioles	Heath 1980
Predominant microfibrillar component in cell wall	microfibrils of chitin (polmer of N-acetylglucosamine)	microfibrils of chitin (polmer of N-acetylglucosamine)	microfibrils of cellulose (β-1,4-glucan)	microfibrils of cellulose (β-1,4-glucan)	Bartnicki-Garcia 1970, 1987
Storage carbohydrate	glycogen (α-1,4-glucan)	glycogen (α-1,4-glucan)	mycolaminaran (β-1,3-glucan)	laminaran (β-1,3-glucan)	Bartnicki-Garcia and Wang 1983
Lysine synthesis pathway	AAA	AAA	DAP	DAP	Vogel 1965, Bartnicki-Garcia 1970
Steroid metabolism	ergosterol and related compounds	ergosterol and related compounds	fucosterol and related compounds	fucosterol and related compounds	Bu'lock and Osagie 1976
Tryptophan biosynthesis enzymes	sedimentation pattern Types I–III	sedimentation pattern Type 1	sedimentation pattern Type IV	na	Hutter and DeMoss 1967, Bartnicki-Garcia 1970
Somatic ploidy	haploid	haploid	diploid	diploid	Sansome 1987
Small-subunit rRNA	divergent from Oomycetes and Chrysophytes	similar to Ascomycetes; divergent from Oomycetes and Chrysophytes	similar to Chrysophytes	similar to Oomycetes	Gunderson et al. 1987, Wolters and Erdmann 1988, Forster et al. 1990
Actin genes	divergent from *Phytophthora* spp.	na	divergent from higher fungi	na	Dudler 1990, Unkles et al. 1991
TRP 1 gene[b]	divergent from *Phytophthora* spp.	na	divergent from higher fungi	na	Karlovsky and Prell 1991
Hsp 70 gene	divergent from *Bremia*	na	divergent from true fungi	na	Judelson and Michelmore 1989
mtDNA sequences	divergent from *Phytophthora* spp.	na	more similar to plants than to the higher fungi	na	Sachay et al. 1993

[a] Ascomycetes, Basidiomycetes, Zygomycetes, and Deuteromycetes

[b] TRP 1 gene, bifunctional enzyme of tryptophan biosynthetic pathway: indole-3-glycerolphosphate synthetase-*N*-(5′-phosphoribosyl)anthranilate isomerase

AAA α-Aminoadipic acid; *DAP* α,ε-diaminopimelic acid; *na* information not available

predominant cell wall microfibrils; differences in a number of biochemical pathways; completion of meiosis immediately prior to gametogenesis leading to a diploid somatic thallus; and close similarities of RNA and DNA sequences with those in the algae and their divergence from those sequences in the true fungi. These affinities have led a number of authors to remove oomycetes from the Kingdom Fungi and place them within the Kingdom Protista (Raven et al. 1981, Villee et al. 1985, Margulis and Schwartz 1988, Dick 1989). The need to recognise the true phylogenetic affinities of oomycete organisms was become increasingly clear in recent years because these affinities have important repercussions not only for taxonomy and phylogeny but also for cellular and molecular studies. They mean, for example, that experimental or practical applications of drugs or fungicides are unlikely to be effective in oomycetes if the compounds used are specific for the higher fungi. In addition, it is likely that the phylogenetic distance of oomycetes from the higher fungi is a major contributing factor to the failure of gene regulatory sequences from higher fungi to support the expression of foreign genes in transformation experiments with *Phytophthora* (Kinghorn et al. 1991, Judelson et al. 1992). Although there are reports of transformation of oomycetes using heterologous promoters (Manavathu et al. 1988; Bailey et al. 1991, 1993), gene expression in these systems may be due to insertion of the foreign genes at fortuitous sites (Judelson et al. 1992). Transient expression and stable transformation of *P. infestans* (Judelson and Michelmore 1991, Judelson et al. 1991) and *P. megasperma* f. sp. *glycinea* (Judelson et al. 1993) have been achieved using a promoter from *Bremia lactucae*, a closely related oomycete species. While research on oomycetes may remain within the realm of the mycologist, it should be done with a full appreciation of the phylogenetic relationships of these organisms.

The oomycetes are an important group of protists if impact on human life and well-being is taken as a measure of importance. Many oomycete genera contain species that cause serious diseases in plants used for food production, horticulture, forestry and natural resource and recreation areas (Erwin et al. 1983); others are pathogens of freshwater fish (Burr and Beakes 1993). In many cases, disease dissemination and host infection are achieved through the production of large numbers of motile asexual zoospores. The zoospores can swim up to potential hosts where they attach, encyst and invade. There is little doubt that successful infection requires the participation of a variety of molecules on the surface of the zoospores and cysts, molecules that as yet we know little about. This situation is, however, changing as new techniques are applied to studies of surface properties of these cells. Surface carbohydrates have been probed with fluorescent- and gold-labelled lectins, and in three studies monoclonal antibodies (MAbs) have been raised against spore components to give markers for molecules on the surface of zoospores and cysts of species of *Phytophthora, Pythium,* and *Saprolegnia.* Information arising from studies of *Saprolegnia* is presented in Burr and Beakes (1994) and will not be reiterated here. In this article we will review recent results of studies of species of two peronosporalean genera, *Phytophthora and Pythium*, with the main focus on *P. cinnamomi.*

P. cinnamomi is a destructive plant pathogen that has a wide host range, infecting close to 1000 species of plants (Zentmyer 1980). It is a soilborne organism that in most cases attacks the roots of plants. *P. cinnamomi*, like other oomycetes, exists for much of its life cycle as vegetative mycelium. When nutrients or other factors become limiting, the mycelium is induced to sporulate, usually asexually. During sporulation, nulcei and cytoplasm flow from subtending hyphae into expanding hyphal tips to form sporangia (Figs. 1 and 5–7). The mature multinucleate sporangium is triggered to undergo cytoplasmic cleavage to form typically 20–30 uninucleate zoospores (Fig. 2). When cleavage is complete, the apical pore of the sporangium ruptures and the zoospores are released (Fig. 3). The biflagellate zoospores (Figs. 4 and 8) can swim through water in the soil and are chemotactically attracted to the roots of potential host plants (Carlile 1983). At the root surface, the zoospores encyst (Figs. 8 and 9). In this process, the cells round up, secrete an adhesive and deposit a cellulosic cell wall. About 20–30 min later the cysts germinate, and the germ tubes grow into the adjacent plant tissues (Fig. 10).

Zoospores are thus key players in the infection of host plants, and molecules on the surface of zoospores and cysts must mediate many of the critical events in the infection process. Structural modifications of the flagellar surface will affect zoospore motility; plasma membrane-bound receptors will initiate signal transduction involved in chemotaxis, the induction of encystment and chemotrophic growth of the germ tube; material secreted onto the cyst surface will serve as an adhesive; and plasma membrane bound enzymes will be responsible for wall formation. Thus, if we are to understand the biology of these protists and the molecular basis of their infection strategies, we must understand the structural and biochemical characteristics of their cell surfaces. Research in this area has now begun and promises to yield much valuable information. These studies, involving as they do the generation of immunological probes for surface components, also promise to make a significant contribution to our understanding of the taxonomy and phylogeny of these protists.

Biology of cell surface antigens

The zoospore surface

Oomycete zoospores lack a cell wall and thus their outer surface is that of the plasma membrane which is coated with carbohydrates (the glycocalyx) as revealed by ruthenium red staining (Fig. 11) (Hardham 1989) or by silver methenamine cytochemistry (Lehnen and Powell 1988, Sadowski and Powell 1990). Surface carbohydrates have been further analysed by lectin cytochemistry. The lectin, Concanavalin A (ConA) binds to both cell body and flagellar surfaces of *P. cinnamomi* zoospores (Fig. 12) (Hardham 1985; Bacic et al. 1985; Hardham and Suzuki 1986, 1990; Hardham et al. 1991 a). Zoospores of a number of other *Phytophthora* spp. (*P. boehmeriae, P. citricola, P. cryptogea, P. meadii, P. megasperma* var. *sojae*, *P. nicotianae*, and *P. palmivora*), *Pythium aphanidermatum*, *Py. butleri*, *Py. debaryanum*, *Py. periplocum*, *Saprolegnia ferax* (Hardham et al. 1991 a; A. R. Hardham unpubl. obs.), *S. ferax* (Lehnen and Powell 1993), *S. diclina* and *S. parasitica* (Burr and Beakes 1994) also react with ConA, indicating that carbohydrates containing glucose and/or mannose sugars are commonly found on the surface of these cells. Lectins with specificities for N-acetyl galactosamine [soybean agglutinin (SBA), *Helix pomatia* agglutinin, peanut agglutinin], N-acetyl glucosamine [wheat germ agglutinin (EGA)], or fucose (*Ulex europaeus* I agglutinin) have been tested against a range of *Phytophthora* species but none of these lectins binds to the zoospore surface (Hardham and Suzuki 1990; A. R. Hardham unpubl. obs.). By contrast, WGA does bind to the surface of *Saprolegnia* zoospores (Lehnen and Powell 1989, 1993; Burr and Beakes 1994).

MAbs have been generated towards components associated with zoospores of *P. cinnamomi* (Hardham et al. 1985, 1986, 1991 b) and *Py. aphanidermatum* (Estrada-Garcia et al. 1989). In both cases, the resultant antibodies include some which bind either to the entire surface or to restricted regions of the surface of the zoospores. In *Phytophthora*, two zoospore surface antigens were identified by the antibodies as occurring over the entire surface; one group of antibodies (Zp) (Fig. 13) reacts strongly with zoospores but not with cysts, while the other group (ZCp) (Fig. 14) reacts with components on the surface of zoospores and cysts. This latter pattern of labelling was also shown by *Py. aphanidermatum* antibodies PA 7 and PA 8 (Estrada-Garcia et al. 1989).

Other MAbs that recognise zoospore surface antigens reveal the occurrence of distinct molecular domains on the zoospore surface. In one case, *P. cinnamomi* MAb Zw 1 reacts only with the membrane associated with the water expulsion vacuole (Fig. 15) (Hardham et al. 1986, 1991 a). Other MAbs add molecular information

---

Fig. 1–10. *P. cinnamomi* hyphae, sporangia, zoospores and cysts

Fig. 1–4. Differential interference contrast micrographs showing mature uncleaved sporangia (Fig. 1); cleaved sporangium (Fig. 2); zoospore emerging from apical pore of sporangium (Fig. 3); and zoospores and cyst (*c*) (Fig. 4). Small arrows, flagella; arrowheads, water expulsion vacuoles; large arrow, ventral groove. Figs. 1–3, × 400; Fig. 4, × 1000

Figs. 5–10. Scanning electron micrographs of frozen hydrated (Figs. 5–7, 9, and 10) and chemically-fixed and dried (Fig. 8) *P. cinnamomi* sporangia, zoospore, and cysts

Fig. 5. Young sporangium and hyphal tip. The sporangium develops as a swelling of the hyphal apex. × 3500

Fig. 6. Developing sporangium. × 1300

Fig. 7. Mature sporangium. × 1300

Fig. 8. Zoospore and cysts on root surface. Two flagella (*a*, anterior and *p*, posterior) emerge from the centre of the groove along the ventral surface of the zoospore. The zoospore is sitting on a sheet of material that has been secreted by the underlying cysts. Arrowhead, water expulsion vacuole. Micrograph courtesy of Ms. K.A. Pickerd of the EM Unit, Division of Entomology, C.S.I.R.O. × 3200

Fig. 9. Newly encysted spores on the surface of a root. Sheets of material that has been secreted by the cysts appear as a network covering the cysts and root. It seems likely that this material has been partially dehydrated during sublimation of ice in the microscope. × 2000

Fig. 10. Germinated cysts on a root of *Eucalyptus sieberi*. The apex of the germ tubes becomes flattened as it grows between the epidermal cells (arrowheads). Material secreted by the cysts is still present and appears as a network of fibers coating the cysts and root surface. × 2200

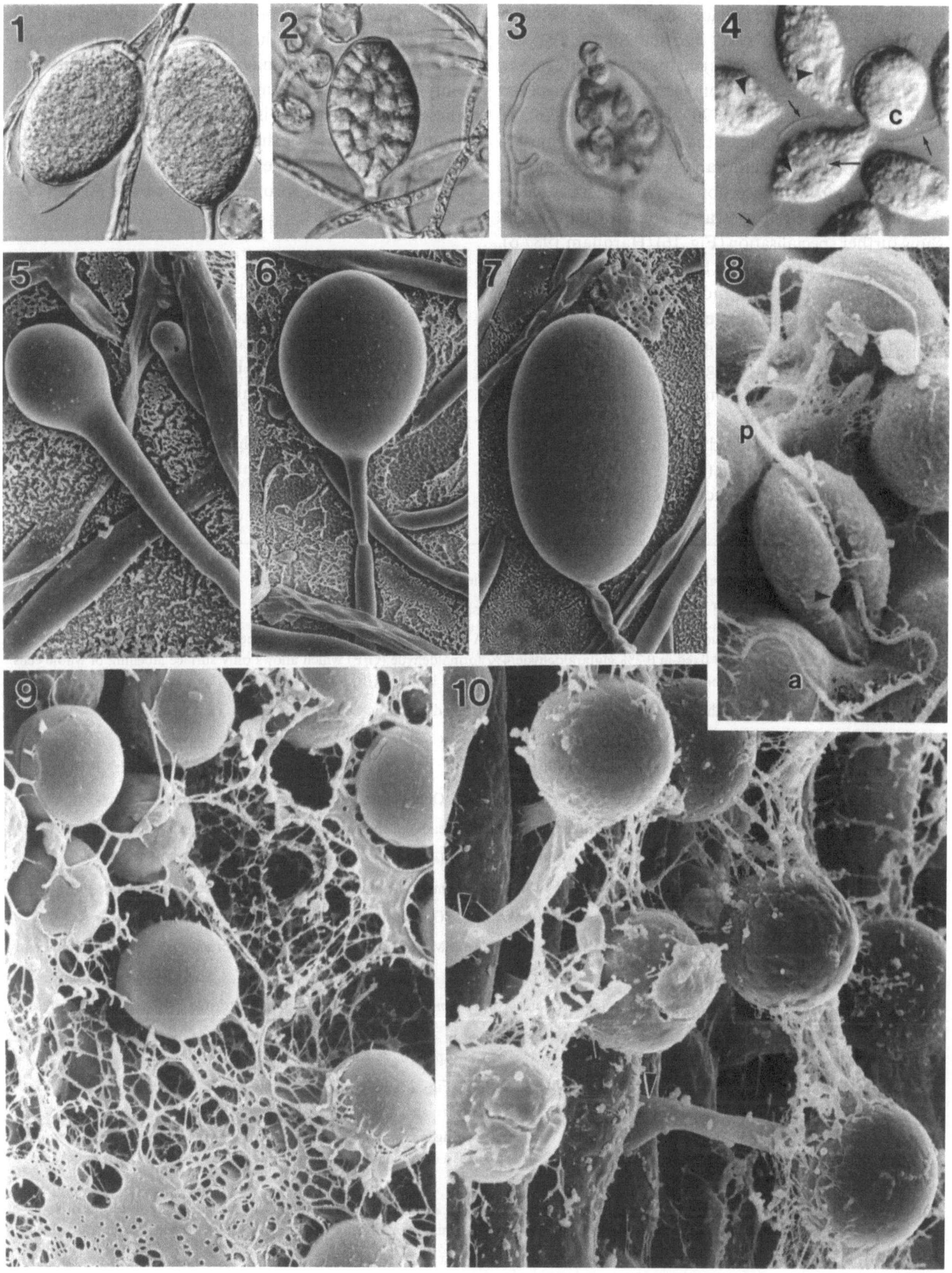
1
2
3
4
c
5
6
7
8
p
a
9
10

to the morphological evidence of differences between the surface of the cell body and that of the flagella. Tubular hairs (mastigonemes) on the anterior flagellum (Fig. 21) and fibrous hairs on the posterior flagellum (Vujicic et al. 1968, Reichle 1969, Hardham 1987 b) are flagella surface appendages that are likely to be associated with plasma membrane specialisations in the form of anchoring or insertion sites for the hairs. Clusters of membrane particles at the base of the flagella show further specialisation (Fig. 16) (Hardham 1987 b). Three groups of antibodies bind to the flagella. One, *P. cinnamomi* MAb Zf, binds to the surface of both flagella (Fig. 17) (Hardham et al. 1986, 1991 a), apparently targeting a component associated with the plasma membrane. The other two groups of antibodies bind to the anterior flagellum only (Hardham et al. 1986, Estrada-Garcia et al. 1989; M. Cope and A. R. Hardham unpubl.). The antibodies in at least one of these groups, Zg in *P. cinnamomi* (Figs. 19 and 20) and PA 2 in *Py. aphanidermatum*, react with the shaft and base of the mastigonemes; antibodies in the other group (Zt) (Fig. 18) may react with a basal segment of the mastigonemes or with plasma membrane components. The only biochemical information available about any of these flagella antigens is derived from periodate and pronase digestions of the Zt antigen. Both treatments destroy antibody-antigen reaction, indicating that the Zt antigen is a glycoprotein (M. Cope and A. R. Hardham unpubl.).

Synthesis and assembly of flagellar components

P. cinnamomi monoclonal antibodies that target the flagella have been used to follow the synthesis and transfer of flagellar surface antigens onto the cell surface (M. Cope and A. R. Hardham unpubl.). The flagellar membrane is assembled during cleavage of the multinucleate sporangia into uninucleate zoospores during zoosporogenesis. Mastigonemes in the oomycetes, as in the heterokont algae (Leedale et al. 1970, Loiseaux 1973, Hill and Outka 1974), are observed

Figs. 11–31. Characteristics and synthesis of zoospore surface components

Fig. 11. Ruthenium red staining of the surface of a *P. cinnamomi* zoospore. × 41,700

Figs. 12–15, 17–20 and **23–31.** Fluorescence micrographs show ConA-FITC or MAbs followed by sheep anti-mouse-FITC (SAM-FITC). Cells in Figs. 12–15 and 17–19 are fixed in 0.2% glutaraldehyde/4% formaldehyde (G/F) which allows only surface labelling of whole cells. Cell in Fig. 20 is fixed in 4% formaldehyde (4F) which allows intracellular labelling

Fig. 12. ConA-FITC labelling of the surface of *P. cinnamomi* zoospores and cyst (*c*). × 800

Fig. 13. The surface of *P. cinnamomi* zoospores labelled with MAb Zp 1. × 600

Fig. 14. *P. citrophthora* zoospores labelled with MAb ZCp 1. × 1000

Fig. 15. The surface of *P. citrophthora* zoospore labelled with MAb Zw 1. Only the water expulsion vacuole is labelled. × 1300

Fig. 16. Freeze-etch micrograph showing intramembrane particles (arrowheads) at the base of *P. cinnamomi* flagellum. × 69,400

Fig. 17. *P. nicotianae* zoospore labelled with MAb Zf 1. Both the anterior (*a*) and posterior (*p*) flagella are labelled. × 1400

Fig. 18. The surface of *P. cinnamomi* zoospores labelled with MAb Zt 2. Only the anterior flagellum of each cell is labelled. × 900

Fig. 19. The surface of a *P. cinnamomi* zoospore labelled with MAb Zg 4. Only the anterior flagellum is labelled. × 800

Fig. 20. Intracellular components of a *P. cinnamomi* zoospore labelled with MAb Zg 4. Rod-shaped packets of mastigonemes are fluorescently labelled. × 1000

Fig. 21. Mastigonemes on the anterior flagellum of a zoospore of *P. cinnamomi* are preserved and/or contrasted by fixation in the presence of Ruthenium red. × 44,000

Fig. 22. Packets of mastigonemes (arrowheads) in the cytoplasm of a zoospore of *P. cinnamomi.* × 23,800

Figs. 23–31. Cryosections of sporangia of *P. cinnamomi* labelled with Zf (Figs. 23, 26, and 29), Zt (Figs. 24, 27, and 30) and Zg (Figs. 25, 28, and 31) MAbs

Figs. 23–25. Mature sporangia before cleavage is induced. Zf 1 (Fig. 23) antigen is seen in the nuclei and as a diffuse labelling of the cytoplasm. Zt 2 (Fig. 24) labels spots in the cytoplasm. Zg 2 (Fig. 25) labels rod-shaped components (packets of mastigonemes) in the cytoplasm. Figs. 23 and 24, × 500; Fig. 25, × 700

Figs. 26–28. Sporangia fixed during early stages of cleavage. Zf (Fig. 26), Zt (Fig. 27), and Zg (Fig. 28) antigens take the form of spots or rods clustered at the ventral surface of the newly formed zoospores. At this stage, flagella axonemes would have fully formed. × 700

Figs. 29–31. Sporangia fixed during later stages of cleavage. Zf (Fig. 29), Zt (Fig. 30), and Zg (Fig. 31) antigens have become localised on the surface of the flagella (arrowheads). × 700

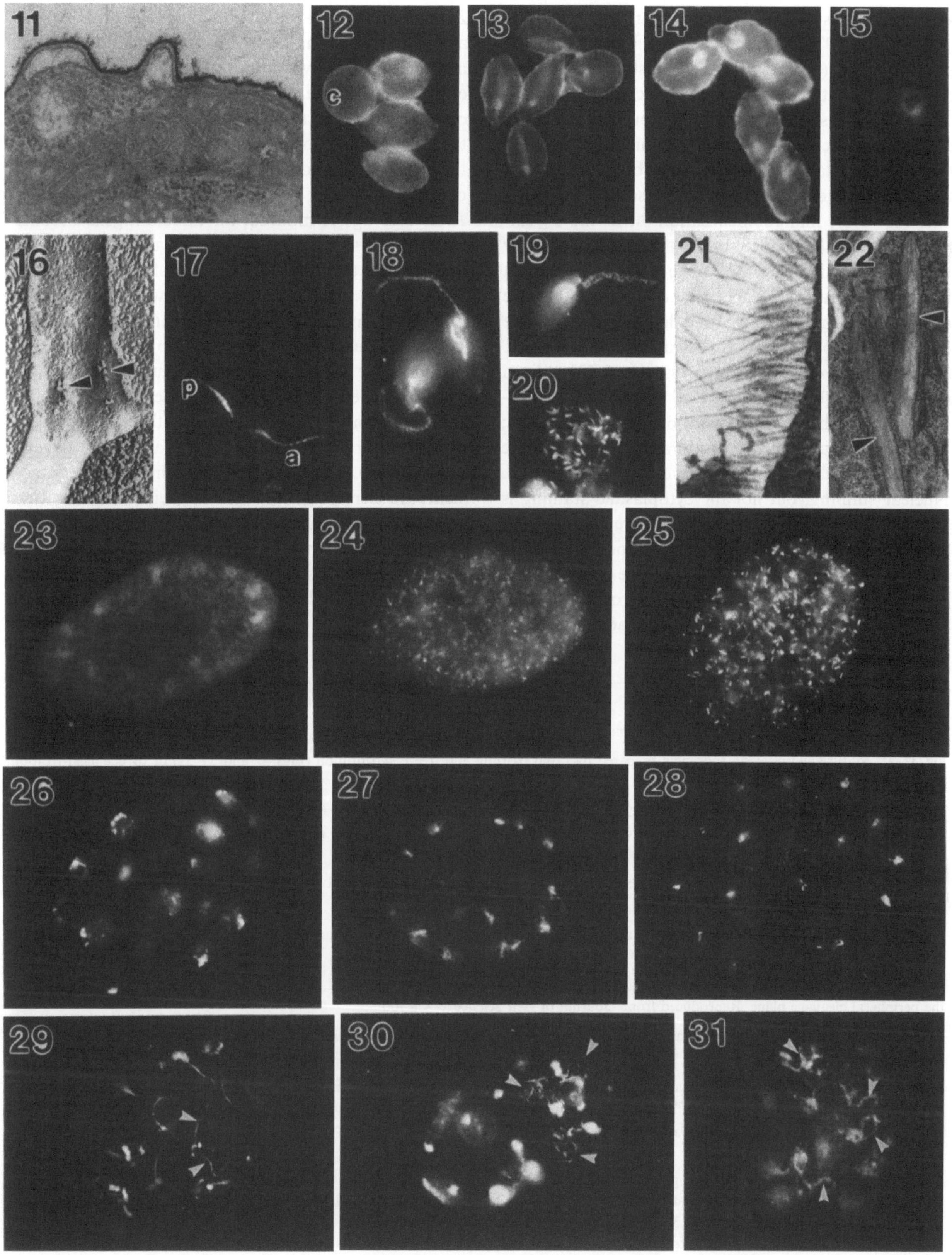
11
12
c
13
14
15
16
17
p
a
18
19
20
21
22
23
24
25
26
27
28
29
30
31

within elements of the ER-perinuclear space (Fig. 22) (Heath et al. 1970, Hardham 1987 b). Unlike flagellar surface scales (Moestrup 1982), mastigonemes have only infrequently been seen in association with the Golgi apparatus in a few algae (e.g., Bouck 1971). Being destined for the flagellar surface, it might be expected that the three flagella antigens would be synthesised at the time of flagella formation during sporangial cleavage. However, immunolocalisations with the *P. cinnamomi* MAbs indicate that this is not the case for any of three flagellar antigens. All three antigens are present in mature sporangia before they are induced to cleave (Figs. 23–25), and have also been detected in vegetative hyphae (M. Cope and A. R. Hardham unpubl. obs.). The situation for the membrane-associated Zf 1 antigen is complicated by the presence of labelling at two different sites within the cytoplasm of mature sporangia; the nucleus is labelled with moderate intensity and the cytoplasm exhibits diffuse low-level labelling (Fig. 23). Nuclear staining, but not the cytoplasmic staining is present in vegetative hyphae. The nature and relationship between the immunoreactive components within the cytoplasm and nucleus remain to be clarified. The antigens could be the same molecule or Zf may be recognising a common epitope on unrelated molecules. Zt and Zg antigens are also present in vegetative hyphae and increase in abundance during sporulation. Within mature sporangia, which have not yet been induced to cleave, the Zt antigen appears as discrete spots throughout the cytoplasm (Fig. 24) and the Zg antigen takes the same rod-like form as seen in zoospores, indicating the presence of bundles of mastigonemes within packets of ER (Fig. 25).

After cleavage is induced, all three antigens become clustered at sites corresponding to the centre of the groove on the ventral surface of the zoospores (Figs. 26–28). This is the region in which the narrow pole of the nucleus, Golgi stacks and the two basal bodies reside (Hyde et al. 1991 a, b). The clustering occurs about 30–45 min after the induction of cleavage at a time when the flagellar axonemes are completely, or almost completely, formed. The three antigens are subsequently transferred to the surface of the flagella (Figs. 29–31). The mechanism by which these components are secreted and sorted into their correct location on the flagellar surface is still not known, although ultrastructural immunocytochemistry with Zf, Zt, Zg, or PA 2 antibodies may help elucidate some details of this intriguing and important process.

Figs. 32–41. Assembly of the surface of *P. cinnamomi* zoospores

Fig. 32–41. Mature and cleaving sporangia prepared by rapid freezing and freeze substitution (Figs. 32 and 34–37) or by chemical fixation (Figs. 33 and 38–41)

Fig. 32. A peripheral nucleus in a mature sporangium before the induction of cleavage. Cleavage vesicles (*cv*) with moderately electron dense, granular contents are clustered around the narrow pole of the nucleus near the basal bodies. *sw* Sporangial wall. Reproduced from "Journal of Cell Science". × 20,000

Fig. 33. Immunolabelling of cleavage vesicles (*cv*) and Golgi cisternae (*GA*) with MAb Cpw 1 and SAM-Au 10. Reproduced with permission from "Mycological Research". × 36,700

Fig. 34. Immunolabelling of the extracellular matrix within the developing cleavage planes (arrowheads) with MAb Cpw 1 and SAM-Au 10. Reproduced from "Journal of Cell Science". × 19,100

Fig. 35. Development of peripheral and internal cleavage planes. During early development, a peripheral layer off cytoplasm is cut off by the peripheral cleavage plane. Reproduced from "Journal of Cell Science". × 8,700

Fig. 36. Golgi apparatus in cleaving sporangium showing trans-Golgi elements (arrowheads) whose appearance resembles that of the developing cleavage planes (*cp*). Reproduced from "Journal of Cell Science". × 25,300

Fig. 37. By completion of cleavage, the layer of peripheral cytoplasm has been re-absorbed and the newly-formed zoospore plasma membrane abuts the sporangial wall. Flagella profiles (arrows) can be seen within the extracellular matrix. Large peripheral vesicles (arrowheads) have become localised within the peripheral cytoplasm of the zoospores. *wev* Water expulsion vacuole. Reproduced with permission from "Journal of Cell Science". × 8,600

Figs. 38–41. Immunofluorescence micrographs of mature but uncleaved (Figs. 38 and 39) and cleaved (Figs. 40 and 41) sporangia labelled with MAbs Cpa 2 (Figs. 38 and 40) and Vsv 1 (Figs. 39 and 41). Both Cpa 2-positive and Vsv 1-positive vesicles are randomly distributed before the induction of cleavage (Figs. 38 and 39, respectively), but become distributed in a polar fashion after cleavage (Figs. 40 and 41). Cpa 2-positive vesicles occur on the dorsal surface of the new zoospores; Vsv 1-positive vesicles are concentrated on the ventral surface near the narrow pole of the nucleus. Figures 39–41 reproduced with permision from "European Journal of Cell Biology". Figures 38 and 39, × 650; Fig. 40, × 850; Fig. 41, × 750

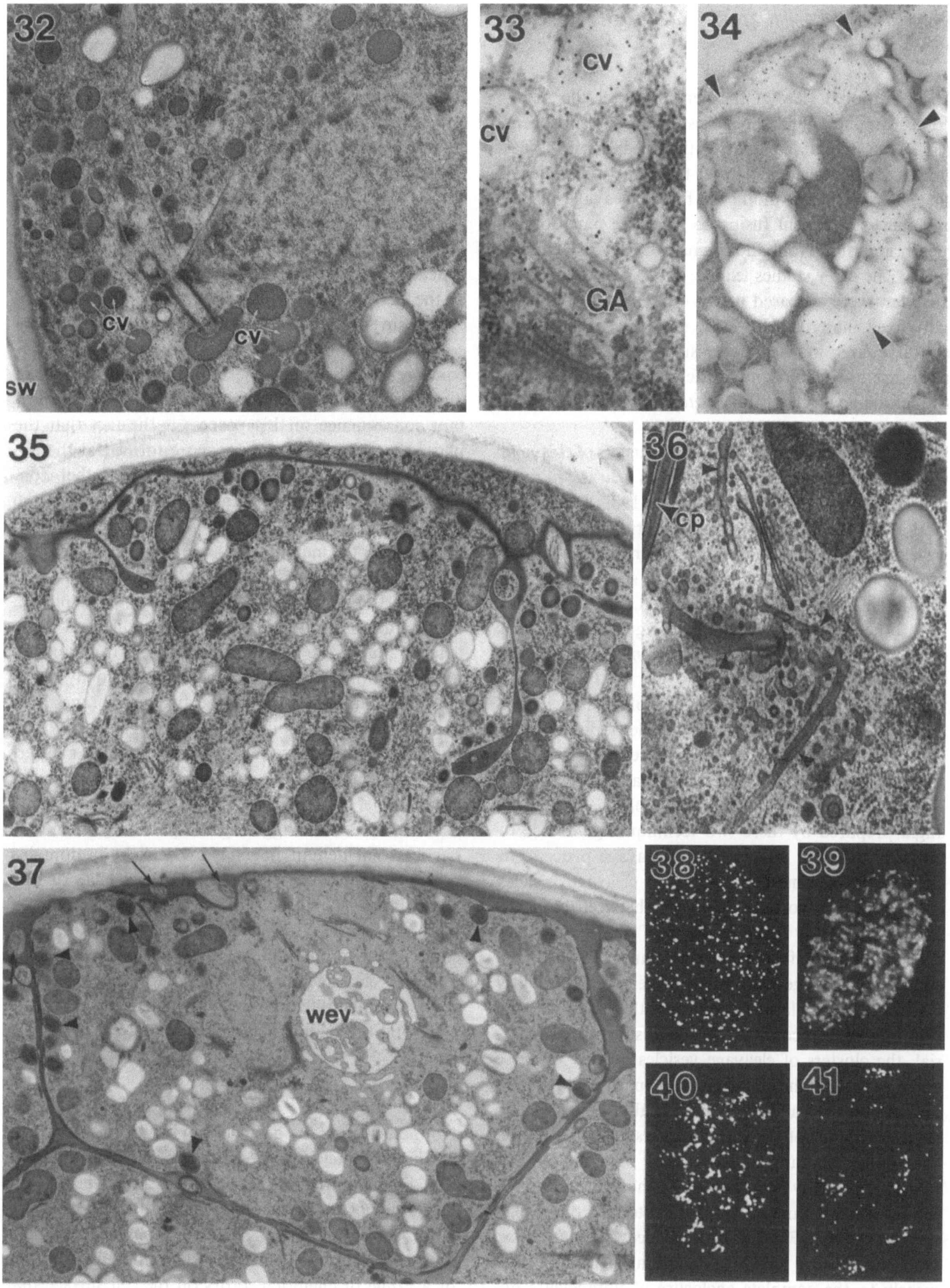
32
cv
cv
sw
33
cv
cv
GA
34
35
36
cp
37
wev
38
39
40
41

Synthesis and assembly of the zoospore plasma membrane

The traditional view of the generation of the membranes that surround zoospores during zoosporogenesis in the Pythiaceae involves the production of specific cleavage vesicles and, after the induction of cleavage, their alignment along the future cleavage planes and their subsequent fusion to form the partitioning membranes (see Hohl and Hamamoto 1967, Hemmes 1983). The early studies leading to the formulation of this view were all based upon observations of chemically-fixed material. However, more recently, examination of rapidly frozen, freeze-substituted material has given rise to a significantly different model of sporangial cleavage at least in *Phytophthora* (Hyde et al. 1991 b).

In mature but uninduced sporangia, clusters of cleavage vesicles occur around the basal bodies near the narrow poles of the nuclei (Fig. 32) (Hemmes 1983; Hyde et al. 1991 a, b). In chemically-fixed material the cleavage vesicles are distinguished from other vesicles in the sporangial cytoplasm by their small size and regular profile; in freeze-substituted material, they stand out from other vesicles because of their granular electron dense contents (Fig. 32). *P. cinnamomi* MAbs in the Cpw group react with the contents of the cleavage vesicles (Fig. 33), providing immunological evidence for the presence of this distinct vesicle population.

While both chemical fixation and freeze-substitution show the presence of clusters of cleavage vesicles in uninduced sporangia, examination of cleaving sporangia prepared by these two techniques gives rise to radically different images of the cleavage process. In chemically fixed material, chains of cleavage vesicles are seen in the cortical cytoplasm between adjacent nuclei, apparently forming sheets of cleavage vesicles distributed in the future cleavage planes. Towards the end of zoosporogenesis, adjacent vesicles apparently fuse to form the plasma membranes that subdivide the sporangium. By contrast, in freeze-substituted material, the clusters of cleavage vesicles quickly disperse and fuse to form the first expanses of partitioning membranes (Figs. 34 and 35) (Hyde et al. 1991 b). At no stage are chains of cleavage vesicles seen in the freeze-substituted sporangia. Expansion of the cleavage planes apparently occurs through the progressive addition of membrane and matrix material from Golgi-derived vesicles (Fig. 36). Internal cleavage planes fuse with the cortical cleavage membranes leading ultimately to the complete enclosure of uninucleate domains of sporangial cytoplasm to form the zoospore precursor cells (Fig. 37). During this time a number of zoospore components, including vesicles and mitochondria (Hyde and Hardham 1993), become reorganised within the sporangial cytoplasm such that they change from a random distribution to polar localisations close to the zoospore plasma membrane (Figs. 38–41).

The internal cleavage planes that form do not fuse with the original plasma membrane of the sporangium, but instead fuse with a cortical cleavage plane that develops in the peripheral cytoplasma, cutting off a layer of cytoplasm adjacent to the sporangial wall (Fig. 35). The components within this "cytoplasmic remnant" become re-absorbed into the main body of cytoplasm during continued development (Fig. 37). One important consequence of this process is the fact that the plasma membrane of the zoospore formed at the periphery of the sporangium is not cobbled together from new and old membranes. Instead, the surface of the plasma membrane of each zoospore is entirely new, arising from the membranes surrounding cleavage vesicles and Golgi vesicles. During zoosporogenesis, therefore, the fungus has the ability to assemble a plasma membrane that potentially has a very different character to that of the plasma membrane surrounding the mature sporangium. No doubt there are many molecules that are common to the two membranes, but considering the very different functions of the two cell types, there must also be significant molecular differences between them. The sporangial plasma membrane, for example, would have no need for proteins involved in chemotaxis or in triggering zoospore encystment, for barriers that delimit flagellar and cell body domains or for membrane insertion sites for flagellar surface appendages. The absence of any reaction of Zf and Zt antibodies with the sporangial plasma membrane is consistent with their absence from this membrane. The mode of cleavage that is observed in *Phytophthora*, seems to constitute an efficient mechanism for rapidly achieving the required transformation.

Function of zoospore surface components

No molecule (or the gene coding for it) on the surface of an oomycete zoospore has yet been isolated and its function unambiguously identified. Nevertheless, the behaviour of the zoospores is indicative of a need for a number of specific surface proteins. One example is that of zoospore chemotaxis which must require one or a range of chemoreceptors on the cell surface that can bind chemoattractant and set in train a signal transduction pathway that regulates the direction of cell movement. A wide range of compounds attract *Phytophthora* and other oomycete zoospores (Carlile 1983, Cahill and Hardham 1993) but as yet no receptor has been identified. It has been discovered recently, however, that the isoflavones, genistein and daidzein, are active chemoattractants for *P. sojae* in the nanomolar range (Morris and Ward 1992). These compounds are produced by soybean roots and only attract *P. sojae* and not other *Phytophthora* spp. that are not pathogenic on soybean. The high levels of specificity in this system make it one in which zoospore chemoreceptors might finally be identified and characterised.

Our lack of information on the character of the chemoreceptor system includes ignorance of its distribution on the cell surface; the location of binding sites for isovaleraldehyde were not determined. It has been suggested that chemoreceptors might occur around the anterior end of the zoospore and be involved in comparisons of attractant concentration across the array of receptors (Carlile 1966). Given the small cell size and the speeds at which they swim (50–200 µm/s), this putative spatial sampling system seems unlikely. It seems more likely that the cells utilise a temporal sampling system as employed by bacteria (Adler 1976). In this case, the chemoreceptor could occur anywhere on the cell surface.

The function of surface appendages on flagella is intriguing. In many cases we have little idea of the advantages cells gain by adding surface hairs or by building intricate patterns of scales on their flagella. We do, however, have circumstantial evidence for a role of the mastigonemes on the anterior flagellum of oomycete zoospores and heterokont algae. Direct observation of zoospore motility indicates that forward thrust is provided by the anterior flagellum which generates an approximitely sinusoidal wave originating at the flagellar base and progressing to its tip. Normally, this wave motion would propel the cell backwards. It has thus been suggested that the two rows of tubular mastigonemes along the anterior flagellum effectively reverse the thrust of flagellar beat (Jahn et al. 1964, Holwill 1982). Wave propagation along the flagellum could swing the mastigonemes back and forth like paddles, resulting in forward motion.

We also have indirect evidence for another function for one or more surface components and that is a role in the induction of encystment. The timing of the induction encystment is likely to be critical for efficient infection as it means the zoospores become immobilised at the surface of a plant root having swum there. When purified Zf 1 Mabs are added to swimming *P. cinnamomi* zoospores, the cells are rapidly induced to encyst (Hardham and Suzuki 1986). The effect is specific for this antibody and is not caused by other antibodies, even though their binding in some cases also includes the flagellar surface (for example, the Zp antibodies which bind to the whole zoospore surface). The same phenomenon was observed with other *Phytophthora* spp., suggesting that in *Phytophthora*, plasma membrane-bound receptors involved in triggering encystment are restricted to the surface of the flagella. In similar experiments with *Py. aphanidermatum* (Estrada-Garcia et al. 1990 b), *S. diclina* and *S. parasitica* (Burr and Beakes 1994), antibodies which target the flagella and cell body surface have a similar effect.

The cyst surface

The properties of the surface of oomycete cysts are quite different from those of the zoospores (Figs. 42–52). During encystment, the zoospores secrete material from vesicles stored in the peripheral cytoplasm, and they rapidly synthesise a cell wall. The secreted material includes an adhesive that bonds the cells to the adjacent root or other surface. The wall contains microfibrils of cellulose and is strong enough to allow turgor pressure to build-up in the cysts by about 10 min after the onset of encystment.

There are three types of storage vesicles within the cortical cytoplasm of the zoospores, as well as a system of flat membranous cisternae (Hardham 1987 a, Hardham and Gubler 1990, Gubler and Hardham 1991, Hardham et al. 1991 a). All four components show distinct polarities in their distribution within the cell cortex. The most prominent of the three peripheral vesicles are large vesicles about 0.4–0.6 µm in diameter (Figs. 42 and 46) (Hardham et al. 1991 a). In the Pythiaceae they are usually called large peripheral vesicles; in *Saprolegnia* they are known as a fibrillar vesicles (see Burr and Beakes 1994). In *P. cinnamomi*, their contents are labelled by MAb Lpv 1 (Figs. 42 and 46). The large peripheral vesicles occur throughout most of the cortical cytoplasm except near the centre of the ventral surface.

The other two types of peripheral vesicles are about 0.3 µm in diameter (Hardham et al. 1991 a). Of these, the dorsal vesicles occur predominantly on the dorsal surface and avoid the groove region on the ventral surface (Figs. 43, 44, 46, and 47). Their high molecular weight glycoprotein contents are labelled by MAb

Cpa 2 (Figs. 43, 46, and 47) and by the lectin SBA (Fig. 44). On immunoblots antibody labelling appears as three bands of over 300 kDa in molecular weight (Gubler and Hardham 1988). The ventral vesicles, on the other hand, occur predominantly along the edges of the groove on the ventral surface of the zoospores (Figs. 45 and 47). Their contents are labelled by MAb Vsv 1 and are greater than 200 kDa in molecular weight (Hardham and Gubler 1990).

Within the first 2 min following the induction of encystment, the contents of the dorsal and ventral vesicles are secreted onto the surface of the cysts (Figs. 43 and 48–50) (Hardham and Gubler 1990). Because the vesicles are restricted in their distribution, the secreted material is, at least initially, also localised to certain parts of the cyst surface (Fig. 50). The polarity of vesicle distribution and secretion have important implications for the infection process. When the zoospores swim up to the root of a potential host, they orient so that the ventral surface faces the root. They then encyst while maintaining this orientation. This means that it is the contents of the ventral vesicles that are secreted onto the root surface. The contents of the dorsal vesicles initially cover the surface of the cyst that faces away from the root surface. With time this material tends to spread out over the surface (see Figs. 8 and 9). The timing of secretion of the dorsal and ventral vesicles coincides with timing of the acquisition of adhesiveness of the encysting cells (Gubler et al. 1989). Our interpretation of these data is that it is the ventral vesicles that contain the adhesive that sticks the zoospore to the root surface.

Cross-reactivity of the Vsv 1 antibody with vesicles in some species of *Pythium* has shown that zoospores of this genus also contain vesicles that are confined to the ventral surface (M. Cope et al. unpubl. obs.), information which is interpreted as evidence of a similar mechanism of adhesion in this genus. *Saprolegnia* zoospores also contain vesicles that occur only near the ventral groove (Lehnen and Powell 1989, 1993; Burr and Beakes 1994). They are called K-bodies because of their location near the kinetosomes. Although the number and morphology of the K-bodies and the composition of their contents differs from that of the ventral vesicles in *Phytophthora*, analysis of their behaviour during encystment suggests that they also serve to store the adhesive used during attachment to potential hosts and other surfaces (Lehnen and Powell 1989, Burr and Beakes 1994).

The contents of the large peripheral vesicles are not secreted during encystment (Figs. 48 and 49) but act as a reserve of proteins that are degraded and utilised to support growth of the germinated spore during the first 4 h or so after cyst germination (Gubler and Hardham 1990). Although there has been som evidence to the contrary (Estrada-Garcia et al. 1990 a), studies using the *P. cinnamomi* MAbs suggest that the large peripheral vesicles are also retained during encystment of *Pythium* zoospores (M. Cope et al. unpubl. obs.).

All three vesicle types are synthesised during fungal

Figs. 42–52. Secretion of surface components during zoospore encystment in *P. cinnamomi*

Figs. 42–45. Fluorescence micrographs of *P. cinnamomi* zoospores and cysts fixed in 4 F and labelled with Lpv 1 (Fig. 42), Cpa 2 (Fig. 43), SBA-FITC (Fig. 44) and Vsv 1 (Fig. 45). Cpa 2- and SBA-positive vesicles occur on the dorsal surface and avoid the ventral groove region (arrows). Vsv 1-positive vesicles are concentrated along the sides of the ventral groove (arrows). Figure 42, × 850; Fig. 43, × 800; Fig. 44, × 750; Fig. 45, × 1000

Fig. 46. Section of a *P. cinnamomi* zoospore labelled with Lpv 1-Au 18 and Cpa 2-Au 10. Lpv 1 labels the large peripheral vesicles. Cpa 2 labels the small dorsal vesicles. × 19,600

Fig. 47. Section of a *P. cinnamomi* zoospore labelled with Cpa 2-Au 11 and Vsv 1-Au 16. Vsv 1 labels small ventral vesicles (*v*) near the groove (*g*). Cpa 2 labels a small dorsal vesicle (*d*). × 20,800

Fig. 48. Section of a *P. cinnamomi* spore fixed 1 min after the induction of encystment. Cpa 2-positive material has been secreted from the dorsal vesicles onto the cysts surface (small gold particles). The large peripheral vesicles, labelled by Lpv 1-Au 18, remain in the cortical cytoplasm. × 24,500

Fig. 49. Section of a *P. cinnamomi* cyst fixed 10 min after the induction of encystment. Cpa 2-positive material (small gold particles) coat the surface of the cyst. The large peripheral vesicles labelled with Lpv 1-Au 15 have moved away from the cell surface. × 21,700

Fig. 50. *P. cinnamomi* cysts on the surface of a *E. sieberi* root labelled with SBA-rhodamine (**a**) and Vsv 1-SAM-FITC (**b**). SBA labels the Cpa antigen. The two secreted components show complementary distributions on the surface of the cysts. The Vsv 1-positive material forms an adhesive pad between the cyst and the root. The SBA- (Cpa-) positive material coats the opposite surface of the cysts. × 400

Fig. 51. The surface of *P. cinnamomi* cysts labelled with ConA-FITC. × 700

Fig. 52. The surface of *P. nicotianae* cysts labelled with Cpw 1 and SAM-FITC. × 1300

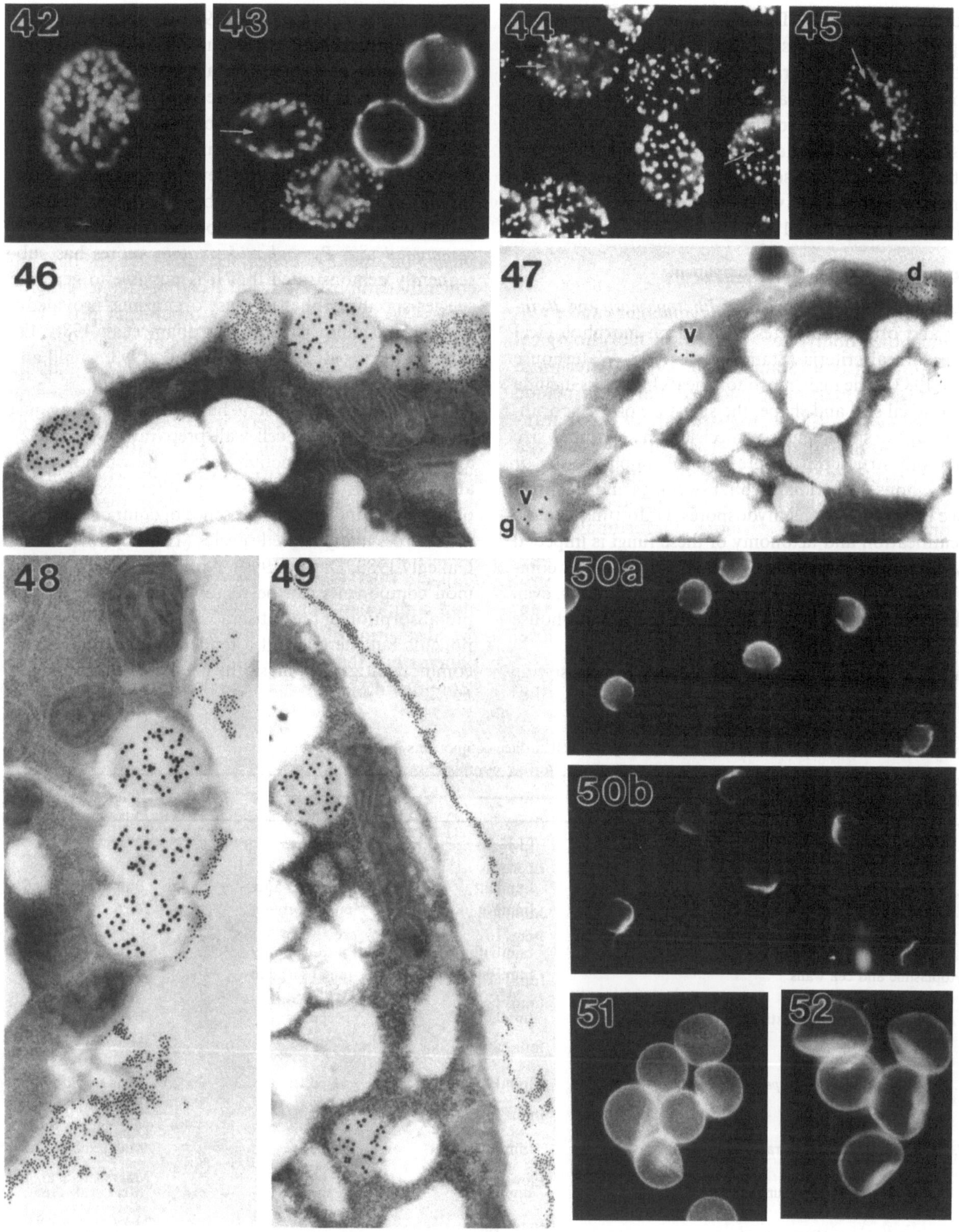
42
43
44
45
46
47
d
v
v
g
48
49
50a
50b
51
52

sporulation (J. D. W. Dearnaley and A. R. Hardham unpubl. obs.). The vesicles are randomly distributed within the mature sporangia (Figs. 38 and 39) but are reorganised during cleavage (Figs. 40 and 41). Movement of the vesicles to the cell cortex occurs in the absence of microtubules or microfilaments, however, without microtubules sorting of the vesicles within the cortex into dorsal and ventral domains does not occur (Hyde and Hardham 1993).

Taxonomy of cell surface components

The identification of species of *Phytophthora* and *Pythium* and other oomycetes is based on morphological and cultural criteria (Stamps et al. 1990, Waterhouse et al. 1983). The main morphological characters include sporangial size and shape, the structure of the sporangium apex, sporangium pedicel length, oogonial size and wall structure, the presence of amphigynous or paragynous antheridia, hyphal swellings and the presence or absence of chlamydospores. Unfortunately, the identification and taxonomy of these fungi is impeded by difficulties in producing sexual stages and is complicated by considerable variation in morphology, even when uniform culture conditions are used (Waterhouse et al. 1983, Gallegly 1983).

In her critique of existing approaches for classifying *Phytophthora* spp., Gallegly (1983) reviewed and suggested improvements to the current system based on morphological characters; she also assessed alternative methods such as protein and serological comparisons. Although the antisera that had been produced (Table 2) all lacked absolute species-specificity, Gallegly foresaw the potential benefits in the development of a central supply of antibody-based probes for species identification (Gallegly 1983). Development of MAbs to *P. cinnamomi* and *Py. aphanidermatum* spores has subsequently demonstrated that it is possible to generate species-specific probes to these organisms, provided a suitable inoculum is used (Hardham et al. 1986, Estrada-Garcia et al. 1989, Gabor et al. 1993, Cahill and Hardham 1994).

All the polyclonal antisera have been raised against mycelium extracts or cell wall preparations (Table 2), and all show cross-reactivity especially when not pre-absorbed with antigens from another species. These observations indicate the presence of components common across many related species (Table 2) (reviewed in Gallegly 1983, Dewey 1990). Antibodies to these common components can be removed to some extent by pre-absorption of the antisera with cross-reacting components but there is still a tendency for antibodies to common antigens to mask the presence of components

Table 2. Specificities of polyclonal antisera raised against cell surface components in oomycete fungi

Inoculum	Assay	Specificity	References
P. fragariae mycelium	ELISA	> species	Werres 1988
P. cactorum, P. cinnamomi and *P. erythroseptica* mycelium extracts	ImmDiff, ImmFlu	⩾ species after pre-absorption	Burrell et al. 1966
P. cinnamomi and *P. drechsleri* cytoplasmic and cell walls	ImmDiff, ImmFlu	⩾ species after pre-absorption	Halsall 1976
9 *Pythium* spp. mycelium extracts	ImmDiff	> species	Krywiencyzk and Dorworth 1979
P. cinnamomi and *P. megasperma* cell walls	ImmFlu	> species	MacDonald and Duniway 1979
P. cinnamomi mycelium extracts and cell wall	ImmFlu	> genus	Malajczuk et al. 1975
6 *Phytophthora* spp. mycelium extracts	ImmDiff	⩾ species	Merz et al. 1969
Py. graminicola mycelium extract	ImmFlu	> species	White 1976
P. fragariae mycelium extract	ELISA	> species	Mohan 1988

ELISA Enzyme-linked immunosorbent assay; *ImmDiff* immunodiffusion; *ImmFlu* immunofluorescence microscopy

Table 3. Specificities of MAbs directed towards *P. cinnamomi* zoospore surface antigens[a]

Antibody	Antigen	Immunofluorescence assay[b]	Specificity
Zp 1, Zp 2	zoospore periphery	2/17 *P. cinnamomi* isolates, 0/18 other *Phytophthora* or *Pythium* spp.	< species
Zt 1	tinsel flagellum	23/23 *P. cinnamomi* isolates, 0/21 other *Phytophthora* or *Pythium* spp.	species specific
Zt 2	tinsel flagellum	20/20 *P. cinnamomi* isolates, 0/20 other *Phytophthora* or *Pythium* spp.	species specific
Zg 1	mastigonemes	23/23 *P. cinnamomi* and 1/3 *P. cryptogea* isolates, 0/21 other *Phytophthora* or *Pythium* spp.	> species but < genus
Zg 2–4	mastigonemes	18–22/23 *P. cinnamomi* isolates, 0/20 other *Phytophthora* or *Pythium* spp.	< species
Zf 1	both flagella	21/21 *Phytophthora* and 6/6 *Pythium* spp.	> genus
Zw 1	WEV	21/21 *Phytophthora* and 6/6 *Pythium* spp.	> genus
ZCp 1	zoospore and cyst periphery	17/18 *Phytophthora* and 5/5 *Pythium* spp.	neither species nor genus specific
ZCp 2	zoospore and cyst periphery	23/23 *Phytophthora* spp., 0/8 *Pythium* spp.	genus specific

[a] All antibodies were raised against a mixture of approximately 95% zoospores and 5% cysts of *P. cinnamomi* fixed in 2% glutaraldehyde in 50 mM Pipes buffer. References: Hardham et al. (1986, 1991 a), Gabor et al. (1993), Cahill and Hardham (1994)
[b] Values (x/y) show the number of positive reactions (x) of the number of isolates or species tested (y)

possessing restricted taxonomic distributions. The production of MAbs, on the other hand, allows the taxonomic specificities of individual epitopes on a molecule to be determined. Although the available data are still frustratingly limited, the MAbs that have been raised to *Phytophthora, Pythium,* and *Saprolegnia* reveal that cell surface components can exhibit a wide range of taxonomic specificities, and that different epitopes on cell surface molecules can possess different taxonomic distributions. Comparisons of the specificities of antibodies raised against *Phytophthora, Pythium,* and *Saprolegnia* surface components reveal that there is a tendency for components associated with the plasma membrane to possess the most restricted taxonomic specificities (Tables 3–5). The only antibodies so far reported that bind to only a small number of isolates of a particular species are the two MAbs Zp 1 and Zp 2 which target the plasma membrane (or its associated glycocalyx) of *P. cinnamomi* zoospores (Table 5) (Hardham et al. 1986, 1991 a). The restricted occurrence of the epitope(s) recognised by these antibodies contrasts with the widespread occurrence of mannosyl and glucosyl residues revealed by ConA labelling to be part of the surface of zoospores of many oomycete species (see discussion above). Other MAbs (Zg 2–4) that react with mastigonemes on the zoospore anterior flagellum bind to most, but not all, isolates of the species, *P. cinnamomi.* To date no antibodies specific for a single pathogenic race or mating type have been found.

Table 4. Specificities of MAbs directed towards *P. cinnamomi* cyst surface antigens[a]

Antibody	Antigen	Immunofluorescence assay[b]	Specificity
Cpa 1, 2	dorsal vesicles, cyst coat	4/15 *Phytophthora* and 2/4 *Pythium* spp.	neither species nor genus specific
Cpa 3, 4, 9–11	dorsal vesicles, cyst coat	23/23 *P. cinnamomi* isolates and 0/13 other *Phytophthora* spp., 0/7 *Pythium* spp.	species specific
Cpa 5–7	dorsal vesicles, cyst coat	23/23 *P. cinnamomi* isolates and 0/12 other *Phytophthora* spp., 0/7 *Pythium* spp.	species specific
Cpa 8	dorsal vesicles, cyst coat	22/22 *P. cinnamomi* isolates and 1/13 other *Phytophthora* spp., 0/7 *Pythium* spp.	neither species nor genus specific
Cpa 12[c]	dorsal vesicles, cyst coat	22/23 *P. cinnamomi* isolates, 0/21 other *Phytophthora* or *Pythium* spp.	< species
Cpw 1	peripheral cisternae, cyst wall	20/20 *Phytophthora* spp., 0/3 *Pythium* spp.	genus specific
Cpw 4	peripheral cisternae, cyst wall	15/15 *Phytophthora* spp. and 1/9 *Pythium* spp.	> genus
Gvv[c]	ventral vesicles, cyst adhesive	23/23 *P. cinnamomi* isolates, 0/20 other *Phytophthora* or *Pythium* spp.	species specific
Vsv 1[c]	ventral vesicles, cyst adhesive	13/14 *Phytophthora* and 2/4 *Pythium* spp.	neither species nor genus specific

[a]Hardham et al. (1986, 1991 a, b), Gabor et al. (1993), Cahill and Hardham (1994)
[b] Values (x/y) show the number of positive reactions (x) of the number of isolates or species tested (y)
[c] These antibodies were raised against *P. cinnamomi* spores fixed in 4% paraformaldehyde in 50 mM Pipes buffer after neonatal tolerization of mice with spores fixed in 2% glutaraldehyde in 50 mM Pipes buffer (Hardham et al. 1991 b)

Species-specific antibodies have been generated to *P. cinnamomi* (Hardham et al. 1986, 1991 a; Gabor et al. 1993; Cahill and Hardham 1994) and *Py. aphanidermatum* (Estrada-Garcia et al. 1989) spore surface components. The species-specific components include molecules on the surface of the anterior flagellum (Zt, PA 2), molecules over the entire zoospore (PA 1), and cyst surface coat and adhesive components (Cpa 3–7, Cpa 9–11, Gvv, PA 3–6). Genus-specific antigens include those associated with the surface of both zoospores and cysts (ZCp 2) and the cyst cell wall (Cpw 1, Cpw 3). Antigens common to more than one genus have been generated in all three studies (Hardham et al. 1986, 1991 a; Estrada-Garcia et al. 1989; Burr and Beakes 1994) and include those associated with zoospore flagella (Zf), the water expulsion vacuole (Zw), the surface of zoospores and cyst (ZCp 1, PA 7, PA 8, HB 5, HB 6), the cyst wall (Cpw 4), and the cyst surface adhesive (Vsv 1).

The reactions of the Cpa group of MAbs raised against *P. cinnamomi* spores illustrate the fact that cell surface antigens in these protists can possess epitopes that exhibit different levels of specificity. An intracellular antigen in the *P. cinnamomi* zoospores, Lpv, also shows a similar range of different epitope specificities (Gabor et al. 1993). Plant carbohydrates are highly antigenic (Anderson et al. 1984, Smith et al. 1984, Evans et al. 1988, Knox and Roberts 1989), and it is thus not surprising that periodate treatments show that many of the antibodies raised in these studies are directed to

Table 5. Specificities of MAbs directed towards cell surface components in oomycete fungi

Inoculum	Ab	Antigen	Assay	Specificity	References
Py. aphanidermatum zoospores and cysts	PA 1	zoospore surface	ImmFlu	species specific	Estrada-Garcia et al. 1989
Py. aphanidermatum zoospores and cysts	PA 2	anterior flagellum	ImmFlu	species specific	Estrada-Garcia et al. 1989
Py. aphanidermatum zoospores and cysts	PA 3–6	cyst surface	ImmFlu	species specific	Estrada-Garcia et al. 1989
Py. aphanidermatum zoospores and cysts	PA 7, 8	zoospore and cyst surface	ImmFlu	> genus	Estrada-Garcia et al. 1989
S. parasitica zoospores and cysts	HB 5, 6	zoospore and cyst surface	ImmFlu	> genus	Burr and Beakes 1994
P. megasperma f.sp. *glycinea* culture filtrate	see ref.	extracellular and mycelial wall Gps	ELISA, ImmBlot	> race, otherwise not determined	Wycoff et al. 1987
P. megasperma f.sp. *glycinea* mycelium walls	see ref.	extracellular and mycelial wall Gps	ELISA, ImmBlot	> race, otherwise not determined	Wycoff and Ayers 1990
AgriDiagnostics kits	na	na	ELISA, ImmBlot	neither species nor genus specific	Petersen et al. 1989, MacDonald et al. 1990, Ali-Shytayeh et al. 1991, Pscheidt et al. 1992

ImmFlu immunofluorescence microscopy; *Gps* glycoproteins; *ELISA* enzyme-linked immunosorbent assay; *ImmBlot* immunoblotting; *na* not available

carbohydrate epitopes on surface glycoproteins (Wycoff et al. 1987, Gubler and Hardham 1988, Hardham and Gubler 1990, Estrada-Garcia et al. 1990 a, Wycoff and Ayers 1990; Cope and Hardham unpubl.). At this stage, however, there are insufficient data to assess the relationship between epitope type (carbohydrate or protein) and its taxonomic sepecificity.

Concluding remarks

MAbs raised against components on the surface of spores of *Phytophthora* and *Pythium* spp. have proved to be of great value for studies of the cell biology and taxonomy of these protists. Immunocytochemical studies have made it possible to identify a number of surface molecules that are involved in the infection process and to uncover new details of their synthesis, assembly and function on the cell surface.

Investigations of the reaction of the MAbs with a wide range of *Phytophthora* and *Pythium* spp. have revealed that spore surface components can possess a range of taxonomic specificities. So far these studies have focused on comparisons at lower levels of taxonomic relationships, looking for isolate-, race-, mating-type-, species- or genus-specific markers. At present it is not known to what extent components on the surface of these protists might include molecules that are sufficiently conserved to be useful markers for higher levels of phylogenetic comparisons. We do, however, already have antibodies that bind to all species of *Phytophthora* and *Pythium* tested (*P. cinnamomi* MAbs Zf 1 and Zw 1). Further screening would determine if these components are diagnostic for the family, Pythiaceae, or even for the order, Peronosporales. What is clear, is that the immunological approach reviewed in this article has the potential to make a significant contribution to our understanding of the cell biology and taxonomy of oomycete protists in the future.

Acknowledgements

I thank Dr. L. R. Lehnen Jr. for providing the inspiration for Table 1, and Mrs J. Maleszka for excellent technical assistance and the production of the cryoSEM micrographs.

References

Adler J (1976) The sensing of chemicals by bacteria. Scient Am 234: 40–47

Ali-Shtayeh MS, MacDonald JD, Kabashima J (1991) A method for using commercial ELISA tests to detect zoospores of *Phytophthora* and *Pythium* species in irrigation water. Plant Dis 75: 305–311

Anderson MA, Sandrin MS, Clarke AE (1984) A high proportion of hybridomas raised to a plant extract secrete antibody to arabinose or galactose. Plant Physiol 75: 1013–1016

Arms K, Camp PS (1979) Biology. Holt, Rinehart and Winston, WB Saunders, New York

Bacic A, Williams ML, Clarke AE (1985) Studies on the cell surface of zoospores and cysts of the fungus *Phytophthora cinnamomi*: nature of the surface saccharides as determined by quantitative lectin binding studies. J Histochem Cytochem 33: 384–388

Bailey AM, Mena GL, Herrera-Estrella L (1991) Genetic transformation of the plant pathogens *Phytophthora capsici* and *Phytophthora parasitica*. Nucleic Acids Res 19: 4273–4278

– – – (1993) Transformation of four pathogenic *Phytophthora* spp. by microprojectile bombardment on intact mycelia. Curr Genet 23: 42–46

Barr DJS (1981) The phylogenetic and taxonomic implications of flagellar rootlet morphology among zoosporic fungi. BioSystems 14: 359–370

– (1983) The zoosporic grouping of plant pathogens. Entity or non-entity? In: Buczacki ST (ed) Zoosporic plant pathogens. A modern perspective. Academic Press, London, pp 43–83

Bartnicki-Garcia S (1970) Cell wall composition and other biochemical markers in fungal phylogeny. In: Harborne JB (ed) Phytochemical phylogeny, Academic Press, London, pp 81–103

– (1987) The cell wall: a crucial structure in fungal evolution. In: Rayner ADM, Brasier CM, Moore D (eds) Evolutionary biology of the fungi. Cambridge University Press, Cambridge, pp 389–403

– Wang MC (1983) Biochemical aspects of morphogenesis in *Phytophthora*. In: Erwin DC, Bartnicki-Garcia S, Tsao P (eds) *Phytophthora*. Its biology, taxonomy, ecology, and pathology. American Phytopathological Society, St. Paul, MN, pp 121–137

Bouck GB (1971) The structure, origin, isolation, and composition of the tubular mastigonemes of the *Ochromonas* flagellum. J Cell Biol 50: 362–384

Bu'Lock JD, Osagie AU (1976) Sterol biosynthesis via cycloartenol in *Saprolegnia*. Phytochemistry 15: 1249–1251

Burr AW, Beakes GW (1994) A comparative study of zoospore and cysts surface structure in saprophytic and fish pathogenic *Saprolegnia* species (oomycetes fungi) using lectins and monoclonal antibodies. Protoplasma 181: 142–163

Burrell RG, Clayton CW, Gallegly ME, Lilly VG (1966) Factors affecting the antigenicity of the mycelium of three species of *Phytophthora*. Phytopathology 56: 422–426

Cahill DM, Hardham AR (1994) Exploitation of zoospore taxis in the development of a novel dipstick immunoassay for the specific detection of *Phytophthora cinnamomi*. Phytopathology 84: 193–200

Carlile MJ (1966) The orientation of zoospores and germ-tubes. In: Madelin MF (ed) The fungal spore, Butterworths, London, pp 175–186

– (1983) Motility, taxis, and tropism in *Phytophthora*. In: Erwin DC, Bartnicki-Garcia S, Tsao P (eds) *Phytophthora*. Its biology, taxonomy, ecology, and pathology. American Phytopathology Society, St. Paul, MN, pp 95–107

Cavalier-Smith T (1981) Eukaryote kingdoms: seven or nine? BioSystems 14: 461–481

– (1986) The kingdom Chromista: origin and systematics. Prog Phycol Res 4: 309–347

Dewey FM (1990) The use of monoclonal antibodies to detect plant invading fungi. In: Schots A (eds) Monoclonal antibodies in agriculture. Pudoc, Wageningen, pp 21–25

Dick MW (1989) Phylum Oomycota. In: Margulis L, Corliss JO, Melkonian M, Chapman DJ (eds) Handbook of Protoctista. Jones and Bartlett, Boston, pp 661–685

Dudler R (1990) The single-copy actin gene of *Phytophthora megasperma* encodes a protein considerably diverged from any other known actin. Plant Mol Biol 14: 415–422

Erwin DC, Bartnicki-Garcia S, Tsao PH (eds) (1983) *Phytophthora*. Its biology, taxonomy, ecology, and pathology. American Phytopathological Society, St. Paul, MN

Estrada Garcia MT, Green JR, Booth JM, White JG, Callow JA (1989) Monoclonal antibodies to cell surface components of zoospores and cysts of the fungus *Pythium aphanidermatum* reveal species-specific antigens. Exp Mycol 13: 348–355

– Callow JA, Green JR (1990 a) Monoclonal antibodies to the adhesive cell coat secreted by *Pythium aphanidermatum* zoospores recognise 200×10^3 M_r glycoproteins stored within large peripheral vesicles. J Cell Sci 95: 199–206

– Ray TC, Green JR, Callow JA, Kennedy JF (1990 b) Encystment of *Pythium aphanidermatum* zoospores is induced by root mucilage polysaccharides, pectin and a monoclonal antibody to a surface antigen. J Exp Biol 41: 693–699

Evans PT, Holaway BL, Malmberg RL (1988) Biochemical differentiation in the tobacco flower probed with monoclonal antibodies. Planta 175: 259–269

Foster H, Coffey MD, Elwood H, Sogin ML (1990) Sequence analysis of the small subunit ribosomal RNAs of three zoosporic fungi and implications for fungal evolution. Mycologia 82: 306–312

Gabor BK, O'Gara ET, Philip BA, Horan DP, Hardham AR (1993) Specificities of monoclonal antibodies to *Phytophthora cinnamomi* in two rapid diagnostic assays. Plant Dis 77: 1189–1197

Gallegly ME (1983) New criteria for classifying *Phytophthora* and critique of existing approaches. In: Erwin DC, Bartnicki-Garcia S, Tsao P (eds) *Phytophthora*. Its biology, taxonomy, ecology, and pathology. American Phytopathological Society, St. Paul, MN, pp 167–172

Gubler F, Hardham AR (1988) Secretion of adhesive material during encystment of *Phytophthora cinnamomi* zoospores, characterized by immunogold labelling with monoclonal antibodies to components of peripheral vesicles. J Cell Sci 90: 225–235

– – (1990) Protein storage in large peripheral vesicles in *Phytophthora* zoospores and its breakdown after cyst germination. Exp Mycol 14: 393–404

– – (1991) The fate of peripheral vesicles in zoospores of *Phytophthora cinnamomi* during infection of plants. In: Mendgen K, Lesemann D-E (eds) Electron microscopy of plant pathogenesis. Springer, Berlin Heidelberg New York Tokyo, pp 197–210

– – Duniec J (1989) Characterising adhesiveness of *Phytophthora cinnamomi* zoospores during encystment. Protoplasma 149: 24–30

Gunderson JH, Elwood H, Ingold A, Kindle K, Sogin ML (1987)

Phylogenetic relationships between chlorophytes, chrysophytes, and oomycetes. Proc Natl Acad Sci USA 84: 5823–5827
Halsall DM (1976) Specificity of cytoplasmic and cell-wall antigens from four species of *Phytophthora*. J Gen Microbiol 94: 149–158
Hardham AR (1985) Studies on the cell surface of zoospores and cysts of the fungus *Phytophthora cinnamomi*: the influence of fixation on patterns of lectin binding. J Histochem Cytochem 33: 110–118
– (1987 a) Ultrastructure and serial section reconstruction of zoospores of the fungus *Phytophthora cinnamomi*. Exp Mycol 11: 297–306
– (1987 b) Microtubules and the flagellar apparatus in zoospores and cysts of the fungus *Phytophthora cinnamomi*. Protoplasma 137: 109–124
– (1989) Lectin and antibody labelling of surface components of spores of *Phytophthora cinnamomi*. Aust J Plant Physiol 16: 19–32
– Gubler F (1990) Polarity of attachment of zoospores of a root pathogen and pre-alignment of the emerging germ tube. Cell Biol Int Rep 14: 947–956
– Suzaki E (1986) Encystment of zoospores of the fungus, *Phytophthora cinnamomi*, is induced by specific lectin and monoclonal antibody binding to the cell surface. Protoplasma 133: 165–173
– – (1990) Glycoconjugates on the surface of the pathogenic fungus *Phytophthora cinnamomi* studied using fluorescence and electron microscopy and flow cytometry. Can J Microbiol 36: 183–192
– – – Perkin JL (1985) The detection of monoclonal antibodies specific for surface components on zoospores and cysts of *Phytophthora cinnamomi*. Exp Mycol 9: 264–268
– – – (1986) Monoclonal antibodies to isolate-, species- and genus-specific components on the surface of zoospores and cysts of the fungus *Phytophthora cinnamomi*. Can J Bot 64: 311–321
– Gubler F, Duniec J (1991 a) Ultrastructural and immunological studies of zoospores of *Phytophthora*. In: Lucas JA, Shattock RC, Shaw DS, Cooke LR (eds) *Phytophthora*. Cambridge University Press, Cambridge, pp 50–69
– – – Elliott J (1991 b) A review of methods for the production and use of monoclonal antibodies to study zoosporic plant pathogens. J Microsc 162: 305–318
Heath IB (1980) Variant mitoses in lower eukaryotes: indicators of the evolution of mitosis? Int Rev Cytol 64: 1–80
– Greenwood AD, Griffiths HB (1970) The origin of flimmer in *Saprolegnia, Dictyuchus, Synura* and *Cryptomonas*. J Cell Sci 7: 445–461
Hemmes DE (1983) Cytology of *Phytophthora*. In: Erwin DC, Bartnicki-Garcia S, Tsao PH (eds) *Phytophthora*. Its biology, taxonomy, ecology, and pathology. American Phytopathological Society, St. Paul, MN, pp 9–40
Hill FG, Outka DE (1974) The structure and origin of mastigonemes in *Ochromonas minute* and *Monas* sp. J Protozool 21: 299–312
Hohl HR, Hamamoto ST (1967) Ultrastructural changes during zoospore formation in *Phytophthora parasitica*. Amer J Bot 54: 1131–1139
Holwill MEJ (1982) Dynamics of eukaryotic flagellar movement. In: Amos WB, Duckett JG (eds) Prokaryotic and eukaryotic flagella. Cambridge University Press, Cambridge, pp 289–312
Hutter R, DeMoss J (1967) Organization of the tryptophan pathway: a phylogenetic study of the fungi. J Bacteriol 94: 1896–1907
Hyde GJ, Hardham AR (1993) Microtubules regulate the generation of polarity in zoospores of *Phytophthora cinnamomi*. Eur J Cell Biol 62: 75–85
– Gubler F, Hardham AR (1991 a) Ultrastructure of zoosporogenesis in *Phytophthora cinnamomi*. Mycol Res 95: 577–591
– Lancelle S, Hepler PK, Hardham AR (1991 b) Freeze substitution reveals a new model for sporangial cleavage in *Phytophthora*, a result with implications for cytokinesis in other eukaryotes. J Cell Sci 100: 735–748
Jahn TL, Landman MD, Fonseca JR (1964) The mechanism of locomotion of flagellates. II. Function of the mastigonemes of *Ochromonas*. J Protozool 11: 291–296
Judelson HS, Michelmore RV (1989) Structure and expression of a gene encoding heat-shock protein Hsp 70 from the oomycete fungus *Bremia lactucae*. Gene 79: 207–217
– – (1991) Transient expression of genes in the oomycete *Phytophthora infestans* using *Bremia lactucae* regulatory sequences. Curr Genet 19: 453–459
– Tyler BM, Michelmore RW (1991) Transformation of the oomycete pathogen, *Phytophthora infestans*. Mol Plant Microbe Interact 4: 602–607
– – – (1992) Regulatory sequences for expressing genes in oomycete fungi. Mol Gen Genet 234: 138–146
– Coffey MD, Arredondo FR, Tyler BM (1993) Transformation of the oomycete pathogen *Phytophthora megasperma* f.sp. *glycinea* occurs by DNA integration into single or multiple chromosomes. Curr Genet 23: 211–218
Karlovsky P, Prell HH (1991) The TRP1 gene of *Phytophthora parasitica* encoding indole-3-glycerolphosphate synthase-N-(5'-phosphoribosyl)anthranilate isomerase: structure and evolutionary distance from homologous fungal genes. Gene 109: 161–165
Kinghorn JR, Moon RP, Unkles SE, Duncan JM (1991) Gene structure and expression in *Phytophthora infestans* and the development of gene-mediated transformation. In: Lucas JA, Shattock RC, Shaw DS, Cooke LR (eds) *Phytophthora*. Cambridge University Press, Cambridge, pp 295–311
Knox JP, Roberts K (1989) Carbohydrate antigens and lectin receptors of the plasma membrane of carrot cells. Protoplasma 152: 123–129
Krywiencyzk J, Dorworth CE (1980) Serological relationships of some fungi of the genus *Pythium*. Can J Bot 58: 1412–1417
Leedale GF, Leadbetter BSC, Massalski A (1970) The intracellular origin of flagellar hairs in the Chrysophyceae and Xanthophyceae. J Cell Sci 6: 710–719
Lehnen LR Jr, Powell MJ (1988) Cytochemical localization of carbohydrates in zoospores of *Saprolegnia ferax*. Mycologia 80: 423–432
– – (1989) The role of kinetosome-associated organelles in the attachment of encysting secondary zoospores of *Saprolegnia ferax* to substrates. Protoplasma 149: 163–174
– – (1993) Characterization of cell surface carbohydrates on asexual spores of the water mold *Saprolegnia ferax*. Protoplasma 175: 161–172
Loiseaux S (1973) Ultrastructure of zoidogenesis in unilocular zoidocysts of several brown algae. J Phycol 9: 277–289
MacDonald JD, Duniway JM (1979) Use of fluorescent antibodies to study the survival of *Phytophthora megasperma* and *P. cinnamomi* zoospores in soil. Phytopathology 69: 436–441
– Stites J, Kabashima J (1990) Comparison of serological and culture plate methods for detecting species of *Phytophthora, Pythium,* and *Rhizoctonia* in ornamental plants. Plant Dis 74: 665–659

Malajczuk N, McComb AJ, Parker CA (1975) An immunofluorescence technique for detecting *Phytophthora cinnamomi* Rands. Aust J Bot 23: 289–309

Manavathu EK, Suryanarayana K, Hasnain SE, Leung W (1988) DNA-mediated transformation in the aquatic filamentous fungus *Achlya ambisexualis*. J Gen Bicrobiol 134: 2019–2028

Margulis L, Schwartz KV (1988) Five kingdoms. An illustrated guide to the phyla of life on earth. WH Freeman, New York

Merz WG, Burrell RG, Gallegly ME (1969) A serological comparison of six homothallic species of *Phytophthora*. Phytopathology 59: 367–370

Moestrup Ø (1982) Flagellar structure in algae: a review, with new observations particularly on the Chrysophyceae, Phaeophyceae (Fucophyceae), Euglenophyceae and *Reckertia*. Phycologia 21: 427–528

Mohan SB (1988) Evaluation of antisera raised against *Phytophthora fragariae* for detecting the red core disease of strawberries by enzyme-linked immunosorbent assay (ELISA). Plant Pathol 37: 206–216

Morris PF, Ward EWB (1992) Chemoattraction of zoospores of the soybean pathogen, *Phytophthora sojae*, by isoflavones. Physiol Mol Plant Pathol 40: 17–22

Petersen FP, Maybroda AM, Grothaus GD, Miller SA (1989) Monoclonal antibodies and methods for fungal pathogen detection. United States Patent no 4,845,197

Prell HH, Karlovsky P, Bahnweg G (1991) Towards transformation in *Phytophthora nicotianae*. In: Lucas JA, Shattock RC, Shaw DS, Cooke LR (eds) *Phytophthora*. Cambridge University Press, Cambridge, pp 312–325

Pscheidt JW, Burkett JZ, Fisher SL, Hamm PB (1992) Sensitivity and clinical use of *Phytophthora*-specific immunoassay kits. Plant Dis 76: 928–932

Raven PH, Evert RF, Curtis H (1981) Biology of plants. Worth Publishers, New York

Ray PM, Steeves TA, Fultz SA (1983) Botany. WB Saunders, Philadelphia, pp 784

Reichle RE (1969) Fine structure of *Phytophthora parasitica* zoospores. Mycologia 61: 30–51

Sachay DJ, Hudspeth DSS, Nadler SA, Hudspeth MES (1993) Oomycete mtDNA: *Phytophthora* genes for cytochrome c oxidase use an unmodified genetic code and encode proteins most similar to those of plants. Exp Mycol 17: 7–23

Sadowski LA, Powell MJ (1990) Cytochemical detection of polysaccharides in zoospores of *Aphanomyces euteiches*. Can J Bot 68: 1379–1388

Sansome ER (1987) Fungal chromosomes as observed with the light microscope. In: Rayner ADM, Brasier CM, Moore D (eds) Evolutionary biology of the fungi. Cambridge University Press, Cambridge, pp 97–113

Smith E, Roberts K, Hutchings A, Galfre G (1984) Monoclonal antibodies to the major structural glycoprotein of the *Chlamydomonas* cell wall. Planta 161: 330–338

Stamps DJ, Waterhouse GM, Newhook FJ, Hall GS (1990) Revised tabular key to the species of *Phytophthora*. CAB International, Wallingford

Unkles SE, Moon RP, Hawkins AR, Duncan JM, Kinghorn JR (1991) Actin in the oomycetous fungus *Phytophthora infestans* is the product of several genes. Gene 100: 105–112

Villee CA, Solomon EP, Martin CE, Martin DW, Berg LR, Davis PW (1985) Biology. WB Saunders, Fort Worth, pp 1412

Vogel HJ (1965) Lysine biosynthesis and evolution. In: Bryson V, Vogel HJ (eds) Evolving genes and proteins. Academic Press, New York, pp 25–40

Vujicic R, Colhoun J, Chapman JA (1968) Some observations on the zoospores of *Phytophthora erythroseptica*. Trans Br Mycol Soc 51: 125–127

Waterhouse GM, Newhook FJ, Stamps DJ (1983) Present criteria for classification of *Phytophthora*. In: Erwin DC, Bartnicki-Garcia S, Tsao P (eds) *Phytophthora*. Its biology, taxonomy, ecology, and pathology. American Phytopathological Society, St. Paul, MN, pp 139–147

Werres S (1988) Enzyme-linked immunosorbent assay (ELISA) as a method for detection of *Phytophthora fragariae* Hickman in strawberry roots. Nachrichtenbl Deutsch Planzenschutz 40: 146–150

White DG (1976) The preparation and use of a fluorescent antibody reagent for the detection of *Phythium graminicola*. Phytopathology 66: 523–525

Wolters J, Erdmann VA (1988) Cladistic analysis of ribosomal RNAs – the phylogeny of eukaryotes with respect to the endosymbiotic theory. BioSystems 21: 209–214

Wycoff KL, Ayers AR (1990) Monoclonal antibodies to surface and extracellular antigens of a fungal plant pathogen, *Phytophthora megasperma* f. sp. *glycinea*, recognize specific carbohydrate epitopes. Physiol Mol Plant Pathol 37: 55–79

– Jellison J, Ayers AR (1987) Monoclonal antibodies to glycoprotein antigens of a fungal plant pathogen, *Phytophthora megasperma* f. sp. *glycinea*. Plant Physiol 85: 508–515

Zentmyer GA (1980) *Phytophthora cinnamomi* and the diseases it causes. American Phytopathological Society, St. Paul, MN

Protoplasma (1994) 181: 233–244

PROTOPLASMA

Structure, composition, and biogenesis of prasinophyte cell coverings

B. Becker, B. Marin, and **M. Melkonian***

Botanisches Institut, Universität zu Köln, Köln

Received August 5, 1993
Accepted October 18, 1993

Summary. The cell body and flagellar surfaces of certain green flagellates are covered by non-mineralized scales. Scale structure has been widely used in the systematics of this group of algae commonly known as the Prasinophyceae. The special importance of the flagellar hairs as a taxonomic marker is discussed. We summarize current knowledge about the structure and chemical composition of these scales with emphasis on thecate flagellates. Scales consist mainly of acidic polysaccharides involving unusual 2-keto sugar acids. Glycoproteins as minor components are mainly involved in mediating scale subunit and scale-membrane interactions and species specific glycosylation patterns exist. In thecate prasinophytes the elaboration of 3-deoxy-*manno*-2-octulosonic acid and galacturonic acid side chains presumably favours a complex of thecal scales with calcium ions and thus extracellular coalescence of the scales to a rigid cell wall. Scales are formed within the Golgi apparatus (GA) and especially in thecate prasinophytes scale formation (i.e., during flagellar regeneration) represents an excellent model system for GA function. Movement of developing scales through the GA requires cisternal progression. Biogenesis of scales involves mainly polysaccharide synthesis, whereas about 50% of the scale-associated glycoproteins are added from a pre-existing pool. Possible functions of prasinophyte scales are briefly discussed.

Keywords: Prasinophyceae; Cell covering; Scales; Cell wall; Cell wall development.

Abbreviations: Dha 3-deoxy-*lyxo*-2-heptulosaric acid; DSA *Datura stramonium* agglutinin; ER endoplasmic reticulum; GA Golgi apparatus; GNA *Galanthus nivalis* agglutinin; Kdo 3-deoxy-*manno*-2-octulosonic acid; MeKdo 3-deoxy-5-*O*-methyl-*manno*-2-octulosonic acid; SDS-PAGE sodium dodecyl sulphate polyacrylamide gel electrophoresis.

Introduction

Scaly green flagellates are often grouped into a separate class of green algae, the Prasinophyceae, consisting of about 14 flagellate and 1 or 2 non-flagellate genera (Moestrup and Throndsen 1988, Throndsen 1988, Eikrem and Throndsen 1990, Melkonian 1990, Guillard et al. 1991, Inouye and Hori 1991, Moestrup 1991). A characteristic feature of the prasinophytes is the presence of non-mineralized scales on the surface of their cell bodies and flagella. Scales are often present in multiple layers and several structural types have been described up till now. Since each prasinophyte taxon is characterized by a special garniture of scales, scales are widely used for identification and classification into orders, families and genera (e.g., Moestrup and Throndsen 1988, Melkonian 1990). In the order Chlorodendrales (sensu Melkonian and Preisig 1986; two genera: *Tetraselmis* and *Scherffelia*) the cell body is covered by a theca (cell wall), which is formed by extracellular fusion of scales, whereas the flagellar surface is covered by layers of individual scales.

In this review, we summarize present knowledge about the structure and chemical composition of the extracellular cell coverings of the prasinophytes with emphasis on the Chlorodendrales. We address special issues such as the possible function of scales and their suitability as taxonomic markers and discuss recent results on scale biogenesis.

The extracellular matrix of prasinophytes consists of scales

Several years before the advent of electron microscopy Chadefaud (1941, 1947) had recognized that some green flagellates differed from others in that flagella emerged from an apical depression (flagellar groove). These flagellates were later shown by I. Manton, M. Parke, and co-workers to be covered by peculiar non-mineralized scales often regularly arranged in several

* Correspondence and reprints: Lehrstuhl I, Botanisches Institut, Universität zu Köln, Gyrhofstrasse 15, D-50931 Köln, Federal Republic of Germany.

layers (Manton and Parke 1960, 1965; Parke and Rayns 1964; Manton et al. 1965; Parke and Hartog-Adams 1965). As described below, prasinophycean scales are synthesized in the Golgi apparatus, again first observed by I. Manton (Manton and Ettl 1965, Manton et al. 1965, Manton 1966) and later studied in detail (Moestrup and Thomsen 1974, Moestrup and Walne 1979, McFadden and Wetherbee 1985, McFadden and Melkonian 1986, Melkonian et al. 1991, Domozych et al. 1992). The ultrastructure of scales has been extensively studied during the last decades for systematic and taxonomic purposes (e.g., Moestrup and Ettl 1979, Melkonian and Robenek 1981, McFadden et al. 1986 a, Moestrup 1990, Moestrup and Hill 1993). Some examples of the complex ultrastructure of scales are presented in Fig. 1 A–F.

Thecate prasinophytes produce a cell wall

The cell body of the quadriflagellate genera *Tetraselmis* and *Scherffelia* is covered by a solid cell wall (theca) instead of distinct scales (Manton and Parke 1965, Parke and Manton 1965). Before flagellar scales were discovered and the mode of theca formation evaluated by electron microscopy, these two genera were often thought to be related to chlamydomonad-type walled flagellates (e.g., *Scherffelia*; see Ettl 1983). The ultrastructural examination of dividing cells proved that the theca develops from the extracellular fusion of special (thecal) scales (see below) and is thus, in ontogenetic terms (and biochemical, see below), quite different from the Chlamydomonas-type cell wall (Roberts et al. 1985). Recent comparative analysis of thecal ultrastructure has led to the recognition of several distinct types of thecae within the Chlorodendrales (e.g., Fig. 2 A–D) (Marin et al. unpubl. obs.). Thecal types can be distinguished by four criteria: (1) the number of layers, which vary from one to five and which can differ in electron density, (2) the behaviour of thecae following shedding during release of progeny cells from the sporangium (Fig. 2 A and C), (3) the presence or absence of "pit hairs", which are attached to the theca in the region next to the flagellar insertion (Manton and Parke 1965; Norris et al. 1980; Hori et al. 1982, 1983, 1986), (4) the presence or absence of fine hairs covering the whole surface of the theca (e.g., *T. convolutae*; Parke and Manton 1967).

Scales and thecae are non-mineralized

The chemical composition of scales and thecae from various prasinophytes has been analyzed (Lewin 1958; Gooday 1971; Manton et al. 1973; Aken and Pienaar 1985; Becker et al. 1989, 1990, 1991; Domozych et al. 1991; Becker and Melkonian 1992). Most studies showed that scales and thecae are non-mineralized, and that carbohydrates are the major constituents (Lewin 1958; Gooday 1971; Aken and Pienaar 1985; Becker et al. 1989, 1990, 1991). A lower carbohydrate content has been reported only for the basket scales of *Mesostigma viride* (24%; Domozych et al. 1991). However, this value is based on a simple photometric assay (phenol/sulfuric acid) and could be an underestimation given the unusual monosaccharide composition of scales (see below).

The protein content of scales and thecae is generally low. For thecal preparations 0.55–1.0% total nitrogen (3–6% protein, using the standard factor 6.3) has been reported (Lewin 1958, Becker et al. 1991). Some differences in the protein content of different scale types appear to exist. A mixture of scales from *Pyramimonas parkeae* showed a protein content of 3–4% (Aken and Pienaar 1985). "Man scales" from the flagellum of *Tetraselmis striata* contained no protein at all, whereas pentagonal scales from the same source yielded a protein content of about 10% (Becker et al. 1990). Finally, a preparation of basket scales from *M. viride* had a protein content of 5% (Domozych et al. 1991).

Other constituents of scales and thecae are calcium, sulphate, and phosphate. Calcium and sulphate were detected using X-ray microanalysis in the thecae of *Tetraselmis tetrathele* (Manton et al. 1973), and in cyst

Fig. 1 A–F. Scales from 6 prasinophycean genera in positively stained whole mount preparations. **A** Spider web scales from *Mantoniella squamata*. Spider web scales consist of 16 radial spokes connected by several rings of concentric fibrils. 8 major radial spokes running from the periphery to the centre of the scale (arrowheads) alternate with others which terminate at the innermost ring of concentric fibrils (lines). A detailed description is given in Moestrup (1990). **B** Three different types of spider web scales from *Mamiella gilva*. Large body scales (*l.b.*) of circular outline with 16 radial spokes. Within the inner ring of fibrils the 8 major radial spokes are slightly enlarged in diameter towards the centre (lines). Small body scales (*s.b.*) of ellipsoid outline with only 8 radial spokes. Flagellar scales (*fl.*) of ellipsoid outline which contain 14 radial spokes (Moestrup 1984). One spoke is always elongated to form a conspicuous spine (►). Another spoke is strengthened also towards the centre of the flagellar scale (>). **C** Spider web scales from *Pterosperma cristatum*. 8 radical spokes are connected by irregular rings of concentric fibrils. The peripheral concentric fibrils are curved concavely between the radial spokes (arrowheads). **D** Body scales from *Pyramimonas* sp. square scales (*sq.*), box scales (*b.*), crown scales (*cr.*), and footprint scales (arrows). **E** Basket scales from *Mesostigma viride*. **F** Stellate scales from *Nephroselmis rotunda*. A constant number of spines (10) is observed

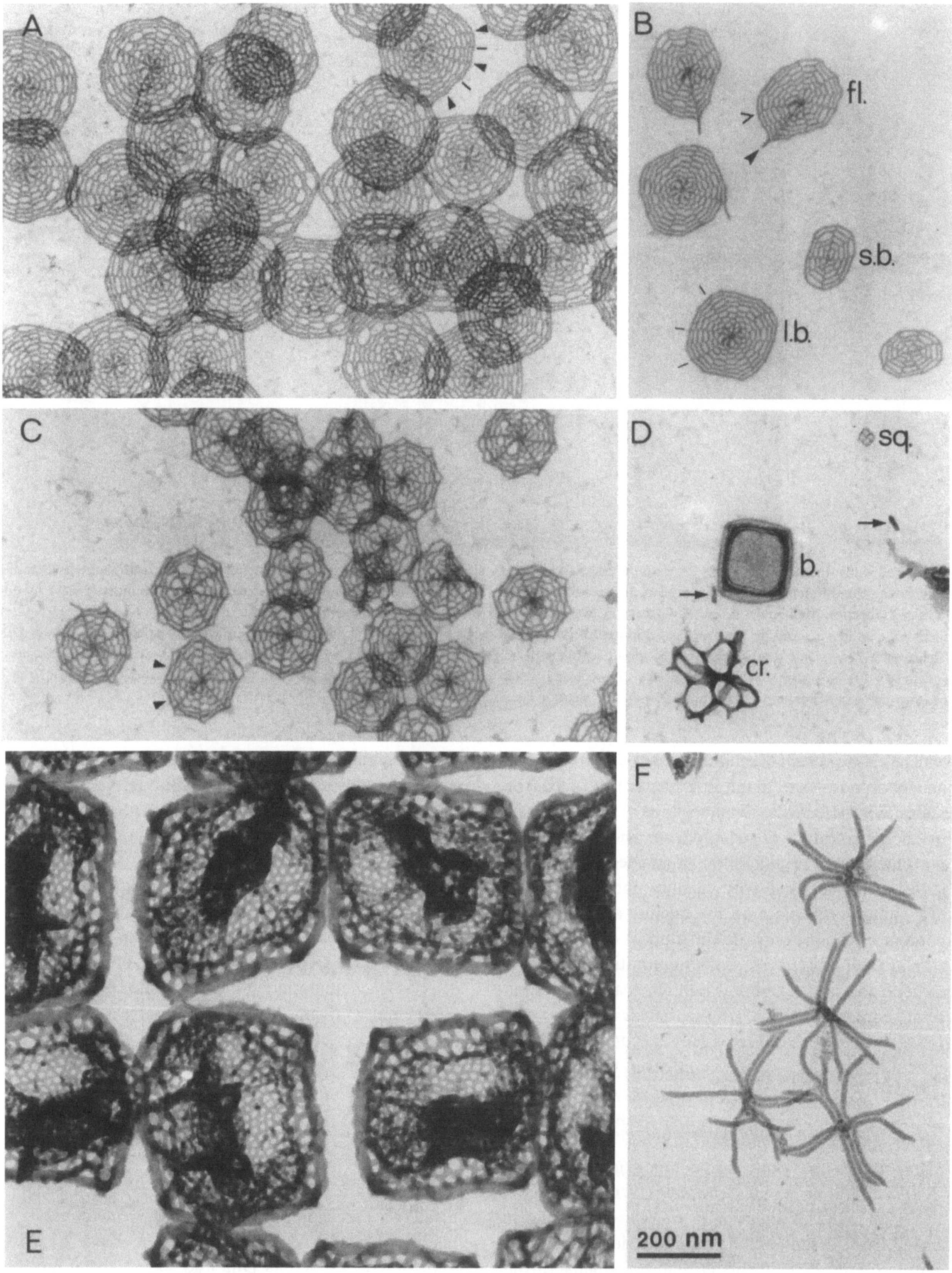
A
B
fl.
s.b.
l.b.
C
D
sq.
b.
cr.
E
F
200 nm

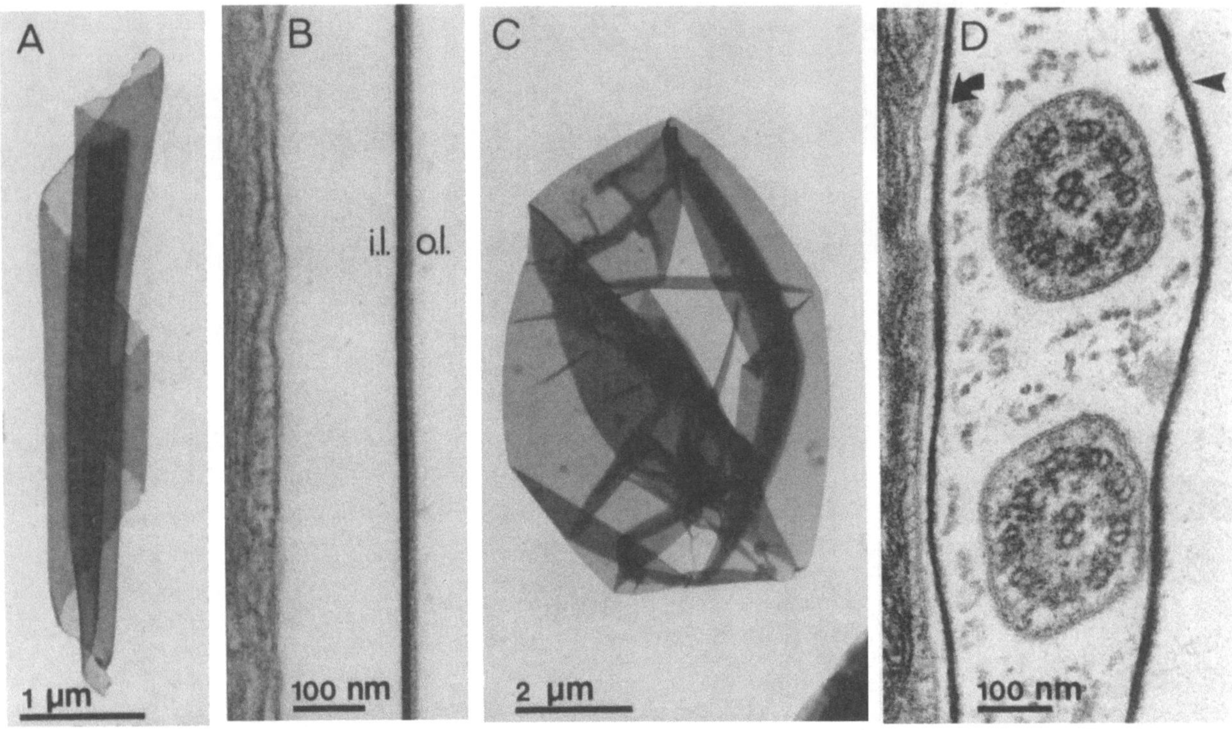

Fig. 2. Cell walls (thecae) of different *Tetraselmis* strains in positively stained whole mount preparations (**A** and **C**) and thin sections (**B** and **D**) showing the differences between thecae of the "scrolls' and the "no scrolls" type (R. Lewin pers. comm.). Strain numbers refer to the Culture Collection Melkonian. **A** and **B** *Tetraselmis chui* (strain M 576). **A** After cell division and release of the progeny cells, the parental theca rolls up like a scroll. **B** The theca is composed of an electron dense inner layer (*i.l*) and less electron dense outer layer (*o.l*). **C** and **D** Thecae of 2 *Tetraselmis* strains belonging to the no scrolls type. **C** Discarded theca of *Tetraselmis* sp. (strain M 836). **D** *Tetraselmis rubens* (strain M 612): A progeny cell and 2 flagella still enclosed within the parental theca. Thecae of the parental (arrowhead) as well as the progeny cell (curved arrow) consist both of only one electron dense layer

walls of *P. parkeae* (Aken and Pienaar 1985). Calcium and phosphate were found in a preparation of basket scales from *M. viride* (Domozych et al. 1991). Based on the apparently low carbohydrate and protein content Domozych et al. (1991) concluded that basket scales are mineralized with calcium phosphate. However, quantitative data on the calcium and phosphate content of the basket scales in support of this conclusion are still lacking. The only quantitative data on the mineral composition of prasinophyte scales are the sulphate contents in thecae of *Tetraselmis subcordiformis* (less than 1%; Lewin 1958) and *T. striata* (6%; Becker et al. 1991), and the calcium content in thecae of three different *Tetraselmis* spp. (4%; Becker et al. 1991).

Scales and thecae contain unusual 2-keto-sugar acids

Early studies on the chemical composition of thecae of *T. subcordiformis* and *T. tetrathele* (Lewin 1958, Gooday 1971), and of scales of *P. parkeae* (Aken and Pienaar 1985) described the polysaccharides as pectic in nature (major constituents galactose and galacturonic acid). Recent studies, however, using advanced analytical techniques and milder hydrolysis conditions, gave strikingly different results. Although galacturonic acid and galactose are present in the thecae of all *Tetraselmis* and *Scherffelia* strains examined so far, and in many scale preparations, the major constituents of prasinophyte scales are unusual 2-keto-sugar acids (Becker et al. 1989, 1990, 1991; Domozych et al. 1991). Thecae contain high amounts of 3-deoxy-*manno*-2-octulosonic acid (Kdo) (> 40%; Table 1) and less 3-deoxy-5-*O*-methyl-*manno*-2-octulosonic acid (MeKdo) (7%) and 3-deoxy-*lyxo*-2-heptulosaric acid (Dha) (10%) (Becker et al. 1989, 1991), whereas scale preparations from various prasinophyte taxa contain less Kdo and more MeKdo and Dha (Table 1) (Becker et al. 1989, 1990, 1991). Conflicting data have been reported for scales of *M. viride*. Domozych et al. (1991) found only glucose (65% of total carbohydrate) and Dha as monosaccharide constituents of the basket scales. Our results on a scale preparation from *M. viride*, consisting mainly of basket scales showed the presence of Kdo and MeKdo as well as Dha and some neutral monosaccharides but no glucose (Becker et al. 1991). The

Table 1. Chemical composition of thecal flagellar scales of *Tetraselmis striata* (Becker et al. 1990, 1991)

	Thecae	p-Scales	m-Scales
Protein content (% d.w.)	3.8[a]	11.0	–
Carbohydrates (% d.w.)	82.0	50–70[b]	
Monosaccharide composition[c]			
Arabinose	1.0	+	5.0
Gulose	4.0	12.0	9.0
Galactose	7.0	+	5.0
Glucose	–	2.0	+
Galacturonic acid	21.0	–	–
Kdo	54.0	23.0	23.0
MeKdo	4.0	35.0	33.0
Dha	8.0	28.0	25.0
Other constituents (% d.w.)			
Calcium	4.0	n.d.	n.d.
Sulfate	6.0	n.d.	n.d.

[a] Based on the total nitrogen content
[b] Not specified for scale type
[c] mol% of total carbohydrate
d.w. Dry weight
n.d. Not determined; + trace; – not detectable

demonstration of glucose in scale preparations obtained by sucrose gradient centrifugation (as in Domozych et al. 1991) is always problematical since, in our experience, it is difficult to completely remove sucrose by dialysis. Consequently, monosaccharide analysis of scales isolated by sucrose gradient centrifugation often shows high amounts of glucose (fructose is destroyed by standard hydrolysis procedures). The differing results regarding the 2-keto-sugar acids are difficult to explain, since complete documentation and experimental details were not provided by Domozych et al. (1991).

Scale structure is maintained by calcium

Several lines of evidence indicate that scale structure is maintained by calcium. In 1973 Manton et al. related the presence of calcium and galacturonic acid to thecal stability and solubilization of cell wall material by hot water extraction was shown to be completely reversible when divalent cations were added (Gooday 1971, Manton et al. 1973). Several years later Melkonian (1982) demonstrated that calcium is needed for scale-scale interactions of flagellar scales in *Tetraselmis cordiformis*. The structure of scales from *P. parkaea* is impaired by treatment with EDTA, whereas the same chelator had no effect on thecae from *T. tetrathele* (Gooday 1971) or on basket scales of *Mesostigma* (Domozych et al. 1991). The structure of the latter was, however, affected by CDTA, which has a stronger affinity to calcium than EDTA. Obviously some differences exist with respect to the affinities of scales and thecae for calcium.

Scales consist of complex polysaccharides

Although the chemical composition of prasinophyte scales and thecae is now well established, almost nothing is known about the structure of scale polysaccharides. Treatment of scales and thecae with hydrolytic enzymes (cellulase, hemicellulase, pectinase) showed no effect on the theca (Gooday 1971; Becker et al. unpubl. results), but scales from *P. parkeae* were dissolved by pectinase treatment (Aken and Pienaar 1985). Hydrolysis of scales of *P. parkeae* with 2 N sulfuric acid yielded a gelatinous residue (uronic acids, which resisted hydrolysis?) and soluble neutral monosaccharides. The gelatinous residue bound alcian blue at pH 2.5 and pH 0.5. These results led the authors to speculate that scales of *P. parkeae* contain a sulphated galacturonan domain, although the residue was not affected by pectinase treatment (Aken and Pienaar 1985). Results in our laboratory indicate that the theca of *T. striata* consists of several polysaccharides of different molecular size with only slight differences in monosaccharide composition (Becker et al. 1994). We isolated and characterized oligosaccharides obtained by partial hydrolysis of the whole cell wall preparation. So far we have established the structure of galacturonic acid and Kdo containing oligosaccharides. The major oligosaccharide obtained is the trisaccharide α-D-GalpA-(1→4)-α-D-GalpA-(1→8)-Kdo (Becker et al. 1994). Interestingly, it appears that at least some of the galacturonic acid residues are sulphated. Preliminary data on the methylation analysis of the cell wall indicates that most of the Kdo is present in 2→4 linked chains (unpubl. results), which are substituted with galacturonic acid side chains.

Some scale types contain glycoproteins

As already discussed, there are considerable differences in the protein content of scales. Only a few scale types have been further analyzed. Basket scales of *M. viride* contain two proteins (Domozych et al. 1991) which have not been further characterized. The best studied scale type are the pentagonal scales (underlayer scales) from the flagellum of *T. striata* (Becker et al. 1990, Becker and Melkonian 1992). Of the four scale types present on the surface of the flagellum, only the two associated with the flagellar membrane contain significant amounts of protein (Becker et al. 1990). Nine out of 13 Coomassie brillant blue-stained protein bands from an SDS-PAGE of a crude scale fraction are associated with the pentagonal scales (Becker et al. 1990). The proteins are glycoproteins containing N-glycosidic carbohydrates of the high-mannose type (Becker and Melkonian 1992) and mediate scale-subunit and scale-plasma membrane interactions (Becker and Melkonian 1992). Remarkably, scale-associated glycoproteins from *Scherffelia dubia* differ from those of *T. striata* as is evident from SDS-PAGE analysis (Fig. 3, lane 1; compare with Becker et al. 1990) and lectin binding studies (Fig. 3, lanes 2 and 3). The binding of the lectin DSA (specific for galactose linked β-1→4 to N-acetylglucosamine) to scale associated glycoproteins from *S. dubia* is the indication of the presence of processed N-glycans in prasinophytes and is very important with respect to the function of the GA in *Scherffelia* (see also below).

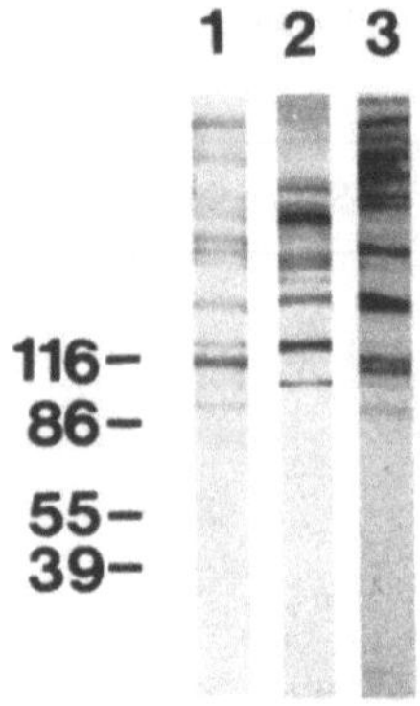

Fig. 3. SDS-PAGE (4–15%) analysis of a crude scale fraction obtained from *Scherffelia dubia*. Proteins were transferred to a polyvinyliden difluoride (PVDF) membrane. *1* Amido black stain; *2* and *3* detection of lectin binding to scale-associated proteins using the digoxigenin labeled lectins and an anti-digoxigenin antibody. *2* GNA, *3* DSA. M_rs (kDa) of marker proteins are indicated on the left. Protein load: 10 µg in each lane

Scales are made in the Golgi apparatus

In 1960, Manton and Parke noted that scales are formed inside the cell. Later studies (Manton and Parke 1965, Manton 1966, Moestrup and Thomsen 1974, Moestrup and Walne 1979, McFadden and Melkonian 1986, McFadden et al. 1986 b, Domozych et al 1992) showed that scales can be detected in the Golgi apparatus and in secretory vesicles. In prasinophytes the Golgi apparatus consists of a defined number of Golgi bodies (e.g., two in *Scherffelia*). In *S. dubia* pentagonal scales are additionally found in a membranous reticulum anterior to and physically separated from the Golgi bodies (McFadden et al. 1986 b). Since for many scale types a maturation process from the proximal (cis) to the distal (trans) face of the Golgi body is observed (Moestrup and Walne 1979, McFadden and Melkonian 1986, Domozych et al. 1992), it is generally accepted that scales are synthesized in the Golgi bodies, transported to the cell surface by the secretory vesicles and released by exocytosis. The function of the Golgi apparatus in scale biogenesis will be discussed later in more detail.

Thecal scales are converted extracellularly into a cell wall

In the order Chlorodendrales, cells produce scales only during cell division (Manton and Parke 1965, McFadden and Melkonian 1986, McFadden et al. 1986 b). Both flagellar and thecal scales are formed in the Golgi apparatus consecutively. Thecal scales resemble the stellate body scales of *Nephroselmis* (Manton and Parke 1965, McFadden and Melkonian 1986). After secretion, scales coalesce extracellularly inside the parental theca to form a new cell wall (Manton and Parke 1965, Domozych et al. 1981, McFadden et al. 1986 b). The molecular basis for the fusion process is presently unknown. However, based on the differences in monosaccharide composition of scales and theca, and our results on the structure of isolated oligosaccharides from the theca (Becker et al. 1994), we can now propose a tentative model for the extracellular wall assembly in the Chlorodendrales. The model is based on the structure of thecal scales in *S. dubia* (McFadden et al. 1986 b). Only the outer layer of thecal scales is incorporated into the following model (Fig. 4):

(1) Theca scales are synthesized in the Golgi apparatus and presumably have a monosaccharide composition similar to other prasinophycean scales, i.e., thecal scales have a lower Kdo and galacturonic acid content than the mature theca (Fig. 4 A).

(2) Kdo and galacturonic acid residues are then added. Our results on the structure of thecal oligosaccharides indicate that presumably homopolymeric Kdo chains with sulfated galacturonic mono- and disaccharide side chains are added. Whether this happens in the Golgi apparatus, extracellularly, or in another post-Golgi compartment is presently unknown (Fig. 4 B).

(3) Finally, fusion of scales takes place by crosslinking individual scales through complexation with calcium (Fig. 4 C). Covalent crosslinks, however, may also be involved. Electron microscope observations indicate that this occurs extracellularly, but prior to release of progeny cells from the parental cell wall.

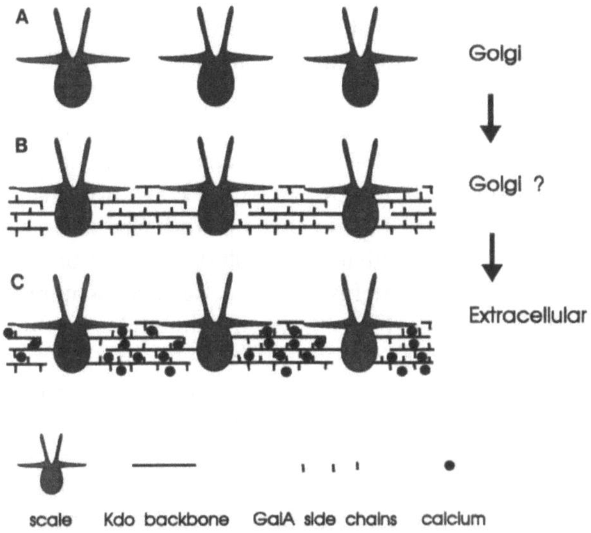

Fig. 4 A–C. Tentative model of cell wall development in *Scherffelia dubia*. For simplicity only the outer scale type is shown. For detailed explanation see text

Although this model is speculative, it might be a good working hypothesis for designing experiments to unravel the mode of theca formation. To test our model, we are currently involved in the production of antibodies towards isolated oligosaccharides (e.g., galacturonic acid and Kdo containing oligosaccharides) and localization of the epitopes during cell wall formation. In addition further data on the structure of the theca and scales are clearly needed and will help in understanding the biogenesis of scales and theca.

Different scale types are produced in one cisterna

Each of the large, partially cellulosic scales of the prymnesiophytes (e.g., *Chrysochromulina chiton*; Manton 1967 a, b) is produced in one cisterna of the single Golgi body. In contrast, prasinophyte scales are often small (less than 100 nm) and cells have often more than one Golgi body. Thus the question arose whether different scale types are synthesized by different Golgi bodies or different cisternae within one Golgi body. All investigations on scale biogenesis in prasinophytes undertaken so far have shown that Golgi bodies are equivalent in their capacity to synthesize all types of scales of a taxon. Furthermore, each cisterna is capable of producing different scales simultaneously (Manton 1966, Moestrup and Thomsen 1974, Moestrup and Walne 1979, McFadden and Melkonian 1986, Domozych et al. 1992). In species of *Pyramimonas*, flagellar and cell body scales were observed within one cisterna (Moestrup and Walne 1979), suggesting that all types of scales in this taxon can be produced at the same time. In thecate prasinophytes, however, it has been found that flagellar and thecal scale formation are temporally separated, the former preceding the latter during cell division. Intracisternal differentiation has often been observed, that is, that specific scale types are formed in certain parts of a cisterna. For example, the large body scales of *Pyramimonas* and *Mesostigma* form at the cisternal periphery (Moestrup and Walne 1979, Domozych et al. 1992); similarly, the pentagonal scales and flagellar hairs of *Scherffelia* (McFadden and Melkonian 1986). A close association of the cisternal membrane with the developing scale was shown for the larger body scales of some prasinophytes (Moestrup and Walne 1979, Domozych et al. 1992), which raises the question of whether the cisternal membrane serves as mold for the formation of this scale type (Domozych et al. 1992).

Prasinophyte cells contain a pool of scale-associated glycoproteins

Much of our knowledge on the biogenesis of cell coverings in eukaryotic cells has been obtained using enzyme inhibitors. One of the inhibitors most often applied is tunicamycin, which blocks the biosynthesis of N-linked oligosaccharides (i.e., the transfer of the first N-acetyl glucosamine to dolichol phosphate is inhibited). Following deflagellation, tunicamycin-treated cells (2 μg/ml) of *Pyramimonas* (McFadden and Wetherbee 1985) and *Tetraselmis* (Reize and Melkonian 1987) regenerate flagella to 50–75% of their original length. A recent reinvestigation of the effect of tunicamycin on flagellar regeneration in *Scherffelia dubia* (1.0 μg/ml) gave similar results (Becker et al. unpubl.). We were able to show with lectin fluorescence microscopy that flagella regenerated in the presence of tunicamycin exhibit the mature glycosylation pattern (unpubl. results). In agreement with this, we detected all of the scale-associated glycoproteins in a Golgi enriched vesicle fraction (Grunow et al. 1993) from non-regenerating cells of *S. dubia*. We therefore conclude that an intracellular pool of scale-associated glycoproteins exists in this species. Since these glycoproteins reveal the mature glycosylation pattern, it will be interesting to localize them at the ultrastructural level.

Glycoprotein versus polysaccharide synthesis

Whereas most polysaccharides of plant cells are synthesized in the Golgi apparatus, including the trans-Golgi-network (see recent review by Driouich et al. 1993), the polypeptide component of glycoproteins is generally made on the rough endoplasmic reticulum

(ER). Core glycosylation of asparagine linked oligosaccharide takes place in the ER (Kornfeld and Kornfeld 1985). High-mannose type glycans undergo no further processing, whereas many N-linked glycans are processed to varying degrees in the GA (Kornfeld and Kornfeld 1985). The general structure of N-linked glycans is conserved throughout the eukaryotic world, with modifications in the processing steps. In contrast, O-glycans are synthesized mainly in the GA in plants and animals. In *T. striata*, only N-glycans of the high-mannose type have been observed (Becker et al. 1990, Becker and Melkonian 1992). In *S. dubia*, however, lectin binding studies showed the presence of processed N-glycans as well (see above; Grunow et al. 1993). Most scale types contain only small amounts of proteins, and cells apparently contain a pool of mature glycoproteins (see above). Golgi function during scale biogenesis in thecate prasinophytes therefore involves mainly polysaccharide synthesis.

Nothing is known about the biosynthesis of scale polysaccharides. Electron microscopical investigations of scale biogenesis in *Scherffelia dubia* showed that scales are first visible as electron-dense granules in the 2nd or 3rd cisterna from the cis face and become more elaborate as they move towards the trans face of the GA (McFadden and Melkonian 1986, McFadden et al. 1986 b). Similar results have been obtained for other scales (Domozych et al. 1992). A recent reinvestigation of GA activity during the first minutes following deflagellation in *S. dubia* showed that mature scales can be recognized after 5 min in the 3rd to 6th cisterna. Between 5 and 20 min, scale-bearing cisternae successively extend to the trans face of the GA. Interestingly, scales are never seen inside the numerous transition vesicles surrounding the Golgi stack. Thus, during scale biogenesis, the GA clearly follows the classical membrane flow, or "cisternal progression", model of Golgi apparatus function (Melkonian et al. 1986, 1991).

The formation of scales: a model system to study Golgi apparatus function

Two models describing the transport of products through the GA have been presented. The "vesicle shuttle" model (Dunphy and Rothman 1985, Rothman and Orci 1992) and the "cisternal progression" model (Morré and Mollenhauer 1974). In the vesicle shuttle model, transport between adjacent cisternae occurs by so-called transition vesicles, whereas in the cisternal progression model new cisternae are formed at the cis-side of the Golgi body and move through the stack. In recent years, the vesicle shuttle model has been favoured, since it was difficult to reconcile cisternal progression with compart mentalization of the GA. Compart mentalization of the GA was indicated by immunolocalization of glycosyltransferases involved in the processing of N-linked glycans to different subcompartments (for review, see Roth 1991). Further support came from experiments of Rothman and colleagues, who showed that intercisternal Golgi transport involved membrane fusion (for review, see Rothman and Orci 1992). However, recent immunolocalization of various resident enzymes of the GA involving double labelings, showed considerable overlap of the antigen distribution to GA subcompartments and also revealed cell type specific differences in distribution (e.g., Velasco et al. 1993, Nilsson et al. 1993). The question of the mode of product transport through the GA is therefore still unsettled (see review by Mellman and Simons 1992). Scale formation in prasinophytes, which clearly involves cisternal progression, represents a model system to study GA function and may shed light on the general mechanisms of transport of secretory products through the GA.

The use of scales as a taxonomic marker in prasinophytes

The classification of prasinophyte taxa into species, genera, families, and orders is largely based on the presence or absence and ultrastructure of the various scale types which exist in the group (e.g., Moestrup and Throndsen 1988, Melkonian 1990). For several reasons scales are regarded as important taxonomic markers in the prasinophytes: (1) in most taxa there are several scale types per cell, arranged in 1–5 layers on the surface of the cell body and flagella, those of each layer having a unique morphology for that taxon, (2) scale structure is genetically controlled and conserved in each taxon and independent of different environmental parameters.

The presence of certain scale types in some genera and the absence in others can be used to define prasinophycean orders, families and genera (Moestrup and Throndsen 1988, Melkonian 1990). For example, square scales (pentagonal or underlayer scales), the most widespread scale type within the green algae, are characteristic of all scaly prasinophytes except for the order Mamiellales (Moestrup 1984, 1990; Eikrem and Throndsen 1990). The occurrence of thecae is a taxonomic marker for the order Chlorodendrales (sensu Melkonian 1990). Maltese cross scales are confined to the order Pseudoscourfieldiales (sensu Melkonian 1990) with the genera *Pseudoscourfieldia* and *Nephro-*

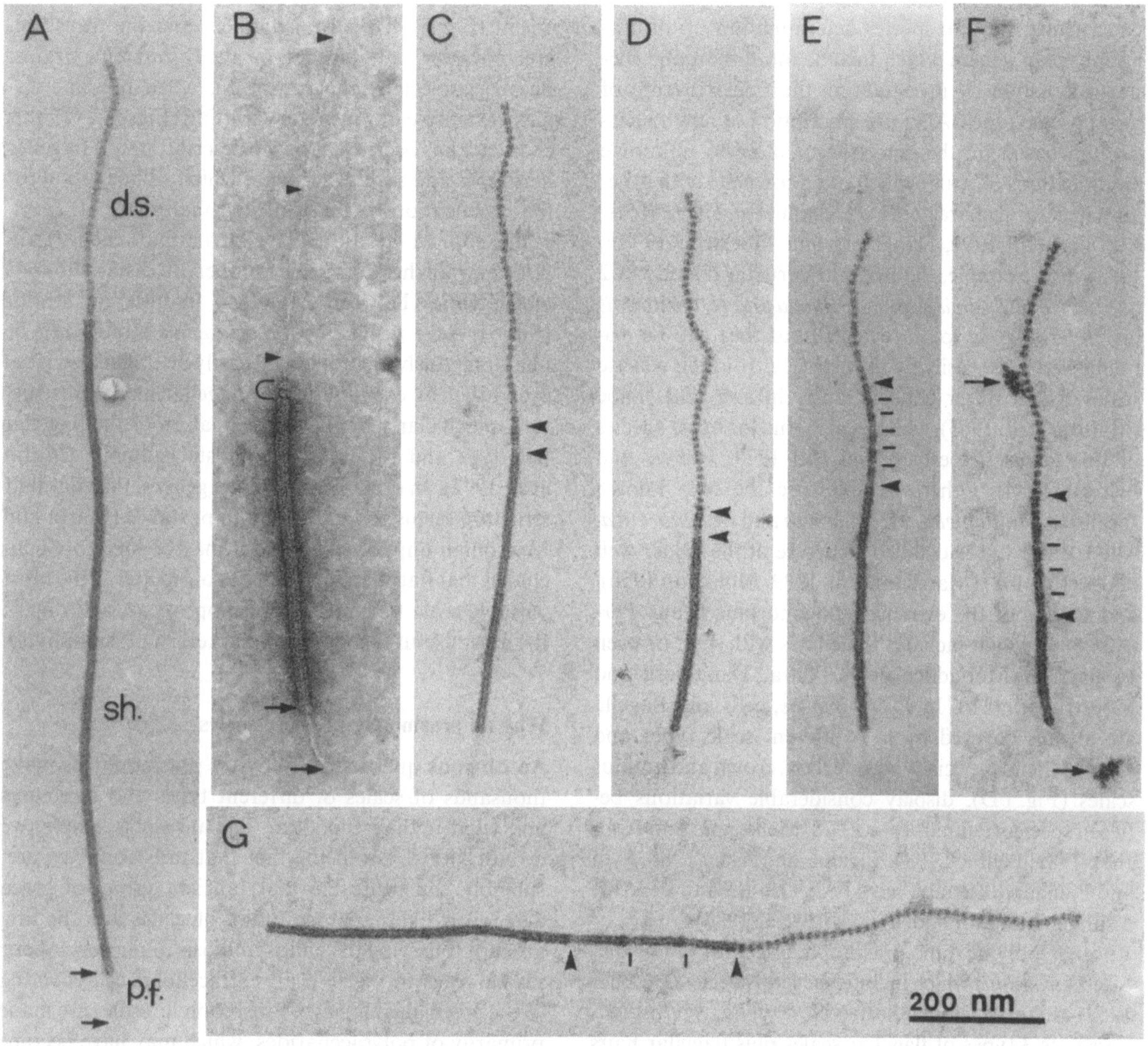

Fig. 5. Flagellar hairs of different strains of *Tetraselmis* in positive and negative contrast. For description of flagellar hair types and details of electron microscopy, see Marin et al. (1993). Strain numbers refer to the Culture Collection Melkonian. **A** and **B** Flagellar hairs of type 1 which consist of a proximal filament (*p.f.*, arrows), a tubular shaft (*sh.*), and a fixed number of distal subunits (*d.s.*) in linear arrangement. **A** strain M 648 in positive contrast. **B** Negatively stained flagellar hair of strain M 625 showing clearly the proximal filament (arrows) and the globular shape of two distal subunits (c) as well as the presence of a long distal filament (arrowheads). **C** and **D** Flagellar hairs of type 2 containing a simple transition inserted between shaft and distal subunits (limits indicated by arrowheads). Flagellar hairs with 37 (**C**) (strain M 653) and 43 (**D**) (strain M 917) distal subunits. **E–G** Flagellar hairs of types 3 and 4 with multiple transition (limits indicated by arrowheads). Segment limits within a multiple transition are indicated by short lines (lines). **E** Flagellar hair of type 3 (strain M 612) with a multiple transition consisting of 6 segments, followed by 19 distal subunits. **F** and **G** Flagellar hairs of type 4 with multiple transition composed of several segments of the same length. The flagellar hair of M 921 (**F**) consists of a short shaft, 5 transition zone-segments of the same length and 31 distal subunits. Arrows, "man scales". The flagellar hair of M 579 contains a transition of 3 very long segments followed by 42 distal subunits (**G**)

selmis. The latter is additionally characterized by stellate scales (Fig. 1 F) which are absent from *Pseudoscourfieldia* and other prasinophytes (Moestrup and Throndsen 1988). As in the case of *Nephroselmis*, many further genera possess peculiar scale types which occur nowhere else. Examples are the crown and footprint scales of *Pyramimonas* and the basket scales of *Mesostigma* (Fig. 1 D, E).

In addition to the presence or absence of scale types, the fine structural variation of scales of the same type can be used for taxonomic purposes, especially the description and distinction of strains or species of the same genus. Indeed, species determination of all prasinophycean genera which include more than one species is possible by investigating the fine structure of certain scale types. Also the description of new species is often based on the detection of hitherto unknown modifications of scales which often correlate with other ultrastructural characters (Marchant et al. 1989, Moestrup and Hill 1991, Daugbjerg and Moestrup 1992 a, 1993). For example, the first ultrastructural study of 2 marine *Nephroselmis* spp., *N. rotunda* and *N. pyriformis* (as *Heteromastix* spp.), revealed that they can be distinguished most easily by ultrastructure of their stellate scales which differ conspicuously in size and shape (Manton et al. 1965), which is also true for other species of this genus (Moestrup and Ettl 1979, Inouye and Pienaar 1984). Other examples are the only known species of *Mantoniella, M. squamata* and *M. antarctica*, which differ in the detailed structure of the spider web scales (Fig. 1 A) (Marchant et al. 1989, Moestrup 1990), and species of the extremely polymorphic genus *Pyramimonas* which includes flagellates with 4, 8, or even 16 flagella (McFadden et al. 1986 a, Daugbjerg and Moestrup 1992 b). In *Pyramimonas*, cells and flagella are always covered by 6–7 different scale types and most of them, especially square, box, crown and limulus scales (Fig. 1 D), display considerable variations between species (Pennick et al. 1978, Melkonian and Robenek 1981, Pennick 1984, Pienaar and Aken 1985, Sym and Pienaar 1991) and have been also used to describe 4 subgenera (McFadden et al. 1986 a, 1987).

Another polymorphic prasinophyte genus, is the thecate *Tetraselmis* which includes several species (Butcher 1959). Like its close relative *Scherffelia, Tetraselmis* possesses 3 types of flagellar scales plus flagellar hairs (hair scales) (Norris et al. 1980, Melkonian 1982, Melkonian and Preisig 1986, Becker et al. 1990). In contrast to *Pyramimonas*, no intrageneric variation of flagellar scales has previously been detected in *Tetraselmis*, and scale morphology thus seemed to be non informative for the taxonomy of thecate prasinophytes (Norris et al. 1980). However, recent ultrastructural examination of flagellar hairs of *Tetraselmis* demonstrated not only the unexpectedly complex fine structure of flagellar hairs but also the wide range of intrageneric variation of this scale type when analyzed with a modified staining technique (Marin et al. 1993). Flagellar hairs consist of distinct parts in linear arrangement: a proximal filament which mediates the contact to the flagellar surface, a tubular shaft, a modification of the shaft called transition (i.e., transition zone of Marin et al. 1993), a linear row of globular distal subunits and a distal filament (Fig. 5). The distal filament and the transition are, however, not present in all *Tetraselmis* strains. Because especially the transition, when present, displays extensive intrageneric variation (Fig. 5 C–G), this character has been used for the description of 4 flagellar hair types and several subtypes which will be useful for the clarification of the confused taxonomy of *Tetraselmis* spp. In addition, flagellar hairs of each *Tetraselmis* strain have a fixed number of distal subunits which varies between 2 and 44 in different strains (Fig. 5). As a result, most *Tetraselmis* strains can be unambiguously identified and easily distinguished from each other by examination of flagellar hair ultrastructure, particularly by the determination of the flagellar hair type and the number of distal subunits (Marin et al. 1993). In other prasinophyte genera, flagellar hair structure seems to be of equal importance (Marin and Melkonian unpubl. obs.) and it may be therefore concluded that flagellar hairs can be considered as the most variable scale type of the Prasinophyceae, and one of the most useful taxonomical markers in prasinophytes.

Why do prasinophytes make scales?

An obvious question is: Why do prasinophytes make thousands of scales of different types and structures and what is their function? The answer is simple, we do not know! Given the fact that prasinophytes presumably constitute the most ancient group of green algae, it is likely that evolution of scales was the first attempt following the endosymbiosis of photosynthetic prokaryotes to develop an extracellular cell covering in the green algal lineage. Prasinophyte scales are made primarily of polysaccharides, which may have become abundant in the ancestral flagellates, once photosynthesis was integrated into the cell's metabolism. A continuous cell wall interferes with flagellation and therefore most eukaryotic flagellates developed other types of cell surface structures. It is likely that prasinophyte scales have functions that have been attributed to cell walls in general. Deep-etch analysis of various prasinophyte cell coverings revealed that individual scales often form a tight fitting extracellular layer resembling a cell wall (e.g., Melkonian and Robenek 1981, McFadden et al. 1986 a). We may therefore conclude that prasinophyte scales perform several functions of which protection against parasites (fungi, protozoa, and bacteria) and viruses may be of special importance. Unusual monosaccharides like the 2-keto sugar acids may have supported this function. The biosynthetic capabilities to synthesize these unusual sugar acids may have been inherited through the endosymbiosis from the cyanobacterial symbiont.

References

Aken ME, Pienaar FN (1985) Preliminary investigations on the chemical composition of the scale boundary and cyst wall of *Pyramimonas pseudoparkeae* (Prasinophyceae). S Afr J Bot 51: 408–416

Becker B, Hard K, Melkonian M, Kamerling JP, Vliegenthart JFG (1989) Identification of 3-deoxy-*manno*-2-octulosonic acid (Kdo), 3-deoxy-5-*O*-methyl-*manno*-2-octulosonic acid and 3-deoxy-*lyxo*-2-heptulosaric acid in the cell wall (theca) of the green alga *Tetraselmis striata* (Butcher). Eur J Biochem 182: 153–160

– Becker D, Kamerling JP, Melkonian M (1991) 2-Keto-sugar acids in green flagellates: a chemical marker for prasinophycean scales. J Phycol 27: 498–504

– Lommerse JPM, Melkonian M, Kamerling JP, Vliegenthart JFG (1994) The structure of an acidic trisaccharide component from a cell wall polysaccharide preparation of the green alga *Tetraselmis striata* (Butcher). Carbohydr Res (submitted)

Becker D, Melkonian M (1992) N-linked glycoproteins associated with flagellar scales in a flagellate green alga: characterization of interactions. Eur J Cell Biol 57: 109–116

– Becker B, Satir P, Melkonian M (1990) Isolation, purification, and characterization of flagellar scales from the green flagellate *Tetraselmis striata* (Prasinophyceae). Protoplasma 156: 103–112

Butcher RW (1959) An introductory account of the smaller algae of the British coastal waters. Part 1: Introduction and Chlorophyceae. Fishery Investigations. Ministry Agriculture Fisheries Food Ser IV 1: 1–74

Chadefaud M (1941) Sur l'organisation et la position systématique des Flagellés du genre *Pyramidomonas*. Rev Scient 79: 113–114

– (1947) Études sur l'organisation de deux Volvocales sédentaires marines: *Prasinocladus lubricus* et *Chlorodendron subsalsum*. Rev Scient 85: 862–865

Daugbjerg N, Moestrup Ø (1992 a) Fine structure of *Pyramimonas cyclotreta* sp.nov. (Prasinophyceae) form Northern Foxe Basin, Arctic Canada, with some observations on growth rates. Eur J Protistol 28: 288–298

– – (1992 b) Ultrastructure of *Pyramimonas* sp.nov. (Prasinophyceae), a species with 16 flagella from northern Foxe Basin, Arctic Canada, including observations on growth rates. Can J Bot 70: 1259–1273

– – (1993) Four new species of *Pyramimonas* (Prasinophyceae) from arctic Canada including a light and electron microscopic description of *Pyramimonas quadrifolia* sp.nov. Eur J Phycol 28: 3–16

Domozych DS, Stewart KD, Mattox KR (1981) Development of the cell wall in *Tetraselmis*: role of the Golgi apparatus and extracellular wall assembly. J Cell Sci 52: 351–371

– Wells B, Shaw PJ (1991) Basket scales of the green alga, *Mesostigma viride*: chemistry and ultrastructure. J Cell Sci 100: 397–407

– – – (1992) Scale biogenesis in the green alga, *Mesostigma viride*. Protoplasma 167: 19–32

Driouich A, Faye L, Staehelin A (1993) The plant Golgi apparatus: a factory for complex polysaccharides and glycoproteins. Trend Biochem Sci 18: 210–214

Dunphy WG, Rothman JE (1985) Compartmental organization of the Golgi stack. Cell 42: 13–21

Eikrem W, Throndsen J (1990) The ultrastructure of *Bathycoccus* gen.nov. and *Bathycoccus prasinos* sp.nov., a non-motile picoplanctic alga (Chlorphyta, Prasinophyceae) from the Mediterrancean and Atlantic. Phycologia 29: 344–350

Ettl H (1983) Chlorophyta I. Phytomonadina. In: Ettl H, Gerloff J, Heynig H, Mollenhauer D (eds) Süßwasserflora von Mitteleuropa, vol 9. G Fischer, Stuttgart

Gooday GW (1971) A biochemical and autoradiographic study of the role of the Golgi bodies in thecal formation in *Platymonas tetrathele*. J Exp Bot 22: 959–971

Grunow A, Becker B, Melkonian M (1993) Isolation and characterization of the Golgi apparatus of a flagellate scaly green alga. Eur J Cell Biol 61: 10–20

Guillard RRL, Keller MD, O'Kelly CJ, Floyd GL (1991) *Pycnococcus provasolii*, gen. et sp. nov., a coccoid prasinoxanthin-containing phytoplankter from the western North Atlantic and Gulf of Mexico. J Phycol 27: 39–47

Hori T, Norris RE, Chihara M (1982) Studies on the ultrastructure and taxonomy of the genus *Tetraselmis* (Prasinophyceae) I. Subgenus *Tetraselmis*. Bot Mag Tokyo 95: 49–61

– – – (1983) Studies on the ultrastructure and taxonomy of the genus *Tetraselmis* (Prasinophyceae) II. Subgenus *Prasinocladia*. Bot Mag Tokyo 96: 385–392

– – – (1986) Studies on the ultrastructure and taxonomy of the genus *Tetraselmis* (Prasinophyceae) III. Subgenus *Parviselmis*. Bot Mag Tokyo 99: 123–135

Inouye I, Pienaar RN (1984) Light and electron microscope observations on *Nephroselmis astigmatica* sp.nov. (Prasinophyceae). Nord J Bot 4: 409–423

– Hori T (1991) High speed video analysis of the flagellar beat and swimming patterns of algae: possible evolutionary trends in green algae. Protoplasma 164: 54–69

Kornfeld R, Kornfeld S (1985) Assembly of asparagine linked oligosaccharides. Annu Rev Biochem 54: 631–664

Lewin RA (1958) The cell walls of *Platymonas*. J Gen Microbiol 19: 87–90

McFadden GI, Melkonian M (1986) Golgi apparatus activity and membrane flow during scale biogenesis in the green flagellate *Scherffelia dubia* (Prasinophyceae). I: Flagellar regeneration. Protoplasma 130: 186–198

– Wetherbee R (1985) Flagellar regeneration and associated scales deposition in *Pyramimonas gelidicola* (Prasinophyceae, Chlorophyta). Protoplasma 128: 31–37

– Hill DRA, Wetherbee R (1986 a) A study of the genus *Pyramimonas* (Prasinophyceae, Chlorophyta) from southeastern Australia. Nord J Bot 6: 209–234

– Preisig HR, Melkonian M (1986 b) Golgi apparatus activity and membrane flow during scale biogenesis in the green flagellate *Scherffelia dubia* (Prasinophyceae). II: Cell wall secretion and assembly. Protoplasma 131: 174–184

– Hill DRA, Wetherbee R (1987) Electron microscopic observations on *Pyramimonas olivacea* N. Carter (Prasinophyceae, Chlorophyta). Phycologia 26: 322–327

Manton I (1966) Observations on scale production in *Pyramimonas amylifera* Conrad. J Cell Sci 1: 429–438

– (1967 a) Further observations on the fine structure of *Chrysochromulina chiton* with special reference to the haptonema, peculiar Golgi structure and scale production. J Cell Sci 2: 265–272

– (1967 b) Further observations on scale formation in *Chrysochromulina chiton*. J Cell Sci 2: 411–418

Manton I, Parke M (1960) Further observations on small green flagellates with special reference to possible relatives of *Chromulina pusilla* Butcher. J Mar Biol Assoc UK 39: 275–298

– – (1965) Observations on the fine structure of two species of *Platymonas* with special reference to flagellar scales and the mode of origin of the theca. J Mar Biol Assoc UK 45: 743–754

– Ettl H (1965) Observations on the fine structure of *Mesostigma viride* Lauterborn. J Linn Soc Bot 59: 175–184

– Rayns DG, Ettl H, Parke M (1965) Further observations on green flagellates with scaly flagella: the genus *Heteromastix* Korshikov. J Mar Biol Assoc UK 45: 241–255

– Oates K, Gooday G (1973) Further observation on the chemical composition of thecae of *Platymonas tetrathele* West (Prasinophyceae) by means of the X-ray microanalyzer electron microscopy (EMMA). J Exp Bot 24: 223–229

Marchant HJ, Buck KR, Garrison DL, Thomsen HA (1989) *Mantoniella* in antarctic waters including the description of *M. antarctica* sp.nov. (Prasinophyceae). J Phycol 25: 167–174

Marin B, Matzke C, Melkonian M (1993) Flagellar hairs of *Tetraselmis* (Prasinophyceae): ultrastructural types and intrageneric variation. Phycologia 32: 213–222

Melkonian M (1982) Effect of divalent cations on flagellar scales in the green flagellate *Tetraselmis cordiformis*. Protoplasma 111: 221–233

– (1990) Phylum Chlorophyta. Class Prasinophyceae. In: Margulis M, Corliss JO, Melkonian M, Chapman DJ (eds) Handbook of Protoctista. Jones and Bartlett, Boston, pp 600–607

– Preisig HR (1986) A light and electron microscopic study of *Scherffelia dubia*, a new member of the scaly green flagellates (Prasinophyceae). Nord J Bot 6: 235–256

– Robenek H (1981) Comparative ultrastructure of underlayer scales in four species of the green flagellate *Pyramimonas*: a freeze-fracture and thin section study. Phycologia 20: 365–376

– McFadden GI, Reize IB, Becker D (1986) Secretion of organic scales in green algae: secretory products are transported through the Golgi apparatus by cisternal progression. Ber Deutsch Bot Ges 99: 263–280

– Becker B, Becker D (1991) Scale formation in algae. J Electron Microsc Techn 17: 165–178

Mellman I, Simons K (1992) The Golgi complex: In vitro veritas? Cell 68: 829–840

Moestrup Ø (1984) Further studies on *Nephroselmis* and its allies (Prasinophyceae). II. *Mamiella* gen.nov., Mamiellaceae fam.nov., Mamiellales ord.nov. Nord J Bot 4: 109–121

– (1990) Scale structure in *Mantoniella squamata*, with some comments on the phylogeny of the Prasinophyceae (Chlorophyta). Phycologia 29: 437–442

– (1991) Further studies of presumably primitive green algae, including the description of Pedinophyceae class.nov. and *Resultor* gen.nov. J Phycol 27: 119–133

– Ettl H (1979) A light and electron microscopical study of the flagellate *Nephroselmis olivacea* (Prasinophyceae). Opera Bot 49: 1–39

– Hill DRA (1991) Studies on the genus *Pyramimonas* (Prasinophyceae) from Australian and European waters: *P. propulsa* sp.nov. and *P. mitra* sp.nov. Phycologia 30: 534–546

– – (1993) Reconstruction of the flagellar surface armour in *Pyramimonas mantoniae* sp.nov. and *P. mitra* Moestrup et Hill (Prasinophyceae). Phycologia 32: 59–67

– Thomsen HA (1974) An ultrastructural study of the flagellate *Pyramimonas orientalis* with particular emphasis on Golgi apparatus activity and the flagellar apparatus. Protoplasma 81: 247–269

– Throndsen J (1988) Light and electron microscopical studies on *Pseudoscourfieldia marina*, a primitive scaly green flagellate (Prasionphyceae) with posterior flagella. Can J Bot 66: 1415–1434

– Walne PL (1979) Studies on scale morphogenesis in the Golgi apparatus of *Pyramimonas tetrarhynchus* (Prasinophyceae). J Cell Sci 36: 437–459

Morré DJ, Mollenhauer HH (1974) The endomembrane concept: a functional integration of endoplasmic reticulum and Golgi apparatus. In: Robards AW (ed) Dynamic aspects of plant ultrastructure. McGraw-Hill, London, pp 84–137

Nilsson T, Pypaert M, Hoe MH, Slusarewicz P, Berger EG, Warren G (1993) Overlapping distribution of two glycosyltransferases in the Golgi apparatus of HeLa cells. J Cell Biol 120: 5–13

Norris RE, Hori T, Chihara M (1980) Revision of the genus *Tetraselmis* (Class Prasinophyceae). Bot Mag Tokyo 93: 317–339

Parke M, Hartog-Adams I (1965) Three species of *Halosphaera*. J Mar Biol Assoc UK 45: 537–557

– Manton I (1965) Preliminary observations on the fine structure of *Prasinocladus miranus*. J Mar Biol Assoc UK 45: 525–536

– – (1967) The specific identity of the algal symbiont in *Convoluta roscoffensis*. J Mar Biol Assoc UK 47: 445–464

– Rayns DG (1964) Studies on marine flagellates VII. *Nephroselmis gilva* sp.nov. and some allied forms. J Mar Biol Assoc UK 44: 209–217

Pennick NC (1984) Comparative ultrastructure and occurrence of scales in *Pyramimonas* (Chlorophyta, Prasinophyceae). Arch Protistenk 128: 3–11

– Clarke KJ, Belcher JH (1978) Studies of the external morphology of *Pyramimonas* 1. *P. orientalis* and its allies in culture. Arch Protistenk 120: 304–311

Pienaar RN, Aken ME (1985) The ultrastructure of *Pyramimonas pseudoparkeae* sp.nov. (Prasinophyceae) form South Africa. J Phycol 21: 428–447

Reize IB, Belkonian M (1987) Flagellar regeneration in the scaly green flagellate *Tetraselmis striata* (Prasinophyceae): regeneration kinetics and effect of inhibitors. Helgol Meeresunters 41: 149–164

Roberts K, Grief C, Hills GJ, Shaw PJ (1985) Cell wall glycoproteins: structure and function. J Cell Sci 2 [Suppl]: 105–127

Roth J (1991) Localization of glycosylation sites in the Golgi apparatus using immunolabeling and cytochemistry. J Electron Microsc Tech 17: 121–131

Rothman JE, Orci L (1992) Molecular dissection of the secretory pathway. Nature 355: 409–415

Sym SD, Pienaar RN (1991) Light and electron microscopy of a punctate species of *Pyramimonas, P. mucifera* sp.nov. (Prasinophyceae). J Phycol 27: 277–290

Throndsen J (1988) *Cymbomonas* Schiller (Prasinophyceae) reinvestigated by light and electron microscopy. Arch Protistenk 136: 327–336

Velasco A, Hendricks L, Moremen KW, Tulsiani DRP, Touster O, Farquhar MG (1993) Cell type-dependent variations in the subcellular distribution of α-mannosidase I and II. J Cell Biol 122: 39–51

Protoplasma (1994) 181: 245–258

PROTOPLASMA

Volvocine cell walls and their constituent glycoproteins: an evolutionary perspective

J. P. Woessner* and **Ursula W. Goodenough**

Department of Biology, Washington University, St. Louis, Missouri

Received August 5, 1993
Accepted November 8, 1993

Summary. Similarities in the composition of the extracellular matrix suggest that only some species of the unicellular *Chlamydomonas* are closely related to the colonial and multicellular flagellated members of the family Volvocaceae. The cell walls from all of the algae in this volvocine group contain a crystalline layer. This lattice structure can be used as a phylogenetic marker to divide *Chlamydomonas* species into distinct classes, only one of which includes the volvocacean algae. Similarly, not all species of *Chlamydomonas* are sensitive to each other's cell wall lytic enzymes, implying divergence of the enzyme's inner wall substrate. Interspecific reconstitution of the crystalline layer is possible between *C. reinhardtii* and the multicellular *Volvox carteri*, but not between *C. reinhardtii* and *C. eugametos*. The hydroxyproline-rich glycoproteins (HRGPs) which make up the crystalline layer in genera which have a similar crystal structure exhibit many homologies. Interestingly, the evolutionarily distant cell walls of *C. reinhardtii* and *C. eugametos* also contain some HRGPs displaying a few morphological and amino acid sequence homologies. The morphological similarities between the flagellar agglutinins (HRGPs responsible for sexual recognition and adhesion during the mating reaction) and the cell wall HRGPs leads to the proposal of a superfamily from which novel HRGPs (designed for self-assembly/recognition) can constantly evolve. Just as variations in the wall HRGPs can lead to unique wall structures, new agglutinins facilitate sexual isolation of new species. Thus, the HRGPs could emerge as valuable phylogenetic markers.

Keywords: Evolution; Hydroxyproline-rich glycoprotein; Cell wall; Phylogenetics; *Chlamydomonas*; Volvocales.

Abbreviations: GLE gametic lytic enzyme; GP glycoprotein; HRGP hydroxyproline-rich glycoprotein; SDS PAGE sodium dodecyl sulfate polyacrylamide gel electrophoresis; VLE vegetative lytic enzyme; VSP vegetative serine/proline-rich; WP wall protein; ZSP zygotic serine/proline-rich.

Introduction

Within the order Volvocales, the volvocine line connects the unicellular green algae in the genus *Chlamydomonas* with the colonial and multicellular green algae comprising the family Volvocaceae. It has been proposed that the conceptual ordering of the volvocine genera (e.g., from *Chlamydomonas*, through *Gonium, Pandorina, Eudorina,* and *Pleodorina*, to *Volvox*), based on traits such as cell number, organismic size, type of sexual reproduction and ratio of somatic to reproductive cells, represents the directional evolutionary history in the group (Schmitt et al. 1992). Indeed, common patterns of subcellular organization, cell wall structure, and asexual and sexual reproduction are found in *Chlamydomonas reinhardtii* and all members of the family Volvocaceae. The colonial members (*Gonium, Pandorina, Pleodorina,* and *Eudorina*) each contain a species-specific number of identical *Chlamydomonas*-like cells, while the multicellular *Volvox* are comprised of non-motile germ cells and flagellated *Chlamydomonas*-like somatic cells. In all these volvocaceans, the cells are held together in a recognizable pattern by a non-cellulosic extracellular matrix with an abundance of hydroxyproline-rich glycoproteins (HRGPs). The similarities and differences between *C. reinhardtii* and volvocacean algae have been recently reviewed in Schmitt et al. (1992).

With an eye towards phylogenetic and evolutionary interpretations, this review compares the various volvocine cell walls, considering first the intact matrices, then in vitro wall assembly, then the isolated wall proteins, and finally the genes encoding the structural components. While marked differences ar observed among the assembled walls of various *Chlamydomonas* spp., some highly conserved amino acid domains in the wall proteins yield HRGP building blocks with common morphological structures (e.g., fibrous rods with glob-

* Correspondence and reprints: Department of Biology, Box 1229, Washington University, St. Louis, MO 63130, U.S.A.

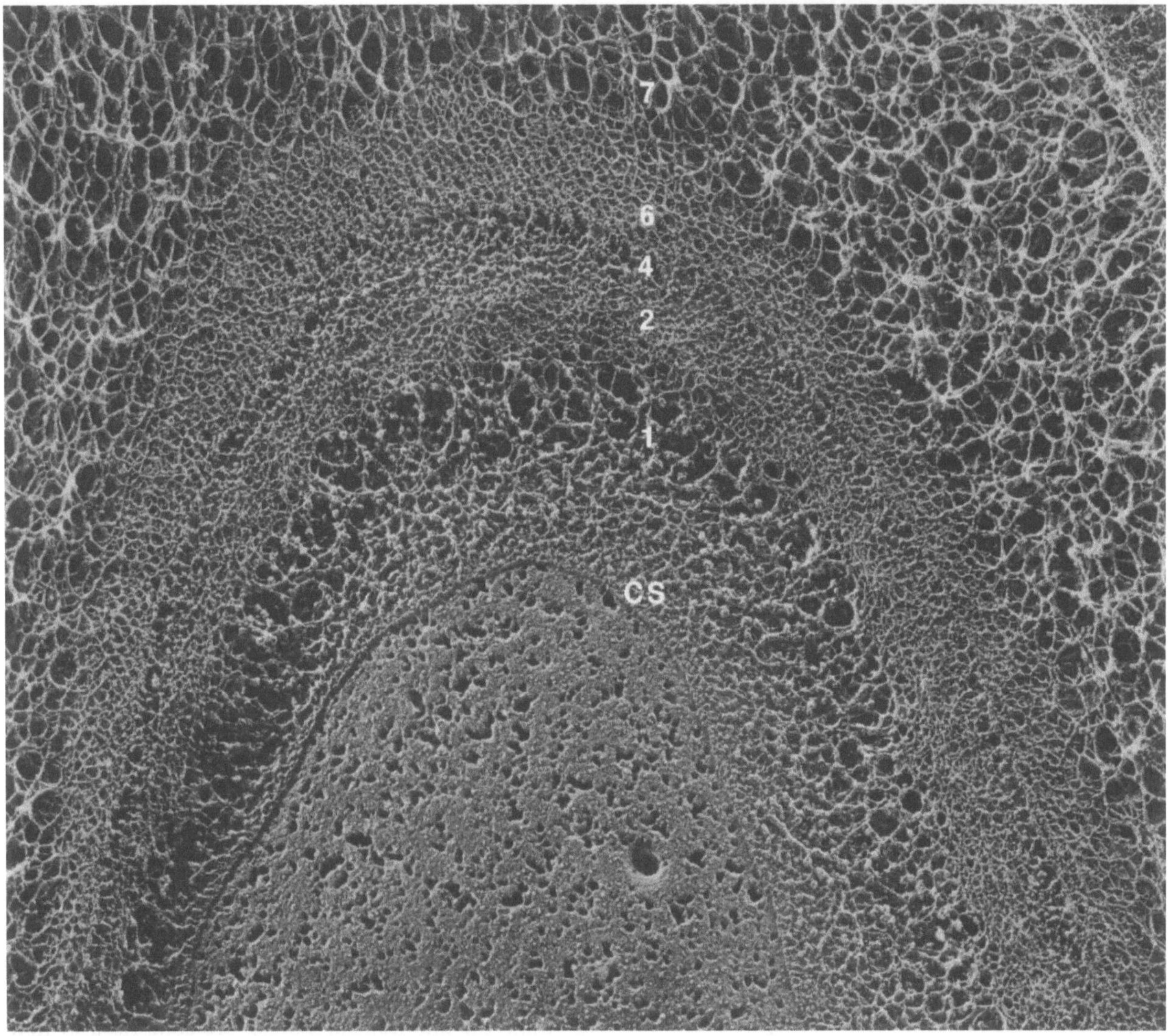

Fig. 1. Cell wall of vegetative *Chlamydomonas reinhardtii* in tangential fracture. *CS* Cell surface; *1, 2, 4, 6,* and *7* major wall layers. × 65,000. This and all of the following micrographs were prepared by John Heuser, Washington University Medical School, St. Louis, MO. Reprinted from Goodenough and Heuser (1985)

ular heads). These assemble into matrices wherein the globular domains are interconnected by the fibrous domains (Roberts et al. 1985 a, Goodenough and Heuser 1988 b). Variability in the assembled matrices comes from differences in two interrelated aspects of the HRGPs, conformation and post-translational modifications. If all of the different cell wall HRGPs derive from a single gene family, as has been proposed (Goodenough 1985), then it should be possible to use these proteins as phylogenetic markers within the volvocine line. An obvious conclusion from the studies presented in this review, reached by other researchers as well, is that the genus *Chlamydomonas* needs to be reorganized. Specifically, it is clear that *C. reinhardtii* has more traits in common with the colonial and multicellular Volvocaceae than with most other *Chlamydomonas* spp., which fall in a group with *C. eugametos*. The closing section will review some molecular phylogenetic studies, involving statistical analysis of aligned rRNA sequences from the Volvocales, which support the monophyletic grouping of *C. reinhardtii* with the Volvocaceae, but indicate that the evolutionary progression from the unicellular *Chlamydomonas* to the multicellular *Volvox* is not unidirectional and monophyletic.

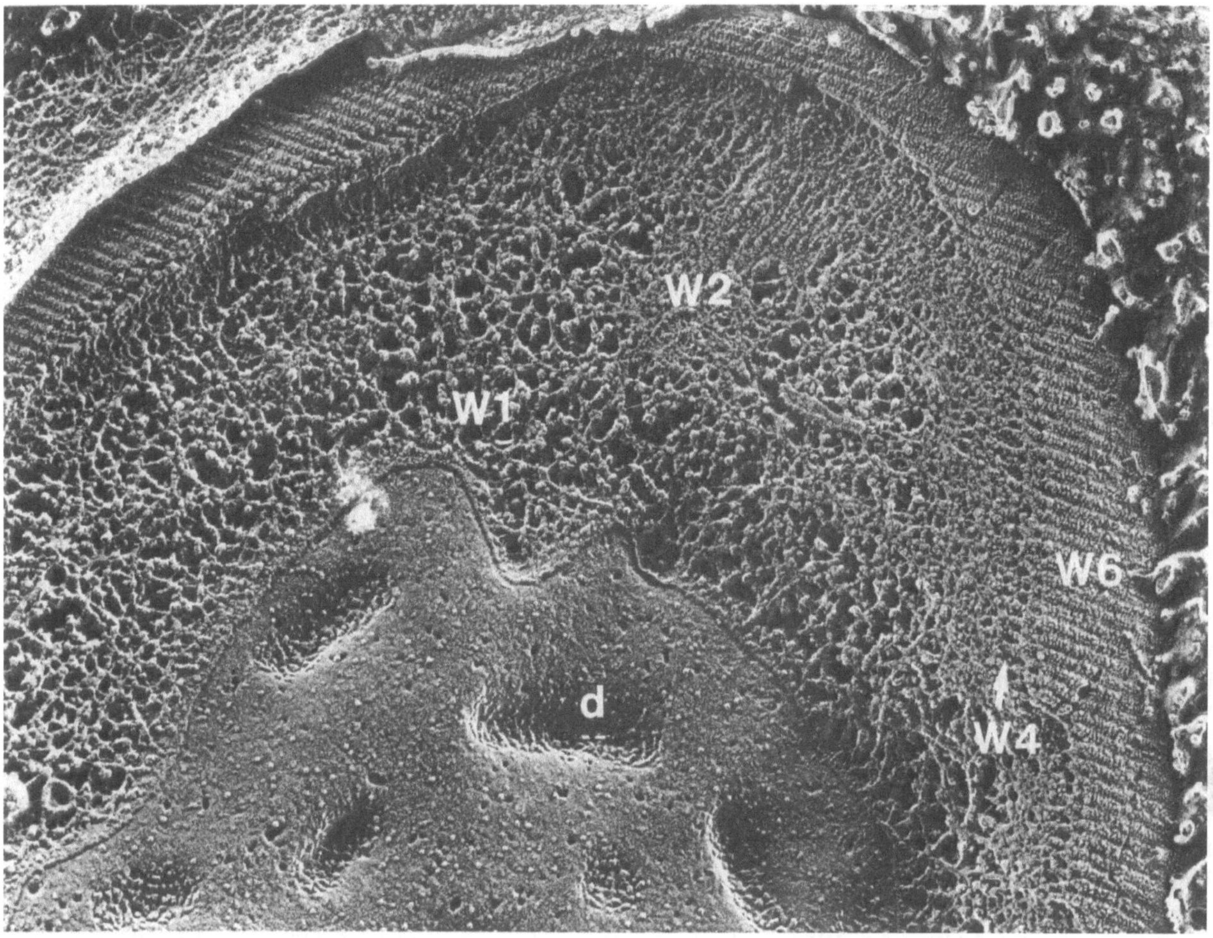

Fig. 2. Cell wall of vegetative *Chlamydomonas eugametos*. The etched face of the plasmalemma shows oblong depressions (*d*) that are absent in *C. reinhardtii*. *W1*, *W2*, *W4*, and *W6* Wall layers. × 80,000. Reprinted from Goodenough and Heuser (1988 b)

Characterization of volvocine cell walls by electron microscopy

Among other morphological traits, the thickness of the cell wall and the overall body shape (which is determined by the cell wall framework; Hills 1973, Imam et al. 1988) were used by Ettl (1976) to define various species of *Chlamydomonas*. The 450 described species of *Chlamydomonas* display a wide array of cell body shapes and sizes. *Chlamydomonas reinhardtii* is the designated type species and has received the most intensive scientific scrutiny (see Harris 1989). The non-cellulosic cell wall of vegetative *C. reinhardtii* cells is constructed from 17–30 types of glycoproteins (Adair and Snell 1990), many of which are HRGPs (Roberts 1974). Each wall is composed of five recognized layers (Goodenough and Heuser 1985, Roberts et al. 1985 a), each of which carries a characteristic set of distinct glycoproteins (Fig. 1). The innermost layer, W1, has a loose network of fibers radiating from the plasma membrane to the central triplet. The central triplet consists of three layers: W2, a tight weave of thick fibers running parallel to the cell surface; W4, a layer of large granules lined up in chains parallel to the cell surface; and the crystalline W6 layer. The outermost layer, W7, also has loose radiating fibers which may or may not be related to those in W1 (see Fig. 3).

The central triplet with its crystalline W6 layer is the most conserved feature of volvocalean cell walls (Roberts 1974). Roberts et al. (1982), using optical diffraction analysis of electron micrographs of cell walls from many different Volvocales, defined four distinct classes of crystalline structure. Class I contains *C. reinhardtii*, *C. angulosa*, *C. cribrum*, *C. inepta*, and the colonial and multicellular Volvocaceae. Class II, considered by Roberts (1974) to be the most primitive wall from which the other types have evolved, is the most predominant type in the unicellular Volvocales, including the majority of *Chlamydomonas* examined. *C. asymmetrica* is the only representative of Class III, while *Vitreochlamys incisa* and *Lobomonas piriformis* make up Class IV. The large number of species which fall into Class II, and the grouping of several *Chlamydomonas* spp. with Volvocaceae indicates that outer-wall ultrastructure has been highly conserved in the Class II lineage and that the colonial and multicellular Volvocaceae probably evolved from a Class I ancestor. Using crystal structure as a phylogenetic marker, therefore, the genus *Chlamydomonas* can be divided into at least three different lineages.

In their quick-freeze deep-etch electron microscopy analysis of the Class II *C. eugametos* wall, Goodenough and Heuser (1988 b) defined W1, W2, W4, and W6 layers (Fig. 2). While W1 and W2 appear structurally similar to the comparable layers found in *C. reinhardtii*, W4 is not granular in *C. eugametos* but probably a subdomain of W2. W6 is a dense, "chainmail" crystal which is radically different from the open W6 weave in *C. reinhardtii*. Interestingly, the Class II version of W6 closely resembles the crystalline pattern of the flagellar collar in the *C. reinhardtii* and *V. carteri* cell walls (Goodenough and Heuser 1988 b, Imam and Snell, 1988, Kirk et al. 1986). Flagellar collars are specialized domains at the cell anterior through which the flagella traverse the wall. In *C. eugametos* and *Chlorogonium elongatum* (both Class II), the flagellar collar appears as a simple infolding of the W6 crystal; whereas in *C. reinhardtii* and *V. carteri* the collar is quite distinct from the other layers of the cell wall. Therefore, expanding on Roberts' (1974) scheme, it has been proposed that the primordial Volvocales produced Class II cell walls from which a line of unicells arose that kept Class II crystals for its flagellar collars, but evolved Class I structure for its cell walls (Goodenough and Heuser 1988 b). This line then gave rise to the colonial and multicellular Volvocaceae.

The colonial and multicellular Volvocaceae have a more complex cell wall than the unicellular green algae. Kirk et al. (1986) have identified four major zones in the *Volvox* wall – the flagellar zone includes the flagellum and the periflagellar region; the boundary zone (with central triplet found in all Volvocales) surrounds the entire colony; the cellular zone encompasses individual cells in the organism; and the deep zone designates the central space of the colony. Both the cellular and deep zones are unique to the Volvocaceae, and these zones make up the majority of the extracellular matrix. Kirk et al. (1986) point out that while comparisons of the *Volvox* boundary zone with the *C. reinhardtii* crystalline layer emphasize the relatedness of these two organisms, the inner zones are what define species-specific extracellular matrix patterns in the family. Extensive modification and elaboration of the amorphous W2 layer of a unicellular ancestor like *C. reinhardtii* may have led to the diversity found in the colonial and multicellular Volvocaceae (Schmitt et al. 1992).

Enzymatic digestion of volvocalean cell walls

Work by Schlösser (1976, 1984) and Matsuda et al. (1987) on species-specific sensitivity to lytic enzymes among the Volvocales lends support to the idea that phylogenetic diversity is reflected by inner wall composition. The vegetative lytic enzyme (VLE) specifically digests the mother cell wall, releasing the daughter cells after mitosis (Schlösser 1966), while the gametic lytic enzyme (GLE) is produced by mating gametes to digest the cell wall prior to zygote formation (Claes 1971). GLE will digest the vegetative or gametic cell wall at any time during the life cycle and has been shown to act specifically on the flagellar collar and within the W2 layer of *C. reinhardtii* (Goodenough and Heuser 1985, Imam and Snell 1988, Monk 1988, Waffenschmidt et al. 1988). Schlösser (1976, 1984) classified 45 species of *Chlamydomonas* into 15 groups based on sensitivity to VLE. There is not enough overlap in the species examined by Schlösser and by Roberts et al. (1982) to make definitive statements about how the VLE groupings match the crystal lattice classes. Nevertheless, *C. dysosmos* and *C eugametos* which have Class II walls, fall into Schlösser's groups 7 and 12, respectively, and *C. reinhardtii, C. angulosa,* and *C. inepta*, which share Class I walls, belong to groups 1, 2 and 2, respectively. The latter two groups must be closely related because the VLE from group 2 members can digest the walls of all group 1 cells; this action, however, is not reciprocal (Schlösser 1976).

GLE from *C. reinhardtii* can digest the vegetative cell walls of all of Schlösser's group 1 strains as well as *C. cribrum* (not grouped by Schlösser), and partially affects the wall of one group 2 member, *C. komma* (Matsuda et al. 1987). No other *Chlamydomonas* spp. showed any sensitivity to this GLE. Hence, *C. reinhardtii* GLE and VLE both exhibit similar group-specific activity. Matsuda et al. (1987) also tested the ability of this GLE to digest the walls of colonial and multicellular Volvocales, and found that six species of *Gonium* and two species of *Astrephomene* were sensitive to GLE but not to VLE; other colonial or multicellular

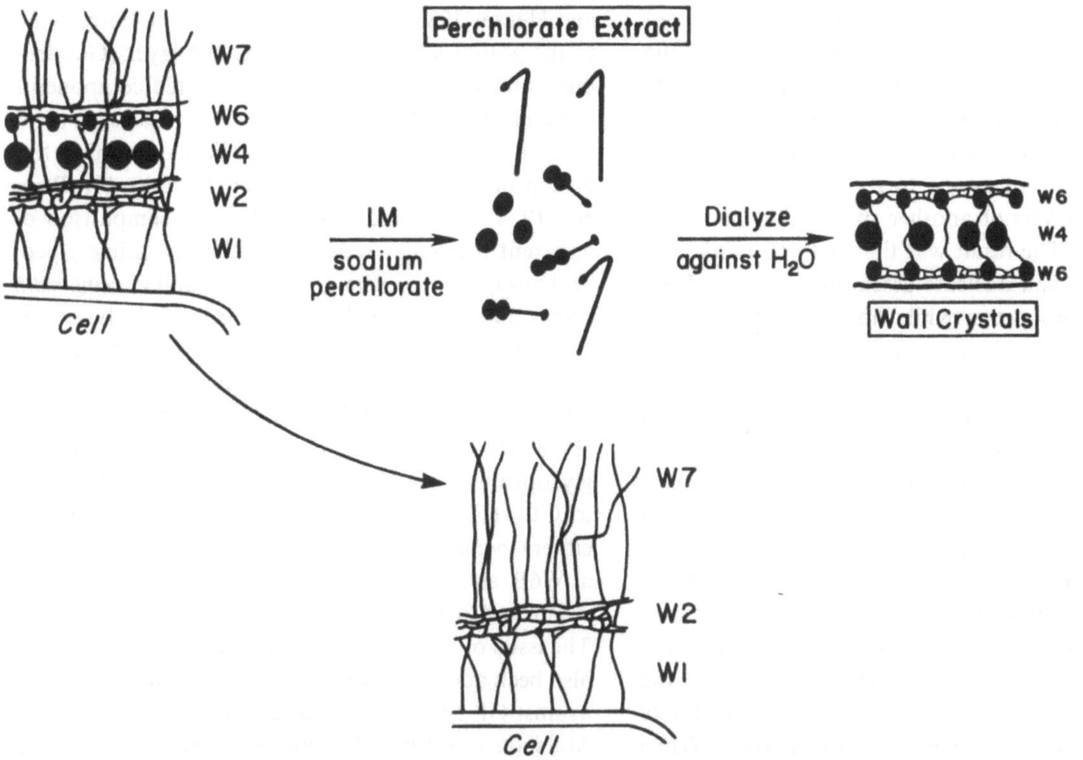

Fig. 3. Proposed sequence of events for *C. reinhardtii* wall crystal formation in vitro. Reprinted from Goodenough et al. (1986 a)

volvocalean algae were not susceptible. Representatives from all four classes of crystalline walls were tested by Matsuda et al. (1987) and only members of Class I showed any sensitivity to GLE from *C. reinhardtii.* While some of the colonial Volvocaceae (Class I) exhibited sensitivity, supporting Roberts' classification, a number of other colonial and multicellular Volvocales (including Class I members) were resistant to this GLE. These observations indicate that there is variation in the inner wall/collar composition such that there can be a common crystalline outer wall (Class I) and a genus-specific inner wall. They also provide additional evidence for an evolutionary sequence from an ancestor resembling *C. reinhardtii* to the colonial Volvocaceae.

In vitro reassembly of volvocine cell walls

In 1973, Hills demonstrated that a portion of the cell wall of *C. reinhardtii* is soluble in chaotropes (LiCl or $NaClO_4$) and that when the chaotrope is dialyzed away, the solubilized proteins reassemble in vitro. This initial observation has now been documented and analyzed in many laboratories (see Adair and Snell 1990). The *C. reinhardtii* cell wall can be divided into two major domains – the salt-soluble outer layers including W4, W6, and W7 (accounting for > 90% of the cell wall mass), and the highly insoluble, covalently cross-linked inner layers W1 and W2 (see Fig. 1). Extraction with chaotropes disassembles the crystalline layer into its component glycoproteins, and if these components are dialyzed in the absence of the insoluble layers they will self-assemble into a crystal structure composed of a W4 layer sandwiched between two W6 layers (Fig. 3) (Goodenough et al. 1986 a). While the crystal structure of the W6 layers in this in vitro sandwich is indistinguishable from the W6 lattice found in situ (Goodenough and Heuser 1988 a), it is only when the chaotrope-extracted components are dialyzed in the presence of the insoluble layers (either wall "shells" (Hills et al. 1975) or salt-extracted cells (Adair et al. 1987)) that the in vitro reassembly creates a single W6 layer as is found in situ. The salt-extracted cells and "shells" retain the characteristic shape of the organism, demonstrating that the inner layers of the wall generate the wide range of cell shapes found in *Chlamydomonas* spp. With the knowledge that the crystalline layers of *C. reinhardtii, Volvox carteri,* and *Gonium pectorale* are all in Class I, and that the W2 layer (boundary and cellular zones of Kirk et al. 1986) of *V. carteri* resembles that of *C. reinhardtii* (Goodenough and Heuser 1985),

Adair et al. (1987) undertook interspecific wall reconstitution analyses. First they demonstrated that the crystalline layers of *V. carteri* and *C. eugametos* could be solubilized in sodium perchlorate and reassembled in vitro. Next, they took the soluble components from each species and attempted to reconstitute a wall on the insoluble layers of another species that had been salt-extracted. The result was that both *C. reinhardtii* and *V. carteri* can nucleate assembly of each other's crystalline components, while no reconstitution was observed between *C. reinhardtii* and *C. eugametos*. Later experiments showed that *C. reinhardtii* and *Gonium pectorale* can each nucleate assembly of the other's outer wall (Adair and Snell 1990). Therefore, association of the outer wall glycoproteins with the W2 layer components is not species-specific and indicates a high degree of functional conservation in Class I cell walls. No wall reconstitution is observed when cell wall-less mutants of *C. reinhardtii* lacking the W2 layer are mixed with extracted crystalline components from a wild-type cell (Hills 1973, Adair et al. 1987), suggesting that the insoluble inner layers are critical to nucleate assembly of the outer wall monomers, at least in vitro. While the insensitivity of *V. carteri* to the *C. reinhardtii* GLE (Matsuda et al. 1987) implies differences in the inner wall layers between these two organisms, the interspecific reassembly data document the persistence of domains in the *Volvox* W2 layer that are homologous to the nucleation sites in the *C. reinhardtii* W2 layer. The inability of *C. reinhardtii* and *C. eugametos* to reconstitute each other's outer wall provides further proof of the evolutionary distance between these two species and between algae with Class I or Class II walls.

Volvocine structural HRGPs

Since the Class I volvocine algae have a similar cell wall crystal structure, are capable of interspecific wall assembly, and exhibit some cross-sensitivity to one another's autolysins, it is reasonable to expect homologies among the proteins which make up their cell walls. Four main glycoproteins are extracted from the outer wall of *C. reinhardtii* (Roberts 1974, Goodenough et al. 1986 a); three are HRGPs (GP 1, GP 2, and GP 3), and one is glycine-rich (GP 1.5). Detailed electron microscopy studies on in vitro crystal assembly using purified glycoproteins have shown that the W6 crystalline layer is actually a bilayer, with the inner layer (W6A) formed by coassembly of GP 2 and GP 3 monomers and the outer layer (W6B) constructed by self-assembly of GP 1 monomers on the GP 2/GP 3 template (Goodenough and Heuser 1988 a). GP 1.5 was shown to correspond to the granules that comprise the W4 layer (Goodenough and Heuser 1985). When extracted components of the crystalline layers of different Volvocales are compared by SDS PAGE, it becomes clear that only a few glycoproteins are involved in outer wall assembly (Roberts 1974, Adair and Appel 1989). A comparison of apparent molecular weights for the crystalline glycoproteins from *V. carteri*, *G. pectorale*, and *C. reinhardtii* shows that the colonial and multicellular Volvocaceae contain components which migrate in the gel similarly to GP 2 (Goodenough and Heuser 1988 b, Adair and Appel 1989). In addition, Goodenough and Heuser (1988 a) have shown that one of the perchlorate-soluble glycoproteins from *V. carteri* is morphologically similar to GP 2 (Fig. 4). There is no GP 1-like component in either *Gonium* or *Volvox*, and neither organism displays a W6B sublayer (Goodenough and Heuser 1988 a, Adair and Appel 1989).

The issue of GP 2 conservation in the Volvocaceae has also been addressed using a polyclonal antibody raised against GP 2 from *C. reinhardtii* (Roberts et al. 1985 b, Matsuda et al. 1987). This antibody proves to recognize sugar side-chain epitopes on the GP 2 glycoprotein, and by immunofluorescence, cross-reacts with the whole

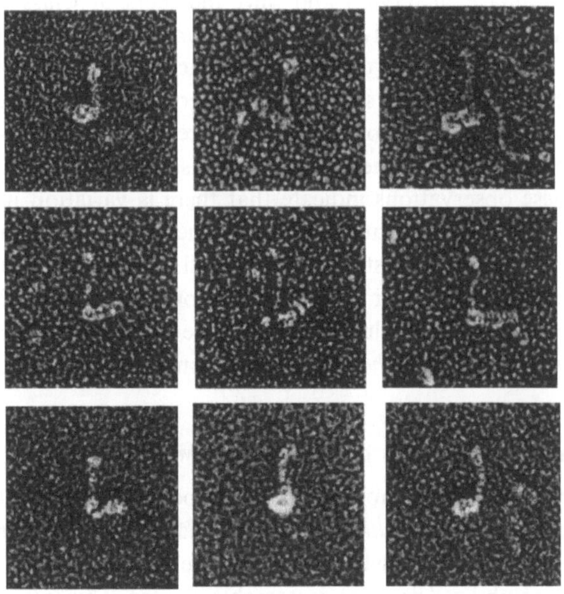

Fig. 4. The top two rows of micrographs show images of the GP 2-like protein found in perchlorate extracts of *Volvox carteri* cell wall crystals. The bottom row shows GP 2 purified from *C. reinhardtii* wall crystals. × 250,000. Reprinted from Goodenough and Heuser (1988 a)

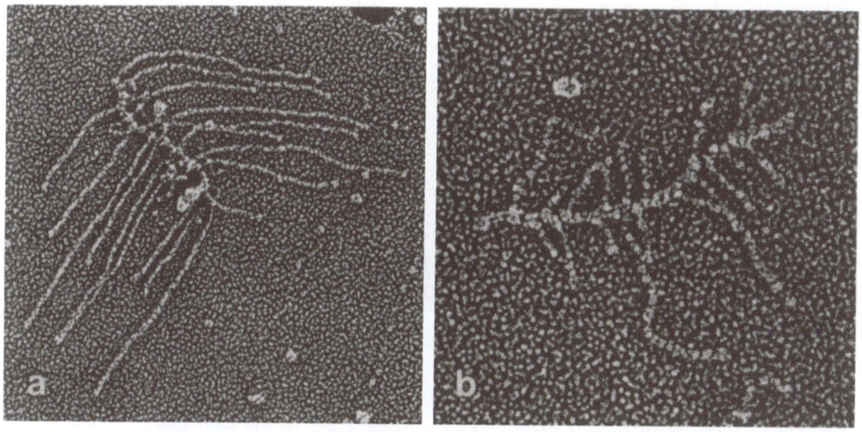

Fig. 5. **a** W 2 fishbone unit from *C. reinhardtii.* × 165,000. Reprinted from Goodenough and Heuser (1985). **b** W 2 fishbone unit from *C. eugametos.* × 274,000. Reprinted from Goodenough and Heuser (1988 b)

cell wall, flagella, and flagellar collars of organisms in Roberts' Class I, but not Class II and III. Roberts et al. (1985 b) found cross-reactivity of the antibody to *Lobomonas piriformis* in Class IV, but Matsuda et al. (1987) could not replicate this result. These antibody data prove that sugar side-chain epitopes are conserved in Class I walls, but reveal nothing about protein backbone conservation among crystalline HRGPs. Adair and Appel (1989) used off-diagonal peptide mapping, amino acid analysis, and Western blots with a polyclonal antibody to deglycosylated GP 2 to demonstrate that there are indeed GP 2 homologs in both *V. carteri* and *G. pectorale*, but not in *C. eugametos* (Class II).

The Class II W6 crystal of *C. eugametos* is composed of two closely migrating glycopolypeptides, designated GP 265 and GP 245 (Goodenough and Heuser 1988 b). Amino acid analysis of this material indicates a preponderance of hydroxyproline, proline and serine, as with the HRGPs which make up the W6 crystal of *C. reinhardtii* (Adair and Appel 1989, Goodenough et al. 1986 a). These *C. eugametos* HRGPs are morphologically dissimilar to GP 2 or GP 3 and, as noted above, assemble into a very different lattice (compare Figs. 1 and 2) (Goodenough and Heuser 1988 b).

The covalent crosslinking of the inner wall layers has confounded the analysis of the individual HRGPs which make up these layers. Boiling mechanically isolated/salt-extracted cell walls in the presence of SDS and β-mercaptoethanol does not solubilize any inner wall components (other than the 225 kDa flagellar collar polypeptide) as determined by SDS PAGE (Imam and Snell 1988). Only when walls released by GLE treatment of *C. reinhardtii* cells are boiled in SDS and β-mercaptoethanol is any solubilization of the inner wall components observed. This solubilization, however, is incomplete and only one distinct band has been identified by SDS PAGE (Adair and Snell 1990). Electron microscopy has been used to compare the solubilized wall components released into the media by *C. eugametos* cells (Goodenough and Heuser 1988 b) and *C. reinhardtii* cells (Goodenough and Heuser 1985). In both cases, fishbone-like units are found, bearing central "spines" and radiating "ribs" (Fig. 5). The main difference between the fishbones from the two species is that the lengths of the long and short ribs in the *C. eugametos* structure are half that of the *C. reinhardtii* ribs. The *C. reinhardtii* fishbones derive from the solubilization of the W2 layer due to the action of GLE released during mating, and while no autolysin-releasing system has yet been identified in *C. eugametos*, the fishbones in the media are proof of its existence. In both species, the fishbone units are believed to interact with the crystalline layer via their spines and thus, they probably serve as the W2-layer nucleation sites for crystal formation.

The presence of similar fishbone structures in these two species of *Chlamydomonas* has some intriguing implications. By the aforementioned classification schemes (Ettl 1976, Roberts et al. 1982, Schlösser 1984) these two species of *Chlamydomonas* are quite distinct, yet both types of walls have radiating and densely packed fibers making up well-defined W1 and W2 layers, both walls have a crystalline lattice which disassembles in perchlorate and reassembles upon dialysis, and some of the components which compose the inner and outer layers of each wall have comparable morphology. This leads to a concept which originated from research on the crystal lattices: there is a common plan for making

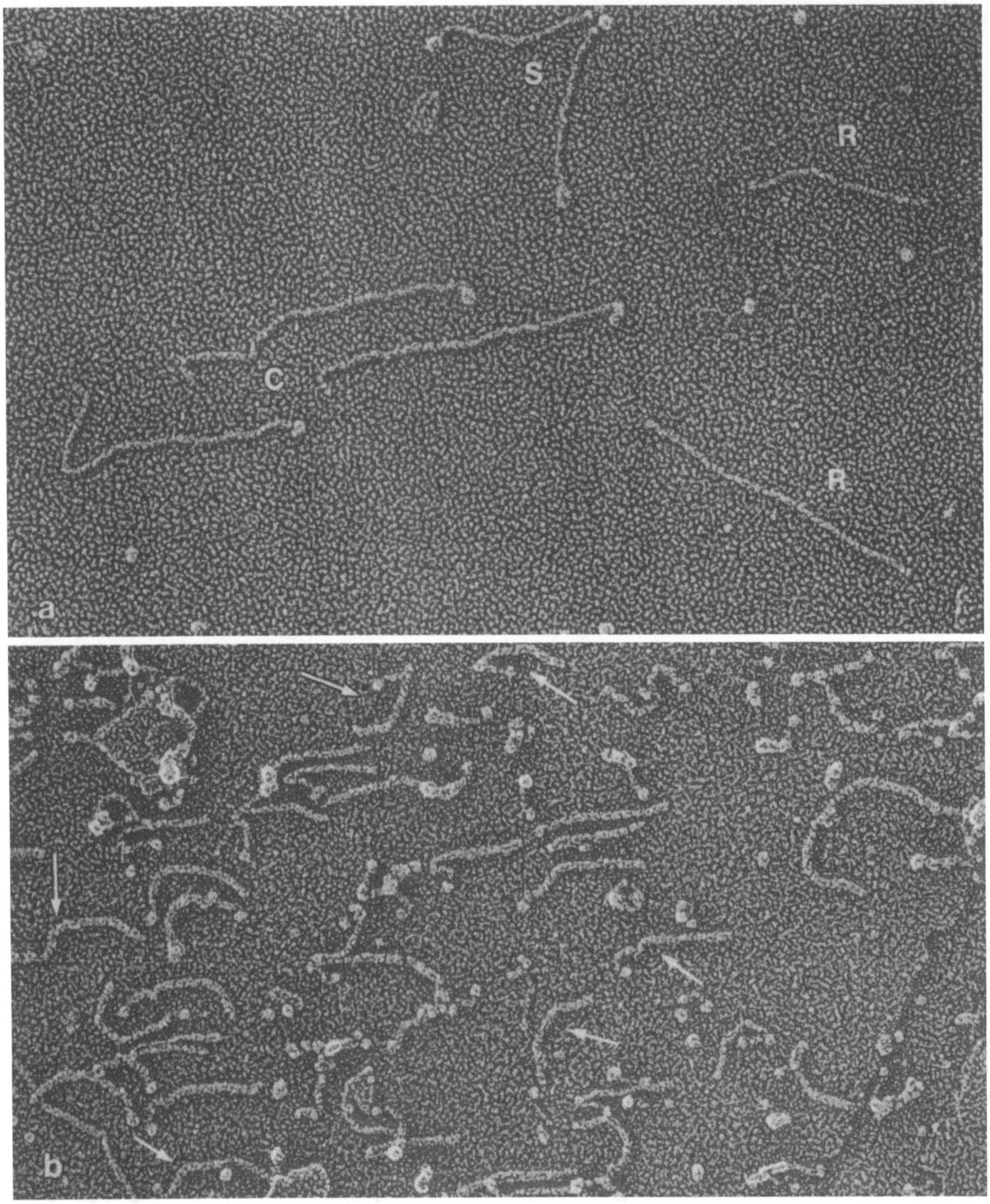

Fig. 6. a A field with three *C. reinhardtii* sexual agglutinins (*C*), two short canes (*S*) that are also found on gametic flagella, and a long (200 nm) and short (100 nm) rib (*R*) which come from the W 2 fishbones. **b** GP 1 monomers (100 nm) with arrows indicating proteins with kinks or bends reminiscent of those seen in agglutinins. × 205,000. Reprinted from Goodenough and Heuser (1985)

a volvocalean cell wall, and modifications in the building blocks led to the diversity exhibited by the assembled matrices (Roberts et al. 1985 a, Goodenough and Heuser 1988 b).

Both Class I and Class II crystals consist of globular and fibrous domains (Goodenough and Heuser 1988 a, b). By transmission electron microscopy, GP 2 (see Fig. 4) and GP 3 from *C. reinhardtii* each carries several globular domains (knobs) interconnected by rods (Goodenough et al. 1986 a). The distant Class II homologs of GP 2 and GP 3, GP 265, and GP 245, are less well differentiated into domains (Goodenough and Heuser 1988 b). The GP 1 monomer of *C. reinhardtii* has the most extreme differentiation (Fig. 6): it carries a terminal knob and a single long rod with two distinct bends (Roberts 1981, Goodenough et al. 1986 a). Models of the assembly suggest that the knobs, rods, and bends all participate in lattice formation (Goodenough and Heuser 1988 a, b).

Chlamydomonas flagellar agglutinins

Studies of the cell wall HRGPs have been complemented by work on the *Chlamydomonas* sexual agglutinins. Agglutinins are the molecules distributed along the flagellar surface of mating type plus and minus gametes that mediate sexual recognition and adhesion during the mating reaction (Adair 1985). The plus and minus agglutinins from both *C. reinhardtii* and *C. eugametos* are HRGPs with a morphology (long fibrous rods with knobs and curves) and amino acid content remarkably similar to that of cell wall HRGPs (see Fig. 6) (Musgrave et al. 1981, Cooper et al. 1983, Samson et al. 1987, van den Ende et al. 1988, Goodenough 1991). When the *C. reinhardtii* and *C. eugametos* agglutinins are compared, they prove to differ in length: the agglutinins in *C. reinhardtii* are both 225 nm, whereas in *C. eugametos* the mating type plus agglutinin is 276 nm and mating type minus molecule is 336 nm. Recall that the long and short ribs of the *C. eugametos* W2 fishbones are half the length of their counterparts in *C. reinhardtii* fishbones. These ribs, like the GP 1 molecule of *C. reinhardtii*, have a basic structural homology to the agglutinin molecules (see Fig. 6). Goodenough and Heuser (1985) proposed that agglutinins originally evolved from a pair of coassembling wall HRGPs that came under control of the mating type locus such that one monomer of the pair was localized only in minus gametic flagella and the other found only on plus flagella. This would imply that agglutination and cell wall assembly are similar processes. To carry this analogy one step further, one can speculate that just as modifications in cell wall HRGPs can lead to unique wall structures, changes in agglutinin HRGPs lead to novel adhesion specificities and hence serve to sexually isolate incipient species (Goodenough 1985, 1991). Thus, the HRGPs could prove to be useful phylogenetic markers.

Volvocine wall genes

The large number of taxonomic groups in the genus *Chlamydomonas* alone implies that there has been tremendous variation in both wall and agglutinin HRGPs. What makes these HRGPs so mutable? In higher plants, where many of the cell wall HRGPs have been sequenced (Showalter 1993), the genes encode proteins with repetitive amino acid motifs (e.g., $Ser(Pro)_4TyrLys$ and ProProValTyrLys). It has been proposed, and later demonstrated, that such repetitive DNA sequences are exceptionally prone to rearrangements (e.g., unequal crossover events and homologous recombination; Goodenough 1985, Raz et al. 1992). Changes in the DNA sequences encoding the repetitive amino acid domains can result in significant alterations in the length, rigidity, post-translational modifications and ultimately, the function of the HRGPs (Kieliszewski and Lamport 1994). The repeating prolines, which are variably hydroxylated and glycosylated depending on the presence of specific neighboring amino acids, contribute to the rigidity of the molecule by adopting an extended polyproline-II helix conformation (Lamport 1977, Kieliszewski et al. 1990). The serines, which can be galactosylated, and the tyrosines, which can form intra- or intermolecular isodityrosine crosslinks, appear only in certain amino acid sequence contexts (Epstein and Lamport 1984, Fry 1986, Kieliszewski and Lamport 1994). Moreover, many of the repeat units are part of peptide palindromes which have been proposed to serve as self-assembly nucleation sites (Kieliszewski et al. 1992). Molecular characterization of volvocine wall genes is just underway and yet there are already some common themes emerging from these studies (repeating amino acid motifs, hydroxylation and glycosylation of prolines, and crosslinking) which mesh neatly with results of work on higher plant cell wall HRGPs. These are reviewed below.

During its life cycle, *C. reinhardtii* erects two biochemically and morphologically distinct cell walls: the vegetative/gametic wall, which has been the focus of this review thus far, and the zygote wall, which is assembled after GLE removes the cell wall from mating gametes. The zygote wall has no obvious crystalline layer, and no perchlorate-soluble components. In deep-etch replicas, the zygote wall appears as an apparently random mass of interconnecting thick fibers (Woessner et al. unpubl.); in thin-sectioned material, several layers can be recognized (Cavalier-Smith 1976, Minami and Goodenough 1978, Grief et al. 1987). Minami and Goodenough (1978) showed that a novel set of glycoproteins is synthesized and secreted by developing zygotes. Ferris and Goodenough (1987) went on to isolate a set of zygote-specific cDNAs whose transcripts were detectable shortly after zygote formation, but absent from gametes or vegetative cells. DNA sequence analysis has demonstrated that at least two of these cDNAs encode wall proteins (Woessner and Good-

1 MRKVSAFGALVAAVAVMSMLTMPTAALAARTSLLATANATATD**APPP**DFSF

52 **EPPP**HEAV**EPPP**AKGAAAF**TPPP**QDLST**EPPP**AEFEQT**SPPP**SASR**SP**

100 **SPSSSASPSPSP**ALVPIPNVGNTA**SPNPSSSPSPSPSPSPSP**AAT**SS**

147 **PSP**IA**SPPP**LPATLTN**QPPP**SAVDQLGGNTQAGSGAAARAGLLLLSCVAA

197 WAFAVL

Fig. 7. Derived amino acid sequence of ZSP-1. The putative signal sequence is underlined and the X(Pro)$_3$ and (SerPro)$_\times$ repeats are in bold face

1 MKSSHRALWASAVLAMAVLACVVPTADAAKK**SPPP**AD**APPP**DSGQ**YPPP**T

51 DWVT**WPPP**VEEHGHH**SPSPSPEDSPSSTPIPTPDILENRDLPSPSPE**

98 **AEDSPSPSPDLEDSPSHSPEPEVESPSPETETESPSPSPEPEVEDS**

144 **ASPSPEPEVEDSPSPSPEPMDESPAPEPSPEPVPSPETPSPAP**YYG

190 A**SPSPSP**YYGGYGNYGPSVPTYG**SSPSPAPSP**YGGYGGYGGYGGYGGYGS

240 YGDDDEEEEDPMSAARRLMLDVPNQGAGRQLSGDAVDLLAGGEEPESEERST

292 RVSRRLLRI

Fig. 8. Derived amino acid sequence of VSP-1. The putative signal sequence is underlined and the X(Pro)$_3$ and (SerPro)$_\times$ repeats are in bold face

enough 1989, 1992). ZSP-1 (originally termed Class IV), which resembles the class of higher plant structural HRGPs called extensins, has a signal sequence, eight X(Pro)$_3$ and two (SerPro)$_\times$ repeats (Fig. 7). The other gene (ZSP-2, originally Class VI) has some similarities to solanaceous lectins (see Showalter 1993) with four lectin-like domains intermixed with two (SerPro)$_\times$ domains (Woessner and Goodenough 1989, and unpubl.). A polyclonal antibody to a ZSP-1/β-galactosidase fusion protein immunoprecipitated 5–10 polypeptides from an in vitro translation of zygote RNA, but detected nothing in immunoprecipitates of secreted zygote proteins from in vivo labelled cells (Woessner and Goodenough 1989, 1992). This result was interpreted to mean that there are several zygote wall components with epitopes similar to those in ZSP-1 but that hydroxylation/glycosylation prevents detection of these proteins in vivo.

To test whether (SerPro)$_\times$ repeats were a common epitope among zygote wall proteins, a polyclonal antibody raised agaist a (SerPro)$_{10}$ oligopeptide was found to immunoprecipitate the identical set of polypeptides from in vitro translations of zygote RNA. In addition, both antibodies were found to recognize a set of five vegetative wall glycopolypeptides from in vitro translations of vegetative RNA, one of which corresponds to the deglycosylated form of GP 2 (Adair and Apt 1990, Woessner and Goodenough 1992). Using clones described by Adair and Apt (1990), we have also isolated several GP 1 cDNAs (none of which is full-length) and determined a partial DNA sequence which encodes a 3′ proline-poor domain preceded by a long domain of (SerPro)$_\times$ (Woessner and Goodenough unpubl.). Therefore, (SerPro)$_\times$ repeats appear to be diagnostic for *C. reinhardtii* wall proteins in much the same way that Ser(Pro)$_4$ repeats indicate extensins in higher plants.

A cDNA expression library made from GLE-treated *C. reinhardtii* vegetative cells, which have high levels of wall-specific messages (Su et al. 1990, Adair and Apt 1990), was screened with the (SerPro)$_{10}$ antibody and five different groups of cDNAs were isolated (VSP-1 to VSP-5; Waffenschmidt et al. 1993). A full-length cDNA for VSP-1 was sequenced and found to encode a signal sequence, four X(Pro)$_3$ repeats, long stretches of (SerPro)$_\times$, a TyrGlyGly domain and an acidic C-terminus lacking both serine and proline (Fig. 8). The amino acid content and abundance of tyrosine distinguish this molecule from any of the outer wall components (Goodenough et al. 1986 a) and thus, VSP-1 is proposed to be an inner wall HRGP. The tyrosines suggested that tyrosine crosslinking may be involved in insolubilization of the inner wall layer, and Waffenschmidt et al. (1993) demonstrated that tyrosine crosslinks in fact mediate wall insolubilization in both vegetative and zygotic cell walls.

Woessner and Goodenough (1992) predicted that (SerPro)$_\times$ motifs might be diagnostic of wall proteins throughout the Volvocales, and a collaborative project with Dr. van den Ende's laboratory supports this proposition. The (SerPro)$_{10}$ antibody was found to immunoprecipitate novel sets of polypeptides from in vitro translations of *C. eugametos* vegetative and zygote RNA and the antibody was used to isolate a vegetative wall gene (WP 6) from a λgt 11 expression library of *C. eugametos* genomic DNA (Woessner et al. in prep.). The DNA sequence of WP 6 has been determined and shown to encode a signal sequence, a 5′ domain rich in cysteine but poor in serine and proline, and a 3′ domain rich in serine and proline with a number of (SerPro)$_\times$ repeats. Interestingly, the (SerPro)$_\times$ repeats are frequently found in a (SerPro)$_3$LysAla motif, a motif repeated as well in *C. reinhardtii* VSP-3, which also has a signal sequence, a 5′ domain rich in cysteine but poor in proline, and a 3′ domain dominated by (SerPro)$_\times$ repeats (Woessner et al. in prep.). Once again, as was pointed out when comparing the fishbone units from these two species, there are homologies in

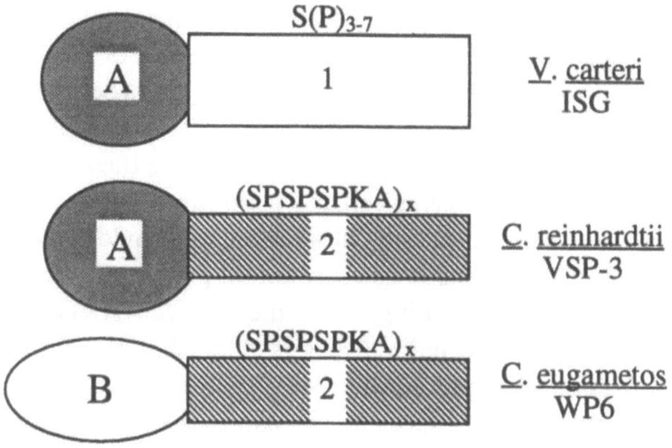

Fig. 9. Domain conservation in volvocine wall proteins. Each protein is represented in cartoon fashion with globular 5′ domain and a rod-like 3′ domain. Homologous 5′ and 3′ domains are shaded and the characteristic amino acid repeat units are presented for each protein

the cell wall structural building blocks, despite the evolutionary distance between *C. eugametos* and *C. reinhardtii*. It is important to point out that the other domains in WP 6 and VSP-3 are quite distinctive, only the (SerPro)$_x$ and (SerPro)$_3$LysAla motifs have been conserved.

Only two other cell wall genes in the Volvocales have been sequenced, both from *Volvox carteri* (Ertl et al. 1989, 1992). SSG 185 is a highly sulfated HRGP which becomes insolubilized in the inner wall layer (cellular zone). There are no extensive (SerPro)$_x$ repeats in this molecule, but an 80 amino acid central domain is almost entirely proline. These prolines are hydroxylated and arabinosylated and give the molecule a rod-like shape as revealed by electron microscopy (Ertl et al. 1989). Covalent cross-linking of this molecule is due to phosphodiester bridges between side chains rather than between peptide chains (Holst et al. 1989). ISG is another sulfated HRGP, expressed only briefly in inverting embryos and sperm cell packets. This glycoprotein localizes to the boundary zone (containing the central triplet) and has two distinct amino acid domains. The 5′ half of the molecule is poor in proline, while the 3′ half is rich in serine and proline, occurring in Ser(Pro)$_{3-7}$ repeats (Ertl et al. 1992). Again, the prolines are both extensively hydroxylated and arabinosylated. Electron microscopy shows ISG to be a rod-like structure (encoded by the 3′ domain) with a globular head (from the 5′ domain). The monomers can aggregate into star-like particles attaching by their knobs with rods radiating outwards (Ertl et al. 1992).

Only a small number of volvocine wall genes have been examined to date, but there is already a striking example of HRGP domain conservation involving *C. reinhardtii, C. eugametos* and *V. carteri* (Fig. 9). The 3′ domains of VSP-3 and WP 6 (which encode the rod-like portion of each protein) are quite similar (88%

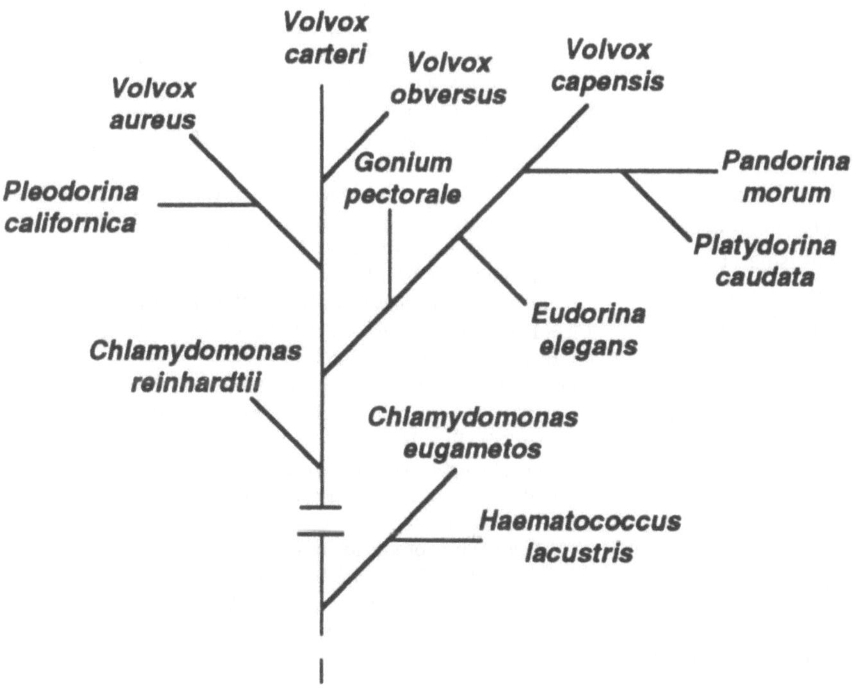

Fig. 10. Phylogenetic tree indicating sister-taxon relationships, as deduced by maximum parsimony analysis of nucleotide sequence from the variable regions of the large and small subunits of nuclear-encoded rRNAs. Reprinted from Schmitt et al. (1992)

similarity over 73 amino acids) while their 5′ domains (encoding the globular portion of each protein) are non-homologous, but the 5′ domains of VSP-3 and ISG (encoding the globular portions of each molecule) are related (52% similarity over 228 amino acids). With a larger pool of characterized volvocine HRGPs it should be possible to define ancestral rod and knob domains and determine the best way to use these proteins as phylogenetic markers.

Conclusions

Recently several groups have constructed consensus trees based on cladistic analyses of aligned nuclear 18 S and 26 S rRNA sequences. Rausch et al. (1989) showed that the volvocine radiation (*Chlamydomonas* to *Volvox*) is recent, probably within the past 50–75 million years. They also demonstrated that the higher plants and the chlorophytes (green algae) last shared a common ancestor about 700 million years ago. In this respect, it is not hard to see why the relatively simple *Chlamydomonas* cell wall, constructed almost exclusively of proteins, is so different from the complex network of numerous proteins and matrix polysaccharides (cellulose, hemicelluloses, lignins, pectins, and waxes) that constitutes the higher plant cell wall (Varner and Lin 1989). Buchheim et al. (1990) compared the results of their cladistic analysis of 18 S rRNA sequences from *Chlamydomonas* spp. with the classification schemes of Ettl (1976), Schlösser (1984), and Roberts et al. (1982). With a few exceptions, all four methods are in close agreement, and all lead to the conclusion that there is no support for a monophyletic or natural genus *Chlamydomonas*. Cladistic analyses also confirm that *C. reinhardtii* is more closely allied with the colonial and multicellular Volvocaceae than with *C. eugametos* (Buchheim and Chapman 1991, Larson et al. 1992) and that there is not a linear evolutionary progression from simple *Chlamydomonas* cells to complex multicellular *Volvox* organisms (Fig. 10). In fact, species of *Volvox* appear on two separate phylogenetic branches of the consensus tree for Volvocales.

In spite of differences seen in crystal structure, lytic enzyme sensitivity, and reassembly of the *C. eugametos* and *C. reinhardtii* cell walls, there are some similarities in the proteins which compose these walls. The accumulation of molecular data about the HRGPs (cell wall proteins and agglutinins) found in the volvocine algae will lend support to the idea of a gene superfamily from which new and different proteins (all specifically designed for self-assembly/recognition) could constantly evolve. The similarities (repeating proline motifs, hydroxylation and arabinosylation of prolines, galactosylation of serines, crosslinking of tyrosines or oligosaccharide side chains, and rod-like molecules) between the volvocine HRGPs and the various types of higher plant cell wall proteins, and the central role these proteins play in organizing cell and plant morphology has led to the proposal that these proteins should make excellent phylogenetic markers for all plants. "The extensin family (including all of the cell wall HRGPs) should mirror the major divisions of the plant kingdom and hence provide another way to assess long-term evolution which, especially in plants, is obscured by extensive homoplasy: gross structures may converge, but molecules probably do not" (Kieliszewski and Lamport 1994).

Acknowledgements

We thank Dr. Marcia Kieliszewski for sending us a preprint of her paper. Part of the work reviewed here was supported by the U.S. Department of Agriculture (Grant no. 92-373304-7936).

References

Adair WS (1985) Characterization of *Chlamydomonas* sexual agglutinins. J Cell Sci [Suppl] 2: 233–260
– Appel H (1989) Identification of a highly conserved hydroxyproline-rich glycoprotein in the cell walls of *Chlamydomonas reinhardtii* and two other Volvocales. Planta 179: 381–386
– Apt KE (1990) Cell wall regeneration in *Chlamydomonas*: accumulation of mRNAs encoding cell wall HRGPs. Proc Natl Acad Sci USA 87: 7355–7359
– Snell WJ (1990) The *Chlamydomonas reinhardtii* cell wall: structure, biochemistry, and molecular biology. In: Adair WS, Mecham RP (eds) Organization and assembly of plant and animal extracellular matrix. Academic Press, San Diego, pp 15–84
– Steinmetz SA, Mattson DM, Goodenough UW, Heuser JE (1987) Nucleated assembly of *Chlamydomonas* and *Volvox* cell walls. J Cell Biol 105: 2373–2382
Buchheim MA, Chapman RL (1991) Phylogeny of the colonial green flagellates: a study of 18 S and 26 S rRNA sequence data. BioSystems 25: 85–100
– Turmel M, Zimmer EA, Chapman RL (1990) Phylogeny of *Chlamydomonas* (Chlorophyta) based on cladistic analysis of nuclear 18 S rRNA sequence data. J Phycol 26: 689–699
Cavalier-Smith T (1976) Electron microscopy of zygospore formation in *Chlamydomonas reinhardtii*. Protoplasma 87: 297–315
Claes H (1971) Autolyse der Zellwand bei den Gameten von *Chlamydomonas reinhardtii*. Arch Mikrobiol 78: 180–188
Cooper JB, Adair WS, Mecham RP, Heuser JE, Goodenough UW (1983) *Chlamydomonas* agglutinin is a hydroxyproline-rich glycoprotein. Proc Natl Acad Sci USA 80: 5898–5901
Epstein L, Lamport DTA (1984) An intramolecular linkage involving isodityrosine in extensin. Phytochemistry 23: 1241–1246
Ertl H, Mengele R, Wenzl S, Engel J, Sumper M (1989) The extracellular matrix of *Volvox carteri*: molecular structure of cellular compartment. J Cell Biol 109: 3493–3501
– Hallmann A, Wenzl S, Sumper M (1992) A novel extensin that may organize extracellular matrix biogenesis in *Volvox carteri*. EMBO J 11: 2055–2062

Ettl H (1976) Die Gattung *Chlamydomonas* Ehrenberg. Nova Hedwigia 49: 1–1122

Ferris PJ, Goodenough UW (1987) Transcription of novel genes, including a gene linked to the mating-type locus, induced by *Chlamydomonas* fertilization. Mol Cell Biol 7: 2360–2366

Fry SC (1986) Cross-linking of matrix polymers in the growing cell walls of angiosperms. Annu Rev Plant Physiol 37: 165–186

Goodenough UW (1985) An essay on the origins and evolution of eukaryotic sex. In: Halvorson HO, Monroy A (eds) The origin and evolution of sex. AR Liss, New York, pp 123–140

– (1991) *Chlamydomonas* mating interactions. In: Dworkin M (ed) Microbial cell-cell interactions. American Society for Microbiology, Washington, DC, pp 71–112

– Heuser JE (1985) The *Chlamydomonas* cell wall and its constituent glycoproteins analyzed by the quick-freeze deep-etch technique. J Cell Biol 101: 1550–1568

– (1988 a) Molecular organization of cell-wall crystals from *Chlamydomonas reinhardtii* and *Volvox carteri*. J Cell Sci 90: 717–733

– (1988 b) Molecular organization of the cell wall and cell-wall crystals from *Chlamydomonas eugametos*. J Cell Sci 90: 735–750

– Gebhart B, Mecham RP, Heuser JE (1986 a) Crystals of the *Chlamydomonas reinhardtii* cell wall: polymerization, depolymerization, and purification of glycoprotein monomers. J Cell Biol 103: 403–417

– Adair WS, Collin-Osdoby P, Heuser JE (1986 b) *Chlamydomonas* cells in contact. In: Gall E, Edelman GM (eds) Cells in contact. Wiley, New York, pp 111–135

Grief C, O'Neill MA, Shaw PJ (1987) The zygote cell wall of *Chlamydomonas reinhardtii*: a structural, chemical, and immunological approach. Planta 170: 433–445

Harris EH (1989) The *Chlamydomonas* sourcebook. Academic Press, San Diego

Hills GJ (1973) Cell wall assembly in vitro from *Chlamydomonas reinhardtii*. Planta 115: 17–23

– Phillips JM, Gay MR, Roberts K (1975) Self-assembly of a plant cell wall in vitro. J Mol Biol 96: 431–444

Holst O, Christoffel V, Frund R, Moll H, Sumper M (1989) A phosphodiester bridge between two arabinose residues as a structural element of an extracellular glycoprotein of *Volvox carteri*. Eur J Biochem 181: 345–350

Imam SH, Snell WJ (1988) The *Chlamydomonas* cell wall degrading enzyme, lysin, acts on two domains within the framework of the wall. J Cell Biol 106: 2211–2221

Kieliszewski M, Lamport DTA (1994) Extensin: repetitive motifs, functional motifs, post-translational codes and phylogeny. Plant J 5: 157–172

– Leykam JF, Lamport DTA (1990) Structure of the threonine-rich extensin from *Zea mays*. Plant Physiol 92: 316–326

– de Zacks R, Leykam JF, Lamport DTA (1992) A repetitive proline-rich protein from the gymnosperm Douglas Fir is a hydroxyproline-rich glycoprotein. Plant Physiol 98: 919–926

Kirk DL, Birchem R, King N (1986) The extracellular matrix of *Volvox*: a comparative study and proposed system of nomenclature. J Cell Sci 80: 207–231

Lamport DTA (1977) Structure, biosynthesis, and significance of cell wall glycoproteins. Rec Adv Phytochem 11: 79–115

Larson A, Kirk MM, Kirk DL (1992) Molecular phylogeny of the volvocine flagellates. Mol Biol Evol 9: 85–105

Matsuda Y, Musgrave A, van den Ende H, Roberts K (1987) Cell walls of algae in the Volvocales: their sensitivity to a cell wall lytic enzyme and labeling with an anti-cell wall glycopeptide of *Chlamydomonas reinhardtii*. Bot Mag 100: 373–384

Minami S, Goodenough UW (1978) Novel glycopolypeptide synthesis induced by gametic cell fusion in *Chlamydomonas reinhardtii* J Cell Biol 77: 165–181

Monk BC (1988) The cell wall of *Chlamydomonas reinhardtii* gametes: composition, structure and autolysin-mediated shedding and dissolution. Planta 176: 441–450

Musgrave A, van Eijk E, te Welscher R, Broekman R, Lens PF, Homan WL, van den Ende H (1991) Sexual agglutination factor from *Chlamydomonas eugametos*. Planta 153: 362–369

Rausch H, Larsen N, Schmitt R (1989) Phylogenetic relationships of the green alga *Volvox carteri* deduced from small-subunit ribosomal RNA comparisons. J Mol Evol 29: 255–265

Raz R, José M, Moya A, Martínez-Izquierdo JA, Puigdomènech P (1992) Different mechanisms generating sequence variability are revealed in distinct regions of the hydroxyproline-rich glycoprotein gene from maize and related species. Mol Gen Genet 233: 252–259

Roberts K (1974) Crystalline glycoproteins of algae: their structure, composition, and assembly. Philos Trans R Soc Lond [Biol] 268: 129–146

– (1981) Visualizing an insoluble glycoprotein. Micron 12: 185–186

– Hills GJ, Shaw P (1982) Structure of algal cell walls. In: Harris JR (eds) Electron microscopy of proteins, vol 3. Academic Press, New York, pp 1–40

– Grief C, Hills J, Shaw PJ (1985 a) Cell wall glycoproteins: structure and function. J Cell Sci [Suppl] 2: 105–127

– Phillips J, Shaw P, Grief C, Smith E (1985 b) An immunological approach to the plant cell wall. In: Brett CT, Hillman JR (eds) Biochemistry of plant cell walls. Cambridge University Press, New York, pp 125–154

Samson MR, Klis FM, Homan WL, van Egmond P, Musgrave A, van den Ende H (1987) Composition and properties of the sexual agglutinins of the flagellated green algae *Chlamydomonas eugametos*. Planta 170: 314–321

Schlösser UG (1966) Enzymatisch gesteuerte Freisetzung von Zoosporea bei *Chlamydomonas reinhardtii dangeard* in Synchronkultur. Arch Mikrobiol 54: 129–159

– (1976) Entwicklungsstadien- und sippenspezifische Zellwand-Lysine bei der Freisetzung von Fortpflanzungszellen in der Gattung *Chlamydomonas*. Ber Deutsch Bot Ges 89: 1–56

– (1984) Species-specific sporangium autolysins (cell-wall dissolving enzymes) in the genus *Chlamydomonas*. In: Irvine DEG, John D (eds) The systematics of green algae. Academic Press, London, pp 409–418

Schmitt R, Fabry S, Kirk DL (1992) In search of molecular origins of cellular differentiation in *Volvox* and its relatives. Int Rev Cytol 139: 189–265

Showalter AM (1993) Structure and function of plant cell wall proteins. Plant Cell 5: 9–23

Su X, Kaska DD, Gibor A (1990) Induction of cytosine-rich poly $(A)^+$ RNAs in *Chlamydomonas reinhardtii* by cell wall removal. Exp Cell Res 187: 54–58

Van den Ende H, Klis FM, Musgrave A (1988) The role of flagella in sexual reproduction of *Chlamydomonas eugametos*. Acta Bot Neerl 37: 327–350

Varner JE, Lin L-S (1989) Plant cell wall architecture. Cell 56: 231–239

Waffenschmidt S, Spessert R, Jaenicke L (1988) Oligosaccharide side chains are essential for cell wall lysis in *Chlamydomonas reinhardtii.* Planta 175: 513–519

Waffenschmidt S, Woessner JP, Beer K, Goodenough UW (1993) Isodityrosine cross-linking mediates cell wall insolubilization in *Chlamydomonas.* Plant Cell 5: 809–820

Woessner JP, Goodenough UW (1989) Molecular characterization of a zygote wall protein: an extensin-like molecule in *Chlamydomonas reinhardtii.* Plant Cell 1: 901–911

– – (1992) Zygote and vegetative cell wall proteins in *Chlamydomonas reinhardtii* share a common epitope, $(SerPro)_x$. Plant Sci 83: 65–76

Protoplasma (1994) 181: 259–268

PROTOPLASMA

Structure and development of the scale case of *Mallomonas adamas* (Synurophyceae)

S. Lavau and **R. Wetherbee***

School of Botany, University of Melbourne, Parkville, Victoria

Received November 22, 1993
Accepted May 2, 1994

Summary. The structure and development of the elaborate scale case of *Mallomonas adamas* (Synurophyceae) was studied throughout the cell cycle. Immediately following division, scale cases normally possess 30 siliceous scales, or 5 rows of 6 scales each. Scale rows overlap one another and spiral around the cell in a clockwise direction when viewed from either end of the cell. Throughout interphase, a duplicate set of 30 scales is manufactured and secreted into the existing scale case in a precise sequence, resulting in a scale case with 60 scales, or 5 rows of 12 scales each. Scales are not added to the scale case in rows, but individually positioned within the cytoplasm and inserted one at a time over the entire surface of the cell. Scales appear to occupy specific positions in the scale case. Scale cases are dynamic cell coverings, their elaborate patterning being maintained throughout cell growth and during division when half the parental scales are inherited by each daughter cell.

Keywords: Cell surface; *Mallomonas*; Synurophyceae; Scales; Scale case.

Introduction

The elaborate cell covering of the two common freshwater genera *Mallomonas* and *Synura* is composed of siliceous scales that are imbricated and arranged in a highly patterned armour over the entire surface (e.g., Fig. 1). In addition, elongate bristles of varying morphology and distribution are attached (hinged) to the distal portion of scales in some species of *Mallomonas* (Takahashi 1978, Wee 1982, Leadbeater 1986, Asmund and Kristiansen 1986). Termed a "scale case", this complex structure, along with a number of other features, resulted in the separation of these algal protists into their own class – the Synurophyceae (Andersen 1987). The overall integrity of the scale case is maintained by distinct adhesive layers located at specific sites on both scales and bristles, anchoring them to the plasma membrane and/or to one another (e.g., Leadbeater 1986, Beech et al. 1990).

The formation of scales/bristles, and the role of cytoskeletal components in their assembly and deployment within the cytoplasm, has been well documented (e.g., Schnepf and Deichgräber 1969, Wujek and Kristiansen 1978, Mignot and Brugerolle 1982, Brugerolle and Bricheux 1984, Beech et al. 1990; and recent reviews of Wetherbee et al. 1989, 1992) and will not be dealt with in detail here. Scales and bristles develop within a silica deposition vesicle (SDV) located along the outer surface of only one of the two chloroplasts. In the species discussed in this report, *Mallomonas adamas* Harris and Bradley, only a single chloroplast exists, though the basic mechanism of scale formation and deployment within the cytoplasm appears the same.

Two conflicting hypotheses attempt to explain scale case morphogenesis in *Synura* and *Mallomonas* (Leadbeater 1990, Siver and Glew 1990). The proposals differ both in the position of scale insertion onto the cell surface and the mechanism of their deployment. To date, little evidence has been presented in support of either hypothesis. Leadbeater (1990) states that his evidence is "circumstantial" and "by no means conclusive", while Siver and Glew (1990) remark that the validity of their proposed mechanism needs to be tested, and that it might be "incorrect or only partially valid". Nevertheless, both papers provide valuable data on scale case construction and development, and have provided the impetus for our own investigations described in this report.

* Correspondence and reprints: School of Botany, University of Melbourne, Parkville, Vic. 3052, Australia.

The mechanism of scale case morphogenesis proposed here for *M. adamas* partially concurs with the Leadbeater (1990) model based on *Synura*, in that scales are exactly positioned and oriented within the cytoplasm prior to being secreted into the scale case. However, the scales of *M. adamas* are inserted into the existing scale case one at a time, and not in rows as suggested in the Leadbeater model. Our ultrastructural studies, as well as in vivo observations using time-lapse video microscopy, have shown that scale cases are not static armour, but dynamic structures that adjust to the addition of new scales during cell growth and maintain their elaborate patterning during division through complex interactions with the cell. Although the mechanism for the construction and maintenance of scale cases in both *Synura* and *Mallomonas* is undoubtedly similar, variations between species do occur, and are discussed below.

Materials and methods

The culture of *M. adamas* was isolated from the reservoir at Mount Macedon, Victoria. Cells were grown in diatom enrichment media (Cohn et al. 1989) at 16 °C in a 14 : 10 h light : dark cycle.

For scanning electron microscopy, cells were fixed in 0.5% OsO_4 in their growth medium and collected on nucleopore filters (pore size 2 µm; Millipore), rinsed with distilled water and air-dried. Dry filters were mounted onto aluminium stubs and sputter-coated with gold (Dynavac sputter coater). Specimens were viewed and photographed on a Jeol JSM-840 scanning electron microscope (Kodak TMax 100 film).

The fixation procedure for transmission electron microscopy was developed in the laboratory by Dr. Kirsten Heimann. Cells were fixed throughout the cell cycle with 1.5% (v/v) glutaraldehyde, 0.4% (w/v) aqueous OsO_4 and 0.01% (w/v) Na-cacodylate buffer (pH 7.0) for 15 min (4 °C). Fixed cells were washed three times in Na-cacodylate buffer (pH 7.0) and once in distilled water, postfixed in 1% (w/v) aqueous OsO_4 (1 h, room temp.), washed ten times in distilled water, and incubated in 0.5% (w/v) tannic acid. After 15 min at room temperature, cells were washed five times in distilled water, incubated in 1% (w/v) NH_4Cl and washed once in distilled water. Dehydration was in a series of acetone concentrations (5%, 10%, 15%, 25%, 35%, 50%, 70%, 90%, and 100%), followed by infiltration with increasing proportions of Spurr's resin in acetone (25%, 50%, 75%, and 100%) and polymerized overnight at 60 °C. Embedded material was sectioned on an Ultracut-E ultramicrotome (Reichert) and collected on Formvar-coated copper or nickel grids. Sections were stained in 2% (w/v) uranyl acetate (aqueous; 15 min, room temp.), washed 3 times in distilled water and allowed to dry. Stained sections were viewed and photographed on a Jeol 1200 transmission electron microscope (Kodak electron microscope film 4489).

Results

Scale and scale case morphology

Cells of *M. adamas* are oval in shape measuring 10–18 µm in length and 8–14 µm in width, and possess a single emergent flagellum plus a non-emergent flagellum/basal body. Cells of *M. adamas* are distinguished by their species-specific scale morphology and the overall structure of the scale case. Scales are arranged in five overlapping spiral rows that form a highly ordered pattern. As with other described species of *Synura* and *Mallomonas*, the scale rows always spiral to the right, or clockwise, when viewed from either end of the cell (Figs. 1–3). The scales in each row overlap one another in a posterior to anterior manner, while the spiral rows overlap one another from the anterior to the posterior of the cell. This arrangement is typical of most *Mallomonas* spp., although exceptions have been reported (Siver and Glew 1990, Siver 1991).

At the anterior of the cell (where the flagellum emerges), each scale row ends in a morphologically distinct "apical scale" that has a domed appearance in *M. adamas*. The five apical dome scales comprise a star-shaped apex to the scale case with a small opening for the flagellum (Fig. 1). The posterior end of each scale row is characterized by at least one distinct "caudal scale", which in *M. adamas* has a protruding flange (Figs. 2–3). Unlike the apical dome scales, the number of caudal scales often varies from the norm of five, though a star-shaped posterior to the scale case is evident (Fig. 3). Body scales make up the remainder of the scale case, and vary slightly in size according to their position within the case, being largest at the middle of the cell. The scale

Fig. 1–3. Scanning electron micrographs (SEMs) of *M. adamas* showing the structure of the scale case, including the orientation and imbrication of the 5 scale rows that spiral around the cell in a clockwise direction (arrows) when viewed from the anterior (Fig. 1) or posterior (Figs. 2 and 3) end of the cell. Note the distinct structure of the apical dome scales (stars) located at the anterior end of each scale row (Fig. 1), and the caudal scale located at the posterior end of each scale row (Figs. 2 and 3). The number of caudal scales often varies throughout the cell cycle (Figs. 2 and 3). Bars: 2 µm

Fig. 4. SEMs of whole scales of *M. adamas* revealing the boomerang, where scales overlap in intact scale cases. Bar: 2 µm

Fig. 5. Thin section showing the "adhesive" layers located between scales (i.e., on the boomerangs; large arrows), and between scales and the plasma membrane (small arrows); Bar: 0.5 µm

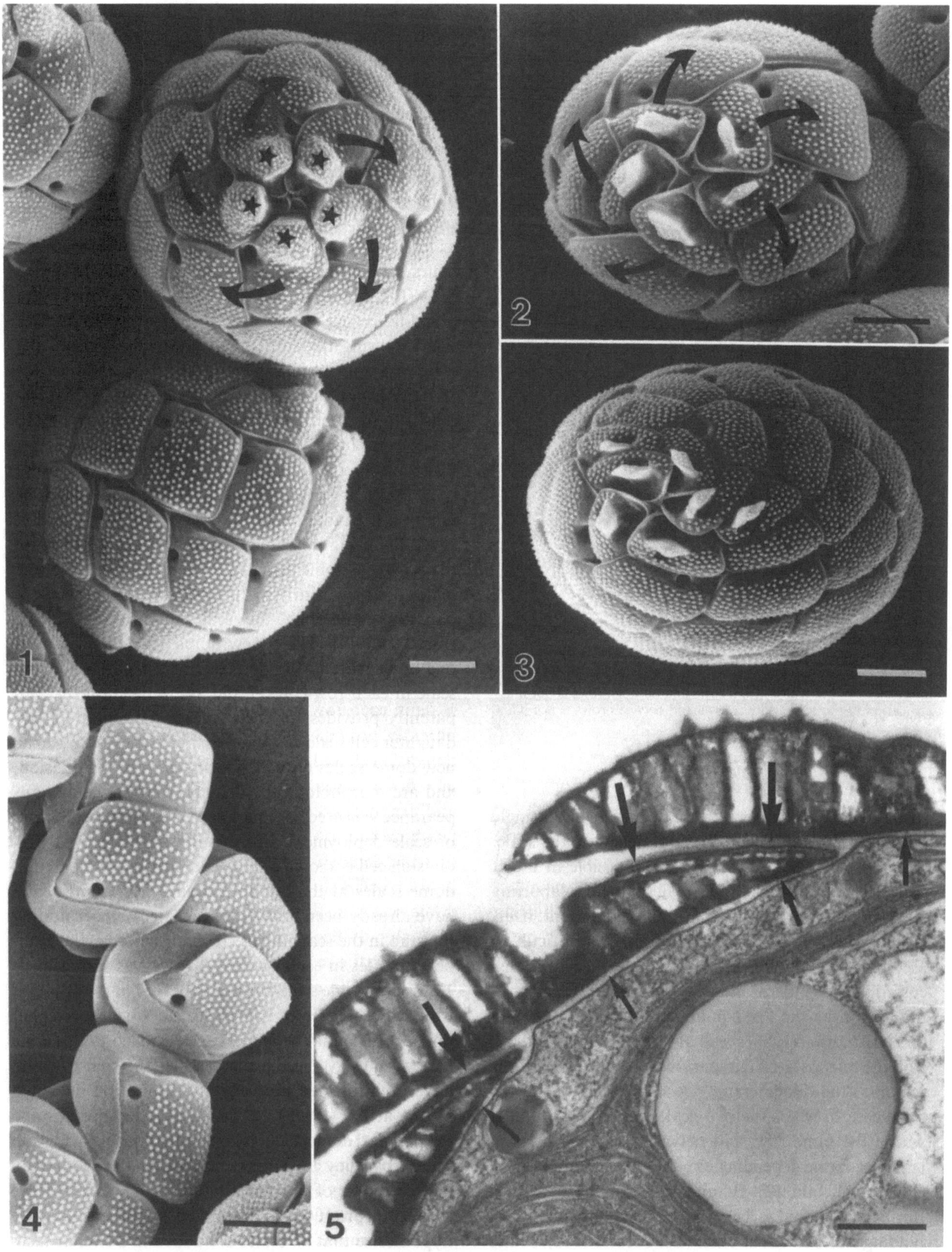
1
2
3
4
5

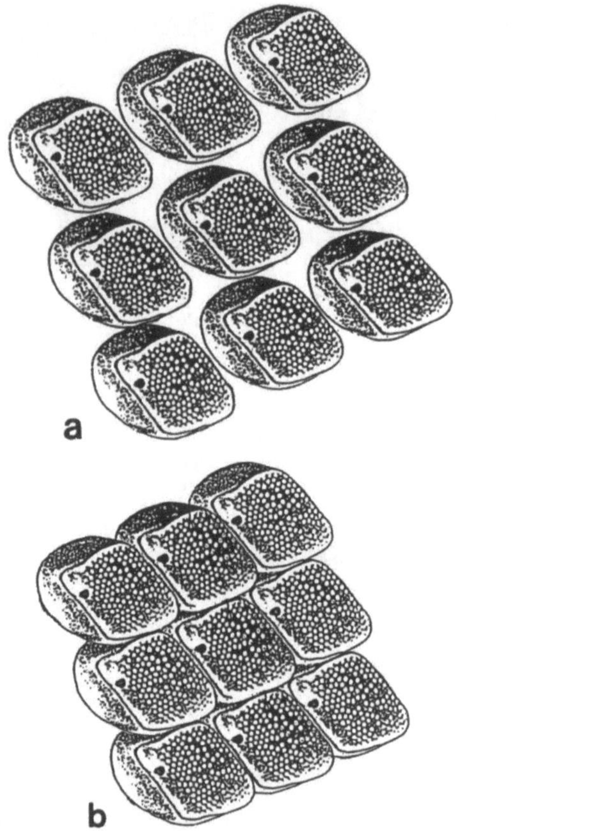

Fig. 6. Diagrams showing how the boomerangs of 9 body scales (**a**) of *M. adamas* overlap one another within a scale case (**b**). The exact imbrication of the scales within rows and between rows is not illustrated

surface bears a number of raised papillae and a single pore-like depression that does not penetrate the scale. Only 50–60% of the scale surface is visible in intact scale cases (Figs. 4 and 6), as scales adhere to abutting scales in the same row and the scales in adjacent, overlapping rows by a V-shaped region. It is difficult to ascertain the homology of this region to structures discussed in the classical literature. Siver (1991) refers to this region as the "proximal rim", whereas Asmund and Kristiansen (1986) speak of the "proximal border". For the purposes of this paper, we shall refer to this region as the "boomerang". A single scale within the scale case overlaps the two scales on either side of it within the same row, and between 2 and 4 scales in adjacent rows depending on their exact orientation (Figs. 6, 11, and 12). Boomerangs are coated with a fibrillar layer that is believed to be adhesive (Fig. 5), binding scales to one another (Leadbeater 1986). Scales are also anchored to the plasma membrane over part of their smooth, proximal surface, though this layer is not as distinctive as the "sticky" boomerangs (Fig. 5).

Development of the scale case

Cells of *M. adamas* divide within a 30 min window 3–4 h before the end of the dark phase of their cell cycle. Immediately following division, recently divided cells are normally surrounded by 30 scales, or 5 rows of 6 scales, with each row ending in 1 apical dome scale (anterior end) and at least 1 caudal scale (posterior end) (Fig. 8). Occasionally cells with 25 scales were found following the division period (5 rows of 5 scales each). During the subsequent interphase period, the scale case doubles in size by the sequential addition of 30 more scales for a total of 60 scales (Fig. 7). In a few cells a total of 50 scales were found, or 5 rows of 10 scales each, indicating that variations exist.

For most of the interphase period, both caudal and body scales are gradually added to the scale case. The number of caudal scales is observed to be somewhat variable during this time, while there are never more than 5 apical dome scales. The number of apical dome scales doubles from 5 to 10 just prior to division, apparently completing the scale duplication process (Fig. 7 and 14). The 5 new dome scales are secreted one body scale to the posterior of the parental dome scale in each row (Figs. 7 and 14), a position that apparently provides for their precise reorientation into daughter cell scale cases at division (see below). As the new dome scales always appear just prior to division, and are morphologically distinct, their sequential appearance was used as a marker to assess the mechanism of scale deployment into existing scale cases. A predivision cell is shown in Fig. 9, and there are 5 parental dome scales at the cell apex, and 2 new dome scales have already been added. By repeatedly reorientating the stub in the scanning electron microscope, the number of scales in each of these 2 rows was found to be 12. A new dome scale has not appeared in the adjacent row, which presently has 11 scales. We believed that the next scale that would have been added to this scale case would have been a dome scale, and that it would have been inserted precisely at the position indicated by the arrow (Fig. 9), or one body scale posterior to the parental dome scale. Cells fixed just prior to division reveal that they are actively involved in the production and secretion of dome scales only (Fig. 13). If dome scales are produced and secreted in this fashion, we hypothesize that all scales are added in a similar fashion, that is one scale at a time, proceeding sequentially

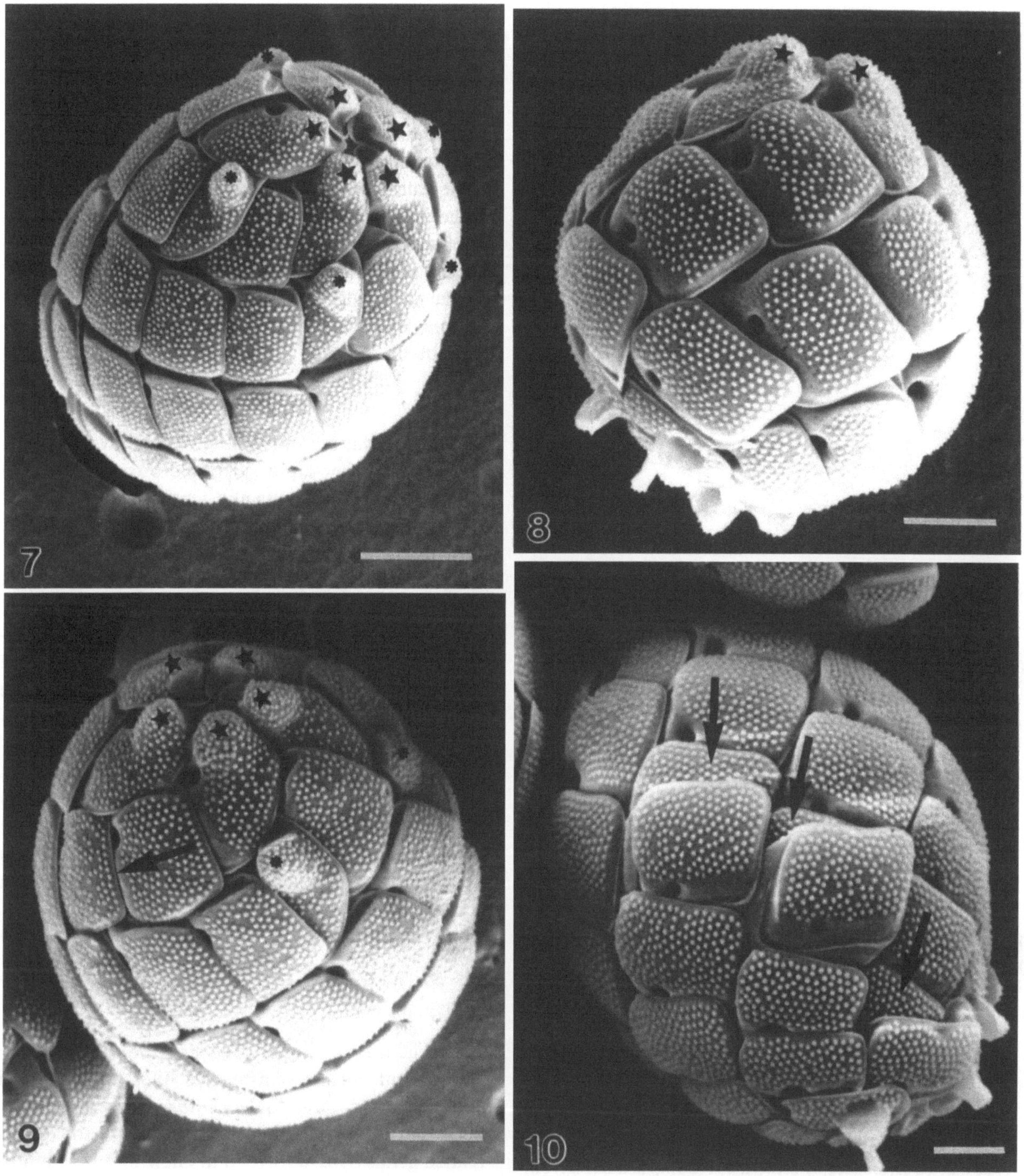

Fig. 7–10. SEMs of developing scale cases of *M. adamas*, with the parental, apical dome scales indicated by stars and each newly inserted apical dome scale by an asterisk

Fig. 7. Fully developed scale case just prior to division. Note that 5 new apical dome scales have been inserted in each row, one body scale to the posterior of each parental, apical dome scale. Bar: 4 μm

Fig. 8. Cell immediately following division. Scale rows contain 6 scales each. Bar: 2 μm

Fig. 9. Scale case completing development. Two scale rows have already obtained their new apical dome scale, while the large arrow indicates where we believe the next apical dome scale would have been inserted. Bar: 3 μm

Fig. 10. Scale case during development, with 3 new body scales (arrows) apparently in the process of being inserted into position within 3 different scale rows. Bar: 2 μm

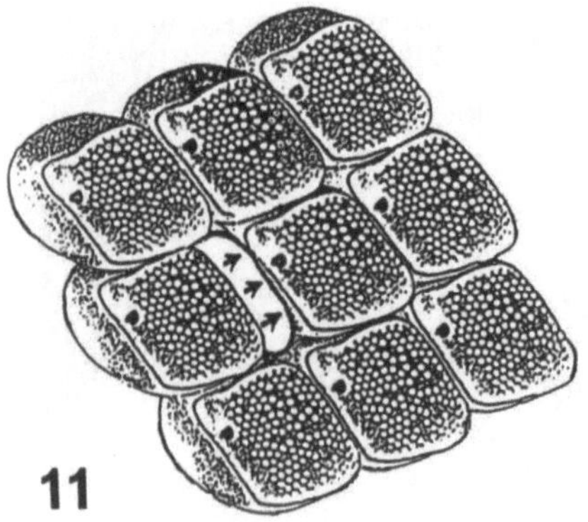

11

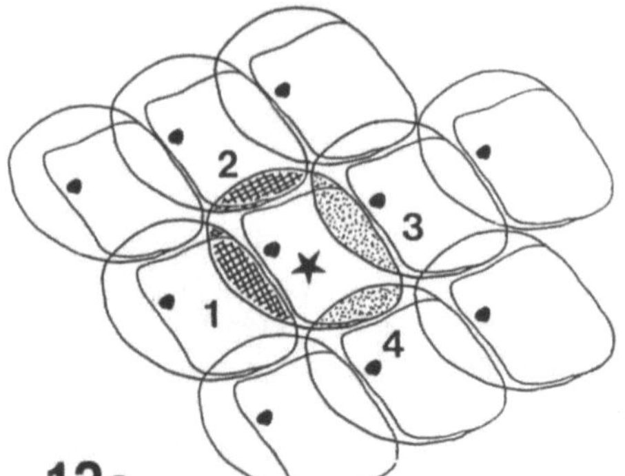

12a

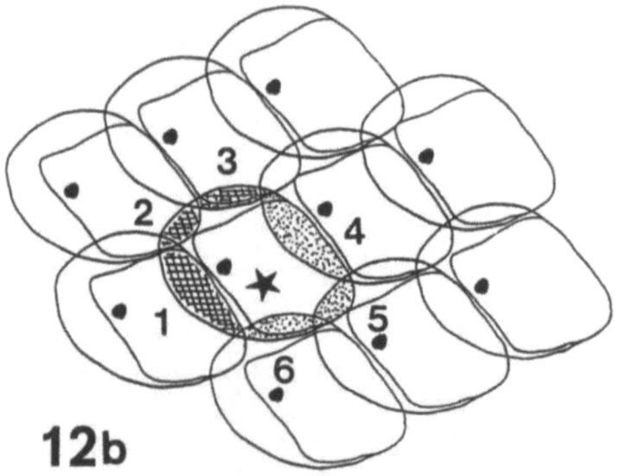

12b

Figs. 11 and **12 a, b.** Diagrams illustrating the realignments required within a scale case following the insertion (arrows, Fig. 11) of a new scale (star, Fig. 12 a, b). Note that the new scale will adhere (i.e., overlap in the region of the boomerang) to the adjacent two scales within the same row, and with either 1 or 2 scales in each of the two adjoining rows (Fig. 12). The region of overlap for the boomerang belonging to the new scale is shown in cross-hatch, while the stippled region shows where the proximal surface of the new scale overlaps the boomerangs of adjoining scales. Depending on its position within the scale case and the stage of development, a single scale may be surrounded by between 4 (Fig. 12 a) and 6 (Fig. 12 b) scales

around the circumference of the cell from one row to the next.

The addition of body scales is less obvious, though scale rows on the same cell are often observed to have unequal numbers of scales, but never varying by more than one. Occasionally, new scales apparently in the process of being secreted into an existing scale case are observed (Fig. 10). Once again, only single scales appear to be added to a row at any one time, with obvious adjustments being made by the existing scales in order to accomodate the new arrival (Figs. 10–12). In Fig. 10, three scales appear to be entering three different rows of the scale case simultaneously. However, images like this are very rare, and normally only a single scale can be seen entering at any one time. The exact sequence for the deployment of all body and caudal scales into growing scale cases is still under investigation, and may relate to the precise positioning of the cytoskeleton (unpubl. obs.).

Division of the scale case

In vivo observations of longitudinal cell division in *M. adamas* using time-lapse video microscopy reveal the complex scale movements and reorientations required to maintain scale case patterning during cytokinesis. Several of these stages are illustrated in Figs. 14–17. The precise size, shape and morphology of scales aids in their re-deployment. In Figs. 14 and 15, in particular, it is possible to envisage how the ten dome scales segregate and re-orientate to form two new star-shaped anteriors as division proceeds. Scale-plasma membrane adhesion is maintained (Fig. 18) while the adhesion of sticky boomerangs appears to be temporarily suspended. In vivo observations have occasionally shown scales to be almost entirely dislodged, only to be retrieved and quickly returned to their proper place in the scale case. Cytokinesis takes approximately 10 min to complete, and mistakes have not been observed in healthy cell cultures.

Discussion

Scales and scale case structure

The structure of the *Mallomonas* scale case has been studied in detail (Takahashi 1978, Wee 1982, Asmund and Kristiansen 1986, Andersen 1987, Beech et al. 1990, Siver and Glew 1990) and compared with the closely related genus *Synura* (Siver and Glew 1990). Although scale case differences between the two genera exist (e.g., presence of bristles in *Mallomonas*), their scale cases are constructed in a similar manner. A set number of scale rows is found in each species (normally 4–7) always spiralling in a clockwise direction when viewed from either the anterior or posterior end of the cell (for reviews, see Siver and Glew 1990, Siver 1991). However, the imbrication of scale cases is such that scales can be in row-like files in a number of different orientations.

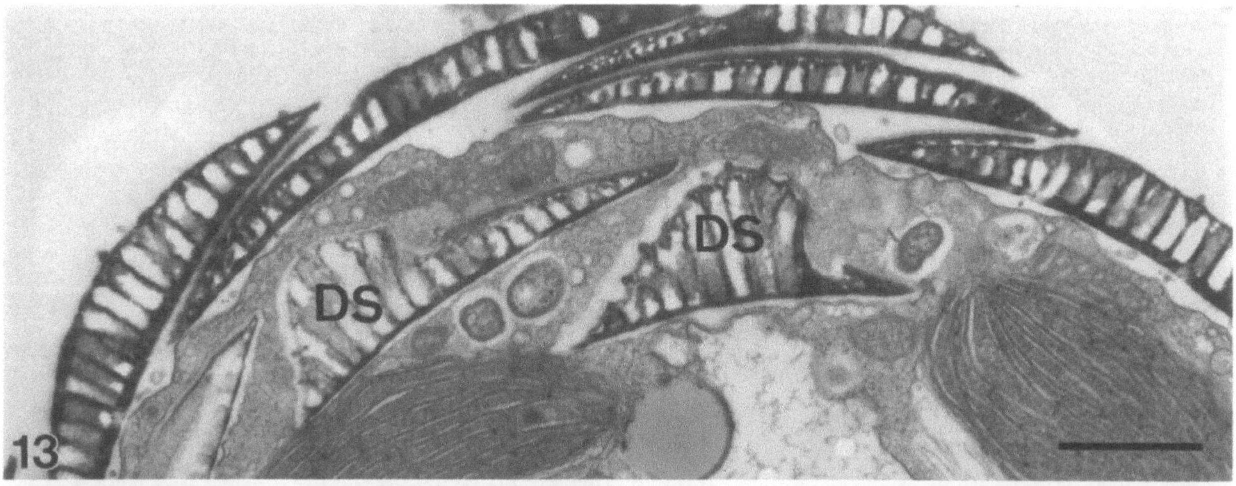

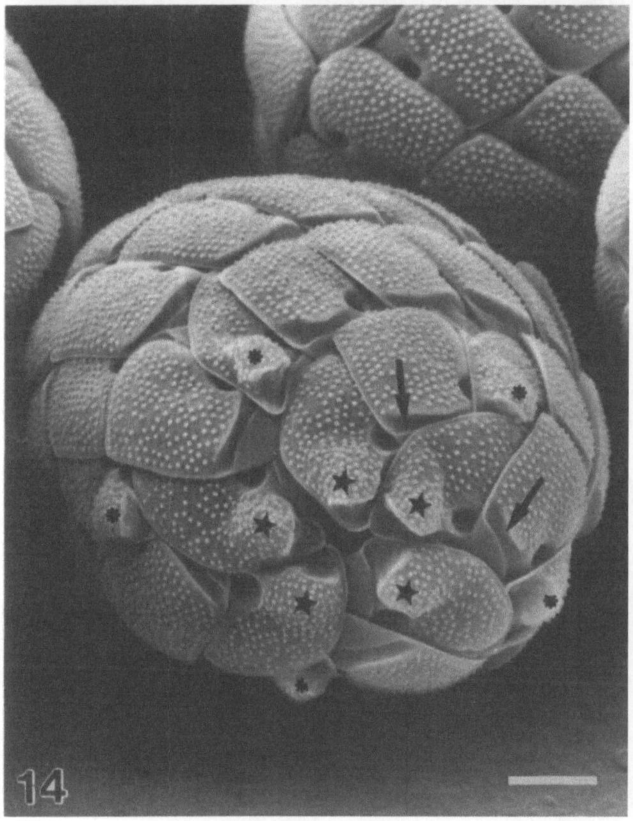

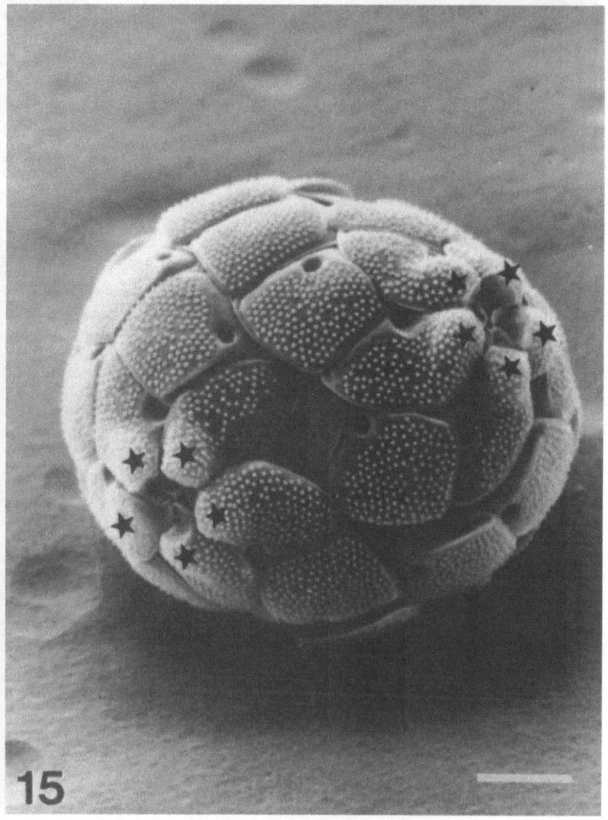

Fig. 13. Thin section of a cell fixed just prior to division. Only apical dome scales (*DS*) are being formed within the cytoplasm at this stage in development. Bar: 1 μm

Figs. 14 and **15.** SEMs of fully developed scale cases in the early stages of division. Bars: 2 μm

Fig. 14. Fully developed scale case with 5 parental apical dome scales (stars) and 5 new apical dome scales (asterisks) just starting to realign at the onset of division. Note the anterior ridges (arrows) that appear to guide scales during their precise movements

Fig. 15. Early stage in the division of the scale case, and already the 10 apical dome scales (stars) have realigned into the anterior ends of two new sibling cells. Note the typical star-shaped apex produced by the 5 dome scales at the anterior end of each scale row

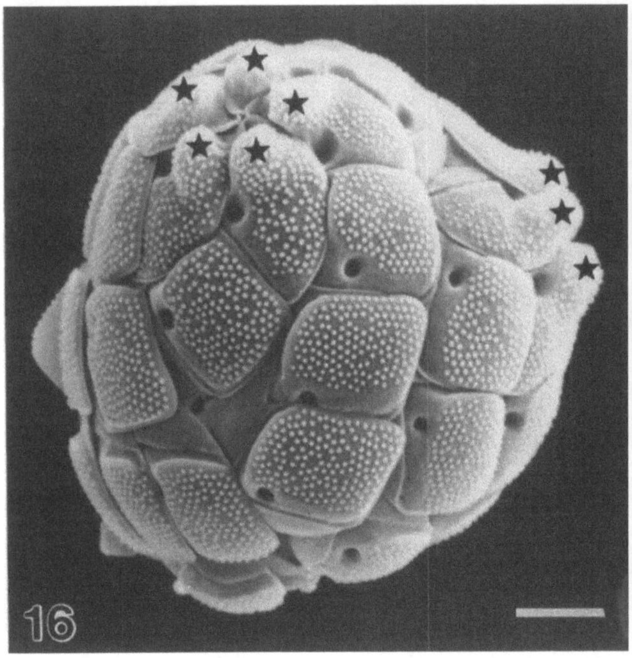
16

17

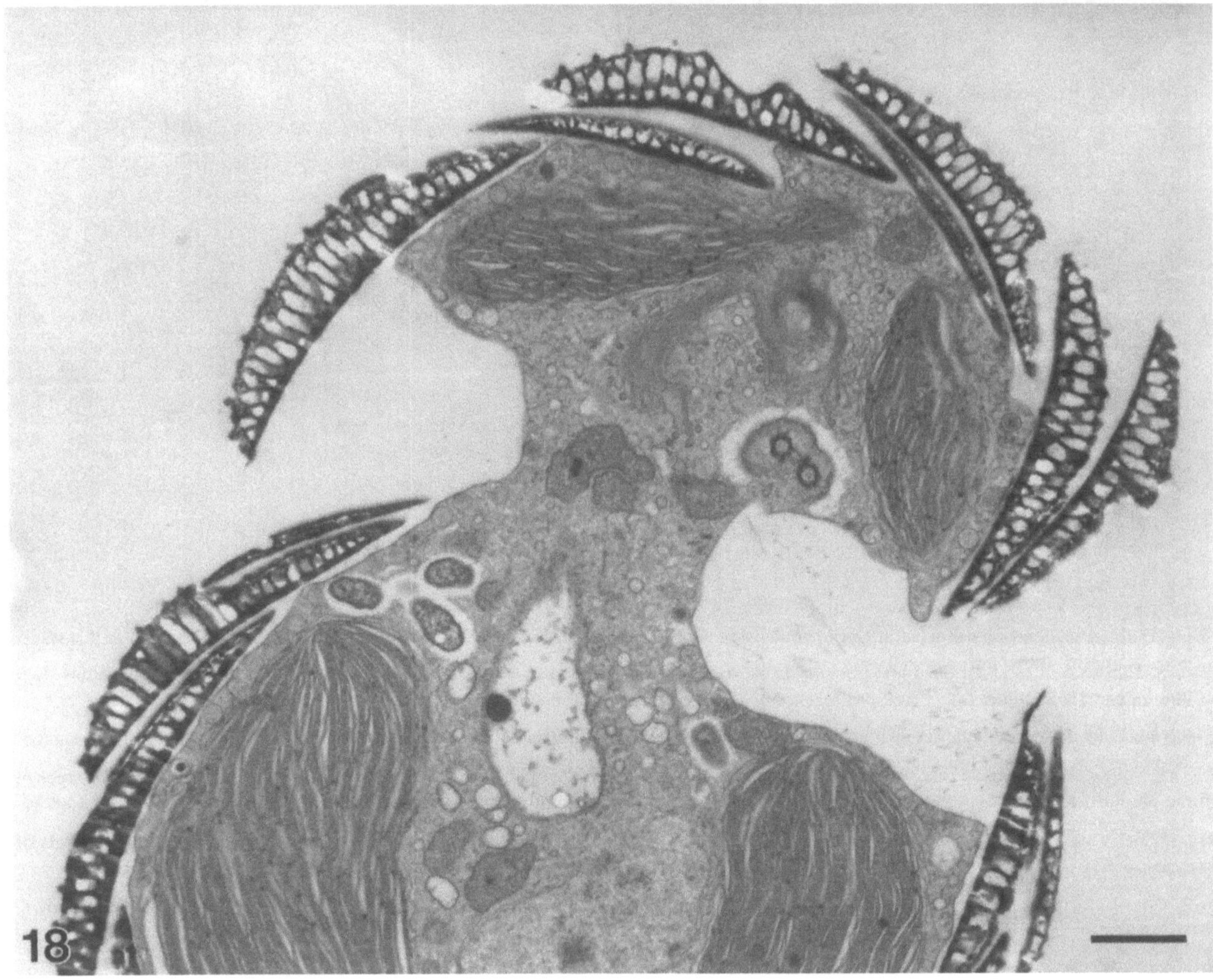
18

For example, in *M. adamas* scales often appear in diagonal rows that spiral counterclockwise, or in the opposite direction from scale rows (e.g., see Figs. 7 and 9). These scale alignments are not the scale rows as described above and discussed by most workers (e.g., Leadbeater 1986, Siver and Glew 1990), and will be referred to here as "scale ranks". Scale ranks are easily distinguished from "scale rows" by the direction of spiralling and by the fact that scale ranks often do not extend the full length of the cell, and may lack either an apical or caudal scale or both. Furthermore, there is a greater number of scale ranks than rows, and the number of ranks increases during the growth of the scale case. As a new scale is inserted into a row, the scales in that row always remain aligned, while scale ranks temporarily lose their alignment. Our perusal of the literature on *Synura* (e.g., Leadbeater 1990), as well as our own preliminary observations, reveal the same pattern of rows and ranks in this genus. In the Leadbeater (1990) model for scale case formation, the diagonal scale "rows" he suggests are added to the scale case during growth, appear to be scale "ranks" by our definition (see below).

Scale case development

With some minor variations, the mechanisms of scale case formation and development should be universal for both *Synura* and *Mallomonas*. In this study, the appearance of apical dome scales is used as a marker to ascertain the order in which scales are inserted into the scale case. The appearance of new apical dome scales only occurs immediately prior to division, and our observations show that they are inserted one at a time into specific positions near the anterior of the existing scale rows. Scale deployment is therefore in a ring around the cell, not in diagonal rows. Sectioned material shows that only dome scales are being assembled at this time. This data partially supports the model of scale case formation proposed for *Synura* (Leadbeater 1990), in that scales are moved around within the cytoplasm, precisely orientated and inserted into the existing scale case in a predetermined position. The alternative model (Siver and Glew 1990), based on a single, posterior site for scale secretion, does not occur in *M. adamas*. New apical dome scales are not accumulated and secreted from a posterior location, nor are they moved to their final position near the cell apex by a process of migration. It is conceivable that the Siver and Glew (1990) model explains how excysted naked cells from resting stomatocysts generate a new scale case. However, though we did not study this process, it seems more likely that only one mechanism exists. We expect that these cells likewise produce scales of predetermined size and shape, and systematically secrete them onto the surface in the position they will occupy in the finished scale case.

According to the model proposed by Leadbeater (1990) and Leadbeater and Green (1993), it is suggested that scales are extruded in rows, and that the scale case increases in size by the addition of complete rows of scales rather than the sequential addition of single scales to existing rows as shown here. *M. adamas* never has more than 5 scale rows during development of the scale case, nor does the number of scale rows increase in any of the 10 species of *Mallomonas* or 4 species of *Synura* we have observed. However, our interpretation of Leadbeater's micrographs and diagrams suggests that he may be referring to an increase in the number of "scale ranks", not "rows". Therefore, our model is not unlike the Leadbeater model, as a sequential addition of scales into different scale rows could result in the appearance of new scale ranks. Although apical dome scales are not added as diagonal rows, but in a ring to the same position within all 5 scale rows, it is possible that caudal and body scales are added in a different pattern, as suggested in Fig. 10. The exact sequence in which caudal and body scales are interpolated into scale cases is presently under investigation. Both the Leadbeater (1990) and Siver and Glew (1990) models are based on the assumption that inserting scales individually into an established scale case would be disruptive to its overall structure, requiring the 4–6 surrounding scales to detach and then re-attach to the new scale (e.g., Fig. 11). However, it is the dynamic aspects of scale case development that are perhaps surprising when considering its primary role as a cell cov-

Figs. 16 and **17.** SEMs of dividing cells showing two stages in the precise reorientation of scales within the scale case during division. Bars: Fig. 16, 2 µm; Fig. 17, 3 µm

Fig. 18. Thin section of a dividing cell of *M. adamas* showing the adhesion of scales to the plasma membrane over part of their proximal surface, and great variation in the relationships of scales to one another as they slide over one another during division. Bar: 1 µm

ering. The adhesion of scales to surrounding scales does not appear to be stronger in rows than between rows (or scale ranks), particularly since cells divide longitudinally by cutting through rows, with the appropriate scales easily separating from one another. In vivo observations of cytokinesis using time-lapse video microscopy have shown a high degree of coordination during the deployment of scales and the maintenance of the scale case patterning during division. Scale rows slide past one another with great precision, while scales that appear misplaced are soon retrieved and integrated into the scale case. Certain structural features of scales, specifically the sticky boomerang and the ridge found anteriorly on each scale (Fig. 14), appear perfectly situated to position scales relative to one another, and to maintain scale case integrity during scale movements.

The overall mechanism of scale case growth described for this species is probably much the same for other species of *Mallomonas* and *Synura*, though differences between species may occur. For example, cells of *M. splendens* (G.S. West) Playfair possess four rows of body scales with the apical and caudal scales in each row, termed base-plate scales, each attaching to an elongate bristle (i.e., each cell has 4 posterior and 4 anterior bristles). The complex duplication and deployment of bristles has been described for this species (Beech et al. 1990) and gives some indication of the dynamic nature of scale cases. The number of scales in the scale case of *M. splendens* does not exactly double prior to division as observed in *M. adamas*. The posterior (caudal) base plate scales and accompanying bristles are duplicated and deployed to the surface prior to division, while the apical base-plate scales and bristles are reduced from 4 to 2 during division. Following division, each daughter cell then duplicates and deploys the 2 new apical base-plate scales and bristles, completing scale case construction. Although these differences seem minor when compared to the overall similarities in scale case construction, we have only studied a few species and cannot discount the possibility that a range of mechanisms are utilized during the deployment of bristles and/or scales, and further investigations are warranted.

Acknowledgements

RW would like to thank the Australian Research Council for financial assistance. We thank Jocelyn Carpenter for technical support and Kirsten Heimann for the fixation of *M. adamas* for transmission electron microscopy. SL thanks LS, AL, and BL for moral support.

References

Andersen RA (1987) Synurophyceae classis nov., as new class of algae. Amer J Bot 74: 337–353

Asmund B, Kristiansen J (1986) The genus *Mallomonas* (Chrysophyceae). Op Bot 85: 1–128

Beech PL, Wetherbee R, Pickett-Heaps JD (1990) Secretion and deployment of bristles in *Mallomonas splendens* (Synurophyceae). J Phycol 26: 112–122

Brugerolle G, Bricheux G (1984) Actin microfilaments are involved in scale formation of the chrysomonad cell *Synura*. Protoplasma 123: 203–212

Cohn SA, Spurck TP, Pickett-Heaps JD, Edgar LA (1989) Perizonium and initial valve formation in the diatom *Navicula cuspidata* (Bacillariophyceae). J Phycol 25: 15–26

Leadbeater BSC (1986) Scale case construction in *Synura petersenii* Korsch. (Chrysophyceae). In: Kristiansen J, Andersen RA (eds) Chrysophytes: aspects and problems. Cambridge University, Cambridge, pp 121–132

– (1990) Ultrastructure and assembly of the scale case in *Synura* (Synurophyceae Andersen). Br Phycol J 25: 117–132

– – Green JC (1993) Cell coverings of microalgae. In: Berner T (ed) Ultrastructure of microalgae. CRC Press, Boca Raton, pp 71–98

Mignot J-P, Brugerolle G (1982) Scale formation in chrysomonad flagellates. J Ultrastruct Res 81: 13–26

Schnepf E, Deichgräber G (1969) Über die Feinstruktur von *Synura petersenii* unter besonderer Berücksichtigung der Morphogenese ihrer Kieselschuppen. Protoplasma 68: 85–97

Siver PA (1991) The biology of *Mallomonas*: morphology, taxonomy and ecology. Kluwer, Dordrecht

– Glew JR (1990) The arrangement of scales and bristles on *Mallomonas* (Chrysophyceae): a proposed mechanism for the formation of the cell covering. Can J Bot 68: 374–380

Takahashi E (1978) Electron microscopical studies of the Synuraceae (Chrysophyceae) in Japan, taxonomy and ecology. Tokai University Press, Tokyo

Wee JL (1982) Studies on the Synuraceae (Chrysophyceae) of Iowa. Bibl Phycol 62: 1–183

Wetherbee R, Koutoulis A, Beech PL (1989) The role of the cytoskeleton during the assembly, secretion and deployment of scales and spines. In: Coleman, AW, Goff LJ, Stein-Taylor JR (eds) Algae as experimental systems. AR Liss, New York, pp 93–108

– – Andersen RA (1992) The microarchitecture of the chrysophycean cytoskeleton. In: Menzel D (ed) The cytoskeleton of the algae. CRC Press, Boca Raton, pp 1–17

Wujek DE, Kristiansen J (1978) Observations on bristle- and scale-production in *Mallomonas caudata* (Chrysophyceae). Arch Protistenk 120: 213–221

Protoplasma (1994) 181: 269–282

PROTOPLASMA

Valve and seta (spine) morphogenesis in the centric diatom *Chaetoceros peruvianus* Brightwell

J. D. Pickett-Heaps*, Jocelyn Carpenter, and A. Koutoulis

School of Botany, University of Melbourne, Parkville, Victoria

Received August 5, 1993
Accepted October 21, 1993

Summary. In *Chaetoceros peruvianus*, the two very long, delicately tapered setae (spine-like processes) of the "lower" valve curve downwards gently until they are often almost parallel, while those emerging from the "upper" valve curve sharply downwards until oriented almost in the same direction as the setae of the lower valve. This curvature creates a deep pit between the bases of the upper valve's setae, where they emerge from the valve. In live cells, extension of setae is rapid and very sensitive to disturbance. After cleavage the new silica deposition vesicle (SDV) appears in the centre of the furrow and expands outwards over it. A distinct microtubule centre (MC) appears directly on top of the SDV. Microtubules (MTs) from the MC ensheath the nucleus, and others fan out over the SDV and plasmalemma. A little later, the MC in the lower daughter cell moves off the SDV, and its MTs now appear to mould the plasmalemma/SDV into the deep pit between the base of the setae. In the upper daughter cell, the MC remains on the SDV. Initiation of setae is first observed as protuberances of bare cytoplasm growing from the sides of the daughter cells, through gaps in the parental valve. Many MTs initially line the plasmalemma of these protuberances as they grow outwards and the SDV also expands over the new surface. As the setae get longer, a unique complex of three organelles appears. Just behind the naked cytoplasm at the tip of the seta, a thin flat layer of fibrous material lines the plasmalemma. This, the first of the complex, is called the "thin band". Immediately behind this is the second, a much thicker, denser fibrous band, the "thick band". At the front edge of the SDV, 5–6 "finger-like outgrowths" of silicified wall grow forwards. These are interconnected by the elements of the thick band which thus apparently dictate the polygonal profile of the seta. These also appear to generate the spinules (tiny spines) that adorn the surface of the seta; the spiral pattern of the spinules indicates that this whole complex might differentiate one after the next, in order. Further back from the tip, evenly spaced transverse ribs are formed. These are connected to the third organelle in the complex, the "striated band"; our interpretation is that the striated band sets up the spacing of the ridges that regularly line the inner surface of the setae. During seta growth, this complex is apparently responsible for controlling the delicate tapering curvature of the very fine silica processes. Since the complex is always seen near the tip of the seta, we conclude that it migrates forwards steadily as the tip grows. While the thin and thick bands could slide continuously over the cell membrane, the striated band must be disassembled and then recycled forward during extension if it is indeed connected to the ridges lining the inside of the setae. We could find no indication that turgor pressure drives extension of the setae, in which event the activity of these organelles is responsible for growth using the just-formed silica tube as the base from which extension is generated.

Keywords: Diatom; Morphogenesis; Seta; Silica; Spine.

Introduction

Diatoms display an extraordinary range of morphological features in their frustules, characteristic for each species. Following cell division, the cytoplast of each daughter cell forms a new valve complete with these features. The silica wall is invariably laid down in a special membranous sac, the silica deposition vesicle (SDV). Many major morphological features of the valve arise by the cell continuously moulding the expanding SDV as silica is deposited in it. The cytoplasmic mechanisms used by diatoms for this moulding are remarkably diverse (reviewed in Pickett-Heaps et al. 1990), involving a range of cell organelles (e.g., MTs, actin filaments, mitochondria, a variety of specialised areas on the cell membrane, etc.) and cytoplasmic activity (e.g., incomplete cytokinesis, controlled plasmolysis, adhesions of the cell membrane, etc.). A few groups of unicellular organisms such as desmids, radiolarians and diatoms display precisely sculptured and

* Correspondence and reprints: School of Botany, University of Melbourne, Parkville, Vic. 3052, Australia.

marvellously diverse range of species. However, the diatoms have revealed a virtuosity in evolving morphogenetic systems using various combinations of organelles and cellular activity unrivalled in eucaryotic cells. This virtuosity is reflected in the great variety of features that diatoms display in their mature valve.
Many diatoms display short or long, fine setae and other spine-like processes. In certain centric diatoms, the labiate process is externally developed into a long straight or curving tube. At the ultrastructural level, a specific organelle, the labiate process apparatus (LPA) (Li and Volcani 1985 a–c), is applied to the SDV at the site of the future labiate process, which then steadily grows out from this point. In *Odontella* and *Ditylum*, the LPA seems to arise from material at the spindle pole (Pickett-Heaps et al. 1988); its exact function is not obvious. It is not clear whether extrusion of these processes is driven by turgor pressure, since they can grow rapidly in cells that are slightly plasmolysed. The lumen of the growing spine is devoid of normal cytoplasmic organelles, so how the diameter and straightness of the growing spine and its SDV is maintained, remains a mystery.
To gain further information on how these and other spine-like processes grow, we investigated morphogenesis of setae of *Chaetoceros peruvianus*. The setae are often extraordinarily long, fine and delicately tapered (e.g., Fryxell 1978) and are thus much less substantial than the spine-like LPs. We find that morphogenesis of setae involves a complex set of fibers that may be unique to this genus, and has nothing in common with extrusion of LPs.

Materials and methods

C. peruvianus is common in marine plankton samples collected from Port Phillip Bay (see von Stosch et al. 1973: table 1) and along the shores of Bass Strait. Single cells were isolated and grown as described for *Ditylum* (Pickett-Heaps et al. 1988). Periodically, sexual reproduction produced large clones of cells. Reisolation and subculture of such cells yielded the most useful material for transmission electron microscopy (TEM).
For light microscopy, actively growing cells were mounted in a variety of ways to attempt measurement of the rate of seta growth, and for time-lapse filming of it. No method worked adequately. While relatively undisturbed cells sometimes displayed extension of setae, our measured rates were highly variable and we cannot tell how representative they are of growth rates in undisturbed cells. After many attempts, some cells recorded on video secreted fully formed setae.
For fixation, cultures of actively dividing cells were fixed initially in 0.5% glutaraldehyde in culture medium, pH 7. After 10–20 s, drops of osmium were added till the final concentration of osmium was about 0.05%. The cells were left in the combined fixative for 10 min, then washed several times in sea water, increasingly diluted with distilled water. After a wash in distilled water, cells were dehydrated slowly in acetone, embedded in Spurr's resin and mounted as a thin layer between teflon-coated microscope slides.
Preselected cells at appropriate stages of division and valve morphogenesis were excised from the thin mounts and remounted on plastic stubs so that setae could be sectioned either longitudinally or transversely. All cells were carefully serially sectioned, with sections mounted on Formvar-coated slot grids. Examination of quite long, unbroken sequences of sections allowed us to reconstruct with confidence the organisation of the growing tips of setae. The sections were stained with uranyl acetate and lead citrate as normal.

Results

C. peruvianus is a common, unicellar marine diatom. Like other members of this genus (Round et al. 1990), it typically has four very long, finely tapered setae per cell. This species is, however, unusual in that all setae are oriented in the same general direction (Rines and Hargraves 1988). Those of the new "upper" cell emerge from the new valve's surface and gently curve downwards and toward each other, until they are often almost parallel; those emerging from the "lower" daughter cell curve sharply downwards at their base and end up almost parallel to the other two (Figs. 1 and 2). Thus, these diatoms are heterovalvate: the two valves in each frustule are morphologically different. The setae are uniformly covered with fine pointed projections (Fig. 3) or "spinules" (Rines and Hargraves 1988) which form a helix (Evensen and Hasle 1975, Round et al. 1990); internally they display regularly spaced ridges (Fig. 4). The diameter of the valve varies between 10–44 μm (Hendey 1964).
In living cells, the lumen of the setae always contained numerous small chloroplast (Fig. 9) prominent also in electron micrographs (Fig. 4). In time-lapse photography, the chloroplasts were seen to move actively along the length of the seta; under strong illumination, they tended to gather back at the base of the setae.

Seta formation: live cells

Extension of setae was initiated soon after cytokinesis (Figs. 7–9) and was very delicate; the slightest disturbance of dividing cells caused extension to stop immediately. Extremely fine cytoplasmic filaments emerged from the tip of cytoplasm of a few setae which had stopped growing; cytoplasmic particles were filmed moving rapidly along these fibres which we suspect were MTs pushing out from the naked cytoplasm at the tip. Eventually, several sequences were obtained of setae growing to their normal length. Immediately after cleavage, the daughter protoplasts were flattened against each other (Fig. 7 a). Over about 15 min, they

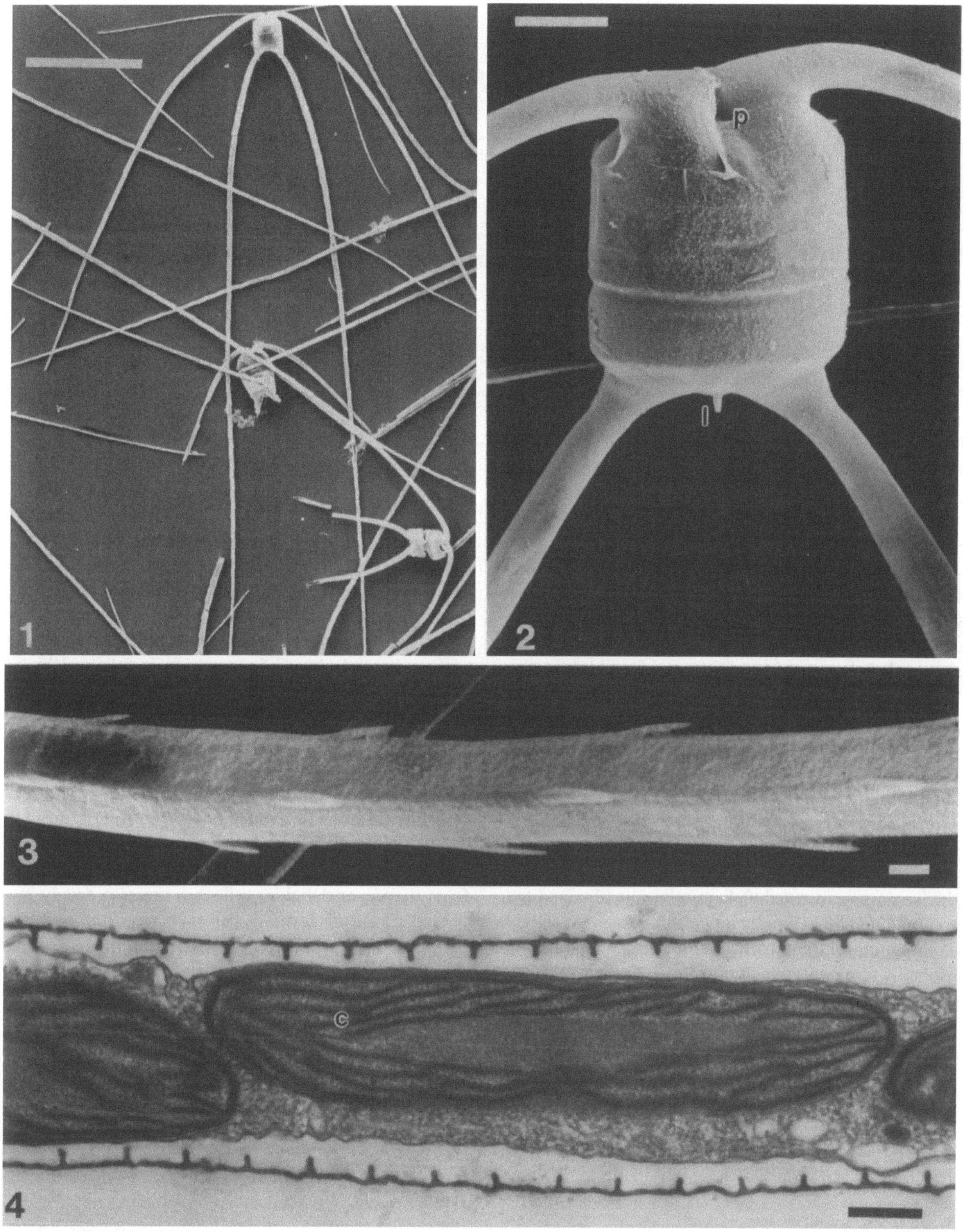

Fig. 1. SEM of whole cells. Bar: 50 μm

Fig. 2. Body of cell; pit (*p*) between base of setae, and labiate process (*l*). Bar: 5 μm

Fig. 3. Seta showing polygonal profile and covering of helical "spinules". Bar: 1 μm

Fig. 4. TEM of seta showing internal ridges of silica, and chloroplasts (*c*) in lumen. Bar: 500 nm

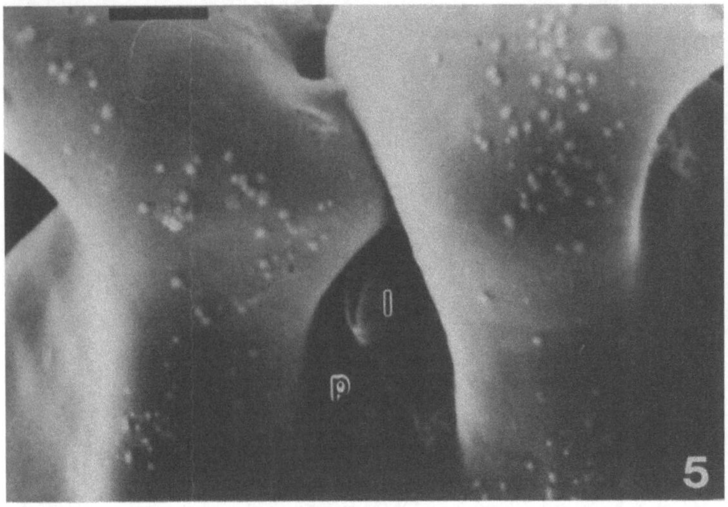

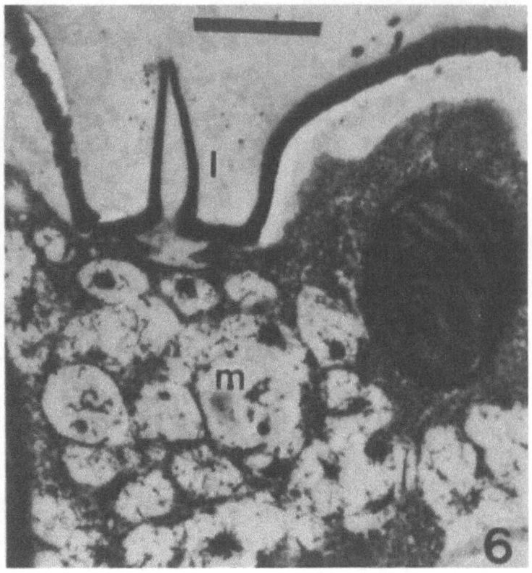

Fig. 5. Pit (*p*) between base of setae, with labiate process (*l*). Bar: 1 μm

Fig. 6. Labiate process with nearby mucilage vesicles (*m*). Bar: 1 μm

shrank slightly as the surface of the furrow became differentiated (Fig. 7 b). Soon, cell expansion separated the girdle bands as setae were initiated (Fig. 7 c) after which, the setae grew out through the holes (Fig. 7 d). The tip of the elongating setae displayed a characteristic shape, slightly constricted in front of a ridge at the leading edge of the SDV. This morphology corresponds very well with that of fixed setae (e.g., as in Fig. 18); there was no indication of bulging at the tip as might be present if extension was osmotically driven (see Discussion).

The rate of seta growth was measured in several batches of cells carefully collected in large chambers for microscopical observation. The rate was highly variable; the maximum observed, 0.4 μm/min, was obtained for several daughter cell pairs in which all four setae grew at the same rate and to the same length. This rate is probably representative of undisturbed cells. It is close to that in *C. didymum* reported by von Stosch et al. (1973) (ca. 31 μm/h) and slightly higher than that in *C. gracilis* (Rogerson et al. 1986; 0.2–0.38 μm/min). Because the setae are so long, it took several hours for them to attain their full length. Rogerson et al. described growth as tri-phasic: after an initial lag phase, a second rapid phase of elongation is followed by a slower, 'plateau' phase.

Valve morphology

The valves were quite fragile and difficult to prepare clean for scanning electron microscopy (SEM) (Figs. 1–3). Between the setae of the lower valve is a small pore with external labia (Fig. 2). When the setae grow out from the upper daughter cell, they create a deeply indented pit in the forming valve, between the base of the setae (Fig. 2). In the centre of this pit is a pore similar to that on the other valve (Fig. 5). By convention, this pore is called a labiate process (LP) (e.g., Li

Figs. 7–9. Live cells, from two time-lapse video sequences

Fig. 7 a–d. Early valve morphogenesis. **a** Post-cytokinesis; cleavage furrow flat (arrowheads), corresponding to Fig. 10. **b** Initial shaping of valve (arrowheads), corresponding to Fig. 14. **c** Initiation of setae as girdle bands open (arrowheads). **d** Extrusion of setae through holes in girdle bands (arrowheads), corresponding to Fig. 16.

Figs. 8 and **9.** Growth of setae

Fig. 8 a, b. Initial growth: shape of rapidly elongating tip corresponds with that in Fig. 18; edge of SDV just detectable (arrowheads)

Fig. 9. Same cells as those in Fig. 8 about 10 min later; numerous chloroplasts in growing setae

Fig. 10. Initiation of valves; nuclei (*n*) with pointed end directed at microtubule centres forming on silica deposition vesicle (arrowheads). Girdle bands overlapped. Bar: 2 μm

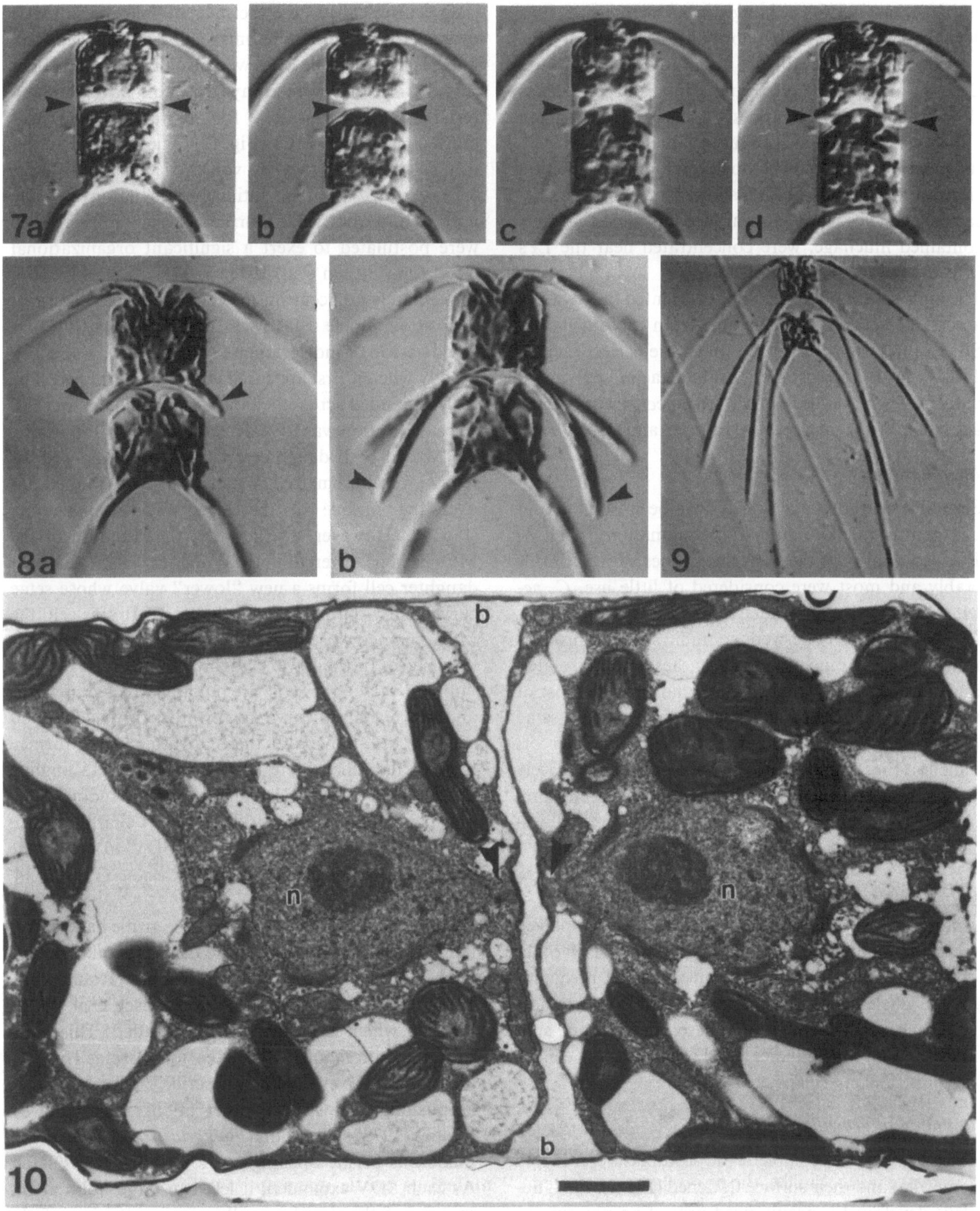
7a
b
c
d
8a
b
9
b
n
n
b
10

and Volcani 1985 b) although in *C. peruvianus*, it is devoid of the internal labia normally diagnostic of LPs. For convenience, we will also adopt this nomenclature. Evensen and Hasle (1975) and Rines and Hargraves (1988) call it a "central process" and say it "varies from a long central tube to a reduced labiate process ...". Its function is unknown; however, our transmission electron microscopy (TEM) showed that on some occasions, vesicles whose appearance indicated that they contained mucilage, were concentrated near the LPs (Fig. 6).

The internal, evenly spaced ridges oriented transversely across the setae were detectable from the outside by SEM (lower right in Fig. 2), and were easiest seen in TEM (Fig. 4). In cross section, the setae are polygonal in form. Their profile in other *Chaetoceros* varies characteristically for each species (Evensen and Hasle 1975).

Interphase cells

Several species of *Chaetoceros* have been isolated and cultured in our laboratory for examination by TEM. Fixation quality of these marine species varied considerably and most were considered of little use. *C. peruvianus*, however, gave adequate to good fixation quality. Our cultures underwent sporadic sexual reproduction producing much larger cells; these, when isolated and cultured, gave good fixation and provide most of the images illustrated here. The lumen of the setae always contained chloroplasts as well as smaller organelles (Figs. 4 and 9). Profiles of MTs were relatively uncommon; in general (see below) MTs were not preserved/stained in the setae although they were well visualised in the rest of the cell.

Cell division

A few dividing cells were examined and they displayed features (not illustrated here) typical of mitosis in other diatoms. The spindle arose from a striated precursor on the nuclear envelope. The typical central spindle stayed compact through anaphase, and did not split up into numerous bundles of MTs as in larger centric diatoms (Pickett-Heaps et al. 1988).

Initiation of growth of setae

Once cleavage was complete, the protoplasts contracted somewhat and their surface flattened (Fig. 7). Each nucleus developed a pronounced projection pointing to the centre of the cleavage furrow (Fig. 10). Numerous MTs lay closely applied to the nuclear envelope, and directed at the cleavage furrow. At the focal point of each MT array appeared a diffuse microtubule organising centre, called here the microtubule centre (MC) by analogy with other diatoms (e.g., Pickett-Heaps and Kowalski 1981). The MCs were usually quite indistinct (Fig. 11) but they were always traceable through many serial sections; a few cells, for unknown reasons had more conspicuous MCs (Fig. 12). They were detected using light microscopy by von Stosch et al. (1973: p. 437) who describe their intimate association with the nucleus and early valve morphogenesis, on which they were postulated to exert a significant organizational effect (von Stosch et al. 1973: p. 441).

Immediately adjacent on the cell membrane, the silica deposition vesicle (SDV) was first seen as a tiny flattened vesicle of indeterminate origin (Figs. 11 and 12); exactly when it first appeared was not determined. The SDV expanded across the face of the plasmalemma and silica deposition was soon detectable within it. Initially, the MC in both daughter cells was superimposed directly on to the small SDV and many MTs fanned out over the surface of the SDV and plasmalemma (Fig. 11). However, a little later, the situation of the MCs became different in the daughter cells. The upper daughter cell forms a new "lower" valve whose setae curve gently out past the girdle bands. In this cell, the MC remained close to the SDV, with MTs spread over its surface. In the lower daughter cell (forming a new "upper" valve), the setae are initially directed upwards and make a sharp curve downwards. In this daughter cell, the MC moved off the SDV after it had formed a small flat silica plate (Fig. 13). Then MTs angling upwards from the MC lay against the expanding SDV/plasmalemma, apparently deflecting it sharply downwards and thereby moulding the SDV to create the deep pit at the base of the mature setae (Figs. 2, 14, and 16).

During and just after cytokinesis, the girdle bands overlap (Figs. 7 and 10). Soon, the cells expand. The last in the series of girdle bands have slots (Round et al. 1990), the "bristle openings" (von Stosch et al. 1973), through which the setae will be extruded. This separation is a prerequisite for extension of setae (Fig. 16) and von Stosch et al. (1973) describe the forming setae as pushing against each other at this stage (Fig. 8). Drebes (1972) showed that *Bacteriastrum hyalinum* has numerous bristle openings, one for each of its setae. As each SDV expanded, it left two large holes at the corners of the daughter cells where extension of the setae was soon to occur (von Stosch et al. 1973: p. 399). The SDV appeared to grow around these areas and then continue to expand along the parental girdle bands to create the form of the valve.

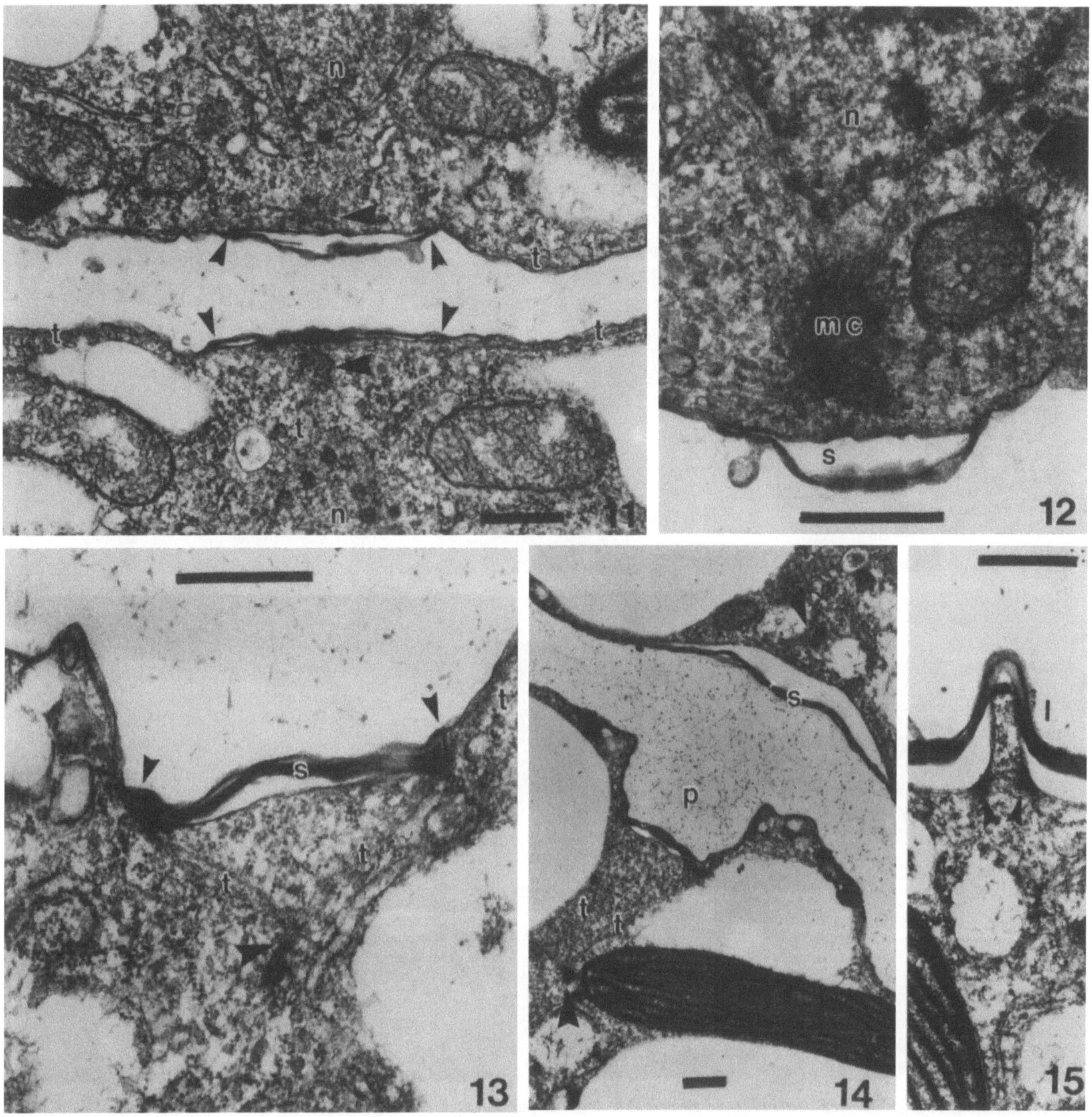

Fig. 11. Microtubule centres (MCs) (large arrowheads) next to silica deposition vesicles (SDVs) whose limits are indicated by small arrowheads. Microtubules (MTs) (*t*) run mostly over surface of cleavage furrow, toward nuclei (*n*). Bar: 500 nm

Fig. 12. An unusually prominent MC (*mc*) on surface of SDV (*s*). Bar: 500 nm

Fig. 13. MC (large arrowhead) moving off SDV (*s*) whose edges are indicated by small arrowheads; MTs (*t*) angling upwards apparently forming pit. Bar: 500 nm

Fig. 14. Pit (*p*) forming at top of one daughter cell; MC at arrowhead. In the other cell where no pit is formed, the MC (arrowhead) stays close to the SDV (*s*). Bar: 500 nm

Fig. 15. Labiate process (*l*) forming with dense material (arrowheads) on SDV. Bar 500 nm

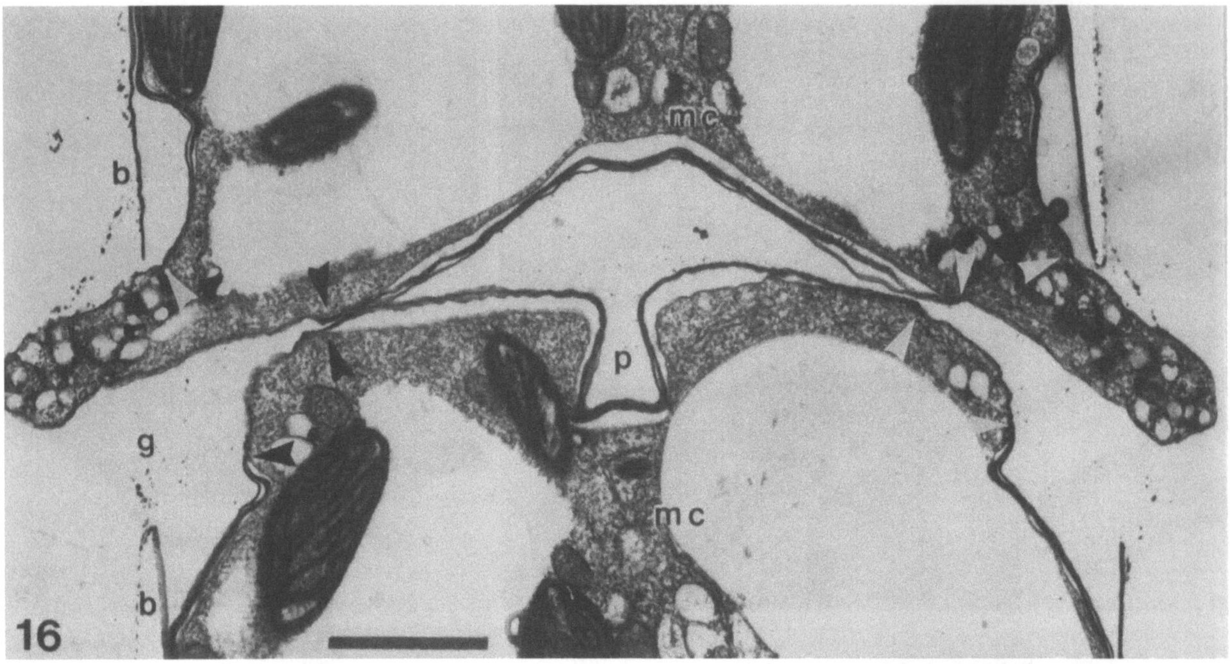

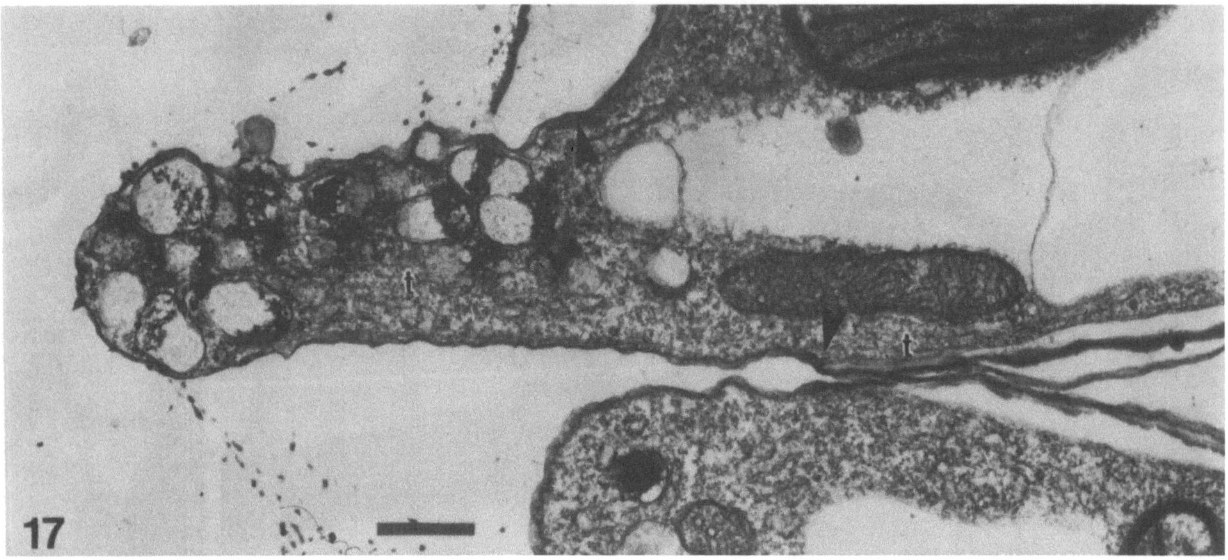

Fig. 16. Initiation of setae through opening (*g*) in girdle bands (*b*); one pair is out of the plane of this section. The expanding SDV in each cell leaves temporary holes (between pairs of arrowheads) through which cytoplasmic expansion occurs. MCs (*mc*) still faintly visible; pit (*p*) formed. Bar: 2 µm

Fig. 17. Detail of cell in Fig. 16; MTs (*t*) running over surface of SDV and plasmalemma into seta. The edge of the SDV is indicated by arrowheads. Bar: 500 nm

At this stage, the small LP was formed on the face of each valve. A dense layer was applied to the SDV at this site (Fig. 15), but it did not resemble the LPA of other centric diatoms (Introduction).

Extension of setae

Extension of the setae was first evident as outgrowths of naked cytoplasm (Iyengar and Subrahmanyan 1944) bulging through the bristle openings in the girdle bands (Fig. 16). Numerous MTs ran into these protrusions, most of them close to the plasmalemma (Fig. 17), and apparently many were associated with the two MCs, still faintly visible. These MTs became less conspicuous as the setae became longer and soon MTs were limited to a relatively few scattered in the lumen of the setae. The cytoplasm was dense and filled with a variety of

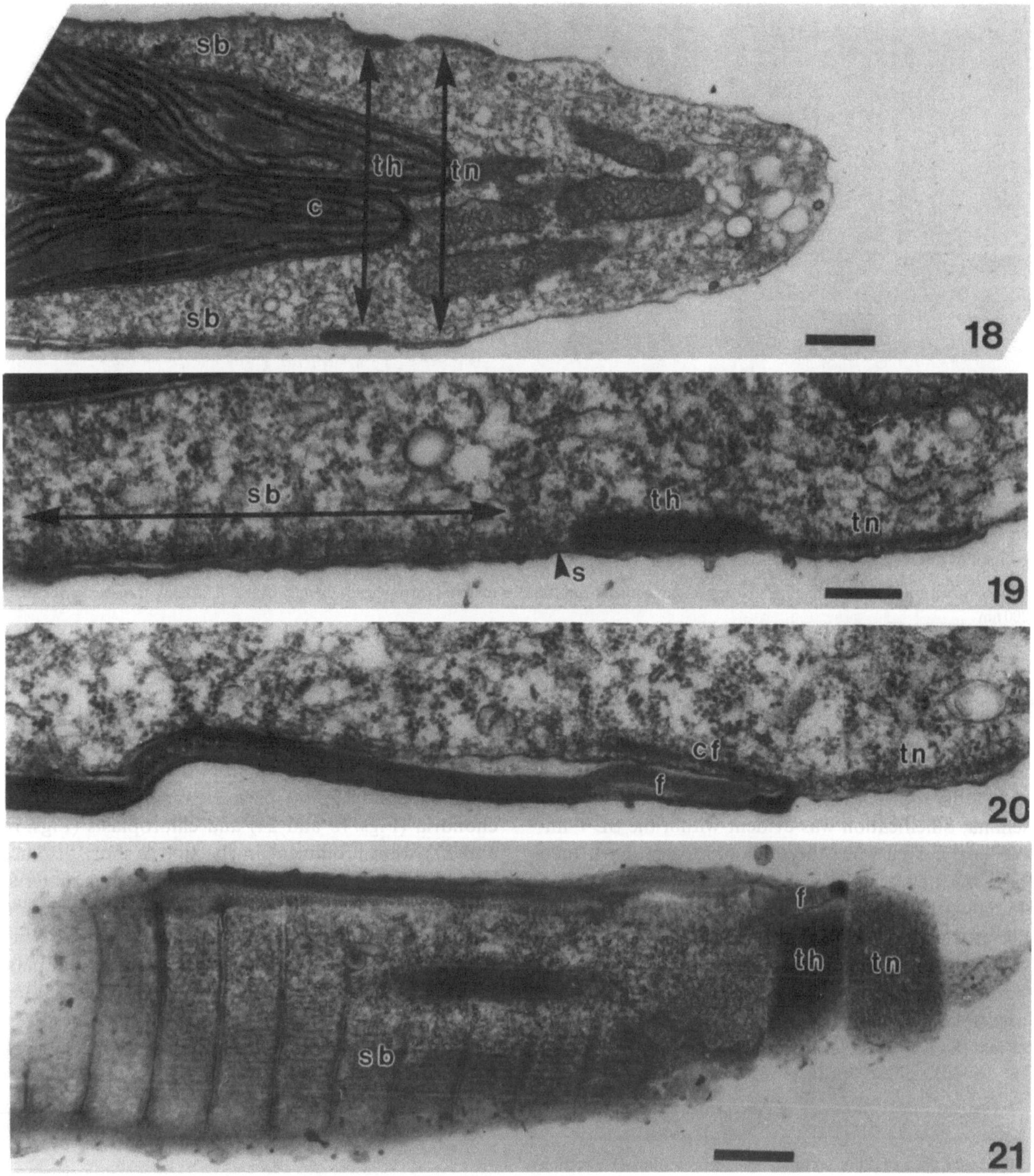

Figs. 18–21. Part of sequence of serial sections, seta in longitudinal section

Fig. 18. Bare tip of seta; then thin band (*tn*), thick band (*th*) and striated band (*sb*). *c* Choroplast. Bar: 500 nm

Fig. 19. Detail of sequence of bands; the outermost edge of the SDV (*s*) indicated by the arrowhead. Bars: 200 nm, for Figs. 19 and 20

Fig. 20. Nearby serial section through one of the finger-like outgrowths (*f*); the thick band is not present because this section goes through the connecting fibre (*cf*), see Figs. 28 and 29

Fig. 21. Section at surface of seta, showing detail of striated band (*sb*). Bar: 500 nm

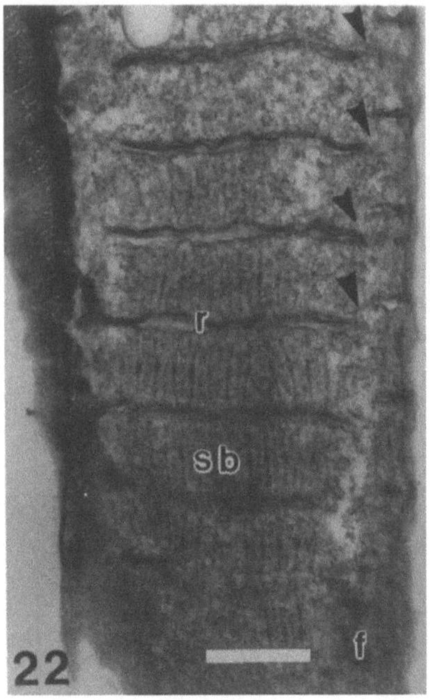

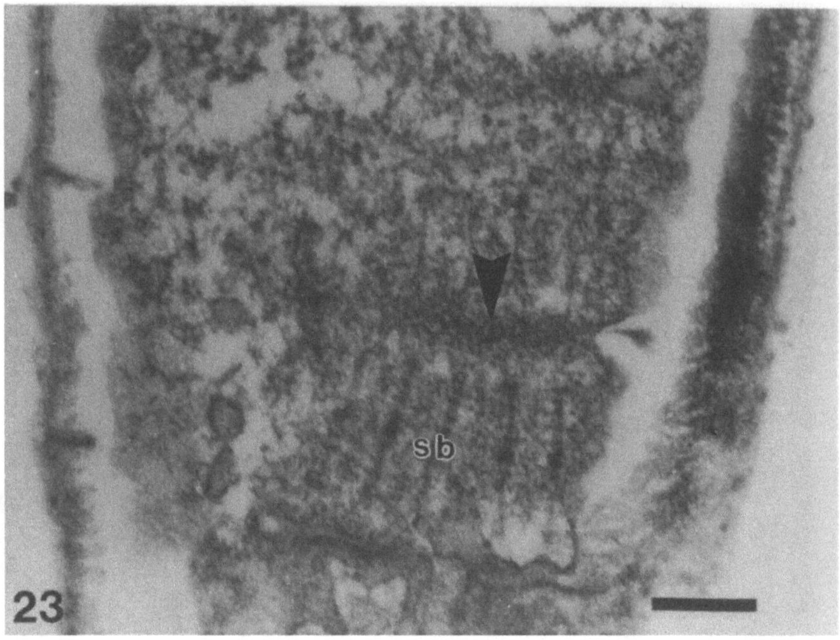

Fig. 22. Fibres of striated band (*sb*) interconnecting forming transverse ribs (*r*) of seta; dislocation across ribs (arrowheads) corresponds to position of finger-like outgrowth (*f*). Bar: 500 nm

Fig. 23. Structure of striated band which appears to connect with fibrous material (arrowhead) directly over the forming transverse ribs. Bar: 200 nm

organelles and small vesicles. The vesicles with dense contents visible in Figs. 16 and 17, may be equivalent to those with a transparent lumen in Fig. 18; their origin and significance is not known. As the cytoplasm grew outwards, the SDV soon extended over the plasmalemma. Silicification was evident within the SDV immediately; even when the setae were very short, their base was appreciably silicified.

Initially, there were no unusual structural features associated with the SDV in the very young setae. However, when the setae had emerged past the separated girdle bands, a remarkable complex and so far, unique assemblage of features was found at their growing tip (Figs. 18–29). These features were present in setae at all subsequent stages of extension and must be involved in their morphogenesis. These features will be described first as seen in longitudinal section.

The cytoplasm at the very tip of the growing seta was enclosed simply by the plasmalemma (Fig. 18); mitochondria (Figs. 18 and 24) and chloroplasts (Figs. 18 and 27) were prominent in the tip. A short distance behind the tip, a thin even layer of fibrous material was applied circumferentially to the plasmalemma. Immediately behind this was a much denser, thick band of fibrous material, again flattened against the plasmalemma and enclosing the cytoplasm (Figs. 18 and 19). The nature of these bands remained to be determined; we call them the "thin band" and the "thick

---

Figs. 24–29. Part of set of serial sections, seta in transverse section

Fig. 24. Thick (*th*) and thin (*tn*) bands near tip; note apparent rigidity of thick band. Bar: 500 nm, for Figs. 24–27

Fig. 25. Further along seta; three of the finger-like outgrowths (*f*) are interconnected by thick band. *tn* Thin band

Fig. 26. Further back still showing developing polygonal profile of seta, becoming silicified around its periphery; elements of striated band (*sb*) visible. *c* Chloroplast

Fig. 27. As for Fig. 26, with a forming "spinule" (see Fig. 3) emerging (top right) from a finger-like outgrowth; below it is the next spinule forming

Figs. 28 and **29.** High magnification of thick bands attached to finger-like outgrowths (*f*) inside SDV, by connecting fibres (*cf*). Bar: 200 nm, for Figs. 28 and 29

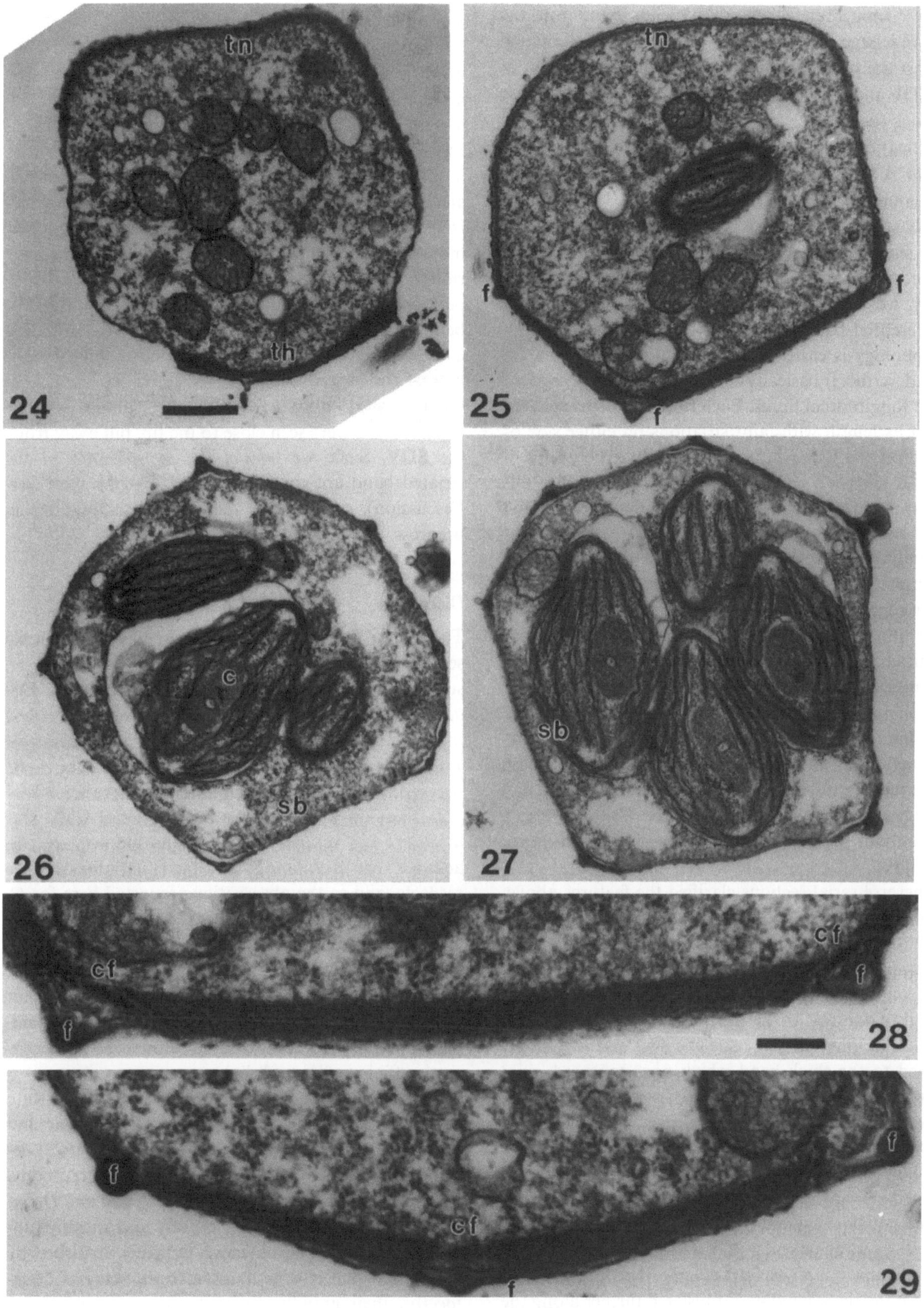
tn
th
24
tn
f
f
f
25
c
sb
26
sb
27
cf
cf
f
f
28
f
f
cf
f
29

band". Immediately behind the thick band was the growing edge of the SDV, very thin and flattened against the plasmalemma (Fig. 19). The association of the SDV and thick band became clearer from transverse sections (below).

Silicification commenced within the SDV very close to the tip. A short distance from the tip where silicification was complete around the periphery of the seta, the evenly-spaced transverse reinforcing ridges (Fig. 4) were starting to form (Fig. 21). In this region, we encountered a third, unique cytoplasmic differentiation which we term the "striated band". Again, this circumferential band was appressed to the SDV and its morphology is easier to see (Figs. 19, 21–23, and 27) than describe. It basically consisted of an evenly spaced set of longitudinal fibres. Each fibre was differentiated along its length with faint striations on either side of a central density; all the fibres were structurally in register (Figs. 22 and 23). Careful examination of adjacent serial sections demonstrated that the striated band was present only along the SDV before and at the initial differentiation of the transverse ridges. Furthermore, a thin layer of slightly dense material on the inner edge of each transverse rib (Fig. 23) appeared to be continuous with a particular region of the sets of longitudinal fibres. The impression created was of a complex fibre whose structure was setting up and establishing the spacing of the transverse ribs of silica.

The lumen of the growing seta contained MTs. Surprisingly, they were not a conspicuous feature at the tip at this slightly later stage of differentiation and may have been relatively poorly preserved. They were scattered about the lumen in no consistent relationship to the SDV.

Transverse serial sections clarified the features above, particularly the relationship of the thick band to the SDV. The SDV was not growing evenly over the plasmalemma. Instead, five or six "finger-like outgrowths" of heavily silicified wall inside the SDV were extending ahead of the rest of the SDV (Fig. 25). Most importantly, the SDV at these specific sites was clearly attached to the thick band which now appeared more accurately as a series of thick bundles of fibres running between and connected to the finger-like outgrowths by short connecting fibres (Figs. 25, 28, and 29). Figure 20 shows one of these finger-like outgrowths; the thin connection fibre in longitudinal section can be compared with the same feature sectioned transversely in Fig. 28. Sections like Fig. 24 confirmed the impression that the thick band was structurally rigid, and the pattern of the thick band was translated further along the seta into its polygonal form (compare Figs. 25–27). Images like Fig. 27 suggest that the finger-like outgrowths also emerge from the surface to form the spinules adorning the surface of the seta, and illustrated in Fig. 3; the spinule emerging at the upper right in Fig. 27 is matched by a less developed spinule below it, as expected since the spinules are regularly spaced and helically arranged. There is no indication of what controls this aspect of morphogenesis. One other minor morphological correlation can be made. The transverse ribs of the seta (Fig. 4) were not continuous around its periphery. This discontinuity seems to arise directly behind some (all?) the finger-like outgrowths (Fig. 22). The striated band is difficult to detect immediately behind the outgrowths (Fig. 20).

Occasionally, fibrous components similar to the striated band were seen close to the tip, but away from the SDV. Since we believe the components of the striated band are continuously recycled forward (see Discussion), such images may represent recycling in progress.

Discussion

The genus *Chaetoceros* is characterised by the fine, long setae extending from the valves. *Chaetoceros* has been sub-divided into two sub-genera: *Phaeoceros* and *Hyalochaete* (Rines and Hargraves 1988), a suggestion first made by Gran in 1897 (cited in Rines and Hargraves 1988). Several morphological distinctions can be made separating these groups. Of immediate relevance, *Phaeoceros* has thick, chloroplast-bearing setae, while *Hyalochaete* has thinner setae, lacking chloroplasts. In addition, the morphology of setae is variable in other respects, and so the observations reported here for *C. peruvianus* may not apply to other species.

The function of setae is not obvious; they probably assist in flotation and would certainly make these cells difficult for predators to ingest. The active movement of chloroplasts along the setae, probably a MT-mediated activity, could also serve to increase photosynthetic efficiency in weak light.

Growth of the setae in this *Chaetoceros* is very delicate and easily disturbed. Earlier papers by Iyengar and Subrahmanyan (1944) and von Stosch et al. (1973) describe their development observed by light microscopy. The spine-like labiate processes in *Ditylum* and *Odontella* (see Introduction) grow rapidly and sturdily (unpubl. obs.) and morphogenesis is rarely disturbed by handling of the cells; in contrast to the setae of *Chaetoceros*, their growth is easy to film. *Chaetoceros* is unusual amongst diatoms in that the cytoplasm of the growing setae is naked and exposed; perhaps this is why they are so sensitive to disturbance. In most diatoms, the new valves are formed while the cytoplast is enclosed and protected within the parent valve.

Tip growth is common in fungi, various algae and green plants. Growth of the setae provides an unusual example of tip growth with a collection of features unlike that found in any other type of cell. The observations reported above show no similarity between extension of the LPs of *Ditylum* and *Odontella* and the setae of *Chaetoceros*. The cytoplasmic system involved in seta morphogenesis is remarkably complex. The only previous report concerning the ultrastructure of seta extrusion is that of Li and Volcani (1985 b). They did not observe MCs in their *Chaetoceros*. Their material suffered more fixation damage than ours and some of their interpretations are, in our opinion, incorrect. In particular, they identified components in the striated band as MTs. They did visualise what we call the thick band but made little comment of the morphogenetic processes at work during seta extension.

This is how we interpret our observations.

Soon after cytokinesis, an MC forms in each daughter cell and immediately it becomes closely apposed to the tiny SDV. This juxtaposition is common in diatoms and related to subsequent morphogenesis of the valve, particularly in raphid pennates (reviewed in Pickett-Heaps et al. 1990). Soon, there is a striking difference between the position of the MCs and the shape of the expanding SDV in the two daughter cells. The MC remains closely applied to the SDV in the upper daughter cell, and its MTs run into the forming setae. However, the MC/MT system in the lower daughter cell moves off the SDV, and by directing its MTs sharply upwards, appears to mould the SDV into the pit between the base of setae that curve downwards. In other species of *Chaetoceros*, the setae extend symmetrically from the cell and we guess that the positioning of the MC/MT complex at this stage of valve morphogenesis might determine the overall orientation of the setae.

MTs from these MCs then appear to set up the position and direction of cytoplasmic outgrowth during initiation of seta morphogenesis since they are numerous and conspicuous at this stage. These MCs are not detectable once setae have started growing. Later, MTs are present in the lumen of the growing tip but our observations do not suggest they play a central, direct role in seta morphogenesis. Perhaps they function in transport along the setae (e.g., generating the chloroplast movement typical of interphase cells). A micrograph by McDonald (published in Pickett-Heaps et al. 1990: fig. 80) shows MTs precisely aligned with the SDV and elements of silica in it, in *Chaetoceros armatum*, but we have not visualised anything equivalent in *C. peruvianus*. During formation of the thicker horns in *Attheya*, MTs are absent (Schnepf et al. 1980); while colchicine inhibits horn formation, this could be an unspecific effect. In contrast, MTs seem to play a central role in morphogenesis of silica spines in *Mallomonas* (Synurophyceae; Mignot and Brugerolle 1982) and the silica strips of choanoflagellates (Leadbeater 1981) and sponges (Garonne et al. 1981) where, however, the SDV remains entirely within the cytoplasm. Rogerson et al. (1986) describe one (?) MT running the length of a seta but they do not illustrate the cytoplasm at the growing tip.

Interpreting our morphological evidence, we suggest that the thick band controls the diameter of the seta by its attachment to the finger-like outgrowths of silica inside the SDV. This explanation also suggests why the setae are not circular in profile as one would expect if they were extended solely under the influence of turgor pressure (see below). If, as we suspect, the finger-like outgrowths extend to form the spinules, then the former probably extend in turn, one after the other. The role of the thin band is not obvious. The striated band may be involved in setting up the spacing of the transverse ribs of silica that reinforce the inner surface or the setae. Specific elements of this band appear to be attached to the faint fibrous layer that surmounts each transverse rib.

The setae grow fast when undisturbed and these three components must move forward as the tip extends. The thin and thick bands could slide continuously ahead of the growing SDV. However, the striated band must behave differently. If it is attached to the forming transverse ribs as appears the case, the rearmost set of fibres must disassemble, be recycled forward and then reassemble when attached to a newly forming rib.

Turgor pressure is used by many plant cells for driving cell expansion. Turgor may function similarly in *C. peruvianus* for extension of the setae. However, these cells sometimes appear faintly plasmolysed during extension – as has also been observed during extension of the spine-like labiate processes of *Ditylum* and *Odontella* (JP-H unpubl. obs.). Even slight increase in internal turgor pressure might be expected to cause ballooning out of the bare cell membrane at the tip of the seta but this has never been observed in actively elongating, live setae (e.g., Figs. 8 and 9). We suspect that extension is actively generated at the tip by the complex of cell organelles described above. Since this complex is structurally attached to the silica tube already deposited just behind the tip (i.e., by the striated band), the whole complex may "crawl" steadily forward on this base – somewhat like a caterpillar tractor moving forward – as the silica tube thickens behind. Just how this complex manages to produce the elegant curvature and the even thinning out of these fine processes is not clear but it seems unimaginable that growth is directly controlled from the distant cell body. Rather, subtle adjustments in the relative positioning of the components from one side to the other could generate curvature.

C. peruvianus is heterovalvate, displaying a relatively unusual ability in diatoms; the daughter cells form valves of different but predetermined morphology, immediately evident in the direction the seta are extruded, and more subtly in the morphology at their base in each new valve. In contrast, most species of *Chaetoceros* display symmetrical setae. In *C. peruvianus*, the altered morphology at the base of the setae is associated with the behaviour of the MC/MT complex which moves off the valve in the upper daughter cell. This behaviour is reminiscent of valve morphogenesis in the monoraphid *Achnanthes coarctata* (Boyle et al. 1984) in which the MC/MTs also move off the partly formed valve in one daughter cell; subsequently, the longitudinal axis of the valve moves laterally and its incipient raphe fills in. As in *C. peruvianus*, this valve ends up having a quite different morphology to the valve produced in the companion daughter cell. Once the morphology of the valves at the base of the setae and the direction of the seta have been established, it is likely that the extrusion machinery functions as it does in other species of *Chaetoceros*, producing the characteristic finely tapered setae.

Acknowledgements

The authors gratefully acknowledge grants from the Australian Research Council (# A18830947, A19030812) supporting this research, and from the Potter Foundation for the equipment used in the video work. We also record our appreciation of the careful review of this manuscript by Drs. A-M. Schmid, S. Cohn, P. Hargraves, and J. Hines who not only provided important background material, but also stimulated further consideration of the morphogenetic events described.

References

Boyle JA, Pickett-Heaps JD, Czarnecki D (1984) Valve morphogenesis in the pennate diatom *Achnanthes coarctata*. J Phycol 20: 563–573

Drebes G (1972) The life history of the centric diatom *Bacteriastrum hyalinum* Lauder. Nova Hedw [Beih] 39: 95–110

Evensen DL, Hasle GR (1975) The morphology of some *Chaetoceros* (Bacillariophyceae) species as seen in electron microscopes. Nova Hedwigia [Beih] 53: 153–184

Fryxell GA (1978) Chain-forming diatoms: three species of Chaetoceraceae. J Phycol 14: 62–71

Garonne R, Simpson TL, Pottu-Boumendil J (1981) Ultrastructure and deposition of silica in sponges. In: Simpson TL, Volcani BE (eds) Sponges and siliceous structures in biological systems. Springer, Berlin Heidelberg New York, pp 495–525

Hendey NI (1964) An introductory account of the smaller algae of British costal waters. Part V. Bacillariophyceae. Fishery Investigations, series IV. O Koeltz, Koenigstein

Iyengar MOP, Subrahmanyan R (1944) On the structure and development of the spines or setae of some centric diatoms. Proc Natl Acad Sci India 14: 114–124

Leadbeater BSC (1981) Ultrastructure and deposition of silica in loricate choanoflagellates. In: Simpson TL, Volcani BE (eds) Sponges and siliceous structures in biological systems. Springer, Berlin Heidelberg New York, pp 295–322

Li C-W, Volcani BE (1985 a) Studies on the biochemistry and fine structure of silica shell formation in diatoms. VIII. Morphogenesis of the cell wall in a centric diatom, *Ditylum brightwellii*. Protoplasma 124: 10–29

– – (1985 b) Studies on the biochemistry and fine structure of silica shell formation in diatoms. IX. Sequential valve formation in a centric diatom *Chaetoceros rostratum*. Protoplasma 124: 30–41

– – (1985 c) Studies on the biochemistry and fine structure of silica shell formation in diatoms. X. Morphogenesis of the labiate process in centric diatoms. Protoplasma 124: 147–156

Mignot JP, Brugerolle G (1982) Scale formation in chrysomonad flagellates. J Ultrastruct Res 81: 13–26

Pickett-Heaps JD, Kowalski SE (1981) Valve morphogenesis and the microtubule-centre of the diatom *Hantzschia amphioxys*. Eur J Cell Biol 25: 150–170

– Wetherbee R, Hill DRA (1988) Cell division and morphogenesis of the labiate process in the centric diatom *Ditylum brightwellii*. Protoplasma 143: 139–149

– Schmid AM, Edgar LA (1990) The cell biology of diatom valve formation. Prog Phycol Res 7: 1–168

Rines JEB, Hargraves PE (1988) The *Chaetoceros* Ehrenberg (Bacillariophyceae) flora of Narragansett Bay, Rhode Island, U.S.A. Bibl Phycol 79: 1–196

Rogerson A, DeFrietas ASW, McInnes AG (1986) Growth rates and ultrastructure of siliceous setae of *Chaetoceros gracilis* (Bacillariophyceae). J Phycol 22: 56–62

Round FE, Crawford RW, Mann DG (1990) The diatoms. Cambridge University Press, Cambridge

Schnepf E, Deichgraber G, Drebes G (1980) Morphogenetic process in *Attheya decora* (Bacillariophyceae, Biddulphiineae). Plant Syst Evol 135: 265–277

von Stosch HA, Theil G, Kowallik KV (1973) Entwicklungsgeschichtliche Untersuchungen an zentrischen Diatomeen. V. Bau und Lebenszyklus von *Chaetoceros didymum*, mit Beobachtungen über einige andere Arten der Gattung. Helgol Wiss Meersunters 25: 384–445

Protoplasma (1994) 181: 283–290

PROTOPLASMA

Protein kinases in protists

M. J. Fazio, J. A. Marrs, and **G. B. Bouck***

Department of Biological Sciences, University of Illinois at Chicago, Chicago, Illinois

Received August 5, 1993
Accepted November 11, 1993

Summary. The vast preponderance of our understanding of protein kinases comes from studies of mammalian or of other higher eukaryotic systems. A survey of the Wilson reference databank yielded 3,807 citations for protein kinases; only nine of these were reports of protein kinases in protists. It is apparent, nonetheless, that this understudied group offers unique opportunities for resolving the mechanisms by which protein kinases mediate a variety of cellular processes. Moreover, generalities about cofactor requirements (e.g., Ca^{2+} alone activates many protist protein kinases), substrate specificity, and the nature of the enzymes themselves (monomeric versus dimeric cyclic-nucleotide dependent protein kinases) will certainly need to be modified.

Keywords: Protein kinases; Protists.

Introduction

Seven years ago Hunter postulated that approximately 1% of all of the different varieties of proteins in mammalian cells would ultimately be found to have protein kinase activity (Hunter 1987). Subsequent findings indicate that this prediction may have been conservative, as protein phosphorylation and dephosphorylation are increasingly documented as the modulators for a wide variety of cellular reactions. Among the most thoroughly studied phosphorylation reactions are those in which external stimuli are perceived at the cell surface and then translated intracellularly into some physiological response. Specific integral proteins in the plasma membrane (receptor protein tyrosine kinases) may be the initial components in a cascade or network of reactions that can mediate growth and cell division (reviewed in Cadena and Gill 1992, Fantl et al. 1993). Loss of this control can lead to unregulated growth as evidenced by the effects of oncogenes which may encode modified versions of the native receptor (Bishop 1987). Normally, as a result of ligand binding to the receptor protein kinase additional protein kinases within the cell are activated or inactivated, thereby propagating the signal. The targets of these cytoplasmic kinases in eukaryotes are generally serine/threonine or tyrosine residues of substrate proteins.

With the exception of certain fungi which have provided powerful systems for studying with genetics the control of protein kinases and phosphatases during cell division (reviewed in Norbury and Nurse 1992), progress in understanding the role of protein kinases in growth and other developmental processes in lower eukaryotes has lagged significantly behind similar studies in other eukaryotes. Among the protists the role and function of protein kinases have received relatively little attention. Nonetheless there is sufficient evidence now available to suggest that protein kinases do provide subtle modulation of cell behavior in ways perhaps equivalent to those found in metazoans. It is the purpose of this review to highlight several different areas in which preliminary findings indicate that the study of protein kinases in protists might contribute significantly to understanding important physiological or developmental events. These include: (1) light regulation of motility, (2) ectoprotein kinases and environmental interactions, (3) the regulation of the cytoskeleton, and (4) the control of membrane fusion during exocytosis.

General properties of protein kinases

As Cozzone (1988) has pointed out, "Research in protein phosphorylation is rather unconventional. The

* Correspondence and reprints: Department of Biological Sciences, University of Illinois at Chicago, Chicago, IL 60607–7060, U.S.A.

problem is most often encountered in the middle, by first targeting a phosphorylated protein in a gel slice or in a chromatography column eluate. Then one proceeds both upstream to characterize the enzyme(s) and downstream to determine the functional consequences of its modification." In many cases work in protists has progressed much in this manner.

The general reaction mediated by a protein kinase can be summarized as follows:

donor + cofactors + substrate + protein kinase ⇒ P-substrate

Donor

The usual donor, ATP, is experimentally administered as [γ-^{32}P]ATP from which the γ-labelled phosphate is transferred and covalently bonded to the substrate protein. γ-Labelled GTP can also be used as a phosphate donor by some protein kinases such as the type II casein kinases (Pinna et al. 1990). [α-^{32}P]ATP has been used as a control for estimating non-enzymatic phosphate transfer, since protein kinases do not catalyze the transfer of the α-phosphate to proteins.

Cofactors

Mg^{2+} or Mn^{2+} is often required for optimal activity of protein kinases (Gealen and Harrison 1990). Other cofactors include Ca^{2+}, Ca^{2+} with calmodulin, cAMP, or cGMP. Protein kinases are often defined by their cofactor requirements, as for example cAMP-dependent protein kinases etc.

Substrate

Most protein kinases are not specific to one protein, and many can be broadly classified by more or less generic substrates. Examples of the latter are the casein kinases and histone kinases for which explicit intracellular functions (where casein is not a natural constituent) are often not well defined. There are two large subclasses of casein kinases, type I and type II which can be separated from one another by phosphocellulose chromatography. Both types phosphorylate casein but each responds differently to inhibitors, and each has different donor specificity (reviewed in Pinna et al. 1990, Tuazon and Traugh 1991). As with the casein kinases substrate identity provides nomenclature for a variety of more specific protein kinases such as the myosin light chain kinase (MLCK). Protein kinases may also be defined by the physiological events they mediate, e.g., the mitogen-activated protein kinases (MAPs) (Pelech and Sanghera 1992, Thomas, 1992). Hanks et al. (1988) divide the protein kinases into two large groups of families based on whether the amino acid tyrosine or serine/threonine is phosphorylated. With the recent discovery of protein kinases that phosphorylate both tyrosine and serine (Featherstone and Russell 1991) this distinction may become less definitive, especially if more such cases are documented.

Autophosphorylation

Many protein kinases, such as the receptor tyrosine kinases that span the plasma membrane, can phosphorylate themselves. In these transmembrane kinases phosphorylation of their cytoplasmic domain is stimulated when the extracellular domain interacts with a specific ligand (reviewed in Fontyl et al. 1993). This tyrosine autophosphorylation provides a signal or target for initiating the next step in the cascade of cellular reactions. Classes of soluble proteins that can recognize and bind to these phosphorylated residues often harbor a particular motif such as an SH2 domain (reviewed in Pawson and Gish 1992). The newly bound proteins are then activated or inactivated, thereby propagating the cascade.

Protein kinases

The catalytic domain of different types of protein kinases is fairly well conserved (Hanks et al. 1988, Taylor et al. 1992) – a property that has facilitated the isolation and identification of genes encoding protein kinases from genomic and cDNA libraries as predicted (Hunter 1987). The three dimensional structure of the catalytic subunit of a cAMP-dependent protein kinase has been determined (Knighton et al. 1991) and is believed to represent a general model for other protein kinases (Taylor et al. 1992).

P-substrate

The phosphorylated protein is the hallmark of protein kinase activity providing suitable controls are exercised. The isolation of protein kinases by biochemical methods frequently relies on the addition of exogenous substrate (e.g., casein, histones) to document stages of kinase purification. A useful method for detecting individual protein kinases in a mixture of proteins is to renaturate proteins after their separation in acrylamide gels and then add cofactors and radiolabelled donors (Kameshita and Fujisawa 1989). Including casein or histones as a substrate in the gel matrix often results in the in situ phosphorylation of the substrate at a specific region (band) in the gel that contains the renatured protein kinase. Moreover, autophosphorylation of a particular renatured protein can readily be distinguished from phosphorylation of the exogenous substrate by comparing the kinase activity in the gel with added substrate to the activity found in gels without added substrate. Polypeptides can also be transferred to PVDF membranes and assayed for autophosphorylation (e.g., Ferrell and Martin 1989).

Protein kinases in protists

In addition to the extensive documentation of protein kinases in higher eukaryotes, protein kinase activity has also been found in the prokaryotes (reviewed in Cozzone 1988), including cyanobacteria (Atkinson et al. 1992). Not surprising there are reports of protein kinase activity in protists in which serine and threonine are the usual phosphorylated residues in substrate proteins. But tyrosine kinases are also present – a finding that was not anticipated in some earlier evaluations, since tyrosine kinases were thought to mediate signal transduction and development in multicellular organisms (Hanks et al. 1988). It seems probable that the relative paucity of reports thus far of protein kinase activity in protists reflects more the limited efforts to identify such enzymes, rather than a lack of enzyme activity in these cells. Even with the evidence now available, it is apparent that novel approaches and important findings will be forthcoming from further studies of kinase activity in members of this group. Several particularly promising areas are considered in detail below.

Light response, motility and protein kinases

In the classical studies of Luck and colleagues using the green unicell, *Chlamydomonas* 250–300 polypeptides were identified in 2-D gels of solubilized flagella after ^{35}S-metabolic labelling of whole cells (reviewed in Luck 1985). The 2-D patterns of polypeptides from wild type cells were then compared with the flagellar polypeptides from mutants that were unable to produce normal flagellar movements; the mutants were also examined by electron microscopy to identify structural lesions. Thus a number of specific polypeptides could be correlated with specific flagellar components involved in motility. In additional studies (Piperno et al. 1981), ^{32}P was also administered to whole cells, the flagellar polypeptides were isolated and 2-D maps prepared. Over 80 polypeptides incorporated phosphate; five of these phosphoproteins were components of the radial spoke stalk which links the central sheath with the peripheral doublet microtubules of the axoneme, and is required for normal ATP-induced flagellar movements. Although the effects of phosphorylation on radial spoke function were not directly determined and protein kinase activity was only inferred, these experiments provided the foundation for further seminal findings using *mbo* mutants (Segal et al. 1984). *Mbo* cells moved backwards only, in a manner similar to the flagellar-like movement induced by an increase in light intensity (photophobic reaction). The mutants no longer responded to light and appeared to lack six polypeptides (b1–b5, 2′) found in wild type cells; 5 of these polypeptides were phosphorylated in vivo. Structurally, beak-like projections on subfiber B of doublet microtubules 5 and 6 of the flagellar axoneme were also absent. Subsequent experiments using isolated flagella and [γ-^{32}P]ATP revealed a Ca^{2+}-dependent protein kinase activity that phosphorylated two proteins, one of which was b4 (Segal and Luck 1985). Since both the photophobic reaction and the protein kinase require calcium ions in the surrounding medium, the broadest interpretation of these experiments is that b4 phosphorylation represents a step in the photophobic response in *Chlamydomonas*.

Clearly, experiments of this kinds incorporating genetics (mutants), biochemistry, physiology, organism behavior and mechanics (structural) would be difficult to carry out in multicellular organisms. The specific kinase associated with b4 phosphorylation in *Chlamydomonas* has yet to be identified but Ca^{2+}-dependent protein kinases have been directly isolated from soluble extracts of several other green algal protists including the filamentous form *Mougeotia* (Roberts 1989, who also reported activity in extracts from *Mesotaenium* and *Spirogyra*) using a synthetic substrate containing the MLCK R subunit phosphorylation sequence, and from *Dunaliella* (Guo and Roux 1990) using histone as a substrate. Evidence for the presence of a myosin light chain kinase (Ca^{2+}-dependent) has recently been documented in *Chara* (McCurdy and Harmon 1992); the authors postulate a possible function for this kinase in cytoplasmic streaming. The phosphorylated amino acid of *Chlamydomonas* (Segal and Luck 1985) is predominately serine ($>$ 93% in isolated flagella), as is the case in *Mougeotia* (Roberts 1989). The function of the calcium dependent kinases purified thus far from green algal protists is not known; it is of interest, however, that as in a number of other protists, none seem to require the presence of calmodulin for Ca^{2+} activation (see below). A recent study using an antiphosphoprotein monoclonal antibody suggested that a 34 kDa (nuclear localization) protein and a 90 kDa (basal body localization) display increased phosphorylation during flagellar regeneration in *Chlamydomonas* (Harper et al. 1993). The kinase inhibitor staurosporine "impaired" regeneration which would be consistent with phosphorylation as a requirement for normal flagellar regrowth.

Parasitic protists and ecto-protein kinases

Protein kinases in the parasitic protozoa such as the trypanosomes, plasmodiums and leishmanias are widely distributed and more thoroughly documented than in other protists (reviewed in Lester et al. 1990). In *Trypanosoma brucei* the pattern of protein phosphorylation changed during different developmental stages. Some of the kinase activity had similarities to casein type II protein kinases although the trypanosome enzyme was not stimulated by polyamines such as spermidine and putrescine (Aboagye-Kwarteng et al. 1991). The phosphoamino acids in casein are presumably threonine/serine, as these are the amino acids generally phosphorylated in casein (Hathaway and Traugh 1982). Tyrosine phosphorylation has also been reported for *T. brucei* (Wheeler and Shapiro 1992) as evidenced by the sensitivity of [γ-^{32}P]ATP incorporation to the tyrosine kinase inhibitor genistein, resistance of the phosphoamino acid to hydrolysis by alkali and the capacity to phosphorylate synthetic poly(Glu^{80}Tyr^{20}). Both protein kinase C and tyrosine kinase revealed distinct patterns of activity during the life cycle of *T. brucei* (Parsons et al. 1990, Keith et al. 1990). Parsons et al. (1993) recently identified at least 10 protein threonine kinases using renaturation assays, and further demonstrated that six of these are regulated during development in *T. brucei.* They also presented additional evidence for the pesence of tyrosine kinases. An epidermal growth factor receptor homologue has also been recorded in trypanosomes (Hide et al. 1989), although the ligand EGF did not appear to have any stimulatory effect on tyrosine kinase activity (Wheeler-Alm and Shapiro 1992). The sum of these findings suggests that as in metazoans, protein kinases play pivotal roles in the regulation of proliferation and differentiation in unicellular protozoa (Parsons et al. 1993).

When *intact* cells of *Leishmania donovani* were incubated with [γ-^{32}P]ATP, 10 different, presumably surface, proteins (externally exposed or ecto-proteins) were radiolabelled (Das et al. 1986). Exogenously added mixed histones, but not H_1, were also phosphorylated and served as an artificial substrate for partial purification of the membrane associated, histone-dependent kinase. Serine was the predominant phosphoamino acid in the labelled histone, and the partially purified kinase activity required only 50 mM NaCl, 10 mM Mg^{2+}, and histones for maximum in vitro activity. In similar assays for ecto-kinase activity intact cells of *L. major* (Lester et al. 1990) incorporated [γ-^{32}P]ATP but not [α-^{32}P]ATP, indicating that phosphorylation was indeed the result of protein kinase activity and not due to degradation of ATP and uptake of labelled phosphate. The ecto-kinase was not purified in these studies, but a number of inhibitors of protein kinases had no effect on in vivo activity. Moreover, three of the endogenously labelled phosphoproteins were released into the medium. Although not unique to the leishmanias (see discussion in Lester et al. 1990), a possible function for ectoprotein kinases in thwarting the defense mechanisms of host cells was suggested from further studies (Hermoso et al. 1991). A membrane-bound protein kinase (LPK-1) was isolated from *L. major* and incubated with components C3, C5, and C9 of the human complement system. The purified kinase phosphorylated the α-chain of C3 and C5, but not C3 b which is a proteolytic cleavage product of C3. Intact cells could, however, phosphorylate both C3 and C3 b suggesting the presence in vivo of more than one ecto-protein kinase. Phosphorylation changes the kinetics of C3 processing which presumably shortcircuits the complement cascade that would normally result in host-induced lysis of the parasite.

Protein kinases active at the cell surface can thus significantly influence host/parasite interactions through modulation of the immune system (complement), or interact with the surrounding environment in yet unknown ways (possibly through the phosphorylation of endogenous surface proteins as reported in *Leishmania*; Lester et al. 1990). In the bacterium *Yersinia pseudotuberculosis,* virulence appears to result from the expression of an endogenous plasmid which secretes a protein that has extensive homology to a eukaryotic serine/threonine protein kinase (Galyov et al. 1993). Further searches for ecto-protein kinases should certainly help clarify not only host-parasite interactions, but perhaps may also reveal how protein kinases mediate other interactions between cells and their environment.

Protein kinases and the protist cytoskeleton

The cytoskeleton and membrane skeleton which generate and maintain the form of many cells consist of a variety of proteins that generally associate non-covalently to assemble flexible meshworks. Because these meshworks are often dynamic and transient, phosphorylation and dephosphorylation would appear to be effective mechanisms for rapidly stabilizing or destabilizing various protein/protein interactions. In the spectrin-based erythrocyte membrane skeleton all the major proteins undergo phosphorylation in vitro (for review, see Mische and Morrow 1988), and cytoskeletal proteins are clearly among the early target sites of protein kinases that are activated during shape changes induced by oncogenic tyrosine kinases (Kellie et al. 1991). In metazoans the peripheral nuclear matrix consists of a meshwork of lamins that disassociates when phosphorylated and reassembles when dephosphorylated (reviewed in Hennekes et al. 1993).

Tyrosine phosphorylation of actin has been correlated with shape changes in the slime mold *Dictyostellium* (Howard et al. 1993). In these experiments starved cells were transferred to growth medium which resulted in profound changes in cell shape and a concomitant increase in vivo of tyrosine phosphorylation of a minor isoform of actin. Actin is known to be a determinant of cell shape in these cells (Shariff and Luna 1992). Overexpression of a tyrosine phosphatase resulted in reduced phosphorylation of actin with diminished cell shape changes. Conversely, cells deficient in this tyrosine phosphatase showed an increase in actin phosphorylation with more rapid and prolonged cell rounding (Howard et al. 1993).

In the unicellular protists *Euglena* the cytoskeleton underlies the plasma membrane. It is therefore analogous to the erythrocyte membrane skeleton, but in *Euglena* the membrane skeleton is constructed of a number of novel proteins. Sonication of whole *Euglena* followed by separation in sucrose gradients yields surface isolates which are enriched in the intact membrane skeleton and free from the bulk of the cell cytosol. The major proteins in these surface isolates are tubulins and the two related articulins (α- and β-articulin) all localized to the submembrane regions (Dubreuil and Bouck 1985). The articulins show no significant sequence similarity to any previously recorded sequences (Marrs and Bouck 1992). The articulins are non-covalently bound to the plasma membrane via a 39 kDa integral membrane protein (IP 39; Rosiere et al. 1991). Incubation of these surface isolates with [γ-^{32}P]ATP and 5 mM Mg^{2+} produced a large number of radiolabelled polypeptides (Fig. 1). When the polypeptides of the surface isolates were separated in acrylamide gels, renatured and incubated with [γ-^{32}P]ATP, it was evident that many of the proteins were autophosphorylated. If casein was included in the gel matrix, additional protein kinase activity was evident (Fazio and Bouck unpubl.). Treatment of surface isolates with urea and NaOH produced protein subfractions that phosphorylated casein in vitro. The significance of the abundant protein kinase activity in surface isolates of *Euglena* is not yet clear. Interestingly, however, IP 39 was phosphorylated in situ on both threonine and tyrosine residues (Fazio and Bouck unpubl.) in domains known to bind the skeletal proteins (articulins; Rosiere et al. 1990). Thus it seems likely, although unproven, that protein kinases are involved in cytoskeletal assembly or stabilization in this protist. Also worth testing is the possibility that phosphorylation may play some role in the shape changes known to be accommodated or powered by this region of the cell (Suzuki and Williamson 1985). The two cAMP-dependent protein kinases that have been partially purified from whole cell extracts of *Euglena* are the only other protein kinases identified from this organism (Carre and Edmunds 1992). These kinases may be of particular interest as potential activators of a variety of cellular activities that display circadian rhythmicities, since cAMP itself exhibits circadian variations (Carre and Edmunds 1992).

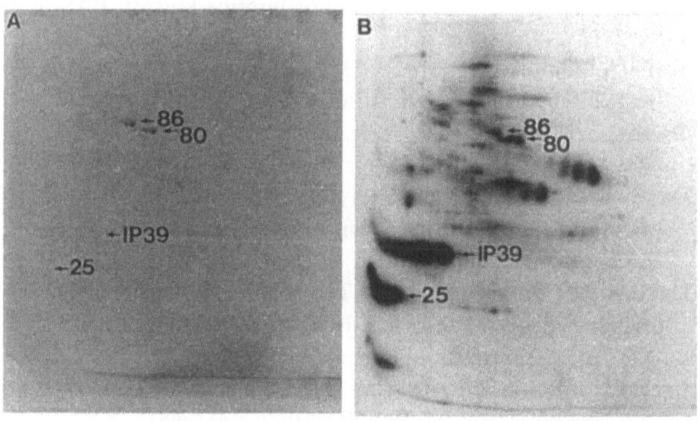

Fig. 1 A, B. Two dimensional gel analysis of proteins phosphorylated by endogenous surface associated protein kinases in *Euglena gracilis*. Surface isolates were incubated with [γ-^{32}P]ATP, rinsed to remove unincorporated label, then separated by NEpHGE in the first dimension and SDS-PAGE in the second dimension. **A** The Commassie blue stained gel. **B** Autoradiograph of **A**. The 80 and 86 kDa articulins (*80* and *86*), the 39 kDa integral membrane protein (*IP 39*) and a 25 kDa membrane skeletal protein (*25*) are indicated

Membrane fusion and protein kinases in Paramecium

Trichocyst discharge in *Paramecium* is a rapid Ca^{2+}-dependent event that can be synchronously induced with a variety of 'secretagogues' such as trinitophenols and aminodextrans (for review, see Satir 1989). One of the requirements for discharge is membrane fusion between the trichocyst vesicle membrane and the plasma membrane. A 63 kDa phosphoprotein termed parafusin appears to be involved in this process. Parafusin is dephosphorylated (activated?) and bound to the vesicle membrane coincident with vesicle fusion, whereas phosphorylated parafusin (inactive) is always found in the supernatant (vesicle-free) fraction. Phosphorylation apparently can be accomplished by two mechanisms: glucose-1-phosphate addition by the action of a phosphotransferase (Subramanian and Satir 1992) and serine phosphorylation by a protein kinase (Murtaugh et al. 1987). Subramanian and Satir (1992) have recently proposed that dephosphoglucosylation may be the physiologically critical event during exocytosis, so the role of the protein kinase activity in this process remains uncertain. The properties of the parafusin associated protein kinase are similar to two Ca^{2+}-dependent protein kinases isolated from soluble extracts of *Paramecium* (Gunderson and Nelson 1987, Son et al. 1993). Threonine/serine of casein was a substrate for both enzymes in vitro, the enzymes autophosphorylated in gel renaturation assays and $^{45}Ca^{2+}$ bound directly to both enzymes in situ. Both enzymes were unusual in that Ca^{2+} alone was sufficient for activation, and neither calmodulin nor lipids had significant effects on enzyme activity. Miglietta and Nelson (1988) reported another protein kinase enriched in cilia of *Paramecium* that was cGMP-dependent. Unlike the dimeric cGMP-dependent protein kinases found elsewhere, the *Paramecium* enzyme appeared to function as a monomer. The authors suggest that this monomeric enzyme may have been an evolutionary precursor to the dimeric enzyme, and that the study of cGMP-dependent protein kinases in these primitive eukaryotes may clarify how the dimeric forms evolved. Monomeric cAMP-dependent protein kinases in *Dictyostelium* lack the amino-terminal domain found in higher eukaryotes which appears to be required for the subunit dimerization (Mutzel et al. 1987). Monomeric cAMP-dependent protein kinases have also been isolated from the marine diatom *Cylindrotheca* (Lin and Volcani 1989).

Outlook

It will be of particular interest in the future to see if kinase modulated signal transduction follows the same pathways as those found in metazoans. For example, are there proteins with SH2 domains? If so, is tryosine phosphorylation of receptors or other proteins a signal for binding by proteins with SH 2 domains? Are there transmembrane receptors that have ligand-induced protein kinase activity? Evidence from studies of the light avoidance response suggests that some responses to environmental cues will involve cascades of phosphorylation and dephosphorylation. Newly discovered methods of posttranslational modification with phosphorylated sugar residues remain to be further explored. The unexpectedly large number of kinase proteins phosphorylated in situ in flagellar axonemes of *Chlamydomonas* and surface isolates of *Euglena* indicates that resolution of the function and properties of protein kinases in protists will be an extraordinarily fertile field in future studies.

Acknowledgements

Some of the work reported in this paper was supported by NSF grant DCB 9105226 to GBB for which the authors are grateful.

References

Aboagye-Kwarteng T, Ole-Moiyoi OK, Lonsdale-Eccles JD (1991) Phosphorylation differences among proteins of bloodstream developmental stages of *Trypanosoma brucei brucei*. Biochem J 275: 7–14

Atkinson M, Allen C, Sequeira L (1992) Tyrosine phosphorylation of a membrane protein from *Pseudomonas solanacearum*. J Bacteriol 174: 4356–4360

Bishop JM (1987) The molecular genetics of cancer. Science 235: 305–311

Cadena DL, Gill GN (1992) Receptor tyrosine kinases. FASEB J 6: 2332–2337

Carré IA, Edmunds LN (1992) cAMP-dependent kinases in the algal flagellate *Euglena gracilis*. J Biol Chem 267: 2135–2137

Cozzone AJ (1988) Protein phosphorylation in prokaryotes. Annu Rev Microbiol 42: 97–125

Das S, Saha AK, Mukhopadhyay NK, Glew RH (1986) A cyclic nucleotide-independent protein kinase in *Leishmania donovani*. Biochem J 240: 641–649

Dubreuil RR, Bouck GB (1985) The membrane skeleton of a unicellular organism consists of bridged, articulating strips. J Cell Biol 101: 1884–1896

Fantl WJ, Johnson DE, Williams LT (1993) Signalling by receptor tyrosine kinases. Annu Rev Biochem 62: 453–481

Featherstone C, Russell P (1991) Fission yeast $p107^{wee1}$ mitotic inhibitor is a tyrosine/serine kinase. Nature 349: 808–811

Ferrell JE, Martin GS (1989) Thrombin stimulates the activities of multiple previously unidentified protein kinases in platelets. J Biol Chem 264: 20723–20729

Galyov EE, Hakansson S, Forsberg A, Wolf-Watz H (1993) A secreted protein kinase of *Yersinia pseudotuberculosis* is an indispensible virulence determinant. Nature 361: 730–732

Geahlen RL, Harrison ML (1990) Protein-tyrosine kinases. In: Kemp BE (ed) Peptides and protein phosphorylation. CRC Press, Boca Raton, pp 239–253

Gundersen RE, Nelson DL (1987) A novel Ca^{2+}-dependent protein kinase from *Paramecium tetraurelia*. J Biol Chem 262: 4602–4609

Guo Y-L, Roux SJ (1990) Partial purification and characterization of a Ca^{2+}-dependent protein kinase from the green alga, *Dunaliella salina*. Plant Physiol 94: 143–150

Hanks SK, Quinn AM, Hunter T (1988) The protein kinase family: conserved features and deduced phylogeny of the catalytic domains. Science 241: 42–52

Harper JDI, Sanders MA, Salisbury JL (1993) Phosphorylation of nuclear and flagellar basal apparatus proteins during flagellar regeneration in *Chlamydomonas reinhardtii*. J Cell Biol 122: 877–886

Hathaway GM, Traugh JA (1982) Casein kinases – multipotential protein kinases. Curr Top Cell Regul 21: 101–127

Hennekes H, Matthias P, Weber K, Nigg EA (1993) Phosphorylation on protein kinase C sites inhibits nuclear import of lamin B_2. J Cell Biol 120: 1293–1304

Hermoso T, Fishelson Z, Becker SI, Hirschberg K, Jaffe CL (1991) Leishmanial protein kinases phosphorylate components of the complement system. EMBO J 10: 4061–4067

Hide G, Gray A, Harrison CM, Tait A (1989) Identification of an epidermal growth factor homologue in trypanosomes. Mol Biochem Parasitol 36: 51–60

Howard PK, Sefton BM, Firtel RA (1993) Tyrosine phosphorylation of actin in *Dictyostelium* associated with cell shape changes. Science 259: 241–244

Hunter T (1987) A thousand and one protein kinases. Cell 50: 823–829

Kameshita I, Fujisawa H (1989) A sensitive method for detection of calmodulin-dependent protein kinase II activity in sodium dodecyl sulfate-polyacrylamide gel. Anal Biochem 183: 139–143

Keith K, Hide G, Tait A (1990) Characterization of protein kinase activities in *Trypanosoma brucei*. Mol Biochem Parasitol 43: 107–116

Kellie S, Horvarth AR, Elmore MA (1991) Cytoskeletal targets for oncogenic tyrosine kinases. J Cell Sci 99: 207–211

Knighton DR, Zheng J, Ten Eyck LF, Ashford VA, Xuong N-H et al (1991) Crystal structure of the catalytic subunit of cAMP-dependent protein kinase. Science 253: 407–414

Lester DS, Hermoso T, Jaffe CL (1990) Extracellular phosphorylation in the parasite, *Leishmania major*. Biochim Biophys Acta 1052: 293–298

Lin PP-C, Volcani BE (1989) Novel adenosine 3′,5′-cyclic monophosphate dependent protein kinases in a marine diatom. Biochemistry 28: 6624–6631

Luck DJL (1984) Genetic and biochemical dissection of the eukaryotic flagellum. J Cell Biol 98: 789–794

Luna EJ, Hitt AL (1992) Cytoskeleton-plasma membrane interactions. Science 258: 955–964

McCurdy DW, Harmon AC (1992) Phosphorylation of a putative myosin light chain in *Chara* By calcium-dependent protein kinase. Protoplasma 171: 85–88

Mann NH, Rippka R, Herdman M (1991) Regulation of protein phosphorylation in the cyanobacterium *Anabaena* strain PCC 7120. J Gen Microbiol 137: 331–339

Marrs JA, Bouck GB (1992) The two major membrane skeletal proteins (articulins) of *Euglena gracilis* define a novel class of cytoskeletal proteins. J Cell Biol 118: 1465–1475

Miglietta LAP, Nelson DL (1988) A novel cGMP-dependent protein kinase from *Paramecium*. J Biol Chem 263: 16096–16105

Mische SM, Morrow JS (1988) Post-translational regulation of the erythrocyte cortical cytoskeleton. Protoplasma 145: 167–175

Murtaugh TJ, Gilligan DM, Satir BH (1987) Purification of and production of an antibody against a 63,000 M_r stimulus-sensitive phosphoprotein in *Paramecium*. J Biol Chem 262: 15734–15739

Mutzel R, Lacombe ML, Simon M-N, deGunsburg J, Vernon M (1987) Cloning and cDNA sequence of the regulatory subunit of cAMP-dependent protein kinase from *Dictyostelium discoideum*. Proc Natl Acad Sci USA 84: 6–10

Norbury C, Nurse P (1992) Animal cell cycles and their control. Annu Rev Biochem 61: 441–47

Parsons M, Valentine M, Deans J, Schieven GL, Ledbetter GL (1990) Distinct patterns of tyrosine phosphorylation during the life cycle of *Trypanosoma brucei*. Mol Biochem Parasitol 45: 241–248

– – Carter V (1993) Protein kinases in divergent eukaryotes: identification of protein kinase activities regulated during trypanosome development. Proc Natl Acad Sci USA 90: 2656–2660

Pawson T, Gish GD (1992) SH2 and SH3 domains: from structure to function. Cell 71: 359–362

Pelech SL, Sanghera JS (1992) Mitogen-activated protein kinases: versatile transducers for cell signaling. Trends Biochem Sci 17: 233–238

Pinna LA, Meggio F, Marchiori F (1990) Type-2 casein kinases: general properties and substrate specificity. In: Kemp BE (ed) Peptides and protein phosphorylation. CRC Press, Boca Raton, pp 145–169

Piperno G, Huang B, Ramanis Z, Luck DJL (1981) Radial spokes of *Chlamydomonas* flagella: polypeptide composition and phosphorylation of stalk components. J Cell Biol 88: 73–79

Roberts DM (1989) Detection of a calcium-activated protein kinase in *Mougeotia* by using synthetic peptide substrates. Plant Physiol 91: 1613–1619

Rosiere TK, Marrs JA, Bouck GB (1990) A 39-kDa plasma membrane protein (IP39) is an anchor for the unusual membrane skeleton of *Euglena gracilis*. J Cell Biol 110: 1077–1088

Satir BH (1989) Signal transduction events associated with exocytosis in ciliates. J Protozool 36: 382–389

Segal RA, Luck DJL (1985) Phosphorylation in isolated *Chlamydomonas* axonemes: a phosphoprotein may mediate the Ca^{2+}-dependent photophobic response. J Cell Biol 101: 1702–1712

– Huang B, Ramanis Z, Luck DJL (1984) Mutant strains of *Chlamydomonas reinhardtii* that move backwards only. J Cell Biol 98: 2026–2034

Son M, Gundersen RE, Nelson DL (1993) A second member of the novel Ca^{2+}-dependent protein kinase family from *Paramecium tetraurelia*. J Biol Chem 268: 5940–5948

Subramanian SV, Satir BH (1992) Carbohydrate cycling in signal transduction: parafusin, a phosphoglycoprotein and possible Ca^{2+}-dependent transducer molecule in exocytosis in *Paramecium*. Proc Natl Acad Sci USA 89: 11297–11301

Suzuki T, Williamson RE (1985) Euglenoid movement in *Euglena fusca*: evidence for sliding between pellicular strips. Protoplasma 124: 137–146

Thomas G (1992) MAP kinase by any other name smells just as sweet. Cell 68: 3–6

Taylor SS, Knighton DR, Zheng J, Ten Eyck LF, Sowadski JM (1992) Structural framework for the protein kinase family. Annu Rev Cell Biol 8: 429–462

Tuazon PT, Traugh JA (1991) Casein kinase I and II – multipotential serine protein kinases: structure, function and regulation. Adv Second Messenger Phosphoprotein Res 23: 123–164

Wheeler-Alm E, Shapiro SZ (1992) Evidence of tyrosine kinase activity in the protozoan parasite *Trypanosoma brucei*. J Protozool 39: 413–416

Verleger: Springer-Verlag KG, Sachsenplatz 4-6, A-1201 Wien. – Herausgeber: Prof. D. J. Morré, Arthur G. Hansen Life Sciences Research Building, Purdue University, West Lafayette, IN 47907, U.S.A. – Redaktion: Sachsenplatz 4-6, A-1201 Wien. – Hersteller: Adolf Holzhausens Nachfolger, Kandlgasse 19-21, A-1070 Wien. – Verlagsort: Wien. – Herstellungsort: Wien. – Printed in Austria.

MIX
Papier aus verantwortungsvollen Quellen
Paper from responsible sources
FSC® C105338

If you have any concerns about our products,
you can contact us on
ProductSafety@springernature.com

In case Publisher is established outside the EU,
the EU authorized representative is:
Springer Nature Customer Service Center GmbH
Europaplatz 3, 69115 Heidelberg, Germany

Printed by Libri Plureos GmbH
in Hamburg, Germany